Books are to be returned on or before
the last date below.

KU-285-066

2 4 SEP 2004

2 1 JAN 2005

2 3 SEP 2005

LIBREX—

LIVERPOOL JMU LIBRARY

3 1111 00717 2214

P T R PRENTICE HALL INTERNATIONAL SERIES
IN THE PHYSICAL AND CHEMICAL ENGINEERING SCIENCES

NEAL R. AMUNDSON, EDITOR, *University of Houston*

ADVISORY EDITORS

ANDREAS ACRIVOS, *Stanford University*
JOHN DAHLER, *University of Minnesota*
THOMAS J. HANRATTY, *University of Illinois*
JOHN M. PRAUSNITZ, *University of California*
L. E. SCRIVEN, *University of Minnesota*

AMUNDSON *Mathematical Methods in Chemical Engineering: Matrices and Their Application*

BALZHISER, SAMUELS, AND ELIASSEN *Chemical Engineering Thermodynamics*

BRIAN *Staged Cascades in Chemical Processing*

BUTT *Reaction Kinetics and Reactor Design*

DENN *Process Fluid Mechanics*

FOGLER *The Elements of Chemical Kinetics and Reactor Calculations: A Self-Paced Approach*

HIMMELBLAU *Basic Principles and Calculations in Chemical Engineering, 4th edition*

HINES AND MADDOX *Mass Transfer: Fundamentals and Applications*

HOLLAND *Fundamentals and Modeling of Separation Processes: Absorption, Distillation, Evaporation, and Extraction*

HOLLAND AND ANTHONY *Fundamentals of Chemical Reaction Engineering*

KUBÍČEK AND HLAVÁČEK *Numerical Solution of Nonlinear Boundary Value Problems with Applications*

KYLE *Chemical and Process Thermodynamics*

LEVICH *Physiochemical Hydrodynamics*

MODELL AND REID *Thermodynamics and Its Applications, 2nd edition*

MYERS AND SEIDER *Introduction to Chemical Engineering and Computer Calculations*

NEWMAN *Electrochemical Systems*

PRAUSNITZ *Molecular Thermodynamics of Fluid-Phase Equilibria*

PRAUSNITZ ET AL. *Computer Calculations for Multicomponent Vapor-Liquid and Liquid-Liquid Equilibria*

RUDD ET AL. *Process Synthesis*

SCHULTZ *Diffraction for Materials Scientists*

SCHULTZ *Polymer Materials Science*

STEPHANOPOULOS *Chemical Process Control: An Introduction to Theory and Practice*

VILLADSEN AND MICHELSEN *Solution of Differential Equation Models by Polynomial Approximation*

WILLIAMS *Polymer Science and Engineering*

Chemical Process Control
An Introduction to Theory and Practice

GEORGE STEPHANOPOULOS

Department of Chemical Engineering
Massachusetts Institute of Technology

Prentice/Hall International, Inc.

Library of Congress Cataloging in Publication Data

Stephanopoulos, George.
  Chemical process control.

  (Prentice-Hall international series in the physical
and chemical engineering sciences)
  Bibliography: p.
  Includes index.
  1. Chemical process control.   I. Title.   II. Series.
TP 155.75.S73   1984   660.281   83-11206
ISBN 0-13-128596-3

Editorial/production supervision: *Anne Simpson* and *Mary Carnis*
Manufacturing buyer: *Anthony Caruso*

This edition may be sold only in those countries
to which it is consigned by Prentice-Hall International.
It is not to be re-exported and it is not for sale
in the U.S.A., Mexico or Canada.

© 1984 by Prentice-Hall, Inc., Englewood Cliffs, New Jersey 07632

All rights reserved. No part of this book may be
reproduced, in any form or by any means,
without permission in writing from the publisher.

Printed in the United States of America

  13 14 15 16 17 18 19 20

ISBN   0-13-128596-3

Prentice-Hall International, Inc., *London*
Prentice-Hall of Australia Pty. Limited, *Sydney*
Editora Prentice-Hall do Brasil, Ltda., *Rio de Janeiro*
Prentice-Hall Canada Inc., *Toronto*
Prentice-Hall of India Private Limited, *New Delhi*
Prentice-Hall of Japan, Inc., *Tokyo*
Prentice-Hall of Southeast Asia Pte. Ltd., *Singapore*
Whitehall Books Limited, *Wellington, New Zealand*
Prentice-Hall, Inc., *Englewood Cliffs, New Jersey*

To Eleni and Nicholas

Contents

11. Dynamic Behavior of Second-Order Systems *186*

12. Dynamic Behavior of Higher-Order Systems *212*

Part IV *Analysis and Design of Feedback Control Systems 239*

13. Introduction to Feedback Control *241*

14. Dynamic Behavior of Feedback-Controlled Processes *258*

Part V   Analysis and Design of Advanced Control Systems *381*

19. Feedback Control of Systems with Large Dead Time or Inverse Response *383*

20. Control Systems with Multiple Loops *394*

21. Feedforward and Ratio Control *411*

22. Adaptive and Inferential Control Systems *431*

31. Process Identification and Adaptive Control   *656*

Preface

As its title suggests, this is an introductory text to the theory and practice of chemical process control. It is intended to cover the needs, as these pertain to the scope of basic chemical engineering education, (1) of a first undergraduate course in process dynamics and control, and (2) of the first part of an advanced undergraduate or graduate course in process control.

During the last ten years, academic research and industrial practice in chemical process control have been shaped by the following important realizations:

1. The structure of chemical processes has become increasingly complex, due to better management of energy and raw materials. As a consequence, the design of control systems for complete plants now constitutes the focal point of engineering interest, rather than controller designs for single processing units. Furthermore, the design of a control system has become intimately related to the design of the process itself.
2. Designing a control system implies identification of control objectives; selection of appropriate measurements and manipulations, as well as the determination of loops connecting these; and identification of the proper control laws. In other words, it is a much more involved question than the traditional one of controller tuning.

3. The advent and rapid growth of digital computers has revolution-
   ized the practice of chemical process control and has allowed the
   industrial implementation of advanced control concepts.

Today, it is widely believed that education in chemical process control has not been adapted to follow modern directions, as these are
depicted above. The present text represents an attempt to bridge the
classical approach to process control problems with the current and
future trends and needs. It is primarily an educational vehicle rather
than a practical guide to the solution of specific industrial problems.
Here, the emphasis is on understanding the nature of process control
problems and their attributes, as well as on systematizing the approach
to their solution. Needless to say, several design tools and methodologies have also been included, but with reduced emphasis. Thus it is
hoped that the following aspects will emerge after studying this book:

1. Chemical process control is a subject of study with its own intrica-
   cies and challenges. It is intimately related to chemical engineering science and practice, and as such it is not the degenerate child
   of any other branch of engineering.
2. The design of a control system is not a mathematical problem, but
   should be perceived as an engineering task, with all its attractive
   challenges and practical shortcomings.
3. A good understanding of physical and chemical phenomena taking place in a chemical process is of paramount importance for
   the design of simple and effective control schemes.
4. Several alternative control configurations are usually possible for
   a given processing unit or a complete plant. The selection of the
   "best" among them is the central question to be resolved.
5. There exist a plethora of analytical tools and design methodologies that one should be familiar with before attempting to tackle
   process control problems.

The text is divided into seven parts. Each part includes a number of
chapters with a common general orientation.

Part I (Chapters 1 through 3) represents a general introduction to the
control aspects of a chemical process. An attempt is made to define
what we mean by process control, to identify the needs and incentives
for process control, to analyze the design questions and formulate the
problems that must be solved, and to provide the rationale for studying
the material that follows in the subsequent chapters.

Part II (Chapters 4 and 5) introduces the reader to the modeling
requirements for process control. It demonstrates how we can construct
useful models, starting from basic principles, and determines the scope
and difficulties of mathematical modeling for process control purposes.

Part III (Chapters 6 through 12) is devoted to the analysis of static and dynamic behavior of processing systems. The emphasis here is on identifying those process characteristics which shape the dynamic response for a variety of processing units. The results of such analysis are used later to design effective controllers. Input–output models have been employed through the use of Laplace transforms.

Part IV (Chapters 13 through 18) covers the analysis and design of feedback control systems, which represent the control schemes encountered most often in a chemical plant. Emphasis has been placed on understanding the effects which various feedback controllers have on the response of controlled processes, and on the selection of the most appropriate among them. The subject of controller tuning has been deemphasized, and as a consequence, the traditional root-locus techniques and frequency response tuning methods have been scaled down.

Part V (Chapters 19 through 22) deals with the description, analysis, and design of more complex control systems, with one controlled output. In particular, Chapter 19 introduces the concept of feedback compensation with Smith's predictor, to cope with systems possessing large dead times or inverse response. Chapter 20 describes and analyzes a variety of multiloop control systems (with one controlled output) often encountered in chemical processes, such as cascade, selective, and split-range. Chapter 21 is devoted exclusively to the analysis and design of feedforward and ratio control systems, while Chapter 22 makes a rather descriptive presentation of adaptive and inferential control schemes: why they are needed and how they can be used.

Chapters 23 through 25 constitute Part VI and are devoted to multivariable control problems. The emphasis here is on generating alternative control configurations in a systematic manner and screening them for the best. It is not meant to cover all aspects of multivariable control, and only one design technique (relative-gain array) is presented for the selection of the least interacting loops. Simple, noninteracting control loops are also designed for low-order systems. Chapters 23 and 25 offer an introduction to the control design problems for complete chemical plants. Also, they outline systematic procedures which can be used to synthesize control schemes for such complex systems.

Part VII (Chapters 26 through 31) is an introduction to process control using digital computers. Initially (Chapter 26), the characteristics of a digital computer control loop are analyzed in terms of the new hardware elements as well as the new control design questions. Chapters 27 through 29 provide the analytical tools for analyzing the response of open-loop and closed-loop discrete-time dynamic systems. Chapter 30 outlines the most popular procedures for designing digital feedback controllers, while Chapter 31 discusses computer-aided, on-line identification of chemical processes and its use for the development of adaptive control systems.

The present book is the culmination of seven years of teaching process control at the University of Minnesota and the National Technical University of Athens. It was designed in such a way as to provide a simple, smooth, and readable account of process control aspects, while providing the interested reader with material, problems, and directions for further study.

With few exceptions, every chapter contains the proper amount of material for just one lecture. In order to maintain continuity and flow of the main text, two mechanisms have been used. First: specific details are usually grouped at the end of each section under the heading Remarks. Second: additional, useful, but not necessary material has been put into appendices at the end of the corresponding chapters. Many examples have been used throughout the text either to explain some concepts or to demonstrate the use of various techniques. Not all of the examples need to be covered during a lecture hour, and some of them can be left for individual study.

To enhance the educational value of the book, a series of Things to Think About at the end of each chapter, as well as a large number of homework problems at the end of each part, have been included. Occasionally, the Things to Think About will direct the reader to find the answers in other books, papers, or handbooks, which he or she can find listed in the sections of annotated bibliography at the end of each part.

I am vastly indebted to many people who have helped and inspired me, in various ways, to start, continue, and complete this book. First and foremost, my gratitude goes to the "Chief," Neal R. Amundson, for supporting me at the conception of this book and strengthening my resolve in so many direct and indirect ways. Rutherford Aris, Arnie Fredrickson, and Skip Scriven may not have realized what an influence their generous presence, "teachings," and friendship have had in shaping this book. Thanks are due to J. Wei for his encouraging words to continue with this project, and to M. M. Denn for being so gracious and helpful when this book was in its embryonic stage. The constructive criticism of J. M. Douglas has been immensely helpful. His generous permission to use passages from his work on the control system design for complete plants is gratefully acknowledged.

A. W. Westerberg and K. Jensen read the manuscript and used it for classroom teaching. Their thoughtful and valuable reviews and suggestions have helped enormously to improve the book. T. Umeda, I. Hashimoto, M. Morari, Y. Arkun, J. Romagnoli, S. Svoronos, M. Nikolaou, and K. Christodoulou contributed many useful remarks, corrections, and suggestions. To all of them, I want to express my sincere gratitude.

I cannot find words to describe the debt I owe to all of my colleagues at the University of Minnesota for having created a stimulating atmosphere of academic excellence, the basic element of any long-lasting endeavor.

Shirley Tabis typed the original manuscript with great care, artistic taste, skill, and dedication, unparalleled in my own experience.

I owe a special debt to my wife, Eleni. She has participated in every stage of this book's development. She copied in her own careful manner my original scribbles (quite often more than once) and a large number of line drawings, proofread the manuscript several times, and edited it for mistakes, while at the same time she was carrying our first son and nurturing him through his first steps. I am without words to thank her.

Finally, to my parents and brothers goes my eternal gratitude for their love, support, and dedication.

George Stephanopoulos
Athens, Greece

The Control of a Chemical Process: Its Characteristics and Associated Problems

The needs are intimately related to the problems, and the problems, as usual, wear a sometimes effective camouflage.

A. S. Foss[*]

The purpose of the following three introductory chapters is:

1. To define what we mean by chemical process control
2. To describe the needs and the incentives for controlling a chemical process
3. To analyze the characteristics of a control system and to formulate the problems that must be solved during its design
4. To provide the rationale for studying the material that follows in subsequent chapters

To achieve the foregoing objectives, we will use a series of examples taken from the chemical industry. These examples are usually simplified and serve only to demonstrate the various qualitative points made.

[*]"Critique of Chemical Process Control Theory," *AIChE J., 19*(2), 209 (1973).

Incentives for Chemical 1
Process Control

A chemical plant is an arrangement of processing units (reactors, heat exchangers, pumps, distillation columns, absorbers, evaporators, tanks, etc.), integrated with one another in a systematic and rational manner. The plant's overall objective is to convert certain raw materials (input feedstock) into desired products using available sources of energy, in the most economical way.

During its operation, a chemical plant must satisfy several requirements imposed by its designers and the general technical, economic, and social conditions in the presence of ever-changing external influences (disturbances). Among such requirements are the following:

1. *Safety:* The safe operation of a chemical process is a primary requirement for the well-being of the people in the plant and for its continued contribution to the economic development. Thus the operating pressures, temperatures, concentration of chemicals, and so on, should always be within allowable limits. For example, if a reactor has been designed to operate at a pressure up to 100 psig, we should have a control system that will maintain the pressure below this value. As another example, we should try to avoid the development of explosive mixtures during the operation of a plant.

2. *Production specifications:* A plant should produce the desired amounts and quality of the final products. For example, we may

require the production of 2 million pounds of ethylene per day, of 99.5% purity. Therefore, a control system is needed to ensure that the production level (2 million pounds per day) and the purity specifications (99.5% ethylene) are satisfied.

3. *Environmental regulations:*   Various federal and state laws may specify that the temperatures, concentrations of chemicals, and flow rates of the effluents from a plant be within certain limits. Such regulations exist, for example, on the amounts of SO_2 that a plant can eject to the atmosphere, and on the quality of water returned to a river or a lake.

4. *Operational constraints:*   The various types of equipments used in a chemical plant have constraints inherent to their operation. Such constraints should be satisfied throughout the operation of a plant. For example, pumps must maintain a certain net positive suction head; tanks should not overflow or go dry; distillation columns should not be flooded; the temperature in a catalytic reactor should not exceed an upper limit since the catalyst will be destroyed. Control systems are needed to satisfy all these operational constraints.

5. *Economics:*   The operation of a plant must conform with the market conditions, that is, the availability of raw materials and the demand of the final products. Furthermore, it should be as economical as possible in its utilization of raw materials, energy, capital, and human labor. Thus it is required that the operating conditions are controlled at given optimum levels of minimum operating cost, maximum profit, and so on.

All the requirements listed above dictate the need for continuous monitoring of the operation of a chemical plant and external intervention (control) to guarantee the satisfaction of the operational objectives. This is accomplished through a rational arrangement of equipment (measuring devices, valves, controllers, computers) and human intervention (plant designers, plant operators), which together constitute the control system.

There are three general classes of needs that a control system is called on to satisfy:

Suppressing the influence of external disturbances
Ensuring the stability of a chemical process
Optimizing the performance of a chemical process

Let us examine these needs using various examples.

1.1   Suppress the Influence of External Disturbances

Suppressing the influence of external disturbances on a process is the most common objective of a controller in a chemical plant. Such disturbances, which denote the effect that the surroundings (external world) have on a reactor, separator, heat exchanger, compressor, and so on, are usually out of the reach of the human operator. Consequently, we need to introduce a control mechanism that will make the proper changes on the process to cancel the negative impact that such disturbances may have on the desired operation of a chemical plant.

Example 1.1: Controlling the Operation of a Stirred Tank Heater

Consider the tank heater system shown in Figure 1.1. A liquid enters the tank with a flow rate F_i (ft³/min) and a temperature T_i (°F), where it is heated with steam (having a flow rate F_{st} lb/min). Let F and T be the flow rate and temperature of the stream leaving the tank. The tank is considered to be well stirred, which implies that the temperature of the effluent is equal to the temperature of the liquid in the tank.

The operational objectives of this heater are:

1. To keep the effluent temperature T at a desired value T_s
2. To keep the volume of the liquid in the tank at a desired value V_s

The operation of the heater is disturbed by external factors such as changes in the feed flow rate and temperature (F_i and T_i). If nothing changed, then after attaining $T = T_s$ and $V = V_s$, we could leave the system alone without any supervision and control. It is clear, though, that this cannot be true since T_i and F_i are subject to frequent changes. Consequently, some form of control action is needed to alleviate the

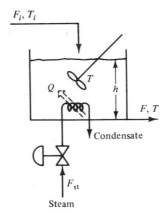

Figure 1.1 Stirred tank heater.

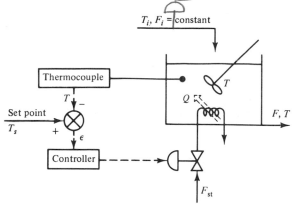

Figure 1.2 Feedback temperature control for a tank heater.

impact of the changing disturbances and keep T and V at the desired values.

In Figure 1.2 we see such a control action to keep $T = T_s$ when T_i or F_i changes. A thermocouple measures the temperature T of the liquid in the tank. Then T is compared with the desired value T_s, yielding a deviation $\epsilon = T_s - T$. The value of the deviation ϵ is sent to a control mechanism which decides what must be done in order for the temperature T to return back to the desired value T_s. If $\epsilon > 0$, which implies that $T < T_s$, the controller opens the steam valve so that more heat can be supplied. On the contrary, the controller closes the steam valve when $\epsilon < 0$ or $T > T_s$. It is clear that when $T = T_s$ (i.e., $\epsilon = 0$), the controller does nothing. This control system, which measures the variable of direct importance (T in this case) after a disturbance had its effect on it, is called the *feedback* control system. The desired value T_s is called the *set point* and is supplied externally by the person in charge of production.

A similar configuration can be used if we want to keep the volume V, or equivalently the liquid level h, at its set point h_s when F_i changes. In this case we measure the level of the liquid in the tank and we open or close the valve that affects the effluent flow rate F, or inlet flow rate F_i (see Figure 1.3). It is clear that the control systems shown in Figure 1.3 are also feedback control systems. All feedback systems shown in Figures 1.2 and 1.3 act post facto (after the fact), that is, after the effect of the disturbances has been felt by the process.

Returning to the tank heater example, we realize that we can use a different control arrangement to maintain $T = T_s$ when T_i changes. Measure the temperature of the inlet stream T_i and open or close the steam valve to provide more or less steam. Such a control configuration is called *feedforward* control and is shown in Figure 1.4. We notice that the feedforward control does not wait until the effect of the disturbances has been felt by the system, but acts appropriately before the external disturbance affects the system, anticipating what its effect will be. The characteristics of the feedback and feedforward control systems will be studied in detail in subsequent chapters.

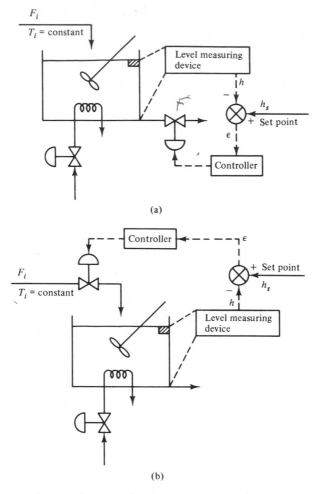

(a)

(b)

Figure 1.3 Alternative liquid-level control schemes.

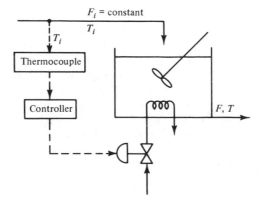

Figure 1.4 Feedforward temperature control for stirred tank heater.

The suppresion of the impact that disturbances have on the operating behavior of processing units is one of the main reasons for the use of control in the chemical industry.

1.2   Ensure the Stability of a Process

Consider the behavior of the variable x shown in Figure 1.5. Notice that at time $t = t_0$ the constant value of x is disturbed by some external factors, but that as time progresses the value of x returns to its initial value and stays there. If x is a process variable such as temperature, pressure, concentration, or flow rate, we say that the process is *stable* or *self-regulating* and needs no external intervention for its stabilization. It is clear that no control mechanism is needed to force x to return to its initial value.

In contrast to the behavior described above, the variable y shown in Figure 1.6 does not return to its initial value after it is disturbed by external influences. Processes whose variables follow the pattern indicated by y in Figure 1.6 (curves A, B, C) are called *unstable* processes and require external control for the stabilization of their behavior. The explosion of a hydrocarbon fuel with air is such an unstable system. Riding a bicycle is an attempt to stabilize an unstable system and we attain that by pedaling, steering, and leaning our body right or left.

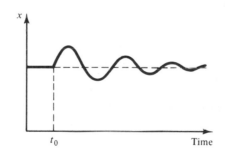

Figure 1.5 Response of a stable system.

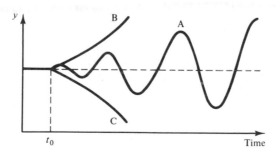

Figure 1.6 Alternative responses of unstable systems.

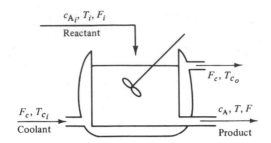

Figure 1.7 CSTR with cooling jacket.

Example 1.2: Controlling the Operation of an Unstable Reactor

Consider a continuous stirred tank reactor (CSTR) in which an irreversible exothermic raction A → B takes place. The heat of reaction is removed by a coolant medium that flows through a jacket around the reactor (Figure 1.7). As is known from the analysis of a CSTR system, the curve that describes the amount of heat released by the exothermic reaction is a sigmoidal function of the temperature T in the reactor (curve A in Figure 1.8). On the other hand, the heat removed by the coolant is a linear function of the temperature T (line B in Figure 1.8). Consequently, when the CSTR is at steady state (i.e., nothing is changing), the heat produced by the reaction should be equal to the heat removed by the coolant. This requirement yields the steady states P_1, P_2, and P_3 at the intersection of curves A and B of Figure 1.8. Steady states P_1 and P_3 are called *stable*, whereas P_2 is *unstable*. To understand the concept of stability, let us consider steady state P_2.

Assume that we are able to start the reactor at the temperature T_2 and the concentration c_{A_2} that corresponds to this temperature. Consider that the temperature of the feed T_i increases. This will cause an increase in the temperature of the reacting mixture, say T_2'. At T_2' the heat released by the reaction (Q_2') is more than the heat removed by the coolant, Q_2'' (see Figure 1.8), thus leading to higher temperatures in the reactor and conse-

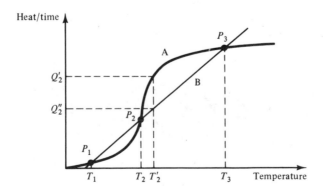

Figure 1.8 The three steady states of a CSTR.

quently to increased rates of reaction. Increased rates of reaction produce larger amounts of heat released by the exothermic reaction, which in turn lead to higher temperatures, and so on. Therefore, we see that an increase in T_i takes the reactor temperature away from steady state P_2 and that the temperature will eventually reach the value of steady state P_3 (Figure 1.9a). Similarly, if T_i were to decrease, the temperature of the reactor would take off from P_2 and end up at P_1 (Figure 1.9b). By contrast, if we were operating at steady state P_3 or P_1 and we perturbed the operation of the reactor, it would return naturally back to point P_3 or P_1 from which it started (see Figure 1.9c,d). (*Note*: The reader should verify this assertion.)

Sometimes we would like to operate the CSTR at the middle unstable steady state, for the following reasons: (1) the low-temperature steady state P_1 causes very low yields because the temperature T_1 is very low, and (2) the high-temperature steady state P_3 may be very high, causing unsafe conditions, destroying the catalyst for a catalytic reactor, degrading the product B, and so on.

In such cases we need a controller that will ensure the stability of the operation at the middle steady state. (*Question*: The reader should suggest a control mechanism to stabilize the operation of the reactor at the unstable steady state P_2. This example demonstrates very vividly the need for stabilizing the operation of a system using some type of control in the presence of external disturbances that tend to take the system away from the desired point.)

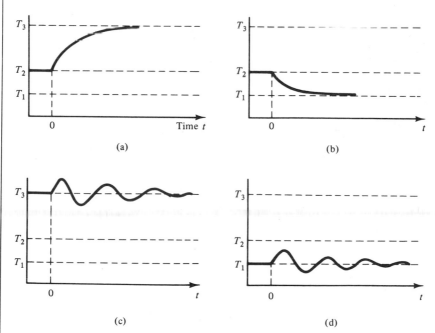

Figure 1.9 Dynamic response of a CSTR: (a) and (b) indicate the instability of the middle steady state, while (c) and (b) demonstrate the stability of the other two.

1.3 Optimize the Performance of a Chemical Process

Safety and the satisfaction of production specifications are the two principal operational objectives for a chemical plant. Once these are achieved, the next goal is how to make the operation of the plant more profitable. Given the fact that the conditions which affect the operation of the plant do not remain the same, it is clear that we would like to be able to change the operation of the plant (flow rates, pressures, concentrations, temperatures) in such a way that an economic objective (profit) is always maximized. This task is undertaken by the automatic controllers of the plant and its human operators.

Let us now see an example from the chemical processing industry where the controller is used to optimize the economic performance of a single unit.

> ### *Example 1.3: Optimizing the Performance of a Batch Reactor*
>
> Consider a batch reactor where the following two consecutive reactions take place:
>
> $$A \xrightarrow{\;1\;} B \xrightarrow{\;2\;} C$$
>
> Both reactions are assumed to be endothermic with first-order kinetics. The heat required for the reactions is supplied by steam which flows through the jacket around the reactor (Figure 1.10). The desired product is B; C is an undesired waste. The economic objective for the operation of the batch reactor is to maximize the profit Φ over a period of time t_R: that is,
>
> $$\text{maximize } \Phi = \int_0^{t_R} \{[\text{revenue from the sales of product B}] -$$
>
> $$+ \text{ cost of steam}\} \; dt + \text{cost of purchasing A} \qquad (1.1)$$
>
> where t_R is the period of reaction.
> The only variable that we can change freely in order to maximize the profit is the steam flow rate Q. The steam flow rate, which can vary with

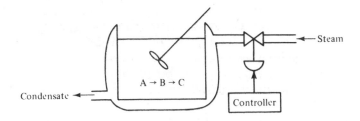

Figure 1.10 Batch reactor with two consecutive reactions.

time, will affect the temperature in the batch reactor and the temperature will, in turn, affect the rates of the desired and undesired reactions. The question is how we should vary $Q(t)$ with time so that the profit Φ is maximized. Let us examine some special policies with respect to $Q(t)$.

1. If $Q(t)$ is given the largest possible value for the entire reaction period t_R, the temperature of the reacting mixture will take the largest value that is possible. Initially, when c_A is large, we will have high yields of B but we will also pay more for the steam. As time goes on and the concentration of B increases, the yield of C also increases. Consequently, toward the end of the reaction period the temperature must decrease, necessitating a decrease in the steam flow rate.
2. If the steam flow rate is kept at its lowest value [i.e., $Q(t) = 0$] for the entire reaction period t_R, we will have no steam cost, but we will also have no production of B.

We see clearly from these two extreme cases that $Q(t)$ will vary between its lowest and highest values during the reaction period t_R. How it should vary to maximize the profit is not trivial and requires the solution of the optimization problem posed above.

In Figure 1.11 we see a general trend that the steam flow rate must follow to optimize the profit Φ. Therefore a control system is needed which will (1) compute the best steam flow rate for every time during the reaction period and (2) adjust the valve (inserted in the steam line) so that the steam flow rate takes its best value [as computed in (1)]. Such problems are known as *optimal control* problems.

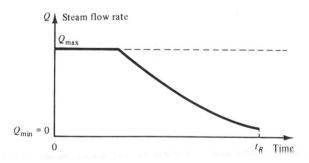

Figure 1.11 Optimal profile of the steam flow rate for the batch reactor of Example 1.3.

This example indicates that the control of the steam flow rate is not used to ensure the stability of the reactor or to eliminate the effect of external disturbances on the reactor, but to optimize its economic performance.

Design Aspects of a Process Control System

2

2.1 Classification of the Variables in a Chemical Process

The variables (flow rates, temperatures, pressures, concentrations, etc.) associated with a chemical process are divided into two groups:

1. *Input* variables, which denote the effect of the surroundings on the chemical process
2. *Output* variables, which denote the effect of the process on the surroundings

---

Example 2.1

For the CSTR reactor discussed in Example 1.2 (Figure 1.7) we have

Input variables: $c_{A_i}, T_i, F_i, T_{c_i}, F_c, (F)$

Output variables: c_A, T, F, T_{c_o}, V

Notice that the effluent flow rate F can be considered either as input or output. If there is a control valve on the effluent stream so that its flow rate can be manipulated by a controller, the variable F is an input, since the opening of the valve is adjusted externally; otherwise, F is an output variable.

Example 2.2

For the tank heater discussed in Example 1.1 (Figure 1.1) we have

Input variables: F_i, T_i, F_{st}, (F)

Output variables: F, V, T

The input variables can be further classified into the following categories:

1. *Manipulated* (or *adjustable*) variables, if their values can be adjusted freely by the human operator or a control mechanism
2. *Disturbances*, if their values are not the result of adjustment by an operator or a control system

The output variables are also classified into the following categories:

1. *Measured* output variables, if their values are known by directly measuring them
2. *Unmeasured* output variables, if they are not or cannot be measured directly

Example 2.3

Suppose that the inlet stream in the CSTR system (Figure 1.7) comes from an upstream unit over which we have no control. Then c_{A_i}, F_i, and T_i are disturbances. If the coolant flow rate is controlled by a control valve, then F_c is a manipulated variable, while T_{c_i} is a disturbance. Also, if the flow rate of the effluent stream is controlled by a valve, F is a manipulated variable; otherwise, it is an output variable.

With respect to the output variables we have the following: T, F, T_{c_o}, and V are measured outputs since their values can be known easily using thermocouples (T, T_{c_o}), a venturi meter (F), and a differential pressure cell (V).

The concentration c_A can be a measured variable if an analyzer (gas chromatograph, infrared spectrometer, etc.) is attached to the effluent stream. In many industrial plants such analyzers are not available because they are expensive and/or have low reliability (give poor measurements or break down easily). Consequently, in such cases c_A is an unmeasured output variable.

Example 2.4

For the tank heater system (Figure 1.1), the inputs F_i and T_i are disturbances, while F_{st} and F are manipulated inputs. The output variables V and T can be measured easily and they are considered measured outputs.

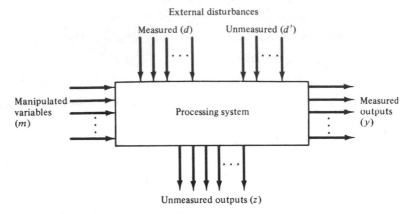

Figure 2.1 Input and output variables around a chemical process.

According to their direct measurability, the disturbances are classified into two categories: *measured* and *unmeasured* disturbances.

Example 2.5

The disturbances F_i and T_i of the stirred tank heater (Figure 1.1) are easily measured; thus they are considered measured disturbances. On the other hand, the feed composition for a distillation column, an extraction unit, reactors, and the like, is not normally measured and consequently is considered an unmeasured disturbance. As we will see later, unmeasured disturbances generate more difficult control problems.

Figure 2.1 summarizes all the classes of variables that we have around a chemical process.

2.2   Design Elements of a Control System

Let us look next at the basic questions that we must ask and try to answer, while attempting to design a control system that will satisfy the control needs for a chemical process.

1 Define control objectives

The central element in any control configuration is the process that we want to control. The first question raised by the control designer is:

Question 1:   *What are the operational objectives that a control system is called upon to achieve?*

The answer to this question determines the *control objectives*. They may have to do with:

> *Ensuring the stability of the process, or*
> *Suppressing the influence of external disturbances, or*
> *Optimizing the economic performance of a plant, or*
> *A combination of the above.*

At the beginning the control objectives are defined qualitatively; subsequently, they are quantified, usually in terms of the output variables.

Example 2.6

For the CSTR system discussed in Example 1.2 (Figure 1.7), the control objective (qualitatively defined) is to ensure the stability of the middle, unstable steady state. But such a qualitative description of the control objectives is not useful for the design of a control system and must be quantified. A quantitative translation of the qualitative control objective requires that the temperature (an output variable) not deviate more than 5% from its nominal value at the unstable steady state.

Example 2.7

For the stirred tank heater of Example 1.1, the control objectives are to maintain the temperature of the outlet (T) and the volume of the fluid (V) in the tank at desired values. For this example the quantification of the control objectives is direct and straightforward: that is,

$$T = T_s$$
$$V = V_s$$

where T_s and V_s are given, desired values.

Example 2.8

For the batch reactor of Example 1.3, the qualitative control objective is the maximization of the profit. The quantitative description of this objective is rather complex. It requires the solution of a maximization problem, which will yield the value of the steam flow rate, $Q(t)$, at each instant during the reaction period.

Select measurements

Whatever our control objectives are, we need some means to monitor the performance of the chemical process. This is done by measuring the values of certain processing variables (temperatures, pressures, concentrations, flow rates, etc.). The second question that arises is:

Question 2:   *What variables should we measure in order to monitor the operational performance of a plant?*

It is self-evident that we would like to monitor directly the variables that represent our control objectives, and this is what is done whenever possible. Such measurements are called *primary* measurements.

Example 2.9

For the tank heater system (Example 1.1) our control objectives are to keep the volume and the temperature of the liquid in the tank at desired levels, that is, keep

$$T = T_s \quad \text{and} \quad V = V_s$$

Consequently, our first attempt is to install measuring devices that will monitor T and V directly. For the present system this is simple by using a thermocouple (for T) and a differential pressure cell (for V).

It sometimes happens that our control objectives are not measurable quantities; that is, they belong to the class of unmeasured outputs. In such cases we must measure other variables which can be measured easily and reliably. Such supporting measurements are called *secondary measurements*.

Then we develop mathematical relationships between the unmeasured outputs and the secondary measurements; that is,

$$\text{unmeasured output} = f(\text{secondary measurements})$$

which allows us to determine the values of the unmeasured outputs (once the values of the secondary measurements are available). In a subsequent chapter we will see that the mathematical relationship between measured and unmeasured outputs results from empirical, experimental, or theoretical considerations.

Example 2.10

Consider a simple distillation column separating a binary mixture of pentane and hexane into two product streams of pentane (distillate) and hexane (bottoms). Our control objective is to maintain the production of a distillate stream with 95 mole % pentane in the presence of changes in the feed composition.

It is clear that our first reaction is to use a composition analyzer to measure the concentration of pentane in the distillate and then using feedback control to manipulate the reflux ratio, so that we can keep the distillate 95% in pentane. This control scheme is shown in Figure 2.2a. An alternative control system is to use a composition analyzer to monitor the concentration of pentane in the feed. Then in a feedforward arrangement we can change the reflux ratio to achieve our objective. This control

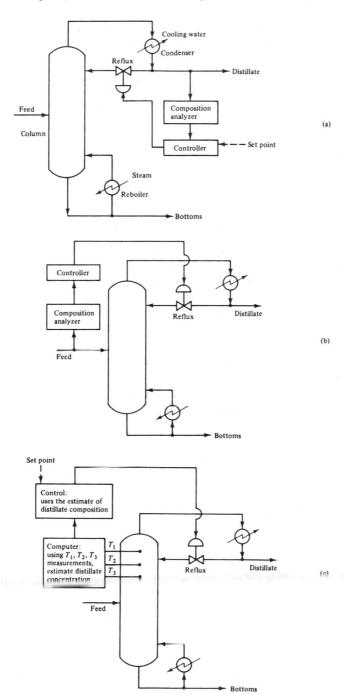

Figure 2.2 Three different systems for the distillate composition control of a simple distillation column: (a) feedback; (b) feedforward; (c) inferential.

scheme is shown in Figure 2.2b. Both of the control systems noted above depend on the composition analyzers. It is possible that such measuring devices are either very costly or of very low reliability for an industrial environment (failing quite often or not providing accurate measurements). In such cases we can measure the temperature of the liquid at various trays along the length of the column quite reliably, using simple thermocouples. Then using the material and energy balances around the trays of the column and the thermodynamic equilibrium relationships between liquid and vapor streams, we can develop a mathematical relationship that gives us the composition of the distillate if the temperatures of some selected trays are known. Figure 2.2c shows such a control scheme that uses temperature measurements (secondary measurements) to estimate or infer the composition of pentane in the distillate (i.e., the value of the control objective).

The third class of measurements that we can make to monitor the behavior of a chemical process includes direct measurement of the external disturbances. Measuring the disturbances before they enter the process can be highly advantageous because it allows us to know *a priori* what the behavior of the chemical process will be and thus take remedial control action to alleviate any undesired consequences. Feedforward control uses direct measurements of the disturbances (see Figure 1.4).

Select manipulated variables

Once the control objectives have been specified and the various measurements identified, the next question relates to how we effect a change in the process:

Question 3:   *What are the manipulated variables to be used to control a chemical process?*

Usually in a process we have a number of available input variables which can be adjusted freely. Which ones we select to use as manipulated variables is a crucial question, as the choice will affect the quality of the control actions we take.

Example 2.11

To control the level of liquid in a tank we can adjust either the flow rate of the inlet stream (Figure 1.3b) or the flow rate of the outlet stream (Figure 1.3a). Which is better is an important question that we will analyze later.

Select the control configuration

After the control objectives, the possible measurements, and the available manipulated variables have been identified, the final problem to be solved is that of defining the control configuration. Before we define a control configuration, let us look at some control systems with different control configurations.

The two feedback control systems in Figure 1.3a and b constitute two different control configurations. Thus the *same information* (measurement of liquid level) *flows to different manipulated variables* [i.e., F (Figure 1.3a) and F_i (Figure 1.3b)]. Similarly, the feedback control system (Figure 1.2) and the feedforward control system (Figure 1.4) for the tank heater constitute two distinctly different control configurations. For these two control systems we use the *same manipulated variable* (i.e., F_{st}) *but different measurements*. Thus for the feedback system of Figure 1.2 we use the temperature of the liquid in the tank, whereas for the feedforward system of Figure 1.4 we measure the temperature of the inlet.

In the examples above we notice that two control configurations differ either in:

1. The information (measurement) flowing to the same manipulated variable, or
2. The manipulated variable to which the same information flows.

Thus for the two feedback control systems in Figure 1.3a and b we use the same information (measurement of the liquid level) but different manipulated variables (F or F_i). On the contrary, for the control systems in Figures 1.2 and 1.4, we have different measurements (T or T_i) which are used to adjust the value of the same manipulated variable (F_{st}).

Later, we will study other types of control configurations, but for the time being we can define a *control configuration* (or *control structure*) as follows:

A control configuration is the information structure that is used to connect the available measurements to the available manipulated variables.

It is clear from the previous examples that normally we will have many different control configurations for a given chemical process, which raises the following question:

Question 4:   *What is the best control configuration for a given chemical process control situation?*

The answer to this question is very critical for the quality of the control system we are asked to design.

Depending on how many controlled outputs and manipulated inputs we have in a chemical process, we can distinguish the control configurations as either *single-input, single-output* (SISO) or *multiple-input, multiple-output* (MIMO) control systems.

For example, for the tank heater system:

(a) If the control objective (controlled output) is to keep the liquid level at a desired value by manipulating the effluent flow rate, we have a SISO system.

(b) On the contrary, if our control objectives are (more than one) to keep the level and the temperature of the liquid at desired values, by manipulating (more than one) the steam flow rate and the effluent flow rate, we have a MIMO system.

In the chemical industry most of the processing systems are multiple-input, multiple-output systems. Since the design of SISO systems is simpler, we will start first with them and progressively cover the design of MIMO systems.

Let us close this section by defining three general types of control configurations.

1. *Feedback control configuration:* uses direct measurements of the controlled variables to adjust the values of the manipulated variables (Figure 2.3). The objective is to keep the controlled vari-

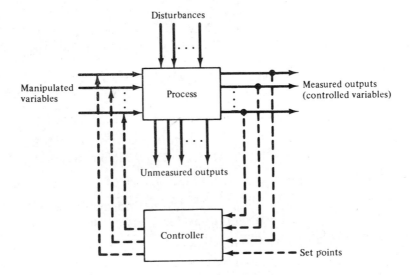

Figure 2.3 General structure of feedback control configurations.

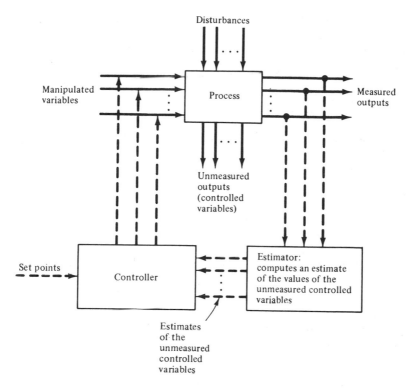

Figure 2.4 General structure of inferential control configurations.

ables at desired levels (set points). Examples of feedback control are shown in Figures 1.2 and 1.3.

2. *Inferential control configuration:*   uses secondary measurements (because the controlled variables cannot be measured) to adjust the values of the manipulated variables (Figure 2.4). The objective here is to keep the (unmeasured) controlled variables at desired levels.

    The estimator uses the values of the available measured outputs, together with the material and energy balances that govern the process, to compute mathematically (estimate) the values of the unmeasured controlled variables. These estimates, in turn, are used by the controller to adjust the values of the manipulated variables. An example of inferential control configuration is shown in Figure 2.2c.

3. *Feedforward control configuration:*   uses direct measurement of the disturbances to adjust the values of the manipulated variables (Figure 2.5). The objective here is to keep the values of the controlled output variables at desired levels. An example of feedforward control configuration is shown in Figure 1.4.

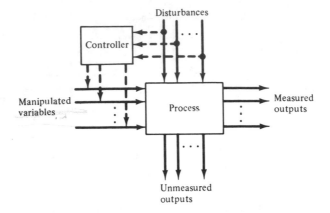

Figure 2.5 General structure of feedforward control configurations.

Design the controller

In every control configuration, the controller is the active element that receives the information from the measurements and takes appropriate control actions to adjust the values of the manipulated variables. For the design of the controller we must answer the following question:

Question 5:   *How is the information, taken from the measurements, used to adjust the values of the manipulated variables?*

The answer to this question constitutes the *control law*, which is implemented automatically by the controller.

Example 2.12

Let us consider the problem of controlling the temperature T of a liquid in a tank (Figure 1.1) when the inlet temperature T_i changes. Assume that the inlet and outlet flow rates are equal. Our measurement will be the liquid temperature and the manipulated variable the rate of heat input, Q, provided by steam. The question is: How should Q change in order to keep temperature T constant, when T_i changes? In other words, we want to develop the control law.

Assume that the heater has been operating for some time and that the liquid temperature has been kept constant at T_s, while the volume of the liquid has remained constant at a value V. We say that the heater has been operating at *steady state* (where nothing changes). Under these conditions the energy balance around the tank yields

$$0 = F\rho c_p(T_{i,s} - T_s) + Q_s \qquad (2.1)$$

where F, ρ, and c_p are the inlet (or outlet) flow rate, density of the liquid,

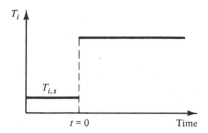

Figure 2.6 Temperature step change.

and the liquid's heat capacity, respectively. T_s, $T_{i,s}$, and Q_s are the corresponding steady-state values. Suppose that suddenly T_i increases as shown in Figure 2.6. If nothing is done on Q, the temperature T will start rising with time. How T changes with time will be given by the transient energy balance around the tank: that is,

$$V\rho c_p \frac{dT}{dt} = F\rho c_p(T_i - T) + Q \tag{2.2}$$

Subtract eq. (2.1) from (2.2) and take

$$V\rho c_p \frac{d(T - T_s)}{dt} = F\rho c_p[(T_i - T_{i,s}) - (T - T_s)] + (Q - Q_s) \tag{2.3}$$

Note that $d(T - T_s)/dt = dT/dt$, since $T_{s_i} = $ constant.

The difference $\epsilon = T - T_s$ denotes the *error* or *deviation* of liquid's temperature from the desired value T_s. We want to drive this error to zero by manipulating appropriately the value of heat input Q. The simplest control law is to require that Q changes proportionally to the error $T - T_s$:

$$Q = -\alpha(T - T_s) + Q_s \tag{2.4}$$

This law is known as *proportional control* and parameter α is called *proportional gain*. Substitute eq. (2.4) into (2.3) and take

$$V\rho c_p \frac{d(T - T_s)}{dt} = F\rho c_p[(T_i - T_{i,s}) - (T - T_s)] - \alpha(T - T_s) \tag{2.5}$$

Equation (2.5) is solved for $(T - T_s)$, and for various values of gain α yields the solutions shown in Figure 2.7. We notice that none of the solutions is satisfactory since $T - T_s \neq 0$. Thus we conclude that the proportional control law is not acceptable.

Considerable improvement in the quality of the resulting control can be obtained if we use a different control law known as *integral control*. In this case Q is proportional to the time integral of $(T - T_s)$:

$$Q = -\alpha' \int_0^t (T - T_s)\, dt + Q_s \tag{2.6}$$

Substitute again Q from eq. (2.6) into (2.3) and take

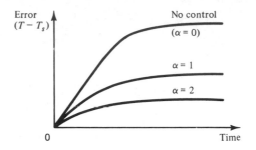

Figure 2.7 Temperature response under proportional feedback control.

$$V\rho c_p \frac{d(T - T_s)}{dt} = F\rho c_p[(T_i - T_{i,s}) - (T - T_s)] - \alpha' \int_0^t (T - T_s)\, dt \quad (2.7)$$

The solution of eq. (2.7) for various values of the parameter α' is shown in Figure 2.8. We notice that integral control is acceptable since it drives the error $T - T_s$ to zero. We also notice that depending on the value of α', the error $T - T_s$ returns to zero faster or slower, oscillates for longer or shorter time, and so on. In other words, the quality of control depends on the value of α'. [*Note:* In Chapter 8 we will learn how to solve integral-differential equations such as (2.7).]

Combining the proportional with the integral action we take a new control law known as *proportional-integral control*. According to this law the value of heat input Q is given by

$$Q = -\alpha(T - T_s) - \alpha' \int_0^t (T - T_s)\, dt + Q_s \quad (2.8)$$

In Part IV (Chapters 13 through 18) we will study the characteristics of various control laws, but it should be remembered that the selection of the appropriate control law is an important question to be answered by the chemical engineer control designer.

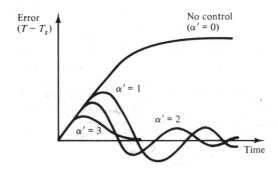

Figure 2.8 Temperature response under integral feedback control.

2.3 Control Aspects of a Complete Chemical Plant

The examples that we discussed in previous sections were concerned with the control of single units such as CSTR, a tank heater, and a batch reactor. It should be emphasized as early as possible that rarely if ever is a chemical process composed of one unit only. On the contrary, a chemical process is composed of a large number of units (reactors, separators, heat exchangers, tanks, pumps, compressors, etc.) which are interconnected with each other through the flow of materials and energy. For such a process the problem of designing a control system is not simple but requires experience and good chemical engineering background.

Without dwelling too much on the control problems of integrated chemical processes, let us see some of their characteristic features which do not show up in the control of single units.

Example 2.13

Consider a simple chemical plant composed of two units: a CSTR and a distillation column (Figure 2.9). The raw materials entering the reactor

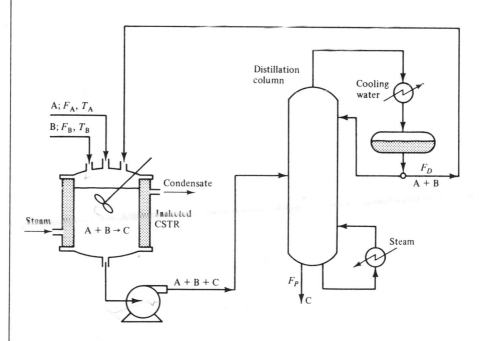

Figure 2.9 The simple chemical plant of Example 2.13.

are A and B with flow rates F_A and F_B and temperatures T_A and T_B, respectively. They react to yield C:

$$A + B \longrightarrow C$$

The reaction is endothermic and the heat is supplied by steam which flows through the jacket of the reactor. The mixture of C and unreacted A and B enters the distillation column, where A + B is separated from the top as the overhead product and C is taken as the bottom product.

The operational objectives for this simple plant are:

1. Product specifications:
   (a) Keep the flow rate of the desired product stream F_P at the specified level.
   (b) Keep the required purity of C in the product stream.
2. Operational constraints:
   (a) Do not overflow the CSTR.
   (b) Do not flood the distillation column or let it go dry.
3. Economic considerations:
   Maximize the profit from the operation of this plant. Since the flow rate and the composition of the product stream are specified, maximizing the profit is equivalent to minimizing the operating costs. It should be noted that the operating cost involves the cost for purchasing the raw materials, the cost of steam used in the CSTR and the reboiler of the distillation column, and the cost of the cooling water used in the condenser.

The disturbances that will affect the foregoing operational objectives are:

(a) The flow rates, compositions, and temperatures of the streams of the two raw materials.
(b) The pressure in the distillation column.
(c) The temperature of the coolant used in the condenser of the distillation column: (For example, if the coolant is water, it will have a different temperature during the day than during the night.)

At first glance the problem of designing a control system even for this simple plant looks very complex. Indeed it is. The basically new feature for the control design of such a system is the interaction between the units (reactor, column). The output of the reactor affects the operation of the column in a profound way and the overhead product of the column influences the conversion in the CSTR. This tight interaction between the two units seriously complicates the design of the control system for the overall process.

Suppose that we want to control the composition of the bottoms product by manipulating the steam in the reboiler. This control action will affect the composition of the overhead product (A + B), which in turn will affect the reaction conversion in the CSTR.

On the other hand, to keep the conversion in the CSTR constant at the desired level, we try to keep the ratio $F_A/F_B =$ constant and the temperature T in the CSTR constant. Any changes in F_A/F_B or T will affect the conversion in the reactor and thus the composition of the feed in the distillation column. A change in the feed composition of the column will affect the purity of the two product streams.

The control of integrated processes is the basic objective for a chemical engineer. Due to its complexity, though, we will start by analyzing the control problems for single units and eventually we will treat the integrated processes.

Hardware for a Process **3**
Control System

In Chapter 2 we examined the various considerations that must be taken into account during the design of a control system and the associated problems that must be resolved. In this chapter we discuss the physical elements (hardware) constituting a control system as it is implemented in practice for the control of real physical processes.

3.1 Hardware Elements of a Control System

In every control configuration we can distinguish the following hardware elements:

1. *The chemical process:* It represents the material equipment together with the physical or chemical operations that occur there.

2. *The measuring instruments or sensors:* Such instruments are used to measure the disturbances, the controlled output variables, or secondary output variables, and are the main sources of information about what is going on in the process. Characteristic examples are:

Thermocouples or resistance thermometers, for measuring the temperature
Venturi meters, for measuring the flow rate
Gas chromatographs, for measuring the composition of a stream

A mercury thermometer is not a good measuring device to be used for control since its measurement cannot be readily transmitted. On the other hand, a thermocouple is acceptable because it develops an electric voltage which can be readily transmitted. Thus transmission is a very crucial factor in selecting the measuring devices.

Since good measurements are very crucial for good control, the measuring devices should be rugged and reliable for an industrial environment.

3. *Transducers:*   Many measurements cannot be used for control until they are converted to physical quantities (such as electric voltage or current, or a pneumatic signal, i.e., compressed air or liquid) which can be transmitted easily. Transducers are used for that purpose. For example, strain gauges are metallic conductors whose electric resistance changes when they are subjected to mechanical strain. Thus they can be used to convert a pressure signal to an electric one.

4. *Transmission lines:*   These are used to carry the measurement signal from the measuring device to the controller. In the past, transmission lines were pneumatic (compressed air or compressed liquids) but with the advent of electronic analog controllers and especially the expanding use of digital computers for control, transmission lines carry electric signals. Many times the measurement signal coming from a measuring device is very weak and cannot be transmitted over a long distance. In such cases the transmission lines are equipped with amplifiers which raise the level of the signal. For example, the output of a thermocouple is of the order of a few millivolts. Before it is transmitted to the controller, it is amplified to the level of a few volts.

5. *The controller:*   This is the hardware element that has "intelligence." It receives the information from the measuring devices and decides what action should be taken. The older controllers were of limited intelligence, could perform very simple operations, and could implement simple control laws. Today, with increasing use of digital computers as controllers, the available machine intelligence has expanded tremendously, and very complicated control laws can be implemented.

6. *The final control element:*   This is the hardware element that implements in real life the decision taken by the controller. For example, if the controller "decides" that the flow rate of the outlet stream should be increased (or decreased) in order to keep the liquid level in a tank at the desired value (see Example 1.1, Figure 1.3a), it is the valve (on the effluent stream) that will implement this decision, opening (or closing) by the commanded amount.

The control valve is the most frequently encountered final control element but not the only one. Other typical final control elements for a chemical process are:

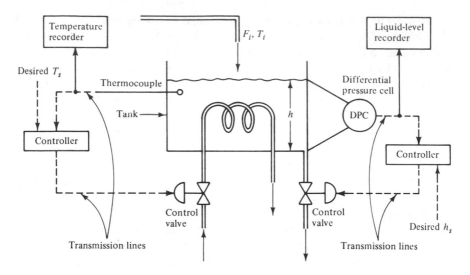

Figure 3.1 Hardware elements for the feedback control of a stirred tank heater.

Relay switches, providing on–off control
Variable-speed pumps
Variable-speed compressors

7. *Recording elements:* These are used to provide a visual demonstration of how a chemical process behaves. Usually, the variables recorded are the variables that are directly measured as part of the control system. Various types of recorders (temperature, pressure, flow rate, composition, etc.) can be seen in the control room of a chemical plant, continuously monitoring the behavior of the process. The recent introduction of digital computers in the process control has also expanded the recording opportunities, through video display units (VDUs).

Figure 3.1 describes the hardware elements used for the control of the stirred tank heater.

3.2 Use of Digital Computers in Process Control

The rapid technological development of digital computers during the last 10 years, coupled with significant reduction of their cost, has had a very profound effect on how the chemical plants are controlled. The expected future improvements, together with the growing sophistication of control design techniques, make the digital computer the centerpiece for the development of a control system for chemical processes.

Already, large chemical plants such as petroleum refineries, ethylene plants, ammonia plants, and many others are under digital computer control. The effects have been very substantial, leading to better control and reduced operating costs.

In the past the control laws that could be implemented by a controller were very simple, such as the proportional or proportional-integral control we discussed in Example 2.12. The fundamental revolution introduced by the digital computer in the practice of process control is the virtually unlimited intelligence that can be exhibited by such units. This phenomenon implies that the control laws that can be used are much more complex and sophisticated. Furthermore, the digital computer, with its easily programmed inherent intelligence, "can learn" as it receives measurements from the process, and it can change the control law that is implementing during the actual operation of the plant.

The digital computers have found very diversified control applications in the process industry. In Part VII (Chapters 26 through 31) we will study both the theoretical and practical aspects associated with the use of digital computers for process control. In the following paragraphs, for the time being, we will discuss some applications characteristic of the diverse usage of digital computers.

1. *Direct digital control (DDC):*   In such applications the computer receives directly the measurements from the process and, based on the control law, which is already programmed and resides in its memory, calculates the values of the manipulated variables. These decisions are now implemented directly on the process by the computer through the proper adjustment of the final control elements (valves, pumps, compressors, switches, etc.). This direct implementation of the control decisions gave rise to the name *direct digital control*, or simply DDC. Figure 3.2 illustrates a typical DDC configuration. The process can be any of the units we have already considered, such as heaters, reactors, or separators. The two interfaces before and after the computer are hardware elements and they are used to provide compatibility in the communication between computer and process. In a later chapter we will discuss the nature of these interfaces. Finally, the human operator can interact with the computer and affect the operation of the DDCs.

Today the chemical industry is moving more and more toward the DDC of the plants. A typical system of DDCs for an ethylene plant can include between 300 and 400 control loops. All the companies that furnish the control systems for the chemical industry rely more and more on DDC.

2. *Supervisory computer control:*   As we discussed earlier, one of the incentives for process control is the optimization of the plant's

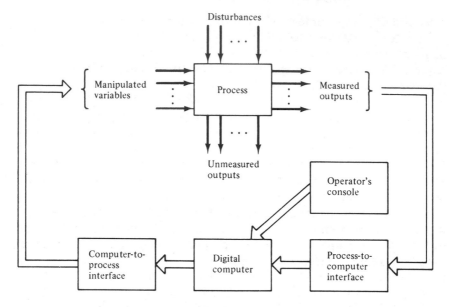

Figure 3.2 Typical DDC configuration.

economic performance. Many times the human operator does not or cannot find the best operating policy for a plant which will minimize the operating cost. This deficiency is due to the enormous complexity of a typical chemical plant. In such cases we can use the speed and the programmed intelligence of a digital computer to analyze the situation

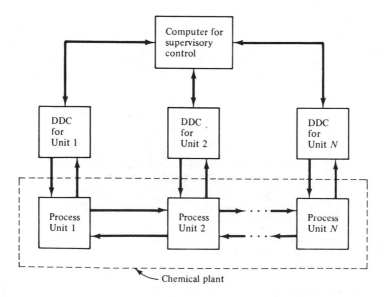

Figure 3.3 Structure of supervisory computer control.

and suggest the best policy. In doing so the computer coordinates the activities of the basic DDC loops (see Figure 3.3).

3. *Scheduling computer control:*   Finally, the computer can be used to schedule the operation of a plant. For example, the conditions in the market (demand, supply, prices) change with time, requiring the management of the chemical plant to change its operational schedule by cutting production to avoid overstocking, increasing production to meet the demand, changing over to a new production line, and so on.

These decisions can be made rationally with the aid of a digital computer, which in turn will communicate these decisions to the supervisory computer controllers. Finally, the supervisory controllers will implement these decisions on the chemical plant through the DDCs.

In subsequent chapters we will deal predominantly with the DDC and a little with supervisory computer control, but we will not concern ourselves with the scheduling computer control, which is the subject matter of a different field.

CONCLUDING REMARKS ON PART I

It is hoped that the reader now has a sketchy outline of:

1. The needs and the incentives for process control
2. The basic questions involved during the design of a control system for a chemical process
3. The hardware elements involved in a control system
4. The importance of the digital computers for the present and future implementation of advanced control techniques

In the remaining chapters we will start a systematic analysis of the various questions raised in this chapter, with our final goal that of being able to design a rational control system for a given process. Thus the subsequent chapters will be less chatty and more rigorous.

THINGS TO THINK ABOUT FOR PART I

1. What is the control objective while you are riding a unicycle or a bicycle? What are the measurements that you instinctively make while riding, and what are the manipulated variables at your disposal?

2. While you are taking a morning shower, what are your control objective, your measurement, and the manipulated variables at your disposal?

3. Compare simple feedback to simple feedforward control configuration (Figure QI.1). Which one would you trust to perform better in achieving your control objective? Why?

4. What factors should you consider in determining what variables to measure for the control of a chemical process? Answer qualitatively.

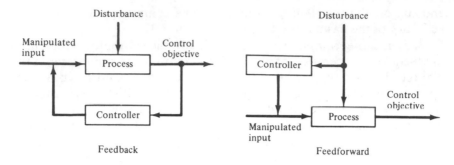

Figure QI.1

5. When is an inferential control configuration needed? What do you think is its primary weakness? Compare it to a simple feedback control configuration. Which one is preferable?

6. Describe the steps that you would go through in designing a control system for maintaining the pH of the liquid in a stirred tank (see Figure QI.2) at a desired value? What questions must you resolve? Develop both feedback and feedforward control configurations for this system.

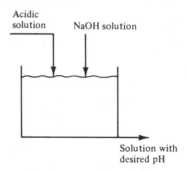

Figure QI.2

7. What is a SISO system and what is a MIMO system? Give examples from the chemical engineering field for both.

8. Define the term "control configuration" and develop three different control configurations for the pH control problem presented in item 6.

9. In the stirred tank heater system shown in Figure 1.1, the flow rate F of the effluent stream is proportional to the square root of the liquid level h in the tank. Show that such a system is self-regulating (i.e., if the inlet flow rate increases or decreases by a unit, the tank will not overflow or empty completely).

10. What is a differential pressure cell, and how does it measure the liquid level in a tank?

11. Is a venturi meter a good measuring device for monitoring and transmitting the flow rate value of a stream?

12. Determine the hardware elements required for the feedback control configuration of the pH in the stirred tank described in item 6.

13. If you were to use a digital computer as your controller in the control configurations of item 8, what new hardware elements would you need?

14. What are the basic and most important advantages offered by digital computers in process control? Discuss the size, capabilities, and prices of the most recent digital microprocessors available in the market. Do you realize the inexpensive potential that they offer for process control?

REFERENCES FOR PART I

Chapter 1. Numerous examples of the needs and incentives for process control can be found in the following books:

1. *Techniques of Process Control*, by P. S. Buckley, John Wiley & Sons, Inc., New York (1964).
2. *Process Control Systems*, 2nd ed., by F. G. Shinskey, McGraw-Hill Book Company, New York (1979).

More on the stability characteristics of CSTRs with exothermic reactions can be found in:

3. *Elementary Chemical Reactor Analysis*, by R. Aris, Prentice-Hall, Inc., Englewood Cliffs, N.J. (1969).

Chapter 2. A good discussion of the problems encountered during the design of a control system can be found in the following papers:

4. "Critique of Chemical Process Control Theory," by A. S. Foss, *AIChE J.*, *19*, 209 (1973).
5. "Advanced Control Practice in the Chemical Process Industry: A View from Industry," by W. Lee and V. W. Weekman, Jr., *AIChE J.*, *22*, 27 (1976).
6. "Design Concepts for Process Control," by A. Kestenbaum, R. Shinnar, and F. E. Thau, *Ind. Eng. Chem., Proc. Des. Dev.*, *15*, 2 (1976).

The reader is encouraged to return to these articles later after he or she has become familiar with the terminology included in the references listed above.

Chapter 3. Details on the characteristics and the design of the measuring devices, transducers, transmitters, controllers, final control elements, and recorders can be found in:

7. *The Chemical Engineer's Handbook*, J. H. Perry (ed.), 5th ed., McGraw-Hill Book Company, New York (1974).

8. *Process Intruments and Controls Handbook*, 2nd ed., D. M. Considine (ed.), McGraw-Hill Book Company, New York (1974).

An excellent reference for the computer control of the chemical processes is the book:

9. *Digital Computer Process Control*, by C. L. Smith, Intext Educational Publishers, Scranton, Pa. (1972).

Applications of computer control can be found in the following articles:

10. "Digital Control of a Distillation System," by E. N. Castellano, C. A. McCain, and F. W. Nobles, *Chem. Eng. Prog.*, *74*(4), 56 (1978).

11. "Energy Conservation via Process Computer Control," by P. R. Latour, *Chem. Eng. Prog.*, *72*(4), 76 (1976).

12. "Computer Control of Ammonia Plants," by L. C. Daigre III and G. R. Nieman, *Chem. Eng. Prog.*, *70*(2), 50 (1974).

13. "Applying Control Computers to an Integrated Plant," by A. E. Nisenfeld, *Chem. Eng. Prog.*, *69*(9), 45 (1973).

PROBLEMS FOR PART I

I.1 Consider the heat exchanger shown in Figure PI.1. Identify:
   (a) The control objectives for this system.
   (b) All the external disturbances that will affect the operation of the exchanger.
   (c) All the available manipulated variables for the control of the exchanger in the presence of disturbances.

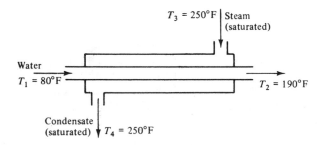

Figure PI.1

I.2 For the same heat exchanger shown in Figure PI.1, consider that the temperature $T_2 = 190°F$ is our basic control objective (i.e., maintain this temperature in the presence of disturbances). Construct two different feedback and two different feedforward control configurations that will satisfy the control objective in the presence of disturbances.

I.3 A steam turbine drives a compressor (Figure PI.2) whose load can change with time. Small variations in the shaft speed of the turbine are controlled through the use of a flyball speed governor. For this system:
(a) Identify all the external disturbances.
(b) Identify all the available manipulated variables.
Also determine the basic control objective and suggest a feedback control system that can be used to satisfy the control objective.

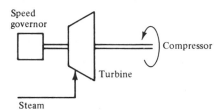

Figure PI.2

I.4 In Figure PI.3 the distillation configuration for the separation of benzene from toluene is given. The feed to the distillation comes from the reactor, where toluene has been hydrodealkylated to produce benzene:

$$\text{toluene} + H_2 \longrightarrow \text{benzene} + CH_4$$

after the excess H_2 and the produced CH_4 have been removed in a flash unit. For the distillation system:

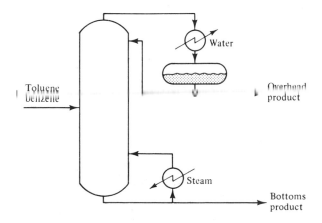

Figure PI.3

      (a) Identify all the control objectives (make sure that you have included all the operational objectives).

      (b) Identify all external disturbances.

      (c) All the available measurements and manipulated variables.

I.5 For the distillation system of Figure PI.3:

      (a) Suggest a feedforward controller that will control the operation of the column in the presence of changes in the feed flow rate.

      (b) Suggest a feedback control configuration to account for changes in the feed flow rate.

      (c) If the control objective is to keep the purity of the overhead product (benzene) constant and the use of concentration measuring devices (gas chromatographs, infrared analyzers, etc.) is not recommended due to their low reliability, suggest an inferential control configuration. What secondary measurements would you use? How would you use them, in principle, to estimate the unmeasured composition of the overhead product?

I.6 Consider the air-heating system used to regulate the temperature in a house (Figure PI.4). The heat is supplied from the combustion of fuel oil.

      (a) Identify the control objectives, the available measurements, and manipulated variables. What are the external disturbances for such a system? Is this a SISO system?

      (b) Develop a feedback control configuration to achieve your control objectives.

      (c) Is a feedforward control configuration possible for achieving your control objectives?

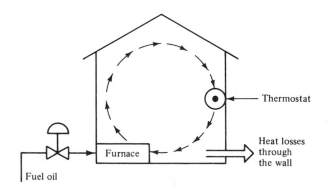

Figure PI.4

I.7 Figure PI.5 shows a system of two tanks which are used for the temporary (tank 1) and longer-term (tank 2) storage of a liquid chemical product. The demand is satisfied from the temporary storage tank, while tank 2 is used to accumulate the liquid product in excess of the demand.

(a) Identify the external disturbances, control objectives, measurements, and manipulated variables available to you. Is this a SISO or a MIMO system?

(b) Develop alternative feedback and/or feedforward control configurations to achieve your control objectives.

(c) Is there any situation that may arise during which you cannot avoid overflowing of the storage tanks?

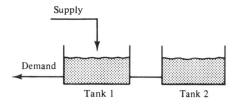

Figure PI.5

I.8   Consider a system of two continuous stirred tank reactors in series (Figure PI.6), where the following endothermic reaction takes place:

$$A + catalyst \longrightarrow B$$

(a) Identify the control objectives for the operation of the two CSTRs.

(b) Classify the variables of the system into inputs and outputs and subsequently classify the inputs into disturbances and manipulated variables and the outputs into measured and unmeasured outputs. Is this a SISO or a MIMO system?

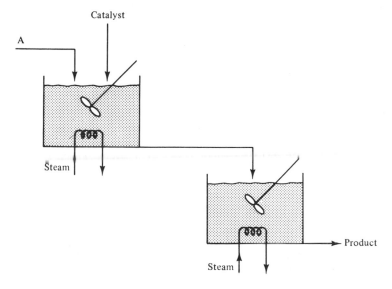

Figure PI.6

   (c) Develop a feedback control configuration that satisfies your objectives using a composition analyzer at the exit stream of the second CSTR.

   (d) Develop an inferential control configuration that uses temperatures and flow rates only, assuming that a composition analyzer is not available.

   (e) Develop a feedforward control configuration that can also use composition analyzers if they are needed.

   (f) In your opinion, which system is easier to control, the two-CSTR system shown in Figure PI.6 or an equivalent one-CSTR system that achieves the same conversion? Explain qualitatively why.

I.9 Consider a tubular catalytic reactor where an endothermic reaction A → B takes place (Figure PI.7a). The reacting mixture is heated with steam flowing in a jacket around the tubular reactor. The stream of the raw material A includes a chemical C which poisons the catalyst over a period of five days. As the catalyst decays, the conversion of A to B decreases. We can make up for this decrease by increasing the temperature of the reacting mixture, which can be achieved by increasing the supply of heat to the reactor through higher pressure steam. Let us assume that the reactor is isothermal along its length. Figure PI.7b

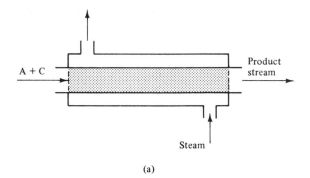

(a)

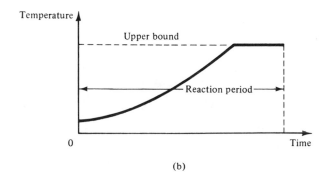

(b)

Figure PI.7

shows the temperature in the reactor during the reaction period that maximizes the profit from the operation of the tubular reactor.

(a) Formulate the optimization problem that yields the temperature profile of Figure PI.7b as its solution.

(b) Develop a feedback control system that will regulate the temperature of the reactor to that shown in Figure PI.7b. What is the control objective here?

(c) Draw a control system that uses a digital computer for the implementation of the feedback system in part (b). Include measuring devices, transmission lines, final control elements, and whatever else is necessary.

I.10 Two liquid streams with flow rates F_1 and F_2 and temperatures T_1 and T_2 flow through two separate pipes which converge at a mixing junction (Figure PI.8). We want to maintain constant the flow rate F_3 and the temperature T_3 of the liquid stream resulting from the mixing of the first two streams.

(a) Identify the control objectives, disturbances, available measurements, and manipulated variables. Is this a SISO or a MIMO system?

(b) Develop a control system that uses only feedback controllers.

(c) Develop a control system that uses only feedforward controllers.

(d) Develop two different control systems that use both feedback and feedforward controllers.

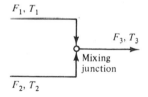

Figure PI.8

Modeling the Dynamic and Static Behavior of Chemical Processes ▐▐

What goes on in the modeller's head is not purely formalizable, either in abstract terms...or in taxonomic views....It has structure, it has technique that can be taught and learned, but involves also a personal touch, not only in trivialities but in deeper considerations of skill and suitability....

R. Aris and M. Penn*

In order to analyze the behavior of a chemical process and to answer some of the questions raised in previous chapters about its control, we need a mathematical representation of the physical and chemical phenomena taking place in it. Such a mathematical representation constitutes the *model* of the system, while the activities leading to the construction of the model will be referred to as *modeling*.

Modeling a chemical process is a very synthetic activity, requiring the use of all the basic principles of chemical engineering science, such as thermodynamics, kinetics, transport phenomena, etc. For the design of controllers for chemical processes, modeling is a very critical step. It should be approached with care and thoughtfulness.

The purpose of the following two chapters is:

1. To explain why we need to develop a mathematical description (model) of a chemical process as a prerequisite to the design of its controller

---

*"The Mere Notion of a Mathematical Model," *Int. J. Math. Modeling*, *1*, 1 (1980).

2. To describe a methodology for the modeling of a chemical process using the balance equations and provide examples of its implementation
3. To determine the scope and the difficulties of the mathematical modeling for process control purposes

It should be noted that the subsequent chapters do not constitute a complete treatment of all the aspects on mathematical modeling, but it is limited to those of interest for process control.

Development   **4**
of a Mathematical Model

Consider a general processing system with its associated variables as shown in Figure 2.1. To investigate how the behavior of a chemical process (i.e., its outputs) changes with time under the influence of changes in the external disturbances and manipulated variables and consequently design an appropriate controller, we can use two different approaches:

1. *Experimental approach*: In this case the physical equipment(s) of the chemical process is available to us. Consequently, we change deliberately the values of various inputs (disturbances, manipulated variables) and through appropriate measuring devices we observe how the outputs (temperatures, pressures, flow rates, concentrations) of the chemical process change with time. Such a procedure is time and effort consuming and it is usually quite costly because a large number of such experiments must be performed

2. *Theoretical approach*: It is quite often the case that we have to design the control system for a chemical process before the process has been constructed. In such a case we cannot rely on the experimental procedure, and we need a different representation of the chemical process in order to study its dynamic behavior. This representation is usually given in terms of a set of mathematical equations (differential, algebraic) whose solution yields the dynamic or static behavior of the chemical process we examine.

In this text we discuss both approaches for the development of a model for a chemical process. Initially, we will examine the theoretical approach, leaving the experimental for subsequent chapters (Chapters 16 and 31).

4.1   Why Do We Need Mathematical Modeling for Process Control?

Let us repeat that our goal is to develop a control system for a chemical process which will guarantee that the operational objectives of our process are satisfied in the presence of ever-changing disturbances. Then, why do we need to develop a mathematical description (model) for the process we want to control?

In the introductory paragraphs earlier we noted that often the physical equipment of the chemical process we want to control have not been constructed. Consequently, we cannot experiment to determine how the process reacts to various inputs and therefore we cannot design the appropriate control system. But even if the process equipment is available for experimentation, the procedure is usually very costly. Therefore, we need a simple description of how the process reacts to various inputs, and this is what the mathematical models can provide to the control designer.

Let us demonstrate now in terms of some examples the need for the development of a mathematical model before we design the control system for a chemical process.

Example 4.1: Design an Integral Controller for a Stirred Tank Heater

Consider the problem of controlling the temperature of a liquid in a tank using integral control (Example 2.12). From Figure 2.8 we notice that the quality of the control depends on the value of the parameter α'. But the question is: How does α' affect the quality of control, and what is its best value? To answer this question we need to know how the value of the liquid temperature T is affected by changes in the value of the inlet temperature T_i or the integral action of the controller. This is given by eq. (2.7), which constitutes the mathematical model of the tank with integral control.

Example 4.2: Design a Feedforward Controller for a Process

In the feedforward control arrangement shown in Figure 4.1 we measure the value of the disturbance and we anticipate what its effect will be on the output of the process that we want to control. In order to keep the

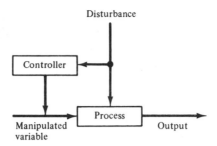

Figure 4.1 Feedforward control configuration.

value of this output at the desired level, we need to change the value of the manipulated variable by such an amount as to eliminate the impact that the disturbance would have on the output. The question is: By how much should we change the manipulated variable in order to cancel the effect of the disturbance? To answer this question we must know the following two relationships:

$$\text{output} = f_1(\text{disturbance})$$

$$\text{output} = f_2(\text{manipulated variable})$$

which are provided by a mathematical model of the process. Indeed, if the output is to remain the same, the manipulated variable must take such a value that

$$f_1(\text{disturbance}) - f_2(\text{manipulated variable}) = 0$$

This example demonstrates very vividly how important mathematical modeling is for the design of a feedforward control system. In fact, without good and accurate mathematical modeling we cannot design efficient feedforward control systems.

Example 4.3: Design of an Inferential Control System

In the inferential control scheme shown in Figure 4.2 we measure the measured output and try to regulate the value of the unmeasured control objective at a desired value. Since the control objective is not measured directly, it can only be estimated from the value of the measured output if a relationship such as the following is available:

$$\text{control objective} = f(\text{measured output})$$

Such a relationship in turn is not possible if we do not have a mathematical representation of the process (mathematical model). Once the value of the control objective can be estimated from a relationship such as the above, it can be compared to the desired value (set point) and the controller can be activated for appropriate action as in feedback control.

We notice, therefore, that the availability of a good mathematical model for the process is indispensable for the design of good inferential control systems.

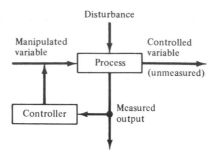

Figure 4.2 Inferential control configuration.

The three examples above indicate very clearly that mathematical modeling of a process is of paramount importance for the design of good and efficient control systems for a chemical process. In the following sections we develop a methodology for the concise modeling of chemical processes.

4.2 State Variables and State Equations for a Chemical Process

In order to characterize a processing system (tank heater, batch reactor, distillation column, heat exchanger, etc.) and its behavior we need:

1. A set of fundamental dependent quantities whose values will describe the natural state of a given system
2. A set of equations in the variables above which will describe how the natural state of the given system changes with time

For most of the processing systems of interest to a chemical engineer there are only three such fundamental quantities: mass, energy, and momentum. Quite often, though, the fundamental dependent variables cannot be measured directly and conveniently. In such cases we select other variables which can be measured conveniently, and when grouped appropriately they determine the value of the fundamental variables. Thus mass, energy, and momentum can be characterized by variables such as density, concentration, temperature, pressure, and flow rate. These characterizing variables are called *state variables* and their values define the *state* of a processing system.

The equations that relate the state variables (dependent variables) to the various independent variables are derived from application of the *conservation principle* on the *fundamental quantities* and are called *state equations*.

The *principle of conservation* of a quantity S states that:

$$\frac{\begin{bmatrix} \text{accumulation of } S \\ \text{within a system} \end{bmatrix}}{\text{time period}} = \frac{\begin{bmatrix} \text{flow of } S \\ \text{in the system} \end{bmatrix}}{\text{time period}} - \frac{\begin{bmatrix} \text{flow of } S \\ \text{out of the system} \end{bmatrix}}{\text{time period}}$$

(4.1)

$$+ \frac{\begin{bmatrix} \text{amount of } S \\ \text{generated within} \\ \text{the system} \end{bmatrix}}{\text{time period}} - \frac{\begin{bmatrix} \text{amount of } S \\ \text{consumed within} \\ \text{the system} \end{bmatrix}}{\text{time period}}$$

The quantity S can be any of the following fundamental quantities:

Total mass
Mass of individual components
Total energy
Momentum

Remark.   It should be remembered that for the physical and chemical processes we will be studying, the total mass and total energy cannot be generated from nothing; neither do they disappear.

Let us review now the forms used most often for the balance equations. Consider the system shown in Figure 4.3. We have:

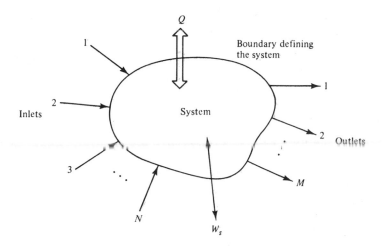

Figure 4.3 A general system and its interactions with the external world.

Total mass balance:

$$\frac{d(\rho V)}{dt} = \sum_{i:\text{inlet}} \rho_i F_i - \sum_{j:\text{outlet}} \rho_j F_j \tag{4.1a}$$

Mass balance on component A:

$$\frac{d(n_A)}{dt} = \frac{d(c_A V)}{dt} = \sum_{i:\text{inlet}} c_{A_i} F_i - \sum_{j:\text{outlet}} c_{A_j} F_j \pm rV \tag{4.1b}$$

Total energy balance:

$$\frac{dE}{dt} = \frac{d(U + K + P)}{dt} = \sum_{i:\text{inlet}} \rho_i F_i h_i - \sum_{j:\text{outlet}} \rho_j F_j h_j \pm Q \pm W_s \tag{4.1c}$$

The variables appearing in the equations above have the following meaning:

ρ = density of the material in the system

ρ_i = density of the material in the ith inlet stream

ρ_j = density of the material in the jth outlet stream

V = total volume of the system

F_i = volumetric flow rate of the ith inlet stream

F_j = volumetric flow rate of the jth outlet stream

n_A = number of moles of component A in the system

c_A = molar concentration (moles/volume) of A in the system

c_{A_i} = molar concentration of A in the ith inlet

c_{A_j} = molar concentration of A in the jth outlet

r = reaction rate per unit volume for component A in the system

h_i = specific enthalpy of the material in the ith inlet stream

h_j = specific enthalpy of the material in the jth outlet stream

U, K, P = internal, kinetic, and potential energies of the system, respectively

Q = amount of heat exchanged between the system and its surroundings per unit time

W_s = shaft work exchanged between the system and its surroundings per unit time

By convention, a quantity is considered positive if it flows *in* the system and negative if it flows *out*.

The state equations with the associated state variables constitute the *mathematical model* of a process, which yields the dynamic or static behavior of the process. The application of the conservation principle as defined by eqs. (4.1) will yield a set of *differential equations* with the fundamental quantities as the dependent variables and time as the independent variable. The solution of the differential equations will determine how the fundamental quantities, or equivalently, the state variables, change with time; that is, it will determine the *dynamic behavior* of the process.

If the state variables do not change with time, we say that the process is at *steady state*. In this case, the rate of accumulation of a fundamental quantity S per unit of time is zero, and the resulting balances yield a set of *algebraic equations*.

Example 4.4: State Variables and State Equations for a Stirred Tank Heater; Its Static and Dynamic Behavior

Consider the stirred tank heater of Example 1.1 (Figure 1.1). The fundamental quantities whose values provide the information about the heater are:

(a) The total mass of the liquid in the tank
(b) The total energy of the material in the tank
(c) Its momentum

The momentum of the heater remains constant even when the disturbances change value and will not be considered further.

Let us now identify the state variables for the tank heater.

Total mass in the tank:

$$\text{total mass} = \rho V = \rho A h \qquad (4.2)$$

where ρ the density of liquid, V the volume of liquid, A the cross-sectional area of the tank, and h the height of the liquid level.

Total energy of the liquid in the tank:

$$E = U + K + P$$

but since the tank does not move, $dK/dt = dP/dt = 0$ and $dE/dt = dU/dt$. For liquid systems,

$$\frac{dU}{dt} \simeq \frac{dH}{dt}$$

where H is the total enthalpy of the liquid in the tank. Furthermore,

$$H = \rho V c_p (T - T_{ref}) = \rho A h c_p (T - T_{ref}) \qquad (4.3)$$

where c_p = heat capacity of the liquid in the tank
T_{ref} = reference temperature where the specific enthalpy of the liquid is assumed to be zero.

From eqs. (4.2) and (4.3) we conclude that the *state variables* for the stirred tank heater are the following;

State variables: h and T

while the

Constant parameters: ρ, A, c_p, T_{ref}

are characteristic of the tank system.

Note. It has been assumed that the density ρ is independent of the temperature.

Let us now proceed to develop the state equations for the stirred tank heater. We will apply the conservation principle on the two fundamental quantities: the total mass and the total energy.

Total mass balance:

$$\frac{\left[\begin{array}{c}\text{accumulation of}\\\text{total mass}\end{array}\right]}{\text{time}} = \frac{\left[\begin{array}{c}\text{input of}\\\text{total mass}\end{array}\right]}{\text{time}} - \frac{\left[\begin{array}{c}\text{output of}\\\text{total mass}\end{array}\right]}{\text{time}}$$

or

$$\frac{d(\rho Ah)}{dt} = \rho F_i - \rho F \tag{4.4}$$

where F_i and F are the volumetric flow rates [i.e., volume per unit of time (ft^3/min, or m^3/min)] for the inlet and outlet streams, respectively. Assuming constant density (independent of temperature), eq. (4.4) becomes

$$A\frac{dh}{dt} = F_i - F \tag{4.4a}$$

Total energy balance:

$$\frac{\left[\begin{array}{c}\text{accumulation of}\\\text{total energy}\end{array}\right]}{\text{time}} = \frac{\left[\begin{array}{c}\text{input of}\\\text{total energy}\end{array}\right]}{\text{time}} - \frac{\left[\begin{array}{c}\text{output of}\\\text{total energy}\end{array}\right]}{\text{time}}$$

$$+ \frac{\left[\begin{array}{c}\text{energy supplied}\\\text{by steam}\end{array}\right]}{\text{time}}$$

or

$$\frac{d[\rho Ah\, c_p(T - T_{ref})]}{dt} = \rho F_i c_p(T_i - T_{ref}) - \rho F c_p(T - T_{ref}) + Q \tag{4.5}$$

where Q is the amount of heat supplied by the steam per unit of time. The equation above can take the following simpler form (assume that $T_{ref} = 0$):

$$A\frac{d(hT)}{dt} = F_i T_i - FT + \frac{Q}{\rho c_p} \tag{4.5a}$$

Additional algebraic manipulations on eq. (4.5a) yield

$$A \frac{d(hT)}{dt} = Ah \frac{dT}{dt} + AT \frac{dh}{dt} = Ah \frac{dT}{dt} + T(F_i - F) = F_iT_i - FT + \frac{Q}{\rho c_p}$$

or

$$Ah \frac{dT}{dt} = F_i(T_i - T) + \frac{Q}{\rho c_p} \qquad (4.5b)$$

Summarizing the modeling steps above, we have:

State equations:

$$A \frac{dh}{dt} = F_i - F \qquad (4.4a)$$

$$Ah \frac{dT}{dt} = F_i(T_i - T) + \frac{Q}{\rho c_p} \qquad (4.5b)$$

The variables in eqs. (4.4a) and (4.5b) can be classified as follows (see also Section 2.1):

State variables: h, T
Output variables: h, T (both measured)
Input variables
  Disturbances: T_i, F_i
  Manipulated variables: Q, F (for feedback control)
                          F_i (for feedforward control)
Parameters: A, ρ, c_p

The state equations (4.4a) and (4.5b), with the state variables, the inputs, and the parameters, constitute the *mathematical model* of the stirred tank heater. We need only solve them in order to find the tank's dynamic or steady-state behavior.

Let us now study the dynamic and static behavior of the stirred tank heater using the state equations (4.4a) and (4.5b). We will assume that initially the tank heater is at steady state (i.e., nothing is changing). This situation is described by the state equations if the rate of accumulation [left-hand sides of (4.4a) and (4.5b)] is set to zero:

$$F_{i,s} - F_s = 0$$

$$F_{i,s}(T_{i,s} - T_s) + \frac{Q_s}{\rho c_p} = 0$$

The subscript s denotes the steady-state value of the corresponding variable.

The system will be disturbed from the steady-state situation if any of the input variables changes value. Let us examine the following two situations:

1. Consider that the inlet temperature T_i decreases by 10% from its steady-state value. The liquid level will remain the same at the

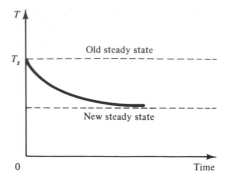

Figure 4.4 Temperature response of a stirred heater to a step decrease in inlet temperature.

steady-state value h_s, since T_i does not influence the total mass in the tank [see also eq. (4.4a)]. On the contrary, the temperature of the liquid will start decreasing with time. How the temperature T changes with time will be determined from the solution of eq. (4.5b) using as *initial condition* the steady-state value of T:

$$T(t = 0) = T_s$$

Figure 4.4 indicates the static and dynamic behavior of the tank for this case. We observe that after a certain time the tank heater again reaches steady-state conditions.

2. Consider that initially the tank heater is at steady state. Then, at time $t = 0$, the inlet flow rate decreases by 10%. It is clear that both the level and the temperature of the liquid in the tank will start changing [notice that F_i is present in both state equations (4.4a) and (4.5b)]. How h and T change with time will be given from the solution of eqs. (4.4a) and (4.5b) using as initial conditions

$$h(t = 0) = h_s \quad \text{and} \quad T(t = 0) = T_s$$

Figure 4.5 summarizes the static and dynamic behavior of the tank heater for this case.

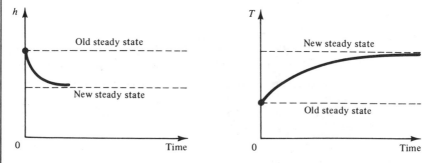

Figure 4.5 Dynamic response of a stirred tank heater to a step decrease in inlet flow rate.

Remark.   It is worth noting that after F_i has changed, the level h and the temperature T reach their new steady states with different speeds. In particular, the level, h, achieves its new steady state faster than the temperature. In a subsequent chapter we will analyze the reasons for such behavior.

4.3   Additional Elements of the Mathematical Models

In addition to the balance equations, we need other relationships to express thermodynamic equilibria, reaction rates, transport rates for heat, mass, momentum, and so on. Such additional relationships needed to complete the mathematical modeling of various chemical and/or physical processes can be classified as follows:

Transport rate equations

They are needed to describe the rate of mass, energy, and momentum transfer between a system and its surroundings. These equations are developed in courses on transport phenomena.

Example 4.5

The amount of heat Q supplied by the steam to the liquid of the tank heater (Example 4.4) is given by the following heat transfer rate equation:

$$Q = UA_t(T_{st} - T)$$

where U = overall heat transfer coefficient
    A_t = total area of heat transfer
    T_{st} = temperature of the steam

Kinetic rate equations

They are needed to describe the rates of chemical reactions taking place in a system. Such equations are developed in a course on chemical kinetics.

Example 4.6

The reaction rate of a first-order reaction taking place in a CSTR is given by

$$r = k_0 e^{-E/RT} c_A$$

where k_0 = preexponential kinetic constant
        E = activation energy for the reaction
        R = ideal gas constant
        T, c_A = temperature and concentration of the reacting fluid.

Reaction and phase equilibria relationships

These are needed to describe the equilibrium situations reached during a chemical reaction or by two or more phases. These relationships are developed in courses on thermodynamics.

Example 4.7

Consider a liquid stream composed of two components A and B at a high pressure p_f and temperature T_f. If the pressure p_f is larger than the bubble-point pressure of the liquid at temperature T_f, no vapor phase will be present. The liquid stream passes through a restriction (valve) and is "flashed" in a drum; that is, its pressure is reduced from p_f to p (Figure 4.6). This abrupt expansion takes place under constant enthalpy. If the pressure p in the drum is smaller than the bubble-point pressure of the liquid stream at the temperature T_f, the liquid will partially vaporize and two phases at equilibrium with each other will be present in the flash drum.

The thermodynamic equilibrium between the vapor and liquid phases imposes certain restrictions on the state variables of the system, and must be included in the mathematical model of the flash drum if it is to be consistent and correct. These equilibrium relationships, as known from chemical thermodynamics, are:

1. Temperature of liquid phase = temperature of vapor phase

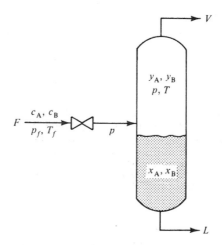

Figure 4.6 Flash drum unit.

2. Pressure of liquid phase = pressure of vapor phase
3. Chemical potential of component i in the liquid phase = chemical potential of component i in the vapor phase

The equilibrium relationships introduce additional equations among the state variables of a system. Care must be exercised so that all the equilibrium relationships are accounted for.

Equations of state

Equations of state are needed to describe the relationship among the intensive variables describing the thermodynamic state of a system. The ideal gas law and the van der Waals equation are two typical equations of state for gaseous systems.

Example 4.8

Let us return to the flash drum system discussed above in Example 4.7. For the vapor phase, from the ideal gas law, we have

$$pV_{vapor} = (\text{moles of A} + \text{moles of B}) \cdot RT \qquad (4.6)$$

But

$$\text{moles of A} + \text{moles of B} = \frac{\text{mass of A} + \text{mass of B}}{\text{average molecular weight}}$$

Therefore, from eq. (4.6) we have

$$\rho_{vapor} = \frac{\text{mass of A} + \text{mass of B}}{V_{vapor}} = \frac{p}{RT} \cdot (\text{average molecular weight})$$

Considering that

$$\text{average molecular weight} = y_A M_A + y_B M_B$$

we have

$$\rho_{vapor} = \frac{p}{RT}[y_A M_A + y_B M_B] \qquad (4.6a)$$

where y_A and y_B are the molar fractions of components A and B and M_A and M_B are the molecular weights of A and B. Equation (4.6a) indicates a relationship among the state variables of the flash drum and must be included in the mathematical model of the flash drum.

Starting from an appropriate equation of thermodynamic state for the liquid phase of the flash drum, we can develop an expression for its density of the form

$$\rho_{liquid} = \rho(T, x_A)$$

where x_A is the molar fraction of component A in the liquid.

4.4   Dead Time

In all of the modeling examples discussed in earlier sections it was assumed that whenever a change takes place in one of the input variables (disturbances, manipulated variables), its effect is instantaneously observed in the state variables and the outputs. Thus whenever the feed composition, c_{A_i}, or the feed temperature, T_i, or the coolant temperature, T_{c_i}, change in the CSTR of Figure 1.7, the effect of the change is felt immediately and the temperature T or concentration c_A of the outlet stream start changing.

The oversimplified picture given above is contrary to our physical experience, which dictates that whenever an input variable of a system changes, there is a time interval (short or long) during which no effect is observed on the outputs of the system. This time interval is called *dead time*, or *transportation lag*, or *pure delay*, or *distance-velocity lag*.

Example 4.9

Consider the flow of an incompressible, nonreacting liquid through a pipe (Figure 4.7a). If the pipe is completely thermally insulated and the heat generated by the friction of the flowing fluid is negligible, it is easy to see that at steady state the temperature T_{out} of the outlet stream will be equal to that of the inlet, T_{in}. Assume now that starting at $t = 0$, the temperature of the inlet changes as shown by curve A in Figure 4.7b. It is

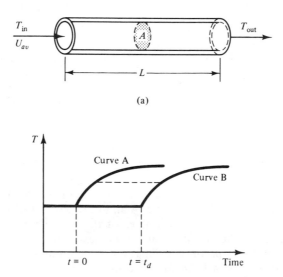

(a)

(b)

Figure 4.7 (a) Pipe flow of Example 4.9; (b) delayed response of exit temperature to inlet temperature change.

clear that the temperature of the outlet, T_{out}, will remain the same until the change reaches the end of the pipe. Then we will observe the temperature of the outlet changing, as shown by curve B in Figure 4.7b. We notice that the change of the outlet temperature follows the same pattern as the change of the inlet temperature with a delay of t_d seconds. t_d is the dead time and from physical considerations it is easy to see that

$$t_d = \frac{\text{volume of the pipe}}{\text{volumetric flow rate}} = \frac{A \cdot L}{A \cdot U_{av}} = \frac{L}{U_{av}} \quad \text{seconds}$$

where U_{av} is the average velocity of the fluid over the cross-sectional area of the pipe, assumed to be constant. Functionally, we can relate T_{in} and T_{out} as follows:

$$T_{out}(t) = T_{in}(t - t_d) \tag{4.7}$$

The dead time is an important element in the mathematical modeling of chemical processes and has a serious impact on the design of effective controllers. As we will see in Chapter 19, the presence of dead time can very easily destabilize the dynamic behavior of a controlled system.

4.5   Additional Examples of Mathematical Modeling

In this section we examine some typical chemical processes and develop their mathematical models.

Example 4.10: Mathematical Model of a Continuous Stirred Tank Reactor (CSTR)

Consider the continuous stirred tank reactor system discussed in Example 1.2 (Figure 1.7). A simple exothermic reaction A → B takes place in the reactor, which is in turn cooled by a coolant that flows through a jacket around the reactor.

The fundamental dependent quantities for the reactor are:

(a) Total mass of the reacting mixture in tank
(b) Mass of chemical A in the reacting mixture
(c) Total energy of the reacting mixture in the tank

Remarks

1. The mass of component B can be found from the total mass and the mass of component A. Therefore, it is not an independent fundamental quantity.
2. The momentum of the CSTR does not change under any operating conditions for the reactor and will be neglected.

Let us apply the conservation principle on the three fundamental quantities:

Total mass balance:

$$\frac{\begin{bmatrix} \text{accumulation} \\ \text{of total mass} \end{bmatrix}}{\text{time}} = \frac{\begin{bmatrix} \text{input of} \\ \text{total mass} \end{bmatrix}}{\text{time}} - \frac{\begin{bmatrix} \text{output of} \\ \text{total mass} \end{bmatrix}}{\text{time}} \pm \frac{\begin{bmatrix} \text{total mass generated} \\ \text{or consumed} \end{bmatrix}}{\text{time}}$$

or

$$\frac{d(\rho V)}{dt} = \rho_i F_i - \rho F \pm 0 \tag{4.8}$$

where ρ_i, ρ = densities of the inlet and outlet streams

F_i, F = volumetric flow rates of the inlet and outlet streams, ft^3/min or m^3/min

V = volume of the reacting mixture in the tank

Mass balance on component A:

$$\frac{\begin{bmatrix} \text{accumulation} \\ \text{of A} \end{bmatrix}}{\text{time}} = \frac{\begin{bmatrix} \text{input} \\ \text{of A} \end{bmatrix}}{\text{time}} - \frac{\begin{bmatrix} \text{output} \\ \text{of A} \end{bmatrix}}{\text{time}} - \frac{\begin{bmatrix} \text{disappearance of A} \\ \text{due to reaction} \end{bmatrix}}{\text{time}}$$

or

$$\frac{d(n_A)}{dt} = \frac{d(c_A V)}{dt} = c_{A_i} F_i - c_A F - rV \tag{4.9}$$

where r = rate of reaction per unit volume

c_{A_i}, c_A = molar concentrations (moles/volume) of A in the inlet and outlet streams and

n_A = number of moles of A in the reacting mixture

Total energy balance:

$$\frac{\begin{bmatrix} \text{accumulation of} \\ \text{total energy} \end{bmatrix}}{\text{time}}$$

$$= \frac{\begin{bmatrix} \text{input of total} \\ \text{energy with feed} \end{bmatrix}}{\text{time}} - \frac{\begin{bmatrix} \text{output of total} \\ \text{energy with outlet} \end{bmatrix}}{\text{time}} - \frac{\begin{bmatrix} \text{energy removed} \\ \text{by coolant} \end{bmatrix}}{\text{time}}$$

In the balance above we have neglected the shaft work done by the impeller of the stirring mechanism. The total energy of the reacting mixture is

$$E = U + K + P$$

where U is the internal energy, K the kinetic energy, and P the potential energy of the reacting mixture. Therefore, assuming that the reactor does not move (i.e., $dK/dt = dP/dt = 0$), the left-hand side of the total energy balance yields

$$\frac{dE}{dt} = \frac{d(U + K + P)}{dt} = \frac{dU}{dt}$$

Since the system is a liquid system, we can make the following approximation:

$$\begin{bmatrix} \text{accumulation of total} \\ \text{energy of the} \\ \text{material in the CSTR} \\ \text{per unit time} \end{bmatrix} = \frac{dU}{dt} \simeq \frac{dH}{dt} = \begin{bmatrix} \text{accumulation of total} \\ \text{enthalpy of the} \\ \text{material in the CSTR} \\ \text{per unit time} \end{bmatrix}$$

Furthermore,

$$\text{input of total energy with feed per unit time} = \rho_i F_i \, h_i(T_i)$$

and

$$\text{output of total energy with the outlet stream per unit time} = \rho F \, h(T)$$

where h_i is the specific enthalpy (enthalpy per unit mass) of the feed stream and h is the specific enthalpy of the outlet stream. Consequently, the total energy balance leads to the equation

$$\frac{dH}{dt} = \rho_i F_i h_i(T_i) - \rho F h(T) - Q \tag{4.10}$$

where Q is the amount of heat removed by the coolant per unit time.

Equations (4.8), (4.9) and (4.10) are not in their final and most convenient form for process control design studies. To bring them to such form *we need to identify the appropriate state variables*.

Characterize Total Mass. We need the density of the reacting mixture, ρ, and its volume, V. The density will be a function of the concentration c_A and c_B and of the temperature T. Quite often the dependence of ρ on c_A, c_B, and T is weak and the density can be considered constant as the reaction proceeds. Therefore, the left-hand side of eq. (4.8) yields

$$\frac{d(\rho V)}{dt} = \rho \frac{dV}{dt} \qquad \text{while} \qquad \rho_i = \rho$$

Under the assumption above, V is the only state variable that is needed to characterize the total mass. Then eq. (4.8) becomes

$$\frac{dV}{dt} = F_i - F \tag{4.8a}$$

Characterize the Mass of Component A. This is simple. From eq. (4.9) we realize that the state variables needed are c_A and V. Algebraic manipulations on eq. (4.9) lead to

$$\frac{d(c_A V)}{dt} = c_A \frac{dV}{dt} + V \frac{dc_A}{dt} = c_{A_i} F_i - c_A F - k_0 e^{-E/RT} c_A V$$

or

$$V \frac{dc_A}{dt} = -c_A(F_i - F) + c_{A_i}F_i - c_A F - k_0 e^{-E/RT} c_A V$$

and finally,

$$\frac{dc_A}{dt} = \frac{F_i}{V}(c_{A_i} - c_A) - k_0 e^{-E/RT} c_A \qquad (4.9a)$$

[*Note.*   We have introduced $r = k_0 e^{-E/RT} c_A$.]

Characterize the Total Energy. We know from thermodynamics that the enthalpy of a liquid system is a function of the temperature and its composition:

$$H = H(T, n_A, n_B)$$

where n_A and n_B are the moles of A and B in the CSTR. Differentiating the expression above, we take

$$\frac{dH}{dt} = \frac{\partial H}{\partial T}\frac{dT}{dt} + \frac{\partial H}{\partial n_A}\frac{dn_A}{dt} + \frac{\partial H}{\partial n_B}\frac{dn_B}{dt} \qquad (4.11)$$

But

$$\frac{\partial H}{\partial T} = \rho V c_p \qquad \frac{\partial H}{\partial n_A} = \tilde{H}_A(T) \qquad \frac{\partial H}{\partial n_B} = \tilde{H}_B(T)$$

where c_p is the specific heat capacity of the reacting mixture and $\tilde{H}_A$ and $\tilde{H}_B$ are the partial molar enthalpies of A and B. Furthermore, from eq. (4.9),

$$\frac{dn_A}{dt} = \frac{d(c_A V)}{dt} = c_{A_i}F_i - c_A F - rV$$

and a similar balance on component B,

$$\frac{dn_B}{dt} = \frac{d(c_B V)}{dt} = 0 - c_B F + rV$$

Substitute the quantities above in eq. (4.11) and take

$$\frac{dH}{dt} = \rho V c_p \frac{dT}{dt} + \tilde{H}_A[c_{A_i}F_i - c_A F - rV] + \tilde{H}_B[-c_B F + rV]$$

Substitute dH/dt by its equal from the total energy balance [eq. (4.10)] and take

$$\rho V c_p \frac{dT}{dt} \qquad\qquad (4.10a)$$

$$= -\tilde{H}_A[c_{A_i}F_i - c_A F - rV] - \tilde{H}_B[-c_B F + rV] + \rho_i F_i h_i - \rho F h - Q$$

Let us now notice that

$$F_i \rho_i h_i(T_i) = F_i[\rho_i h_i(T) + \rho_i c_{p_i}(T_i - T)] = F_i[c_{A_i}\tilde{H}_A(T) + \rho_i c_{p_i}(T_i - T)]$$

and

$$F\rho h(T) = F[c_A \tilde{H}_A(T) + c_B \tilde{H}_B(T)]$$

Consequently, eq. (4.10a) becomes

$$\rho V c_p \frac{dT}{dt} = -\tilde{H}_{AC}\cancel{c_{A_i}F_i} + \cancel{\tilde{H}_{AC}c_A F} + \tilde{H}_A rV + \cancel{\tilde{H}_{BC}c_B F} - \tilde{H}_B rV$$

$$+ F_i \cancel{c_{A_i}\tilde{H}_A} + F_i \rho_i c_{p_i}(T_i - T) - F\cancel{c_A \tilde{H}_A} - \cancel{Fc_B \tilde{H}_B} - Q$$

or

$$\rho V c_p \frac{dT}{dt} = F_i \rho_i c_{p_i}(T_i - T) + (\tilde{H}_A - \tilde{H}_B)rV - Q$$

Finally, since $(\tilde{H}_A - \tilde{H}_B) = (-\Delta H_r) = $ heat of reaction at temperature T, and $\rho = \rho_i$, $c_p = c_{p_i}$,

$$V \frac{dT}{dt} = F_i(T_i - T) + \frac{(-\Delta H_r)rV}{\rho c_p} - \frac{Q}{\rho c_p} \qquad (4.10b)$$

From eq. (4.10b) we conclude that temperature T is the state variable that characterizes the total energy of the system.

Summarizing all the steps above in the mathematical modeling of a CSTR, we have the following:

State variables: V, c_A, T
State equations:

$$\frac{dV}{dt} = F_i - F \qquad (4.8a)$$

$$\frac{dc_A}{dt} = \frac{F_i}{V}(c_{A_i} - c_A) - k_0 e^{-E/RT} c_A \qquad (4.9a)$$

$$\frac{dT}{dt} = \frac{F_i}{V}(T_i - T) + Jk_0 e^{-E/RT} c_A - \frac{Q}{\rho c_p V} \qquad (4.10b)$$

where $J = (-\Delta H_r)/\rho c_p$.

Output variables: V, c_A, T
Input variables: c_{A_i}, F_i, T_i, Q, F (when feedback control is used)

Among the input variables the most common disturbances are:

Disturbances: c_{A_i}, F_i, T_i

while the usual manipulated variables are:

Manipulated variables: Q, F (occasionally F_i or T_i)

The remaining variables are parameters characteristic of the reactor system:

Constant parameters: ρ, c_p, $(-\Delta H_r)$, k_0, E (activation energy), R

In the presence of changes in the input variables, the state variables change. Integration of eqs. (4.8a), (4.9a), and (4.10b) yields $V(t)$, $c_A(t)$, and $T(t)$ as functions of time.

The steady-state behavior of the CSTR is given by eqs. (4.8a), (4.9a), and (4.10b) if their left-hand sides are set equal to zero.

Example 4.11: Mathematical Model of a Mixing Process

Two streams 1 and 2 are being mixed in a well-stirred tank, producing a product stream 3 (Figure 4.8). Each of the two feed streams is composed of two components, A and B, with molar concentrations c_{A_1}, c_{B_1} and c_{A_2}, c_{B_2}, respectively. Also let F_1 and F_2 be the volumetric flow rates of the two streams (ft^3/min, m^3/min) and T_1 and T_2 their corresponding temperatures. Finally, let c_{A_3}, c_{B_3}, F_3, and T_3 be the concentrations, flow rate, and temperature of the product stream. A coil is also immersed in the liquid of the tank and it is used to supply heat to the system with steam, or remove heat with cooling water.

The fundamental quantities needed to describe the mixing process are:

(a) Total mass in the tank
(b) Amounts of components A and B in the tank
(c) Total energy
(d) Momentum of the material in the tank

Remarks

1. The momentum does not change under any operating conditions and it will be neglected in further treatment.
2. We only need to consider two of the following three quantities: total mass, mass of A, mass of B. The third can be computed from the other two.

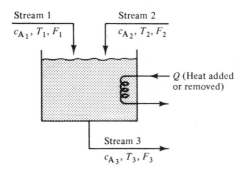

Figure 4.8 Mixing process.

Consider now the balances on the fundamental quantities:

Total mass balance:

$$\dfrac{\left[\begin{array}{c}\text{accumulation of total}\\ \text{mass in the tank}\end{array}\right]}{\text{time}} = \dfrac{\left[\begin{array}{c}\text{input of total}\\ \text{mass in the tank}\end{array}\right]}{\text{time}} - \dfrac{\left[\begin{array}{c}\text{output of total}\\ \text{mass from the tank}\end{array}\right]}{\text{time}}$$

or

$$\frac{d(\rho V)}{dt} = (\rho_1 F_1 + \rho_2 F_2) - \rho_3 F_3 \qquad (4.12)$$

where ρ_1, ρ_2, and ρ_3 are the densities of streams 1, 2, and 3, respectively. Since the content of the tank is well mixed, the density of the product stream ρ_3 is equal to the density of the material in the tank, ρ (i.e., $\rho_3 = \rho$). V is the volume of the material in the tank and is equal to the product of the cross-sectional area of the tank, A, and the height, h, of the liquid level:

$$V = A \cdot h$$

In general, the densities, ρ, ρ_1, and ρ_2 depend on the corresponding concentrations and temperatures:

$$\rho = \rho_3 = f(c_{A_3}, c_{B_3}, T_3) \qquad \rho_1 = f(c_{A_1}, c_{B_1}, T_1) \qquad \rho_2 = f(c_{A_2}, c_{B_2}, T_2)$$

Usually (but not always) the dependencies above are weak and we assume that the densities are independent of the concentrations and temperatures. Therefore, we assume that

$$\rho_1 = \rho_2 = \rho_3 = \rho$$

This transforms eq. (4.12) to the following:

$$\frac{dV}{dt} = A \frac{dh}{dt} = (F_1 + F_2) - F_3 \qquad (4.12a)$$

Balance on component A:

$$\dfrac{\left[\begin{array}{c}\text{accumulation of}\\ \text{component A in}\\ \text{the tank}\end{array}\right]}{\text{time}} = \dfrac{\left[\begin{array}{c}\text{total input of}\\ \text{component A}\\ \text{in the tank}\end{array}\right]}{\text{time}} - \dfrac{\left[\begin{array}{c}\text{total output of}\\ \text{component A}\\ \text{from the tank}\end{array}\right]}{\text{time}}$$

or

$$\frac{d(c_A V)}{dt} = (c_{A_1} F_1 + c_{A_2} F_2) - c_{A_3} F_3$$

or

$$V \frac{dc_A}{dt} + c_A \frac{dV}{dt} = (c_{A_1} F_1 + c_{A_2} F_2) - c_{A_3} F_3 \qquad (4.13)$$

Substituting dV/dt by its equal from eq. (4.12a), we take

$$V\frac{dc_A}{dt} + c_A[(F_1 + F_2) - F_3] = (c_{A_1}F_1 + c_{A_2}F_2) - c_{A_3}F_3$$

and since $c_A = c_{A_3}$ due to the well-stirred assumption,

$$V\frac{dc_{A_3}}{dt} = (c_{A_1} - c_{A_3})F_1 + (c_{A_2} - c_{A_3})F_2 - \overset{= 0}{\underset{}{(c_{A_3} - c_{A_3})F_3}} \qquad (4.13a)$$

Total energy balance:

$$\frac{\begin{bmatrix}\text{accumulation of} \\ \text{total energy}\end{bmatrix}}{\text{time}} = \frac{\begin{bmatrix}\text{input of total energy} \\ \text{with feed streams}\end{bmatrix}}{\text{time}} - \frac{\begin{bmatrix}\text{output of total energy} \\ \text{with product stream}\end{bmatrix}}{\text{time}}$$

$$\pm \frac{\begin{bmatrix}\text{heat added or removed} \\ \text{with the coil}\end{bmatrix}}{\text{time}}$$

The total energy of the material in the tank is

$$E = U \text{ (internal)} + K \text{ (kinetic)} + P \text{ (potential)}$$

Since the tank is not moving, $dK/dt = dP/dt = 0$. Thus $dE/dt = dU/dt$ and for liquid systems,

$$\frac{dU}{dt} \simeq \frac{dH}{dt}$$

where H is the total enthalpy of the material in the tank. Furthermore,

$$\begin{pmatrix}\text{input of total energy} \\ \text{with feed streams} \\ \text{per unit time}\end{pmatrix} = \rho(F_1 h_1 + F_2 h_2)$$

and

$$\begin{pmatrix}\text{output of total energy} \\ \text{with product stream} \\ \text{per unit time}\end{pmatrix} = \rho F_3 h_3$$

where h_1, h_2, and h_3 are the specific enthalpies (enthalpy per unit mass) of streams 1, 2, and 3. Due to the perfect stirring assumption, the specific enthalpy of the material in stream 3 is the same as the specific enthalpy of the material in the tank. Thus

$$H = \rho V h_3$$

Consequently, the total energy balance yields

$$\frac{d(\rho V h_3)}{dt} = \rho(F_1 h_1 + F_2 h_2) - \rho F_3 h_3 \pm Q \qquad (4.14)$$

The question now is how to characterize h_1, h_2, and h_3 in terms of other variables (i.e., temperatures, concentrations, etc.). We know that

$$h_3(T_3) = h_3(T_0) + c_{p3}(T_3 - T_0) \tag{4.15a}$$

$$h_1(T_1) = h_1(T_0) + c_{p1}(T_1 - T_0) \tag{4.15b}$$

$$h_2(T_2) = h_2(T_0) + c_{p2}(T_2 - T_0) \tag{4.15c}$$

where T_0 is the reference temperature. At this temperature

$$ph_3(T_0) = c_{A_3}\tilde{H}_A + c_{B_3}\tilde{H}_B + c_{A_3}\,\Delta\tilde{H}_{S_3}(T_0) \tag{4.16a}$$

$$ph_1(T_0) = c_{A_1}\tilde{H}_A + c_{B_1}\tilde{H}_B + c_{A_1}\,\Delta\tilde{H}_{S_1}(T_0) \tag{4.16b}$$

$$ph_2(T_0) = c_{A_2}\tilde{H}_A + c_{B_2}\tilde{H}_B + c_{A_2}\,\Delta\tilde{H}_{S_2}(T_0) \tag{4.16c}$$

where $\tilde{H}_A$ and $\tilde{H}_B$ are the molar enthalpies (enthalpy per mole) of components A and B at temperature T_0. $\Delta\tilde{H}_{S_1}$, $\Delta\tilde{H}_{S_2}$, and $\Delta\tilde{H}_{S_3}$ are the heats of solution for streams 1, 2, and 3 per mole of A at temperature T_0. Substituting eqs. (4.15a,b,c) and (4.16a,b,c) into the total energy balance eq. (4.14), we take

$$\frac{d[V(c_{A_3}\tilde{H}_A + c_{B_3}\tilde{H}_B + c_{A_3}\,\Delta\tilde{H}_{S_3}) + \rho V c_{p3}(T_3 - T_0)]}{dt}$$

$$= F_1(c_{A_1}\tilde{H}_A + c_{B_1}\tilde{H}_B + c_{A_1}\,\Delta\tilde{H}_{S_1}) + \rho F_1 c_{p1}(T_1 - T_0)$$

$$+ F_2(c_{A_2}\tilde{H}_A + c_{B_2}\tilde{H}_B + c_{A_2}\,\Delta\tilde{H}_{S_2}) + \rho F_2 c_{p2}(T_2 - T_0)$$

$$- F_3(c_{A_3}\tilde{H}_A + c_{B_3}\tilde{H}_B + c_{A_3}\,\Delta\tilde{H}_{S_3}) - \rho F_3 c_{p3}(T_3 - T_0) \pm Q$$

or

$$\overset{\displaystyle = 0 \text{ (balance on A)}}{\rho c_{p3}\frac{d[V(T_3 - T_0)]}{dt} + \tilde{H}_A\left[\frac{d(V c_{A_3})}{dt} - c_{A_1}F_1 - c_{A_2}F_2 + c_{A_3}F_3\right]}$$

$$+ \tilde{H}_B\underset{\displaystyle = 0 \text{ (balance on B)}}{\left[\frac{d(V c_{B_3})}{dt} - c_{B_1}F_1 - c_{B_2}F_2 + c_{B_3}F_3\right]} + \Delta\tilde{H}_{S_3}\frac{d(V c_{A_3})}{dt}$$

$$= F_1 c_{A_1}\,\Delta\tilde{H}_{S_1} + \rho F_1 c_{p1}(T_1 - T_0) + F_2 c_{A_2}\,\Delta\tilde{H}_{S_2} + \rho F_2 c_{p2}(T_2 - T_0)$$

$$- F_3 c_{A_3}\,\Delta\tilde{H}_{S_3} - \rho F_3 c_{p3}(T_3 - T_0) \pm Q$$

or

$$\rho c_{p3} V \frac{dT_3}{dt} + \rho c_{p3}(T_3 - T_0)\frac{dV}{dt} + \Delta\tilde{H}_{S_3}[c_{A_1}F_1 + c_{A_2}F_2 - c_{A_3}F_3]$$

$$= F_1 c_{A_1}\,\Delta\tilde{H}_{S_1} + \rho F_1 c_{p1}(T_1 - T_0) + F_2 c_{A_2}\,\Delta\tilde{H}_{S_2} + \rho F_2 c_{p2}(T_2 - T_0)$$

$$- F_3 c_{A_3}\,\Delta\tilde{H}_{S_3} - \rho F_3 c_{p3}(T_3 - T_0) \pm Q$$

and finally,

$$\rho c_{p_3} V \frac{dT_3}{dt} = c_{A_1} F_1 [\Delta \tilde{H}_{S_1} - \Delta \tilde{H}_{S_3}] + c_{A_2} F_2 [\Delta \tilde{H}_{S_2} - \Delta \tilde{H}_{S_3}]$$

$$+ \rho F_1 [c_{p_1}(T_1 - T_0) - c_{p_3}(T_3 - T_0)]$$

$$+ \rho F_2 [c_{p_2}(T_2 - T_0) - c_{p_3}(T_3 - T_0)] \pm Q$$

If we assume that $c_{p_1} = c_{p_2} = c_{p_3} = c_p$, we have

$$\rho c_p V \frac{dT_3}{dt} = c_{A_1} F_1 [\Delta \tilde{H}_{S_1} - \Delta \tilde{H}_{S_3}] + c_{A_2} F_2 [\Delta \tilde{H}_{S_2} - \Delta \tilde{H}_{S_3}]$$

$$+ \rho F_1 c_p (T_1 - T_3) + \rho F_2 c_p (T_2 - T_3) \pm Q \qquad (4.14a)$$

Summarizing the steps above, we have:

State variables: V, c_{A_3}, T_3
State equations:

$$\frac{dV}{dt} = (F_1 + F_2) - F_3 \qquad (4.12a)$$

$$V \frac{dc_{A_3}}{dt} = (c_{A_1} - c_{A_3}) F_1 + (c_{A_2} - c_{A_3}) F_2 \qquad (4.13a)$$

$$\rho c_p V \frac{dT_3}{dt} = c_{A_1} F_1 [\Delta \tilde{H}_{S_1} - \Delta \tilde{H}_{S_3}] + c_{A_2} F_2 [\Delta \tilde{H}_{S_2} - \Delta \tilde{H}_{S_3}]$$

$$+ \rho F_1 c_p (T_1 - T_3) + \rho F_2 c_p (T_2 - T_3) \pm Q \qquad (4.14a)$$

Input variables: F_1, c_{A_1}, T_1, F_2, c_{A_2}, T_2, F_3 (for feedback control)
Output variables: V (or equivalently the height of liquid level, h),
c_{A_3}, T_3
Parameters (constant): ρ, c_p, $\Delta \tilde{H}_{S_1}$, $\Delta \tilde{H}_{S_2}$, $\Delta \tilde{H}_{S_3}$

Remarks

3. Usually, a mixing tank is equipped with a cooling or heating coil or jacket through which flows a coolant (if heat is released during the mixing of the two solutions) or a heating medium (if heat is absorbed during mixing) in an attempt to keep the mixing isothermal.
4. If the heats of solution are strong functions of concentration (i.e., if $[\Delta \tilde{H}_{S_1} - \Delta \tilde{H}_{S_3}]$ and $[\Delta \tilde{H}_{S_2} - \Delta \tilde{H}_{S_3}]$ are not small quantities), then from the total energy balance, eq. (4.14a), we notice that temperature T_3 depends strongly on the concentrations of the feed streams and their temperatures. If on the other hand, $[\Delta \tilde{H}_{S_1} - \Delta \tilde{H}_{S_3}]$ and $[\Delta \tilde{H}_{S_2} - \Delta \tilde{H}_{S_3}]$ are nearly zero, then T_3 depends basically only on T_1 and T_2.

Example 4.12: Mathematical Model of a Tubular Heat Exchanger

Consider the shell-and-tube heat exchanger shown in Figure 4.9. A liquid flows through the inner tube and it is being heated by steam that flows countercurrently around the tube. The temperature of the liquid not only changes with time but also changes along the axial direction z from the value T_1 at the entrance to the value T_2 at the exit. We will assume that the temperature does not change along the radius of the pipe. Consequently, we have two independent variables, z and t. The state variable of interest for the heat exchanger is the temperature T of the heated liquid. Therefore, we need the energy balance for the characterization of the temperature. To perform this balance, consider the element of length Δz defined in Figure 4.9 by the dashed lines. For this system and over a period of time Δt, we have:

Energy balance:

$$\rho c_p A \; \Delta z \; [(T)|_{t+\Delta t} - (T)_t] = \rho c_p v A (T)|_z \; \Delta t \qquad - \rho c_p v A (T)|_{z+\Delta z} \; \Delta t$$

$$\begin{bmatrix} \text{accumulation of} \\ \text{enthalpy during} \\ \text{the time period} \\ \Delta t \end{bmatrix} \begin{bmatrix} \text{flow in of} \\ \text{enthalpy during} \\ \text{the time period} \\ \Delta t \end{bmatrix} \begin{bmatrix} \text{flow out of} \\ \text{enthalpy during} \\ \text{the time period} \\ \Delta t \end{bmatrix}$$

$$+ Q \; \Delta t \; (\pi D \; \Delta z) \tag{4.17}$$

$$\begin{bmatrix} \text{enthalpy transferred} \\ \text{from the steam to the} \\ \text{liquid, through the} \\ \text{wall, during the time} \\ \text{period } \Delta t \end{bmatrix}$$

where Q = amount of heat transferred from the steam to the liquid per unit of time and unit of heat transfer area
  A = cross-sectional area of the inner tube
  v = average (assumed constant) velocity of the liquid
  D = external diameter of the inner tube

Dividing both sides of eq. (4.17) by $\Delta z \; \Delta t$ and letting $\Delta z \to 0$ and $\Delta t \to 0$,

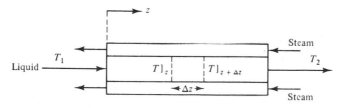

Figure 4.9 Tubular heat exchanger.

we take

$$\rho c_p A \frac{\partial T}{\partial t} + \rho c_p vA \frac{\partial T}{\partial z} = \pi D \, Q \tag{4.18}$$

In eq. (4.18) we can substitute Q by its equal,

$$Q = U(T_{st} - T)$$

and take

$$\rho c_p A \frac{\partial T}{\partial t} + \rho c_p vA \frac{\partial T}{\partial z} = \pi D \, U(T_{st} - T) \tag{4.19}$$

This is the equation of state that models the behavior of liquid's temperature (state variable) along the length of the exchanger. Since eq. (4.19) is a partial differential equation we say that the exchanger has been modeled as a *distributed parameter system*. Note that U is the overall heat transfer coefficient between steam and the liquid in the tube, and T_{st} is the temperature of saturated steam.

Example 4.13: Mathematical Model of an Ideal Binary Distillation Column

Consider a binary mixture of components A and B, to be separated into two product streams using conventional distillation. The mixture is fed in the column as a saturated liquid (i.e., at its bubble point), onto the feed tray f (Figure 4.10), with a molar flow rate (mol/min) F_f and a molar fraction of component A, c_f. The overhead vapor stream is cooled and completely condensed, and then it flows into the reflux drum. The cooling of the overhead vapor is accomplished with cooling water. The liquid from the reflux drum is partly pumped back in the column (top tray, N) with a molar flow rate F_R (*reflux* stream) and is partly removed as the *distillate product* with a molar flow rate F_D. Let us call M_{RD} the liquid holdup in the reflux drum and x_D the molar fraction of component A in the liquid of the reflux drum. It is clear that x_D is the composition for both the reflux and distillate streams.

At the base of the distillation column, a liquid product stream (the *bottoms product*) is removed with a flow rate F_B and a composition x_B (molar fraction of A). A liquid stream with a molar flow rate V is also drawn from the bottom of the column and after it has been heated with steam, it returns to the base of the column. The composition of the recirculating back to column stream is x_B. Let M_B be the liquid holdup at the base of the column.

The column contains N trays numbered from the bottom of the column to the top. Let M_i be the liquid holdup on the ith tray. The vapor holdup on each tray will be assumed to be negligible.

In Figure 4.11a we see the material flows in and out of the feed tray. Similarly, Figure 4.11b and c show the material flows for the top (Nth) and bottom (first) trays, while Figure 4.11d refers to any other tray.

To simplify the system, we will make the following assumptions:

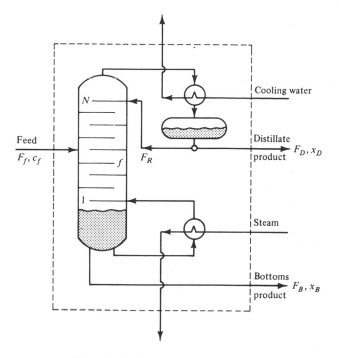

Figure 4.10 Binary distillation column.

1. Vapor holdup on each tray will be neglected.
2. The molar heats of vaporization of both components A and B are approximately equal. This means that 1 mol of condensing vapor releases enough heat to vaporize 1 mol of liquid.
3. The heat losses from the column to the surroundings are assumed to be negligible.
4. The relative volatility α of the two components remains constant throughout the column.
5. Each tray is assumed to be 100% efficient (i.e., the vapor leaving each tray is in equilibrium with the liquid on the tray).

The first three assumptions imply that

$$V = V_1 = V_2 = \cdots = V_N$$

and there is no need for energy balance around each tray.

The last two assumptions imply that a simple vapor–liquid equilibrium relationship can be used to relate the molar fraction of A in the vapor leaving the ith tray (y_i) with the molar fraction of A in the liquid leaving the same tray (x_i):

$$y_i = \frac{\alpha x_i}{1 + (\alpha - 1)x_i} \tag{4.20}$$

where α is the relative volatility of the two components A and B.

The final assumptions that we will make are the following:

6. Neglect the dynamics of the condenser and the reboiler. It is clear that these two units (heat exchangers) constitute processing systems on their own right and as such they have a dynamic behavior (see Example 4.12). Therefore, accurate modeling should include the state equations, which describe the dynamic behavior of condenser and reboiler.

7. Neglect the momentum balance for each tray and assume that the molar flow rate of the liquid leaving each tray is related to the liquid holdup of the tray through the Francis weir formula:

$$L_i = f(M_i) \qquad i = 1, 2, \ldots, f, \ldots, N \qquad (4.21)$$

Let us now develop the state equations that will describe the dynamic behavior of a distillation column. The fundamental quantities are total mass and mass of component A. But the question is: What is the system around which we will make the balances? From a practical point of view,

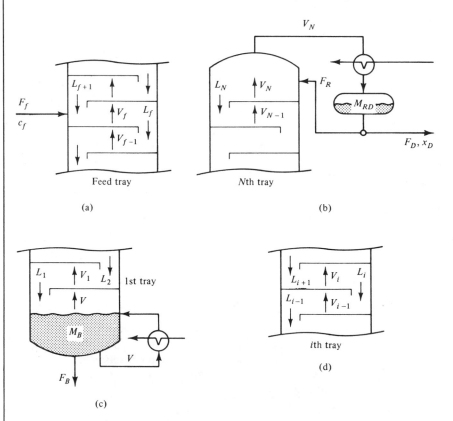

Figure 4.11 Modeling details of the binary distillation column: (a) feed section; (b) top section and overhead accumulator; (c) bottom section and reboiler; (d) general ith tray.

the boundary of the system of interest is outlined by dashed lines in Figure 4.10. Such a boundary clearly identifies the inputs and outputs of practical significance for the overall system. It is also evident that unless we can describe how the concentrations and liquid holdups on each tray change with time, we cannot find how the variables of practical significance, such as x_D and x_B, change with time. Therefore, we are forced to consider the balances around each tray. Thus we have (see also Figure 4.11):

Feed tray ($i = f$):

Total mass:
$$\frac{d(M_f)}{dt} = F_f + L_{f+1} + V_{f-1} - L_f - V_f = F_f + L_{f+1} - L_f \quad (4.22a)$$

Component A:
$$\frac{d(M_f x_f)}{dt} = F_f c_f + L_{f+1} x_{f+1} + V_{f-1} y_{f-1} - L_f x_f - V_f y_f \quad (4.22b)$$

Top tray ($i = N$):

Total mass:
$$\frac{d(M_N)}{dt} = F_R + V_{N-1} - L_N - V_N = F_R - L_N \quad (4.23a)$$

Component A:
$$\frac{d(M_N x_N)}{dt} = F_R x_D + V_{N-1} y_{N-1} - L_N x_N - V_N y_N \quad (4.23b)$$

Bottom tray ($i = 1$):

Total mass:
$$\frac{d(M_1)}{dt} = L_2 - L_1 + V - V_1 - L_2 - L_1 \quad (4.24a)$$

Component A:
$$\frac{d(M_1 x_1)}{dt} = L_2 x_2 + V y_B - L_1 x_1 - V_1 y_1 \quad (4.24b)$$

ith tray ($i = 2, \ldots, N - 1$ and $i \neq f$):

Total mass:
$$\frac{d(M_i)}{dt} = L_{i+1} - L_i + V_{i-1} - V_i = L_{i+1} - L_i \quad (4.25a)$$

Component A:
$$\frac{d(M_i x_i)}{dt} = L_{i+1} x_{i+1} + V_{i-1} y_{i-1} - L_i x_i - V_i y_i \quad (4.25b)$$

Reflux drum:

Total mass:
$$\frac{d(M_{RD})}{dt} = V_N - F_R - F_D \quad (4.26a)$$

Component A:
$$\frac{d(M_{RD} x_D)}{dt} = V_N y_N - (F_R + F_D) x_D \quad (4.26b)$$

Column base:

Total mass:
$$\frac{d(M_B)}{dt} = L_1 - V - F_B \quad (4.27a)$$

Component A:
$$\frac{d(M_B x_B)}{dt} = L_1 x_1 - V y_B - F_B x_B \quad (4.27b)$$

All the equations above are *state equations* and describe the dynamic behavior of the distillation column. The *state variables* of the model are:

Liquid holdups: $M_1, M_2, \ldots, M_f, \ldots, M_N; M_{RD}$ *and* M_B
Liquid concentrations: $x_1, x_2, \ldots, x_f, \ldots, x_N; x_D$ *and* x_B

To complete the modeling of the column, in addition to the state equations, we need the following relationships:

1. Equilibrium relationships:

$$y_i = \frac{\alpha x_i}{1 + (\alpha - 1)x_i} \qquad i = 1, 2, \ldots, f, \ldots, N, B \qquad (4.20)$$

2. Hydraulic relationships (Francis weir formula):

$$L_i = f(M_i) \qquad i = 1, 2, \ldots, f, \ldots, N \qquad (4.21)$$

When all the modeling equations above are solved, we find how the flow rates and concentrations of the two product streams (distillate, bottom) change with time, in the presence of changes in the various input variables.

The modeling steps outlined above indicate that the overall procedure may be tedious and full of simplifying assumptions. At times the resulting model is overwhelming in size and the solution of the corresponding equations may be cumbersome. For the binary distillation column we have to solve a system of

$2N + 4$ nonlinear differential equations (state equations)

and

$2N + 1$ algebraic equations (equilibrium, and hydraulic relationships)

4.6 Modeling Difficulties

The modeling examples discussed in previous sections of this chapter should have alerted the reader to a series of difficulties that we may encounter in our efforts to develop a meaningful and realistic mathematical description of a chemical process.

Example 4.14: Difficulties in the Modeling of a CSTR

Considering the mathematical modeling of a CSTR (Example 4.10), the following difficulties arise:

1. Determine with the desired accuracy the values of various parameters such as the preexponential kinetic constant k_0, the activation energy E, and the overall heat transfer coefficient U.
2. Although the specific heat capacities, c_p and c_{p_i}, have been considered constant, they are in general functions of the temperature T

and the concentration c_A. How do we decide that this dependence is weak (so that we can use constant values as in the example) or strong (in which case the modeling becomes very complicated)? The same questions arise for the densities ρ and ρ_i and the heat of reaction $(-\Delta H_r)$.

3. During the operation of the CSTR, scaling, fouling, and so on, will alter the value of the overall heat transfer coefficient. How can we account for this effect in the mathematical model?

4. We have considered first-order kinetics to describe the reaction rate. Is this correct?

We can classify the difficulties encountered during the mathematical modeling of a process in three categories:

1. Those arising from poorly understood chemical or physical phenomena

2. Those caused from inaccurate values of various parameters

3. Those caused from the size and the complexity of the resulting model

Poorly understood processes

To understand completely the physical and chemical phenomena occurring in a chemical process is virtually impossible. Even an acceptable degree of knowledge is at times very difficult. Typical examples include:

Multicomponent reaction systems with poorly known interactions among the various components and imprecisely known kinetics

Vapor–liquid or liquid–liquid thermodynamic equilibria for multicomponent systems

Heat and mass transfer interactions in distillation columns with nonideal multicomponent mixtures, azeotropic mixtures, and so on.

Example 4.15

Consider the fluidized catalytic cracking process shown in Figure 4.12. An oil feed composed of heavy hydrocarbon molecules is mixed with catalyst and enters a fluidized bed reactor. The long molecules react on the surface of the catalyst and are cracked into lighter product molecules (such as gasoline) which leave the reactor from the top. While cracking is taking place, carbon and other heavy uncracked organic materials are deposited on the surface of the catalyst, leading to its deactivation. The catalyst is then taken into a regenerator, where the material deposited on its surface is burned with air. Then, the regenerated catalyst returns to the reactor after it has been mixed with fresh feed.

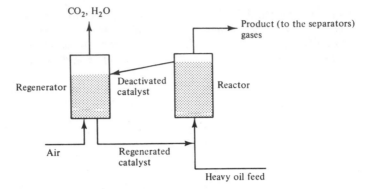

Figure 4.12 Fluid catalytic cracking (FCC) system.

To model the two units, the following information must be available:

1. The reaction rate of the cracking process
2. The rate with which carbon and heavy material are deposited on the catalyst (this will determine the rate of catalyst deactivation)
3. The dependence of the two rates above on the temperature of the reactor and the quality of the feed (light or heavy)
4. The rate with which carbonaceous material deposited on the catalyst is burned off in the regenerator, and its dependence on temperature

All of the foregoing information is not only difficult to acquire, but at times it leads to contradicting contentions. For example, in Figure 4.13 we see two models that describe the effect of the heavy oil feed rate on the reactor temperature. We notice that the qualitative behavior predicted by the two models is quite different.

Finally, the two units (reactor, regenerator) are fluidized beds and it is well known how poorly understood the fluid mechanical characteristics of such units are.

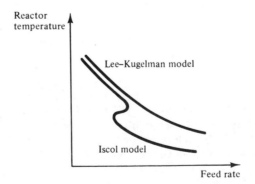

Figure 4.13 Two different models to describe the effect of heavy oil feed rate on reactor temperature for the FCC unit.

Imprecisely known parameters

The availability of accurate values for the parameters of a model is indispensable for any quantitative analysis of the behavior of a process. Unfortunately, this is not always possible. Typical examples include the preexponential constant of a kinetic rate expression.

It should also be pointed out that the values of the parameters do not remain constant over long periods of time. Therefore, for effective modeling we need not only accurate values but also some quantitative description of how the parametric values change with time. Typical examples of changing parameters are the activity of a catalyst and the overall heat transfer coefficient of heat transfer systems (heat exchangers, jacketed reactors, etc.).

The dead time is also a critical parameter whose value is usually imprecisely known and varying. As we will see in a later section, poor knowledge of the dead time can lead to serious stability problems for the process.

When no reliable values for the parameters are available, we resort to experiments on the real process in an effort to estimate some "good" values for them. The experimental procedures will be discussed further in Chapter 31.

Size and complexity of a model

In an effort to develop as accurate and precise a mathematical model as possible, its size and complexity increase significantly.

Example 4.16

Consider a distillation column with 20 trays, a reboiler, and a condenser. The feed is a two-component mixture. Then, as we have seen in Example 4.13, the mathematical model is composed of

$$2N + 4 = 2(20) + 4 = 44 \text{ differential equations}$$

and

$$2N + 1 = 2(20) + 1 = 41 \text{ algebraic equations}$$

The size of the model for such a simple system is already prohibitive. Since the common distillation systems include feeds with more than two components and possess larger numbers of trays, it is clear that such an extensive modeling would lead to cumbersome and hard-to-use models.

Therefore, care must be exercised that the size and complexity of a model do not exceed certain manageable levels, beyond which the model loses its value and becomes less attractive.

THINGS TO THINK ABOUT

1. What is a mathematical model of a physical process, and what do we mean when we talk about mathematical modeling?

2. In Figure 4.13 we see two different curves that relate the temperature and the feed rate of the reactor for the fluid catalytic cracking unit discussed in Example 4.15. Is the term "model" appropriate for each of these curves?

3. Let us recall that the steam tables give the temperature at which water liquid and water vapor are at equilibrium for a given pressure. They also give the specific values for enthalpy, entropy, and volume of both liquid and vapor phases. Do these tables of values constitute a mathematical model?

4. Consider the graphs shown in Figure Q4.1. These graphs were produced by measuring the concentration of B in the reaction A → B, over time, and at various temperatures. Do these graphs represent a mathematical model?

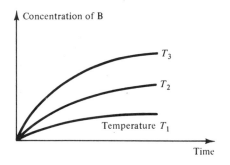

Figure Q4.1

5. Why do you need to develop the mathematical model of a process you want to control?

6. What are the state variables, and what are the state equations? What are they used for?

7. How many state variables do you need to describe a system that is composed of M phases and N components?

8. We know that when two phases are at thermodynamic equilibrium, the chemical potential $\mu_{i,I}$ of every component (i) in phase I is equal to the chemical potential $\mu_{i,II}$ of the same component in phase II:

$$\mu_{i,I} = \mu_{i,II}, \qquad i = 1, 2, \ldots, N$$

Express the equilibrium relationship above in terms of the mole concentrations of the N components in the two phases. The answer to this question will demonstrate to you that we do not need the concentrations of the N components in both phases in order to describe the system.

9. Write a relationship that will give you the molar or the specific enthalpy of

a multicomponent liquid at temperature T and pressure p, with known composition for the N components.

10. Repeat question 9, but with a gas instead of a liquid.

11. Consider the flash drum of Examples 4.7 and 4.8. Develop an expression for the density of the vapor phase, using the van der Waals equation of state. State also an expression for the density of the liquid phase.

12. When is a system at steady state?

13. What is the main reason for the presence of dead time in a process?

14. Do you know of any systems that do not possess dead time?

15. How would you find the dead time of a system?

16. In Figure Q4.2 we see the behavior of the concentration at the outlet of two processes after concentration at the inlets and at time $t = 0$ was increased by 10%. Which process possesses dead time?

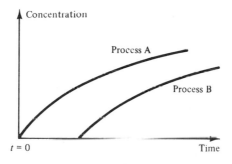

Figure Q4.2

17. What are the assumptions leading to equimolar vapor flow rates (i.e., $V_1 = V_2 = \cdots = V_N = V$) for a binary distillation column?

18. Why have we neglected the energy balances for the binary ideal distillation column of Example 4.13?

19. What are the assumptions leading to the equilibrium relationship (4.20), and how is it derived?

20. Could you have dead time between the overhead vapor and the distillate product? If yes, why?

21. Consider again Example 4.9. Show that the dead time can be computed from the following equation

$$\int_{t-t_d}^{t} F(\theta)d\theta = \text{volume of the pipe}$$

where $F(\theta)$ is the volumetric flow rate of the liquid through the pipe as a function of time. The above equation is more general than that of Example 4.9, where the volumetric flow rate was assumed to be constant.

Modeling Considerations for Control Purposes **5**

The mathematical modeling of physical and chemical phenomena, encountered in processing systems of interest to chemical engineers, is a form of scientific art. Like every type of art it does not conform to, nor obey, clearly specified rules and recipes. It is an expression of all the educational background and previous practical experience of the modeler. Therefore, if previous sections have generated more questions to the reader than answers to specific modeling problems, he or she should not despair. Good and efficient modeling is acquired slowly with ever-increasing ability.

In this chapter we attempt to focus the mathematical modeling on control purposes and needs. Thus we will examine the following issues:

1. Starting from the state-variables model, how one can develop an input–output model which is very convenient for control purposes?
2. Using the mathematical model of a process, how can one determine the degrees of freedom inherent in the process, and consequently, identify the extent of the control problem to be solved?

We will close this chapter with some general guidelines that will help the control designer to formulate the scope of modeling for control purposes.

5.1   The Input–Output Model

Every chemical process and its associated variables can be described pictorially as shown in Figure 5.1. The main block represents the process, while the arrows indicate the inputs and outputs of the process.

A mathematical model that is convenient and useful to a control system designer should conform with the picture above, (i.e., be such that, given the values of the inputs, it provides directly the values of the outputs). In particular, the model should have the following general form for every output;

$$\text{output} = f(\text{input variables})$$

Using Figure 5.1, the relationship above implies that

$$y_i = f(m_1, m_2, \ldots, m_k; d_1, d_2, \ldots, d_\ell) \qquad \text{for } i = 1, 2, \ldots, m$$

Such a model, describing directly the relationship between the input and output variables of a process, is called an *input–output model*. It is a very convenient form since it represents directly the cause-and-effect relationship in processing systems. For this reason it is also appealing to process engineers and control designers.

The mathematical models we learned to develop in Chapter 4 using state variables *are not* of the direct input–output type. Nevertheless, they constitute the basis for the development of an input–output model. This is particularly easy and straightforward *when the state variables coincide completely with the output variables* of a process. In such a case we can integrate the state model to produce the input–output model of the process.

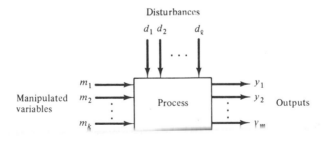

Figure 5.1 A chemical process and its associated inputs and outputs.

Example 5.1: Input–Output Model for the Stirred Tank Heater

Consider again the stirred tank heater discussed in Example 4.4. Assume that $F_i = F$, which yields $dV/dt = 0$, leaving the total energy balance as the only equation of the state model,

$$V \frac{dT}{dt} = F_i(T_i - T) + \frac{Q}{\rho c_p} \tag{4.5b}$$

The amount of heat Q supplied by the steam is given by

$$Q = UA_t(T_{st} - T)$$

where U is the overall heat transfer coefficient, A_t the area of heat transfer, and T_{st} the temperature of the steam. In eq. (4.5b) replace Q by its equal and take

$$V \frac{dT}{dt} + \left(F_i + \frac{UA_t}{\rho c_p} \right) T = F_i T_i + \frac{UA_t}{\rho c_p} T_{st}$$

or

$$\frac{dT}{dt} + aT = \frac{1}{\tau} T_i + KT_{st} \tag{5.1}$$

where

$$a = \frac{1}{\tau} + K \qquad \frac{1}{\tau} = \frac{F_i}{V} \qquad K = \frac{UA_t}{V\rho c_p}$$

Equation (5.1) is the mathematical model of the stirred tank heater with T the state variable, while T_i and T_{st} are the input variables. Let us see how we can develop the corresponding input–output model.

At steady state, eq. (5.1) yields

$$0 + aT_s = \frac{1}{\tau} T_{i,s} + KT_{st,s} \tag{5.2}$$

where T_s, $T_{i,s}$, and $T_{st,s}$ are the steady-state values of the corresponding variables. Subtract (5.2) from (5.1) and take

$$\frac{d(T - T_s)}{dt} + a(T - T_s) = \frac{1}{\tau}(T_i - T_{i,s}) + K(T_{st} - T_{st,s})$$

or

$$\frac{dT'}{dt} + aT' = \frac{1}{\tau} T_i' + KT_{st}' \tag{5.3}$$

where $T' = T - T_s$, $T_i' = T_i - T_{i,s}$, and $T_{st}' = T_{st} - T_{st,s}$ indicate the deviations from the corresponding steady-state values.

The solution of (5.3) is

$$T'(t) = c_1 e^{-at} + e^{-at} \int_0^t e^{at} \left[\frac{1}{\tau} T_i' + KT_{st}' \right] dt \tag{5.4}$$

Assuming that initially the heater is at steady state, [i.e., at $t = 0$, $T' = 0$], we easily find that $c_1 = 0$. Therefore, eq. (5.4) gives

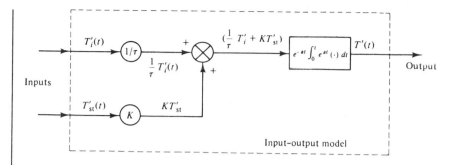

Figure 5.2 Input–output model of a stirred tank heater.

$$T'(t) = e^{-at} \int_0^t e^{at} \left[\frac{1}{\tau} T_i' + KT_{st}' \right] dt \qquad (5.5)$$

Equation (5.5) expresses the relationship between the inputs (T_i', T_{st}') and the output (T'), and constitutes the input–output model for the tank heater. This relationship is also depicted pictorially in Figure 5.2.

Example 5.2: Input–Output Model for a Mixing Process

Consider again the process of mixing of two streams discussed in Example 4.11 (Figure 4.8). Assume that $F_1 + F_2 = F_3$, which implies that $dV/dt - 0$ (i.e., V – constant). The heats of solutions are independent of the concentration, which implies that

$$[\Delta \tilde{H}_{s_1} - \Delta \tilde{H}_{s_3}] = [\Delta \tilde{H}_{s_2} - \Delta \tilde{H}_{s_3}] = 0$$

Then the state equations [eqs. (4.13a) and (4.14a)] are reduced to the following:

$$\frac{dc_{A_3}}{dt} + \left(\frac{F_1}{V} + \frac{F_2}{V} \right) c_{A_3} = \frac{F_1}{V} c_{A_1} + \frac{F_2}{V} c_{A_2} \qquad (5.6a)$$

and

$$\frac{dT_3}{dt} + \left(\frac{F_1}{V} + \frac{F_2}{V} \right) T_3 = \frac{F_1}{V} T_1 + \frac{F_2}{V} T_2 \pm \frac{Q}{\rho c_p V} \qquad (5.6b)$$

At steady state eqs. (5.6a) and (5.6b) yield

$$0 + \left(\frac{F_1}{V} + \frac{F_2}{V} \right) c_{A_3,s} = \frac{F_1}{V} c_{A_1,s} + \frac{F_2}{V} c_{A_2,s} \qquad (5.7a)$$

and

$$0 + \left(\frac{F_1}{V} + \frac{F_2}{V} \right) T_{3,s} = \frac{F_1}{V} T_{1,s} + \frac{F_2}{V} T_{2,s} \pm \frac{Q_s}{\rho c_p V} \qquad (5.7b)$$

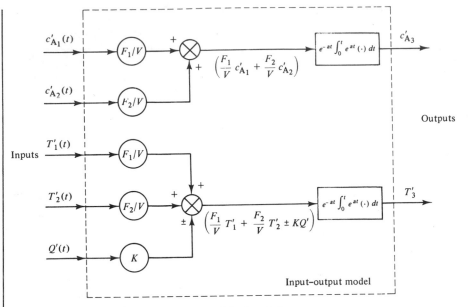

Figure 5.3 Input–output model of a mixing process.

Subtract (5.7a) from (5.6a) and (5.7b) from (5.6b) and take

$$\frac{dc'_{A_3}}{dt} + ac'_{A_3} = \frac{F_1}{V} c'_{A_1} + \frac{F_2}{V} c'_{A_2} \tag{5.8a}$$

$$\frac{dT'_3}{dt} + aT'_3 = \frac{F_1}{V} T'_1 + \frac{F_2}{V} T'_2 \pm KQ' \tag{5.8b}$$

where $c'_{A_1}, c'_{A_2}, c'_{A_3}, T'_1 T'_2, T'_3,$ and Q' are deviation variables defined as follows:

$$c'_{A_1} = c_{A_1} - c_{A_{1,s}} \qquad c'_{A_2} = c_{A_2} - c_{A_{2,s}} \qquad c'_{A_3} = c_{A_3} - c_{A_{3,s}}$$

and

$$T'_1 = T_1 - T_{1,s} \qquad T'_2 = T_2 - T_{2,s} \qquad T'_3 = T_3 - T_{3,s} \qquad Q' = Q - Q_s$$

Also,

$$a = \frac{F_1}{V} + \frac{F_2}{V} \qquad \text{and} \qquad K = \frac{1}{\rho c_p V}$$

The solution of (5.8a) and (5.8b) yields

$$c'_{A_3}(t) = c_1 e^{-at} + e^{-at} \int_0^t e^{at} \left[\frac{F_1}{V} c'_{A_1} + \frac{F_2}{V} c'_{A_2} \right] dt \tag{5.9a}$$

and

$$T'_3(t) = c_2 e^{-at} + e^{-at} \int_0^t e^{at} \left[\frac{F_1}{V} T'_1 + \frac{F_2}{V} T'_2 \pm KQ' \right] dt \tag{5.9b}$$

If the system is initially (i.e., at $t = 0$) at steady state, then

$$c'_{A_3}(t = 0) = 0 \quad \text{and} \quad T'_3(t = 0) = 0$$

and it results in $c_1 = c_2 = 0$.

Equations (5.9a) and (5.9b) represent the input–output model for the mixing process, and is shown schematically in Figure 5.3.

Remarks

1. In Examples 5.1 and 5.2 the output variables coincide with the state variables of the two processes. Consequently, in order to develop the input–output model we need only solve the differential equations of the mass and energy balances. This is not always true. Take as an example the binary distillation column model (Example 4.13 and Figure 4.10). For this system we have:
   State variables:

   Liquid holdups: $M_1, M_2, \ldots, M_f, \ldots, M_N, M_{RD}$ and M_B
   Liquid concentrations: $x_1, x_2, \ldots, x_f, \ldots, x_N, x_D$ and x_B

   Output variables:

   Distillate product flow rate and composition: F_D and x_D
   Bottoms product flow rate and composition: F_B and x_B

   We notice that we have many more state variables than outputs. For such systems, the development of the input–output model is quite involved and difficult. Figure 5.4 depicts pictorially the input–output model that we would like to develop for the binary ideal distillation column.
2. In subsequent chapters we will study the method of Laplace transforms, which allows a much simpler development of input–output models from the corresponding state models.

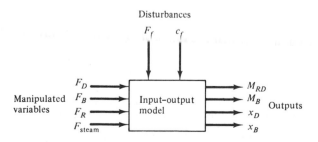

Figure 5.4 Inputs and outputs of a binary distillation column.

5.2   Degrees of Freedom

The degrees of freedom of a processing system are the independent variables that must be specified in order to define the process completely. Consequently, *the desired control of a process will be achieved when and only when all the degrees of freedom have been specified.*

A good understanding of how many degrees of freedom are inherent in a process, and which they are, is very crucial for the design of effective controllers. For a specified system, its mathematical model is the basis for finding the degrees of freedom under both dynamic and static conditions. Let us start with two characteristic examples.

Example 5.3: Degrees of Freedom in a Stirred Tank Heater

The mathematical model of a stirred tank heater (Example 4.4) is given by

$$A \frac{dh}{dt} = F_i - F \qquad (4.4a)$$

$$Ah \frac{dT}{dt} = F_i(T_i - T) + \frac{Q}{\rho c_p} \qquad (4.5b)$$

When eqs. (4.4a) and (4.5b) are solved simultaneously, we can find how h (liquid level) and T (liquid temperature) change with time when the inputs (T_i, F_i, Q) change. Let us ask though the following two questions:

1. Is solution of the equations possible?
2. If solution is possible, how many solutions exist?

To answer these questions, let us count equations and variables:

number of equations = 2;   eqs. (4.4a) and (4.5b)

number of variables = 6; h, T, F_i, F, T_i, and Q

We have assumed that A, ρ, and c_p are parameters with given constant values.

We notice that

number of variables > number of equations

Consequently, the answer to the first question is: yes, there exists at least one solution to the equations modeling the tank heater. With respect to the second question we easily conclude that there is an infinite number of solutions since we can specify arbitrarily the values of four variables $(4 = 6 - 2)$ and solve eqs. (4.4a) and (4.5b) for the remaining two variables.

The arbitrarily specified variables are the *degrees of freedom* and their number is given by the following obvious relationship:

f = (number of variables) − (number of equations)

Suppose that we specify the values of the four variables F_i, T_i, F, and Q. Then we can integrate eqs. (4.4a) and (4.5b) and find how h and T

change with time. If we give different values to F_i, or T_i, or F or Q, we find that h and T change differently than before. Consequently, if we want h and T to change in a prescribed manner, we should not have any degrees of freedom (i.e., variables that can take arbitrary values). This leads us to the conclusion that *in order to specify a process completely, the number of degrees of freedom should be zero.*

Example 5.4: Degrees of Freedom in an Ideal Binary Distillation Column

Consider the model for an ideal binary distillation column developed in Example 4.13. We have:

Number of equations	Origin
$N + 1$	Equilibrium relationships [eq. (4.20)]
N	Hydraulic relationships [eq. (4.21)]
2	Balances around the feed tray [eq. (4.22a), (4.22b)]
2	Balances around the top tray [eqs. (4.23a), (4.23b)]
2	Balances around the bottom tray [eqs. (4.24a), (4.24b)]
$2(N - 3)$	Balances around the ith tray; $i \neq 1, N, f$ [eqs. (4.25a), (4.25b)]
2	Balances around the reflux drum [eqs. (4.26a), (4.26b)]
2	Balances around the column base [eqs. (4.27a), (4.27b)]
Total = $4N + 5$	

Number of variables	Type		
$N + 2$	x_i	$i = 1, 2, \ldots, f, \ldots, N, D, B$	liquid compositions
$N + 1$	y_i	$i = 1, 2, \ldots, f, \ldots, N, B$	vapor compositions
$N + 2$	M_i	$i = 1, 2, \ldots, f, \ldots, N, RD, B$	liquid holdups
N	L_i	$i = 1, 2, \ldots, f, \ldots, N$	liquid flows
6	$F_f, c_f, F_D, F_B, F_R, V$		
Total = $4N + 11$			

The number of degrees of freedom for the ideal binary distillation column is

$$f = (4N + 11) - (4N + 5) = 6$$

That is, we need to specify the values of six variables before we can solve the model of the binary distillation.

The observations made and the conclusions drawn from the two examples above can now be generalized for any processing system described by a set of E independent equations (differential and/or algebraic) containing V independent variables. The number of degrees of freedom for such a system is given by

$$f = V - E$$

According to the value of f, we can distinguish the following cases:

Case 1. If $f = 0$, we have a system of equations with equal number of variables. The solution of the E equations yields unique values for the V variables. In this case we say that the process is *exactly specified*.

Case 2. If $f > 0$, we have more variables than equations. Multiple solutions result from the E equations since we can specify arbitrarily f of the variables. In this case we say that the process is *underspecified by f equations* (i.e., we need f additional equations to have a unique solution).

Case 3. If $f < 0$, we have more equations than variables and in general there is no solution to the E equations. In this case we say that the system is *overspecified by f equations* (i.e., we need to remove f equations to have a solution for the system).

Remarks

1. It is clear from the analysis above that sloppy modeling of a process may lead to a model that does not include all the relevant equations and variables or includes redundant equations and variables. In either case we have an erroneous determination of the degrees of freedom, which may imply incorrectly that we have an infinite number of solutions or no solution at all.
2. The presence of a control loop in a chemical process introduces an additional equation between the corresponding measured and manipulated variables, thus reducing by one the initial number of degrees of freedom for the process.

Example 5.5

The stirred tank heater is modeled by two equations containing six variables, thus yielding four degrees of freedom (Example 5.3). This is true if the effluent flow rate F is determined by a pump, valve, and so on. Let us suppose that this is not the case and that the liquid flows out from the tank freely under the hydrostatic pressure of the liquid in the tank. In this case there is an additional equation relating F to h (e.g., $F = \beta\sqrt{h}$), which reduces the number of degrees of freedom by one.

Example 5.6

Consider again the stirred tank heater, but now under feedback control (Figure 5.5). Control loop 1 maintains the liquid level at a desired value by measuring the level of the liquid and adjusting the value of the effluent flow rate. Therefore, control loop 1 introduces a relationship between F and h. Similarly, control loop 2 maintains the temperature of the liquid at the desired value by manipulating the flow of steam and thus the flow of heat Q. Consequently, control loop 2 introduces a relationship between Q and T.

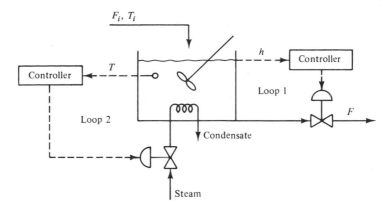

Figure 5.5 Feedback loops of a stirred tank heater.

It is clear from the analysis above that the two control loops introduce two additional equations, thus reducing the degrees of freedom by two.

5.3 Degrees of Freedom and Process Controllers

In general, a carefully modeled process will possess one or more degrees of freedom. Since for $f > 0$ the process will have an infinite number of solutions, the following question arises:

> How do you reduce the number of degrees of freedom to zero so that you can have a completely specified system with unique behavior?

It is clear that for an underspecified system with f degrees of freedom, we need to introduce f additional equations to make the system completely specified. There are two sources which provide the additional equations: (1) the external world and (2) the control system. Let us examine them closer using the stirred tank heater as our example.

Recall from Example 5.3 that the stirred tank heater possesses four degrees of freedom. Therefore, we need four additional relationships, independent of the modeling equations [eqs. (4.4a), (4.5b)]. These are provided from the following considerations:

1. The feed flow rate F_i and feed temperature T_i are the main two disturbances for the stirred tank heater and they are both specified by the external world (e.g., the unit that precedes the tank heater). Although the equations that specify F_i and T_i may not be known to us, nevertheless they exist and remove two degrees of freedom. Thus we have $4 - 2 = 2$ remaining degrees of freedom.

2. Acceptable operation of the tank heater requires that the liquid level and liquid temperature in the tank heater be maintained at desired values. These two control objectives can be achieved with the two control loops shown in Figure 5.5 and discussed in Example 5.6. But the introduction of the two control loops adds two equations (see Example 5.6), thus removing the remaining two degrees of freedom.

Summarizing the observations above we conclude the following:

1. The external world, by specifying the values of the disturbances, removes as many degrees of freedom as the *number of disturbances*.
2. The control system required to achieve the control objectives removes as many degrees of freedom as the *number of control objectives*.

During the reduction in the number of degrees of freedom for a chemical process, care must be exercised not to specify more control objectives than it is possible for the particular system. Thus we can have at most *two* control objectives for the stirred tank heater. When we attempt to have three control objectives, we are led to an overspecified system with $f < 0$.

Example 5.7: Reduce the Degrees of Freedom of an Ideal Binary Distillation Column

Return to the ideal binary distillation column (Figure 4.10). The system possesses six degrees of freedom (see Example 5.4), which are specified as follows:

Specification of the Disturbances. Two are the main disturbances for the binary distillation column: the feed flow rate F_f and the feed composition c_f. Their values are specified by the external world (e.g., a reactor whose effluent stream is the feed to the distillation column). Although the equations specifying F_f and c_f are not known to us, they do exist and remove two degrees of freedom, leaving four for additional specifications.

Specification of the Control Objectives. We can have up to four control objectives since there are four remaining degrees of freedom. Acceptable operation of the binary column requires that the following variables be maintained at desired values:

1. Composition of the distillate stream, x_D
2. Composition of the bottoms stream, x_B
3. Liquid holdup in the reflux drum, M_{RD}
4. Liquid holdup at the base of the column, M_B

Specifications 1 and 2 characterize the two product streams. Specifications 3 and 4 are required for operational feasibility (i.e., we do not want

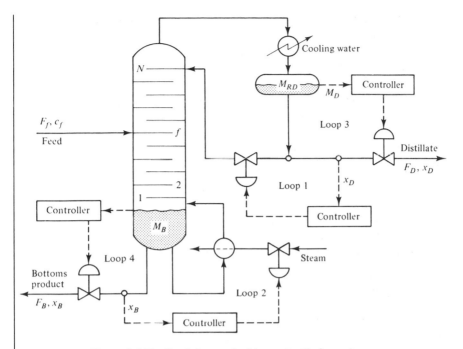

Figure 5.6 Feedback loops of a binary distillation column.

to flood or dry up the reflux drum or the base of the column). Figure 5.6 shows the four control loops that satisfy the foregoing four objectives.

We must note that these four specifications of the control objectives may differ, according to the particular operating objectives. For example, in a different application we may impose the following control objectives:

> Keep at the desired values the distillate flow rate F_D, its composition x_D, and the two liquid holdups M_{RD} and M_B

or

> Keep at the desired values the bottoms flow rate F_B, its composition x_B, and the two liquid holdups, M_{RD}, and M_B.

Care must be exercised *not to specify more control objectives than the available number of degrees of freedom*. In such a case the system becomes overspecified and it is impossible to design a control system that satisfies all the desired control objectives. Thus it is impossible to design a control system for the ideal binary distillation column that can satisfy the following six operational (control) objectives:

> Keep at the desired values the F_D, x_D; F_B, x_B; M_{RD} and M_B.

Example 5.8: Degrees of Freedom of a Mixing Process

Consider the nonisothermal mixing of two streams discussed in

Example 4.11 (Figure 4.8). The mathematical model is given by the equations

$$\frac{dV}{dt} = (F_1 + F_2) - F_3 \tag{4.12a}$$

$$V\frac{dc_{A_3}}{dt} = (c_{A_1} - c_{A_3})F_1 + (c_{A_2} - c_{A_3})F_2 \tag{4.13a}$$

$$\rho c_p V \frac{dT_3}{dt} = c_{A_1}F_1[\Delta\tilde{H}_{S_1} - \Delta\tilde{H}_{S_3}] + c_{A_2}F_2[\Delta\tilde{H}_{S_2} - \Delta\tilde{H}_{S_3}]$$

$$+ \rho F_1 c_p(T_1 - T_3) + \rho F_2 c_p(T_2 - T_3) \pm Q \tag{4.14a}$$

number of variables   *= 16:* $V, F_1, F_2, F_3, c_{A_1}, c_{A_2}, c_{A_3}$
                                    $T_1, T_2, T_3, Q, \rho, c_p, \Delta\tilde{H}_{S_1}, \Delta\tilde{H}_{S_2}, \Delta\tilde{H}_{S_3}$
number of state equations = 3
initial degrees of freedom = 16 – 3 = 13

Further limitations of the degrees of freedom are as follows:

Physical properties of the liquids are specified (i.e., ρ and c_p).
The heats of solution $\Delta\tilde{H}_{S_1}$, $\Delta\tilde{H}_{S_2}$, and $\Delta\tilde{H}_{S_3}$ are functions of the corresponding concentrations, and the reference temperature T_0:

$$\Delta\tilde{H}_{S_1} = f_1(c_{A_1}, c_{B_1}, T_0) \quad \Delta\tilde{H}_{S_2} = f_2(c_{A_2}, c_{B_2}, T_0) \quad \Delta\tilde{H}_{S_3} = f_3(c_{A_3}, c_{B_3}, T_0)$$

These three equations reduce the degrees of freedom by 3. Consequently, after the specifications above we have left,

$$13 - 5 = 8 \quad \text{degrees of freedom}$$

The eight degrees of freedom are now specified as follows:

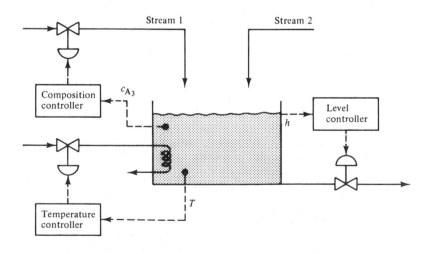

Figure 5.7 Feedback loops of a mixing process.

Specification of the Disturbances. There are five main disturbances coming from the two feed streams:

Feed stream 1: F_1, T_1 Feed stream 2: F_2, T_2 and c_{A_2}

The values of the disturbances are specified by the external world.

Specification of the Control Objectives. Considering five disturbances specified by the external world, we have only $8 - 5 = 3$ degrees of freedom left. Therefore, we can specify up to *three* control objectives. These are:

Keep the volume (V) of the mixture in the tank as well as the temperature (T) and composition (c_{A_3}) of the effluent stream at desired values.

Figure 5.7 shows three possible feedback control loops that satisfy the objectives above.

5.4 Formulating the Scope of Modeling for Process Control

It should be clear by now that efficient modeling of a chemical process is a nontrivial task but also very crucial for the design of a control system. Before closing the present chapter, let us emphasize some of the factors that will determine the scope of modeling for control purposes.

Before attempting to model a process, we must pose the following questions and try to understand well their implications:

1. What are the control objectives that we must satisfy?
2. What are the expected disturbances and their impact?
3. What are the dominant physical and chemical phenomena taking place in the process to be controlled?

A clear understanding of these questions and their answers will help greatly to define and simplify:

1. The system that we will attempt to model
2. The mass, energy, and momentum balances that we should develop
3. The additional equations that will be needed to complete the mathematical model of the process (i.e., transport and kinetic rate expressions, reaction and phase equilibria relationships, etc.)

They will also help to identify:

1. The state variables
2. The input variables (manipulated and disturbances) and
3. The output variables

that the mathematical model should include.

Let us now examine each of the three questions posed above and see how they affect the modeling of a process for control purposes through a series of examples.

Control objectives

As discussed in Chapter 2, the objectives that a control system is called on to satisfy may have to do with:

Ensuring the stability in the operation of a process, or
Suppressing the influence of external disturbances, or
Optimizing the economic performance of a plant, or usually
A combination of the above

All the objectives above are translated into quantitative expressions in terms of flow rates, temperatures, pressures, compositions, volumes, and so on, of the form

$$\text{variable } x = \text{desired value}$$

or

$$\text{variable } x > \text{ or } < \text{ bounding value}$$

where variable x = flow rate, temperature, pressure, volume, composition, and so on.

It is clear, therefore, that if we have identified the variables x which define quantitatively our control objectives, *the mathematical model that we will develop must describe how these variables change with time.* Also, it will help us determine what balances are needed for the development of the mathematical model.

Example 5.9

Consider the stirred tank heater discussed in Example 4.4.

1. If our control objective is to keep the liquid level at a desired value, the only state variable of interest is the volume of the liquid in the tank (or equivalently the height of the liquid level) and consequently we need only consider the total mass balance. The disturbance of interest is the flow rate of the inlet stream, F_i, while the manipulated variables to be considered are the outlet flow rate F and the inlet flow rate F_i.

2. If, on the other hand, our control objective is to keep the temperature of the outlet stream, T, at a desired value, we must consider both state variables: the temperature and the level of the liquid in the tank. This implies that we need write both total mass and energy balances. The disturbances of interest are the temperature and the flow rate of the inlet stream, while the available manipulated variables are F_i, F and Q.

3. Finally, if our control objectives are to keep the temperature of the effluent stream and the liquid level at desired values, we have a situation similar to case 2 above.

Example 5.10

Consider the continuous mixing process discussed in Example 4.11 (Figure 4.8). We can distinguish the following control situations:

1. If our control objective is to keep the concentration of the effluent stream in A at a desired value, the state variables of interest are the volume of the mixture in the tank and its concentration in A. The relevant balances are those on total mass and on component A. The disturbances of interest are c_{A_1} (or c_{B_1}), F_1, c_{A_2} (or c_{B_2}), and F_2. The available manipulated variables are F_1, F_2, F_3, or the ratio F_1/F_2.
2. If, on the other hand, our control objectives are to keep the composition *and* the temperature of effluent stream at desired values we need consider all three state variables (c_{A_3}, V, T_3) and formulate all three balances (total mass, component A, total energy). In this case the important disturbances are c_{A_1}, F_1, T_1, c_{A_2}, F_2, and T_2. The available manipulated variables are F_1, F_2, F_3, the ratio F_1/F_2, and Q.

Expected disturbances and their impact

The external disturbances which are expected to appear and affect the operation of a process will influence the mathematical model that we need to develop. Furthermore, disturbances with a very small impact on the operation of the process can be neglected, whereas disturbances with significant impact on the process must be included in the model. This will determine the complexity of the model needed: that is, what balances and what state variables should be included in the model.

State variables that are affected very little by the expected disturbances can be eliminated from the model and, together with them, the corresponding balances.

Example 5.11

Let us return to the stirred tank heater (Example 4.4). If the feed flow rate (disturbance) is not expected to vary significantly, the volume of the liquid in the tank will remain almost constant. In this case $dV/dt = A\, dh/dt \approx 0$ and we can neglect the total mass balance and the associated state variable h. The mathematical model of interest for control purposes is given by the total energy balance alone [eq. (4.5b)], with temperature the only state variable.

Remarks.   Note that if the feed temperature T_i is not expected to vary significantly, but the feed flow rate F_i is expected to change substantially, the mathematical model cannot be simplified as above, but it will be given by both eqs. (4.4a) and (4.5b).

Similar results to those above can be deduced for the CSTR system (Example 4.10). If the feed flow rate is not expected to change significantly, the model can be simplified (i.e., $dV/dt = 0$) and it is given by eqs. (4.9a) and (4.10b).

Physical–chemical phenomena in a process

A good understanding of the physical–chemical phenomena taking place in a process can lead to significant model simplifications for control purposes. Such simplifications can be done by excluding from the balances (model) those terms that have small contributions.

Example 5.12

Let us return to the continuous mixing process discussed in Example 4.11. Assume that for the particular components A and B of the mixture, the heat of solution does not depend significantly on the composition of A and B. In this case (see Example 4.11)

$$[\Delta \tilde{H}_{S_1} - \Delta \tilde{H}_{S_3}] \approx' [\Delta \tilde{H}_{S_2} - \Delta \tilde{H}_{S_3}] \approx 0$$

and the total energy balance [eq. (4.14a)] can be simplified to the following:

$$\rho c_p V \frac{dT_3}{dt} = \rho F_1 c_p (T_1 - T_3) + \rho F_2 c_p (T_2 - T_3) \pm Q$$

In other words, we see that the nature of the mixing phenomenon leads to a simplification of the model.

Furthermore, assuming that from all possible disturbances only the feed compositions c_{A_1} and c_{A_2} are expected to change significantly whereas the feed flow rates F_1 and F_2 and feed temperatures T_1 and T_2 are expected to remain almost the same, we can omit from the mathematical model the total mass and energy balances and from the set of state variables volume V and temperature T_3. Thus the simplified model is given only by the balance on component A [eq. (4.13a)].

Example 5.13

Consider again the CSTR discussed in Example 4.10. If the heat of reaction for the particular reaction A $\longrightarrow$ B is very small and the temperature of the feed stream is not expected to change significantly, the temperature of the reacting mixture will not change appreciably. In this case the reactor can be assumed *isothermal*. We can exclude the total energy balance from the mathematical model and the temperature from the set of state variables.

Examples 5.9 through 5.13 demonstrate very simply but vividly how the mathematical model of a process can be simplified when we take into account various considerations related to the nature of the process and the characteristics of the control problems.

The control designer always looks out for such simplifications.

THINGS TO THINK ABOUT

1. What is an input–output model, and how can you develop it from a state model? When is this possible?

2. Describe a procedure that would allow you to develop the input–output model for an ideal, binary distillation column.

3. Define the concept of degrees of freedom and relate it to the solution of E equations with V variables.

4. How many degrees of freedom do you have in a system composed of P phases with C components? (Recall Gibbs' rule.)

5. How many degrees of freedom do you have in a system composed of P phases with C components if the mass of each phase is given (i.e., M_1, M_2, ..., M_P)? (Recall Duhem's rule.)

6. How does the number of degrees of freedom affect the number and the selection of the control objectives in a chemical process?

7. Why do we claim that d disturbances reduce the number of degrees of freedom by d?

8. Why can a control system not be designed for an overspecified process?

9. Can you have the desired operation for an underspecified process? If yes, explain why. If no, explain how can you lift the underspecification.

10. Consider a system modeled by the following set of state equations:

$$\frac{dx_1}{dt} = f_1(x_1, x_2, m_1, m_2, m_3, d_1, d_2)$$

$$\frac{dx_2}{dt} = f_2(x_1, x_3, m_1, d_2)$$

$$\frac{dx_3}{dt} = f_3(x_1, x_2, x_3, m_2, m_3, d_1, d_2, d_3)$$

where x_1, x_2, and x_3 are the state variables, m_1, m_2, and m_3 are the manipulated variables, and d_1, d_2, and d_3 are the external disturbances.
(a) How many degrees of freedom does the system possess?
(b) How many control objectives can you specify at most?
(c) Consider this system at steady state. How many degrees of freedom does it possess?

11. A system is described by the following set of state equations:

$$\frac{dx_1}{dt} = f_1(m_1, m_2, d_1, d_2) \quad \text{and} \quad \frac{dx_2}{dt} = f_2(m_1, m_2, d_1)$$

Find the degrees of freedom for the system at its dynamic state and steady state. Are they equal? If not, why? What are the implications on control in this case?

12. What are the principal control considerations that affect the scope of mathematical modeling of a chemical process?

13. In what sense do the control considerations affect the mathematical modeling of a chemical process?

14. What are the usual, general quantitative representations of the control objectives? In terms of what variables are they expressed?

15. How can the expected impact of the disturbances simplify the model of a process? Give an example other than that discussed in the text.

16. Give examples to demonstrate how you can simplify the model of a process by disregarding physical and chemical phenomena with a limited impact on the behavior of the process.

17. Outline the steps that you should take during the development of a mathematical model for chemical process control purposes.

REFERENCES FOR PART II

Chapter 4. Three exceptional references with a large number of process modeling examples are:

1. *Process Dynamics and Control*, Vol. 1, by J. M. Douglas, Prentice-Hall, Inc., Englewood Cliffs, N.J. (1972).
2. *Process Modeling, Simulation, and Control for Chemical Engineers*, by W. L. Luyben, McGraw-Hill Book Company, New York (1973).
3. *Dynamic Behavior of Processes*, by J. C. Friedly, Prentice-Hall, Inc., Englewood Cliffs, N.J. (1972).

For the development of the dynamic material and energy balances, the reader could also consult the following book, from which Examples 4.10 and 4.11 are adapted:

4. *Introduction to Chemical Engineering Analysis*, by T. W. F. Russell and M. M. Denn, John Wiley & Sons, Inc., New York, (1972).

Additional references for material and energy balances are:

5. *Basic Principles and Calculations in Chemical Engineering*, 3rd ed., by D. M. Himmelblau, Prentice-Hall, Inc., Englewood Cliffs, N.J. (1974).
6. *Elementary Principles of Chemical Processes*, by R. M. Felder and R. W. Rousseau, John Wiley & Sons, Inc., New York (1978).

For the modeling of specific unit operations and reactors, there exist a large number of textbooks that the reader could use. Not all models included in these books are convenient for process control purposes, but they could help you to develop simplified and useful models. Among all the available references the following constitute a partial list:

For the modeling of chemical reactors:

7. *Chemical Reaction Engineering*, by O. Levenspiel, John Wiley & Sons, Inc., New York, (1962).
8. *An Introduction to Chemical Engineering Kinetics and Reactor Design*, by C. G. Hill, Jr., John Wiley & Sons, Inc., New York (1977).
9. *Elementary Chemical Reactor Analysis*, by R. Aris, Prentice-Hall, Inc., Englewood Cliffs, N.J. (1969).
10. *Chemical and Catalytic Reaction Engineering*, by J. J. Carberry, McGraw-Hill Book Company, New York (1976).

For the modeling of transport processes:

11. *Transport Phenomena*, by R. B. Bird, W. E. Stewart, and E. N. Lightfoot, John Wiley & Sons, Inc., New York (1960).
12. *Mass-Transfer Operations*, 2nd ed., by R. E. Treybal, McGraw-Hill Book Company, New York (1968).
13. *Heat and Mass Transfer*, 2nd ed., by E. R. G. Eckert and R. M. Drake, Jr., McGraw-Hill Book Company, New York (1959).

For more on reaction equilibria and phase equilibria the reader may consult the following books:

14. *Introduction to Chemical Engineering Thermodynamics*, 3rd ed., by J. M. Smith and H. C. Van Ness, McGraw-Hill Book Company, New York (1975).
15. *Chemical Engineering Kinetics*, 2nd ed., by J. M. Smith, McGraw-Hill Book Company, New York, (1970).

For an extensive discussion of the mathematical modeling of an ideal, binary distillation column and of a nonideal multicomponent column, the reader can consult the books by Douglas [Ref. 1], Luyben [Ref. 2], and Friedly [Ref. 3]. An interesting discussion of the difficulties encountered during the modeling of chemical processes can be found in:

16. "Critique of Chemical Process Control Theory," by A. S. Foss, *AIChE J.*, *19*, 209 (1973).
17. "Advanced Control Practice in the Chemical Process Industry: A View from Industry," by W. Lee and V. W. Weekman, Jr., *AIChE J.*, *22*, 27 (1976).

Chapter 5. For additional study on the degrees of freedom and their impact on the design of process control systems, the reader should consult the following references:

18. *Automatic Control of Processes*, by P. W. Murrill, Intext Educational Publishers, Scranton, Pa. (1967).
19. *The Chemical Engineer's Handbook*, 5th ed., J. H. Perry (ed.), McGraw-Hill Book Company, New York (1974).

PROBLEMS FOR PART II

Chapter 4

II.1 Consider the two systems shown in Figure PII.1. System 1 differs from system 2 by the fact that the level of liquid in tank 2 does not affect the effluent flow rate from tank 1, which is the case for system 2.

(a) Develop the mathematical model for each of the two systems.

(b) What are the state variables for each system, and what type of balance equations have you used?

(c) Which mathematical model is easier to solve, that for system 1 or that for system 2? Why?

Assume that the flow rate of an effluent stream from a tank is proportional to the hydrostatic liquid pressure that causes the flow of liquid. The cross-sectional area of tank 1 is A_1 (ft^2) and of tank 2 is A_2 (ft^2)(for both systems). The flow rates F_1, F_2, and F_3 are in ft^3/min.

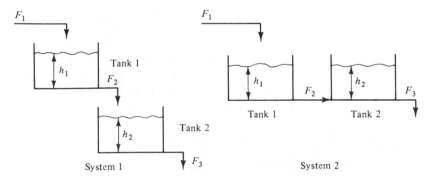

Figure PII.1

II.2 Develop the mathematical model for the system shown in Figure PII.2. What are the state variables for this system and what type of balance equations have you used? All the flow rates are volumetric, and the cross-sectional areas of the three tanks are A_1, A_2, and A_3 (ft^2), respectively. The flow rate F_5 is constant and does not depend on h_3, while all other effluent flow rates are proportional to the corresponding hydrostatic liquid pressures that cause the flow.

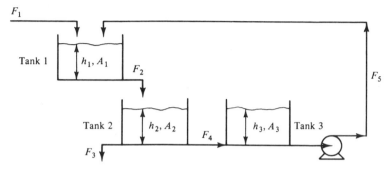

Figure PII.2

II.3 Consider the two stirred tank heaters shown in Figure PII.3.
(a) Identify the state variables of the system.
(b) Determine what balances you should perform.
(c) Develop the state model that describes the dynamic behavior of the system.
(d) How would you express the heats given by the two steam flows in terms of other variables?

The flow rates of the effluent streams are assumed to be proportional to the liquid static pressure that causes the flow of the liquid. The cross-sectional areas of the two tanks are A_1 and A_2 (ft^2) and the flow rates are volumetric. No vapor is produced either in the first or the second tank. A_{t1} and A_{t2} are the heat exchange areas for the two steam coils.

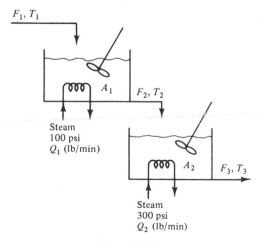

Figure PII.3

LIVERPOOL JOHN MOORES UNIVERSITY
LEARNING SERVICES

II.4 Do the same work as in Problem II.3 for the stirred tank heaters' system shown in Figure PII.4. For tank 1, the steam is injected directly in the liquid water. Water vapor is produced in the second tank. A_1 and A_2 are the cross-sectional areas of the two tanks. Assume that the effluent flow rates are proportional to the liquid static pressure that causes their flow. A_t is the heat transfer area for the steam coil.

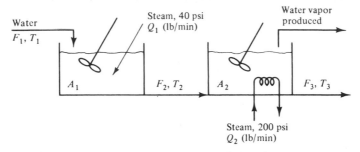

Figure PII.4

II.5 Consider the mixing process taking place in a two-tank system (Figure PII.5).
(a) Identify the state variables of the system.
(b) Determine what balances you should perform.
(c) Develop the state model that describes the dynamic behavior of the process, assuming that the heats of solution are strong functions of the composition.
(d) How is the model simplified if the heats of solution are very weak functions of the composition?
Assume that the flow rates are volumetric and the compositions are in moles per volume. The effluent flow rates are proportional to the liquid static pressure that causes their flow. A_1 and A_2 are the cross-sectional areas of the two tanks and A_t is the heat transfer area for the steam coil.

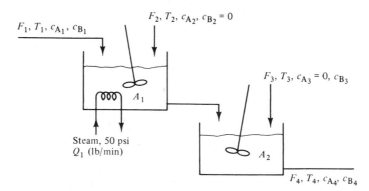

Figure PII.5

II.6 Develop the state model for the batch mixing of two solutions (Figure PII.6). Initially, the tank is empty. The volume of the tank is V

(ft^3). The flow rates are volumetric and the concentrations are in moles per volume.

(a) How long does it take to fill up the tank?

(b) Show how would you find the composition and temperature of the mixture in the tank during the time that the tank is being filled up.

Assume that the flow rates are volumetric, the compositions in moles per volume, and that the heat of solution depends on the composition.

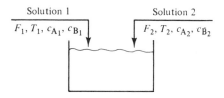

Figure PII.6

II.7 Develop the state model for a batch reactor where the following reactions take place:

$$A \xrightarrow{k_1} B \xrightarrow{k_3} C$$
$$A \xrightarrow{k_2} D$$

All reactions are endothermic and have first-order kinetics. The reacting mixture is heated by steam of 150 psig, which flows through a jacket around the reactor with a rate of Q (lb/min).

II.8 Consider the continuous stirred tank reactor system shown in Figure PII.7. Stream 1 is a mixture of A and B with composition c_{A_1} and c_{B_1} (moles/volume) and has a volumetric flow rate F_1 and a temperature T_1. Stream 2 is pure R. The reactions taking place are:

$$\text{Reaction 1: } A + R \xrightarrow{k_1} P_1$$
$$\text{Reaction 2: } B + 2R \xrightarrow{k_2} P_2$$

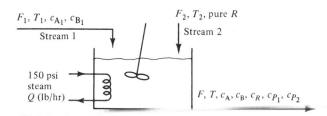

Figure PII.7

Both reactions are endothermic and have second-order (reaction 1) and third-order (reaction 2) kinetics. Heat is supplied to the reaction mixture by steam which flows through a coil, immersed in the reactor's content, with a heat transfer area A_t.

(a) What are the state variables describing the natural state of the system?

(b) What are the balances that you should consider?

(c) Develop the state model for the CSTR system.

(d) How can you simplify the state model if $k_2 \simeq 10^{-7}k_1$ for a large range of temperatures?

(e) Define the assumptions that should be made in order to have an isothermal reactor.

II.9 Develop the state model for the two CSTRs system of Figure PII.8. A simple reaction A $\longrightarrow$ B with first-order kinetics takes place. Assume isothermal conditions. Flow rates F_1 and F_2 are determined by variable speed pumps and thus are independent of the corresponding liquid levels.

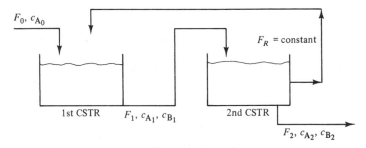

Figure PII.8

II.10 Assuming plug flow conditions for a jacketed tubular reactor (Figure PII.9), develop its state model. A simple exothermic reaction A $\longrightarrow$ B with first-order kinetics takes place. Due to the very large heat of reaction, molten salt, which flows countercurrently to the reaction mixture around the tube of the reactor, is used as coolant to keep the reaction temperature at acceptable levels. Assume constant temperature for the molten salt along the length of the reactor. The reaction takes place in the gaseous phase. The flow rates are volumetric and the compositions are in moles per volume. The internal diameter of the tube is d (in.) and its length l.

(a) Is the system a lumped parameter (described by ordinary differential equations) or a distributed parameter system? Develop the model.

(b) Does the reactor possess dead time between inputs and outputs? Explain.

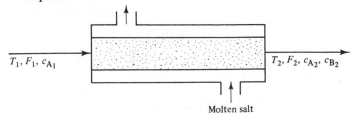

Figure PII.9

II.11 Figure PII.10 shows a simplified representation of a drum boiler. Feed water enters the boiler with a flow rate F_1 (mass/hr) and a temperature T_1 and it is heated by an amount of heat Q (Btu/hr) which is supplied by burned fuel. The generated steam flows out from the top of the boiler, with a flow rate F_2 (mass/hr) and a pressure p (psig). A simple feedback control system has been installed to keep the level of the water in the drum boiler constant by manipulating the flow rate of the feedwater stream.
(a) What are the state variables describing the system?
(b) What balances are appropriate for the drum boiler system?
(c) Develop the state model of the system. For the feedback control system, use a relationship of the form $F_1 = f(h - h_{desired})$.

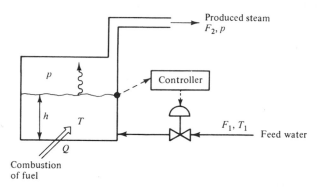

Figure PII.10

II.12 Consider a pipe of length L (in.) with an internal diameter D (in.). Water flows through the pipe with a volumetric flow rate F. Let p_1 be the pressure at the entrance of the pipe and p_2 the pressure at the exit.
(a) Identify the appropriate state variables to describe the system.
(b) What are the relevant balances for the system?
(c) Develop the state model for this laminar flow system.

II.13 Consider turbulence in the flow system of the Problem II.12. At time $t = 0$ a stream containing pure component A is mixed with the entering water, causing a concentration c_A (moles/volume) as the water enters the pipe. Assuming that A does not diffuse along the length of the pipe:
(a) Develop a state model that describes how the concentration of A changes with time and along the length of the pipe.
(b) Show that the system possesses dead time between input and output and compute the value of the dead time.

II.14 A liquid stream is a mixture of two components A and B and has a volumetric (volume/hr) flow rate F, temperature T_f, and pressure p_f. Let c_A and c_B be the mole fractions of A and B in the liquid stream. It is assumed that the pressure p_f is larger than the bubble point pressure of the mixture A and B, so that there is no vapor present.
  The liquid stream passes through an isenthalpic expansion valve

and is "flashed" into a flash drum (Figure 4.6). The pressure p in the drum is assumed to be lower than the bubble point pressure of the liquid mixture at T_f. As a result, two phases at equilibrium with each other appear in the flash drum: a vapor phase with a composition y_A and y_B (molar fractions) which is drawn with a volumetric flow rate V, and a liquid phase with a composition x_A and x_B (molar fractions) drawn with a volumetric flow rate L. Let T be the temperature of the two phases at equilibrium in the flash drum.

(a) What are the fundamental dependent quantities whose values describe the natural state of the flash drum?
(b) What are the boundaries of the system(s) around which you will perform the various balances?
(c) What are the relevant balances?
(d) Besides the balance equations, what additional relationships do you need to complete the state model for the flash drum?
(e) Identify the state variables and the input variables (manipulations, disturbances) of the system.
(f) Develop the complete state model of the system.

Chapter 5

II.15 Starting with the state models for the two systems of Problem II.1 (Figure PII.1), develop the corresponding input–output models. Also:
(a) Find the degrees of freedom for each system.
(b) Specify how many control objectives you can have for each system.

II.16 For each of the following systems, find:
(a) The number of degrees of freedom.
(b) The number of independent control objectives you can specify.
(c) The input–output model.

   System A: The three-tank system of Figure PII.2 (Problem II.2).
   System B: The two stirred tank heaters system of Figure PII.4 (Problem II.4).
   System C: The two-tank mixing process of Figure PII.5 (Problem II.5).

II.17 For the CSTR system of Figure PII.7 (Problem II.8), find:
(a) The number of degrees of freedom.
(b) The number of independent control objectives that you can specify.
(c) For the number of control objectives you have specified, do you have an equal number of manipulated variables so that you can achieve your control objectives?
(d) Identify what you would use for control objectives.
(e) Show how you would form the corresponding input–output model, but do not develop its analytical form.

II.18 Do the same work as in Problem II.17 for the two CSTRs system of Figure PII.8 (Problem II.9).

II.19 For the flash drum system of Problem II.14 (see also Figure 4.6), find:
   (a) The number of degrees of freedom.
   (b) The number of independent control objectives you can specify.
   (c) Identify a set of control objectives that are meaningful from a practical operation point of view.
   (d) Show how you would develop an input–output model for the flash drum, but do not derive its complete analytical form.

II.20 A simple chemical reaction, A $\longrightarrow$ B, with first-order kinetics, takes place in a CSTR. The effluent of the reactor enters an ideal binary distillation column where the unreacted A is taken as the overhead product with a composition y_A (molar fraction) and is recycled back to the reactor after it has been mixed with fresh feed (Figure PII.11). Assume that the mixer and the CSTR are both isothermal.
   (a) Identify the fundamental dependent variables that describe the natural state of the plant.
   (b) What are the relevant balances, and what are the boundaries of the systems around which you will perform the balances?
   (c) Formulate all the relevant balance equations.
   (d) Identify the state variables of the plant.

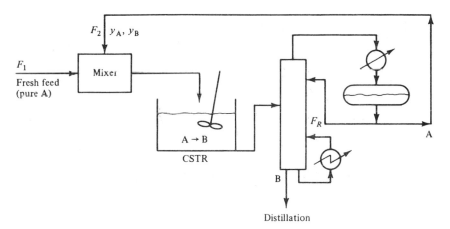

Figure PII.11

II.21 Consider the small plant described in Problem II.20 (Figure PII.11).
   (a) Determine the number of degrees of freedom for the plant.
   (b) If (number of degrees of freedom) > 0, how would you specify the additional equations needed to render an exactly specified system (i.e., number of degrees of freedom = 0)?
   (c) How many disturbance specifications do you have, and how many control objectives can you identify?

(d) Specify the control objectives that have a practical meaning for the plant.

(e) Develop a simple input–output model for the plant without performing the required analytic computations.

II.22 Develop the state model for an ideal binary batch distillation column with N ideal plates (Figure PII.12). At $t = 0$, the composition of the initial mixture is c_A and c_B (molar fractions), and its total mass is M (moles).

(a) List all assumptions that you make for modeling the system.

(b) Identify the relevant balance equations.

(c) In addition to the balance equations, what other relationships do you need to complete the state model?

(d) Show how you can find the time when the composition of the overhead product has reached a composition y_A (molar fraction).

(e) How many degrees of freedom does the system possess?

(f) How many control objectives can you specify?

Additional questions:

(g) Does the vapor flow rate remain constant if the heat input Q remains constant with time?

(h) What about F_R; does it remain constant? Under what conditions would you change it?

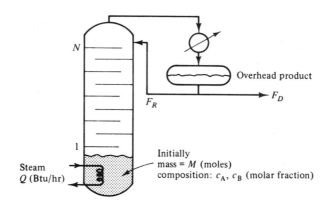

N

Overhead product

F_R

F_D

1

Steam
Q (Btu/hr)

Initially
mass = M (moles)
composition: c_A, c_B (molar fraction)

Figure PII.12

II.23 Develop the state model for a gas absorption column (Figure PII.13) with N ideal transfer units. The flow rate (mol/hr) of the entering gas stream is F_g with a composition (molar fraction) y_A, while the composition of the effluent gas stream should be y'_A (molar fraction). The flow rate of the liquid absorbing stream is F_ℓ (mol/hr) and at its entrance it is free of component A.

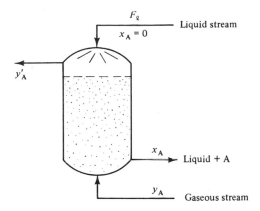

Figure PII.13

II.24 Develop the state model for a multicomponent (C components) nonideal distillation column with N trays. Use the general nomenclature developed in Example 4.13 for the ideal binary distillation.

(a) State your assumptions.

(b) Include the dynamics of the condenser and reboiler considering them as a perfectly stirred tank cooler and heater, respectively.

(c) What relationships do you need in addition to the balance equations?

(d) How many degrees of freedom does the system possess?

(e) How many control objectives can you specify?

Analysis of the Dynamic Behavior of Chemical Processes

In process control, the processes are analytically complex but relatively simple to control in practice...These processes act quantitatively like nothing we know, but qualitatively (and for certain quantitative purposes), they can be modeled in terms of simple gains, deadtimes, lags or combinations thereof....

*E. H. Bristol**

In Part III we study the dynamic and static behavior of several, simple processing systems. Understanding the dynamics of such simple systems allows us to analyze the behavior of more complex systems such as the chemical processes.

The analysis is limited to linear dynamic systems. This may seem incompatible with the fact that most of the chemical engineering processes are modeled by nonlinear equations. However, linear techniques are very valuable and of great practical importance for the following reasons: (1) There is no general theory for the analytic solution of nonlinear differential equations, and consequently no compre hensive analysis of nonlinear dynamic systems. (2) A nonlinear system can be adequately approximated by a linear system near some operating conditions. (3) Significant advances in the linear control

*Industrial Needs and Requirements for Multivariable Control," in *Chemical Process Control*, A. S. Foss and M. M. Denn (eds.), AIChE Symp. Ser. 72, No. 159 (1976).

theory permit the synthesis and design of very effective controllers even for nonlinear processes.

Fundamental, therefore, is the concept of linearization and the procedure for approximating nonlinear systems with linear ones which will be studied in Chapter 6.

The Laplace transforms, which we will discuss in Chapter 7, offer a very simple and elegant procedure for the solution of linear differential equations (Chapter 8) and consequently for the analysis of the behavior of linear systems.

In the remaining chapters of Part III we will discuss:

1. The development of simple input–output models for chemical processes, using the Laplace transforms (Chapter 9)
2. The dynamic analysis of various typical processes (Chapters 10 through 12)

Computer Simulation **6**
and the Linearization
of Nonlinear Systems

In order to find the dynamic behavior of a chemical process, we have to integrate the state equations used to model the process. But most of the processing systems that we will be interested in are modeled by nonlinear differential equations, and it is well known that there is no general mathematical theory for the analytical solution of nonlinear equations. Only for linear differential equations are closed-form, analytic solutions available.

When confronted with the dynamic analysis of nonlinear systems, there are several things we can do, such as:

1. Simulate the nonlinear system on an analog or digital computer and compute its solution numerically, or
2. Transform the nonlinear system into a linear one by an appropriate transformation of its variables, or
3. Develop a linear model that approximates the dynamic behavior of a nonlinear system in the neighborhood of specified operating conditions.

Alternative 2 can be used in very few cases, whereas alternatives 1 and 3 are, in principle, always feasible. In this chapter we discuss the computer simulation of nonlinear processes very briefly because it is a subject to be covered primarily in a course on numerical analysis. More emphasis will be given on the approximation of nonlinear models by linear ones. It should be noted that all the theory for the design of control systems, available from past work, is based on linear systems, and that very small advances have been made toward the development of a control theory for nonlinear systems.

6.1 Computer Simulation of Process Dynamics

Nonlinear differential and/or algebraic equations cannot, in general, be solved analytically, and computer-aided numerical solutions are required. Numerical solutions are also preferred for the equations which can be solved analytically, when the analytic solutions are very complex and provide little insight in the behavior of the system.

Let us consider two processes we have already modeled: the continuous stirred tank reactor and the ideal, binary distillation column.

The model for the CSTR (see Example 4.10) is given by eqs. (4.8a), (4.9a), and (4.10b). These constitute a set of nonlinear equations for which there is no analytic solution available. Therefore, in order to study the dynamic behavior of the CSTR, we must solve the modeling equations numerically using a computer.

The model for the ideal, binary distillation column (see Example 4.13) is composed of

$$2N + 4 \quad \text{nonlinear differential equations}$$

and

$$2N + 1 \quad \text{nonlinear algebraic equations}$$

It is not only the nonlinearity of the equations but also the size of model (24 differential and 21 algebraic equations for a modest 10-tray column) that necessitates a numerical solution in order to study the dynamic behavior of the column.

Today, computer simulation is used extensively to analyze the dynamics of chemical processes or aid in the design of controllers and study their effectiveness in controlling a given process. Analog and digital computers have been used for this purpose, with the emphasis having shifted almost entirely in favor of the digital computers.

Historically, analog computers were the first to be used to simulate the dynamics of chemical processes with or without control. They permitted a rapid solution of the modeling equations, thus providing useful insight as to how a process would react to external disturbances or how effective was the control of the process using various measurements, manipulated variables, and control configurations. The analog computers have several serious drawbacks: (1) Significant time is required to set up the problem and get it running; (2) the need of one hardware element per mathematical operation prohibits the simulation of large, complex systems; (3) the nonlinear terms are simulated by rather expensive hardware elements (function generators) with limited flexibility; and (4) they do not possess memory like the digital computers. The subsequent revolution, brought about by the digital computers, made the analog computers obsolete. Today, they are still used on a small scale and primarily to train operators on the dynamic operation of chemical plants.

The computational power introduced with the digital computers, together with the resulting low cost of computations, has expanded tremendously the scope and the practical significance of computer simulation for process dynamics and control. The availability of sophisticated equation-solving routines for almost every digital computer system available has simplified the required groundwork for process simulation and has relieved the engineer of the need to be an expert in numerical analysis.

Digital computer simulation of process dynamics involves the solution of a set of differential and algebraic equations, which describe the process. There are several categories of numerical methods which can be used to integrate differential equations and solve algebraic ones. Let us examine briefly the simplest and most popular among them.

Numerical solution of algebraic equations

At steady state, the state equations turn to simple algebraic equations, since the rate of accumulation becomes zero. Therefore, in order to determine the steady-state behavior of a process under given conditions, we should be able to solve sets of algebraic equations. All available methods use an iterative trial-and-error procedure, which approaches (hopefully) closer and closer to the solution with each iteration. The key question is to select the appropriate method, which for the given set of equations converges rapidly to the correct solution. Unfortunately, this is a very difficult task and in all but a few instances is impossible to know a priori how successful a method will be in finding the solution to a particular set of equations. Quite often, a method will not converge to the solution, or in other instances it approaches the solution very slowly. Among the most commonly used techniques are the following: (1) interval halving, (2) successive substitution, and (3) Newton–Raphson.

Numerical integration of differential equations

Here again we have a very large number of available techniques. Numerical integration implies an approximation of the continuous differential equations with discrete finite-difference equations. The various integration methods differ in the way they implement this approximation. Thus we have *explicit* methods which march on in time yielding the solution in one pass, or we have *implicit* methods with *predictor–corrector* capabilities. The key questions for an integration technique are the stability of the procedure and the speed with which it reaches the solution. But again, these are questions which, in general, cannot be answered to our satisfaction ahead of time. Among the most popular integration methods is the explicit fourth-order *Runge–Kutta*, which provides satisfactory accuracy and stability of computations as well as low cost.

Digital computer simulation of the chemical process dynamics is used extensively at the present. It allows the engineer to anticipate the behavior of a process not only qualitatively but also quantitatively. It has helped to design more complex and sophisticated control systems. The major drawback of computer simulation is that "it only gives you numbers" and not a general analytic solution in terms of arbitrary, unspecified parameters, which in turn you taylor to your particular problem. Therefore, the results of computer simulation are of ad hoc nature, and you will have to make several runs with different values for the input variables and parameters before you can establish a good understanding of the process dynamics for a range of operating conditions.

6.2   Linearization of Systems with One Variable

Linearization is the process by which we approximate nonlinear systems with linear ones. It is widely used in the study of process dynamics and design of control systems for the following reasons:

1. We can have closed-form, analytic solutions for linear systems. Thus we can have a complete and general picture of a process's behavior independently of the particular values of the parameters and input variables. This is not possible for nonlinear systems, and computer simulation provides us only with the behavior of the system at specified values of inputs and parameters.
2. All the significant developments toward the design of effective control systems have been limited to linear processes.

First, we will study the linearization of a nonlinear equation with one variable and then we will extend it to multivariable systems.

Consider the following nonlinear differential equation, modeling a given process:

$$\frac{dx}{dt} = f(x) \tag{6.1}$$

Expand the nonlinear function $f(x)$ into a Taylor series around the point x_0 and take

$$f(x) = f(x_0) + \left(\frac{df}{dx}\right)_{x_0} \frac{x - x_0}{1!} + \left(\frac{d^2f}{dx^2}\right)_{x_0} \frac{(x - x_0)^2}{2!}$$

$$+ \cdots + \left(\frac{d^n f}{dx^n}\right)_{x_0} \frac{(x - x_0)^n}{n!} + \cdots \tag{6.2}$$

If we neglect all terms of order two and higher, we take the following approximation for the value of $f(x)$:

$$f(x) \approx f(x_0) + \left(\frac{df}{dx}\right)_{x_0} (x - x_0) \tag{6.3}$$

It is well known that the error introduced in the approximation (6.3) is of the same order of magnitude as the term

$$I = \left(\frac{d^2f}{dx^2}\right)_{x_0} \frac{(x - x_0)^2}{2!} \tag{6.4}$$

Consequently, the linear approximation (6.3) is satisfactory only when x is very close to x_0, where the value of the term I is very small.

In Figure 6.1 we can see the nonlinear function $f(x)$ and its linear approximation around x_0. From the same picture it is also clear that the linear approximation depends on the location of the point x_0 around which we make the expansion into a Taylor series. Compare the linear approximation of $f(x)$ at the points x_0 and x_1 (Figure 6.1). *The approximation is exact only at the point of linearization.*

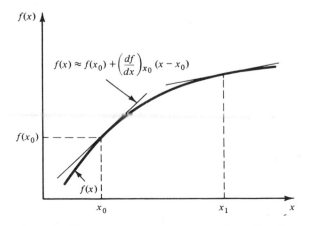

Figure 6.1 Linear approximation of a nonlinear function.

In eq. (6.1), replace $f(x)$ by its linear approximation given by eq. (6.3) and take

$$\frac{dx}{dt} = f(x_0) + \left(\frac{df}{dx}\right)_{x_0} (x - x_0) \tag{6.5}$$

This equation is the linearized approximation of the initial dynamic system given by eq. (6.1). In later chapters the design of the process controller will be based on such approximate linearized models.

Example 6.1

Consider the tank system shown in Figure 6.2a. The total mass balance yields

$$A \frac{dh}{dt} = F_i - F_o \tag{6.6}$$

where A is the cross-sectional area of the tank and h the height of the liquid level. If the outlet flow rate F is a linear function of the liquid's level,

$$F_o = \alpha h \qquad \text{where } \alpha = \text{constant}$$

we take

$$A \frac{dh}{dt} + \alpha h = F_i$$

which is a linear differential equation (modeling a linear dynamic system) and no approximation is needed.

If, on the other hand,

$$F_o = \beta \sqrt{h}$$

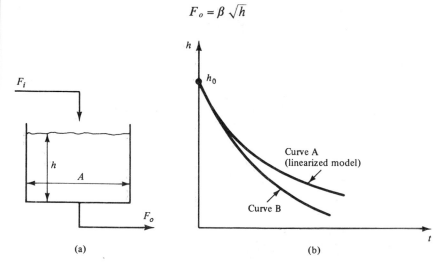

(a)   (b)

Figure 6.2 (a) Tank system of Example 6.1; (b) approximation of liquid level response.

the resulting total mass balance yields a nonlinear dynamic model,

$$A \frac{dh}{dt} + \beta \sqrt{h} = F_i \tag{6.7}$$

Let us develop the linearized approximation for this nonlinear model. The only nonlinear term in eq. (6.7) is $\beta \sqrt{h}$. Take the Taylor series expansion of this term around a point h_0:

$$\beta\sqrt{h} = \beta\sqrt{h_0} + \left[\frac{d}{dh}(\beta\sqrt{h})\right]_{h=h_0}(h - h_0) + \left[\frac{d^2}{dh^2}(\beta\sqrt{h})\right]_{h=h_0}\frac{(h - h_0)^2}{2!} + \cdots$$

$$= \beta\sqrt{h_0} + \frac{\beta}{2\sqrt{h_0}}(h - h_0) - \frac{\beta}{8\sqrt[3]{h_0^2}}(h - h_0)^2 + \cdots$$

Neglecting terms of order two and higher, we take

$$\beta \sqrt{h} \approx \beta \sqrt{h_0} + \frac{\beta}{2\sqrt{h_0}}(h - h_0)$$

which, if introduced in the nonlinear dynamic system (6.7), yields the following linearized approximate model:

$$A\frac{dh}{dt} + \frac{\beta}{2\sqrt{h_0}}h = F_i - \frac{\beta}{2}\sqrt{h_0} \tag{6.8}$$

Let us compare the linearized, approximate model given by eq. (6.8) to the nonlinear one, given by eq. (6.7). Assume that the tank is at steady state with a liquid level h_0. Then at time $t = 0$, we stop the supply of liquid to the tank, while we allow the liquid to flow out. Thus at $t = 0$ the liquid level is at the steady-state value [i.e., $h(t = 0) = h_0$]. Curve A in Figure 6.2b is the solution of eq. (6.8) and curve B in the same figure is the solution of eq. (6.7). We notice that the two curves are very close to each other for a certain period of time. This indicates that the linearized model approximates at the beginning very well the nonlinear model.

As the time increases and the liquid level continues to fall, its value h deviates more and more from the initial value h_0 around which the linearized model was developed. Figure 6.2b indicates very clearly that as the difference $h_0 - h$ increases, the linearized approximation becomes progressively less accurate, as was expected.

6.3 Deviation Variables

Let us now introduce the concept of the *deviation variable*, which we will find very helpful in later chapters for the control of processing systems.

Suppose that x_s is the steady-state value of x describing the initial dynamic system (6.1). Then

$$\frac{dx_s}{dt} = 0 = f(x_s) \tag{6.9}$$

Consider x_s the point of linearization for eq. (6.1) (i.e., $x_0 \equiv x_s$). Then eq. (6.1) yields the following linearized model:

$$\frac{dx}{dt} = f(x_s) + \left(\frac{df}{dx}\right)_{x_s} (x - x_s) \tag{6.10}$$

Subtract eq. (6.9) from (6.10) and take

$$\frac{d(x - x_s)}{dt} = \left(\frac{df}{dx}\right)_{x_s} (x - x_s) \tag{6.11}$$

If we define the *deviation variable x'* as

$$x' = x - x_s$$

then eq. (6.11) takes the form

$$\frac{dx'}{dt} = \left(\frac{df}{dx}\right)_{x_s} x' \tag{6.12}$$

Equation (6.12) is the linearized approximation of the nonlinear dynamic system (6.1), expressed in terms of the deviation variable x'.

The notion of the deviation variable is very useful in process control. Usually we will be concerned with maintaining the value of a process variable (temperature, concentration, pressure, flow rate, volume, etc.) at some desired steady state. Consequently, the steady state becomes a natural candidate point around which to develop the approximate linearized model. In such cases the deviation variable describes directly the magnitude of the dislocation of a system from the desired level of operation. Furthermore, if the controller of the given process has been designed well, it will not allow the process variable to move far away from the desired steady-state value. Consequently, the approximate linearized model expressed in terms of deviation variables will be satisfactory to describe the dynamic behavior of the process near the steady state.

In subsequent chapters we will make extensive use of the linearized forms of differential equations, in terms of deviation variables.

Example 6.2

Consider the linearized model of the tank system given by eq. (6.8) of Example 6.1. Let h_s be the steady-state value of the liquid level for a given value, $F_{i,s}$ of the inlet flow rate. Then the linearized model around h_s

gives

$$A\frac{dh}{dt} + \frac{\beta}{2\sqrt{h_s}}h = F_i - \frac{\beta}{2}\sqrt{h_s} \tag{6.13}$$

At steady state from eq. (6.7) we also have

$$A\frac{dh_s}{dt} + \beta\sqrt{h_s} = F_{i,s} \tag{6.14}$$

$$0 =$$

Subtract eq. (6.14) from (6.13),

$$A\frac{d(h - h_s)}{dt} + \frac{\beta}{2\sqrt{h_s}}(h - h_s) = F_i - F_{i,s}$$

Defining the deviation variables

$$h' = h - h_s \quad \text{and} \quad F_i' = F_i - F_{i,s}$$

we take the following linearized form in terms of deviation variables:

$$A\frac{dh'}{dt} + \frac{\beta}{2\sqrt{h_s}}h' = F_i' \tag{6.15}$$

6.4   Linearization of Systems with Many Variables

In previous sections we developed the linearized approximation of a nonlinear dynamic system that had only one variable. Let us now extend that approach to systems with more than one variable.

Consider the following dynamic system:

$$\frac{dx_1}{dt} = f_1(x_1, x_2) \tag{6.16}$$

$$\frac{dx_2}{dt} = f_2(x_1, x_2) \tag{6.17}$$

Expand the nonlinear functions $f_1(x_1, x_2)$ and $f_2(x_1, x_2)$ into Taylor series around the point $(x_{1,0}, x_{2,0})$ and take

$$f_1(x_1, x_2) = f_1(x_{1,0}, x_{2,0}) + \left[\frac{\partial f_1}{\partial x_1}\right]_{(x_{1,0}, x_{2,0})}(x_1 - x_{1,0}) + \left[\frac{\partial f_1}{\partial x_2}\right]_{(x_{1,0}, x_{2,0})}(x_2 - x_{2,0})$$

$$+ \left[\frac{\partial^2 f_1}{\partial x_1^2}\right]_{(x_{1,0}, x_{2,0})}\frac{(x_1 - x_{1,0})^2}{2!} + \left[\frac{\partial^2 f_1}{\partial x_2^2}\right]_{(x_{1,0}, x_{2,0})}\frac{(x_2 - x_{2,0})^2}{2!}$$

$$+ \left[\frac{\partial^2 f_1}{\partial x_1 \partial x_2}\right]_{(x_{1,0}, x_{2,0})}(x_1 - x_{1,0})(x_2 - x_{2,0}) + \cdots$$

and

$$f_2(x_1, x_2) = f_2(x_{1,0}, x_{2,0}) + \left[\frac{\partial f_2}{\partial x_1}\right]_{(x_{1,0}, x_{2,0})} (x_1 - x_{1,0}) + \left[\frac{\partial f_2}{\partial x_2}\right]_{(x_{1,0}, x_{2,0})} (x_2 - x_{2,0})$$

$$+ \left[\frac{\partial^2 f_2}{\partial x_1^2}\right]_{(x_{1,0}, x_{2,0})} \frac{(x_1 - x_{1,0})^2}{2!} + \left[\frac{\partial^2 f_2}{\partial x_2^2}\right]_{(x_{1,0}, x_{2,0})} \frac{(x_2 - x_{2,0})^2}{2!}$$

$$+ \left[\frac{\partial^2 f_2}{\partial x_1 \partial x_2}\right]_{(x_{1,0}, x_{2,0})} (x_1 - x_{1,0})(x_2 - x_{2,0}) + \cdots$$

Neglect terms of order two and higher and take the following approximations:

$$f_1(x_1, x_2) \approx f_1(x_{1,0}, x_{2,0}) + \left[\frac{\partial f_1}{\partial x_1}\right]_{(x_{1,0}, x_{2,0})} (x_1 - x_{1,0}) + \left[\frac{\partial f_1}{\partial x_2}\right]_{(x_{1,0}, x_{2,0})} (x_2 - x_{2,0})$$

and

$$f_2(x_1, x_2) \approx f_2(x_{1,0}, x_{2,0}) + \left[\frac{\partial f_2}{\partial x_1}\right]_{(x_{1,0}, x_{2,0})} (x_1 - x_{1,0}) + \left[\frac{\partial f_2}{\partial x_2}\right]_{(x_{1,0}, x_{2,0})} (x_2 - x_{2,0})$$

Substitute the foregoing linear approximations of $f_1(x_1, x_2)$ and $f_2(x_1, x_2)$ into eqs. (6.16) and (6.17) of the initial nonlinear dynamic system and take

$$\frac{dx_1}{dt} = f_1(x_{1,0}, x_{2,0}) + \left[\frac{\partial f_1}{\partial x_1}\right]_{(x_{1,0}, x_{2,0})} (x_1 - x_{1,0}) + \left[\frac{\partial f_1}{\partial x_2}\right]_{(x_{1,0}, x_{2,0})} (x_2 - x_{2,0}) \quad (6.18)$$

$$\frac{dx_2}{dt} = f_2(x_{1,0}, x_{2,0}) + \left[\frac{\partial f_2}{\partial x_1}\right]_{(x_{1,0}, x_{2,0})} (x_1 - x_{1,0}) + \left[\frac{\partial f_2}{\partial x_2}\right]_{(x_{1,0}, x_{2,0})} (x_2 - x_{2,0}) \quad (6.19)$$

These last two equations are linear differential equations and constitute the linearized, approximate model of the initial nonlinear system described by eqs. (6.16) and (6.17).

The comments made earlier for the one-dimensional case apply also here:

1. The approximation deteriorates as the point (x_1, x_2) moves away from the point $(x_{1,0}, x_{2,0})$ of linearization.
2. The linearized approximate model depends on the point $(x_{1,0}, x_{2,0})$ around which we make the Taylor series expansion.

Let us now express the linearized system in terms of deviation variables. Select the steady state $(x_{1,s}, x_{2,s})$ as the point around which you will make the linearization [i.e., in eqs. (6.18) and (6.19) put $x_{1,0} \equiv x_{1,s}$ and $x_{2,0} \equiv x_{2,s}$]. At the steady state, eqs. (6.16) and (6.17) yield

$$0 = f_1(x_{1,s}, x_{2,s}) \quad (6.20)$$

$$0 = f_2(x_{1,s}, x_{2,s}) \qquad (6.21)$$

Subtract eq. (6.20) from (6.18) and (6.21) from (6.19) and take

$$\frac{d(x_1 - x_{1,s})}{dt} = \left[\frac{\partial f_1}{\partial x_1}\right]_{(x_{1,s}, x_{2,s})} (x_1 - x_{1,s}) + \left[\frac{\partial f_1}{\partial x_2}\right]_{(x_{1,s}, x_{2,s})} (x_2 - x_{2,s}) \quad (6.22)$$

and

$$\frac{d(x_2 - x_{2,s})}{dt} = \left[\frac{\partial f_2}{\partial x_1}\right]_{(x_{1,s}, x_{2,s})} (x_1 - x_{1,s}) + \left[\frac{\partial f_2}{\partial x_2}\right]_{(x_{1,s}, x_{2,s})} (x_2 - x_{2,s}) \quad (6.23)$$

Defining the deviation variables by

$$x_1' = x_1 - x_{1,s} \qquad \text{and} \qquad x_2' = x_2 - x_{2,s}$$

eqs. (6.22) and (6.23) take the following form in terms of deviation variables:

$$\frac{dx_1'}{dt} = a_{11} x_1' + a_{12} x_2'$$

$$\frac{dx_2'}{dt} = a_{21} x_1' + a_{22} x_2'$$

where

$$a_{11} = \left[\frac{\partial f_1}{\partial x_1}\right]_{(x_{1,s}, x_{2,s})} \qquad a_{12} = \left[\frac{\partial f_1}{\partial x_2}\right]_{(x_{1,s}, x_{2,s})}$$

$$a_{21} = \left[\frac{\partial f_2}{\partial x_1}\right]_{(x_{1,s}, x_{2,s})} \qquad a_{22} = \left[\frac{\partial f_2}{\partial x_2}\right]_{(x_{1,s}, x_{2,s})}$$

A final comment is in order. In the previous and present sections we considered the presence of state variables only in the nonlinear functions. Thus for systems with one variable, we had only the state x, and for systems with two variables we had only states x_1 and x_2. The formulation above should not be perceived as restrictive, but it is easily expanded to include the presence of input variables, like the manipulated variables and the disturbances. The following example demonstrates this point.

Example 6.3

Consider a dynamic system described by two state variables x_1 and x_2 and the following state equations:

$$\frac{dx_1}{dt} = f_1(x_1, x_2, m_1, m_2, d_1)$$

$$\frac{dx_2}{dt} = f_2(x_1, x_2, m_1, m_2, d_2)$$

where m_1 and m_2 are two manipulated variables and d_1 and d_2 are two disturbances affecting the system.

Linearization of the equations above around the nominal values ($x_{1,0}$, $x_{2,0}$, $m_{1,0}$, $m_{2,0}$, $d_{1,0}$, $d_{2,0}$) will yield

$$\frac{dx_1}{dt} = f_1(x_{1,0}, x_{2,0}, m_{1,0}, m_{2,0}, d_{1,0}) + \left(\frac{\partial f_1}{\partial x_1}\right)_0 (x - x_{1,0}) + \left(\frac{\partial f_1}{\partial x_2}\right)_0 (x_2 - x_{2,0})$$

$$+ \left(\frac{\partial f_1}{\partial m_1}\right)_0 (m_1 - m_{1,0}) + \left(\frac{\partial f_1}{\partial m_2}\right)_0 (m_2 - m_{2,0}) + \left(\frac{\partial f_1}{\partial d_1}\right)_0 (d_1 - d_{1,0}) \qquad (6.24)$$

and

$$\frac{dx_2}{dt} = f_2(x_{1,0}, x_{2,0}, m_{1,0}, m_{2,0}, d_{2,0}) + \left(\frac{\partial f_2}{\partial x_1}\right)_0 (x_1 - x_{1,0}) + \left(\frac{\partial f_2}{\partial x_2}\right)_0 (x_2 - x_{2,0})$$

$$+ \left(\frac{\partial f_2}{\partial m_1}\right)_0 (m_1 - m_{1,0}) + \left(\frac{\partial f_2}{\partial m_2}\right)_0 (m_2 - m_{2,0}) + \left(\frac{\partial f_2}{\partial d_2}\right)_0 (d_2 - d_{2,0}) \qquad (6.25)$$

where all the derivatives have been computed at the point of linearization (denoted by the subscript 0).

Assuming that the point of linearization corresponds to the steady-state operation of the system, we can define the deviation variables by

$$x_1' = x_1 - x_{1,0} \qquad m_1' = m_1 - m_{1,0} \qquad d_1' = d_1 - d_{1,0}$$

$$x_2' = x_2 - x_{2,0} \qquad m_2' = m_2 - m_{2,0} \qquad d_2' = d_2 - d_{2,0}$$

Introducing these deviation variables in eqs. (6.24) and (6.25), we take

$$\frac{dx_1'}{dt} = a_{11}x_1' + a_{12}x_2' + b_{11}m_1' + b_{12}m_2' + c_1 d_1' \qquad (6.26)$$

and

$$\frac{dx_2'}{dt} = a_{21}x_1' + a_{22}x_2' + b_{21}m_1' + b_{22}m_2' + c_2 d_2' \qquad (6.27)$$

where the constants a_{ij}, b_{ij}, and c_i are the appropriate derivatives in the eqs. (6.24) and (6.25):

$$a_{11} = \left(\frac{\partial f_1}{\partial x_1}\right)_0 \qquad a_{12} = \left(\frac{\partial f_1}{\partial x_2}\right)_0 \qquad b_{11} = \left(\frac{\partial f_1}{\partial m_1}\right)_0$$

$$b_{12} = \left(\frac{\partial f_1}{\partial m_2}\right)_0 \qquad c_1 = \left(\frac{\partial f_1}{\partial d_1}\right)_0$$

and

$$a_{21} = \left(\frac{\partial f_2}{\partial x_1}\right)_0 \qquad a_{22} = \left(\frac{\partial f_2}{\partial x_2}\right)_0 \qquad b_{21} = \left(\frac{\partial f_2}{\partial m_1}\right)_0$$

$$b_{22} = \left(\frac{\partial f_2}{\partial m_2}\right)_0 \qquad c_2 = \left(\frac{\partial f_2}{\partial d_2}\right)_0$$

It should be noted that the modeling eqs. (6.26) and (6.27) are in the form that we would like to have for process control purposes (i.e., linearized approximation of the nonlinear state equations), in terms of deviation variables.

Example 6.4: Linearization of a Nonisothermal CSTR

The modeling equations for a CSTR were given in Example 4.10 by eqs. (4.8a), (4.9a), and (4.10b). Assume that the volume V of the reacting mixture remains constant. Then the dynamic model of the reactor is reduced to the following:

$$\frac{dc_A}{dt} = \frac{1}{\tau}(c_{A_i} - c_A) - k_0 e^{-E/RT} c_A \tag{6.28}$$

and

$$\frac{dT}{dt} = \frac{1}{\tau}(T_i - T) + Jk_0 e^{-E/RT} c_A - \frac{UA_t}{\rho c_p V}(T - T_c) \tag{6.29}$$

where, $1/\tau = F_i/V$. This model is nonlinear due to the presence of the nonlinear term $e^{-E/RT} c_A$, while all the other terms are linear. Thus to linearize eqs. (6.28) and (6.29), we need only to linearize the nonlinear term around some point $(c_{A,0}, T_0)$.

$$e^{-E/RT} c_A \simeq e^{-E/RT_0} c_{A,0} + \left[\frac{\partial[e^{-E/RT} c_A]}{\partial T}\right]_{(T_0, c_{A,0})} (T - T_0)$$

$$+ \left[\frac{\partial[e^{-E/RT} c_A]}{\partial c_A}\right]_{(T_0, c_{A,0})} (c_A - c_{A,0})$$

$$= e^{-E/RT_0} c_{A,0} + \left(\frac{E}{RT_0^2} e^{-E/RT_0} c_{A,0}\right)(T - T_0) + e^{-E/RT_0}(c_A - c_{A,0})$$

Substituting the approximation above into eqs. (6.28) and (6.29), we take the following linearized model for a nonisothermal CSTR:

$$\frac{dc_A}{dt} = \frac{1}{\tau}(c_{A_i} - c_A) \tag{6.30}$$

$$- k_0 \left[e^{-E/RT_0} c_{A,0} + \frac{E}{RT_0^2} e^{-E/RT_0} c_{A,0}(T - T_0) + e^{-E/RT_0}(c_A - c_{A,0}) \right]$$

$$\frac{dT}{dt} = \frac{1}{\tau}(T_i - T) + Jk_0 \left[e^{-E/RT_0} c_{A,0} + \frac{E}{RT_0^2} e^{-E/RT_0} c_{A,0}(T - T_0) \right.$$

$$\left. + e^{-E/RT_0}(c_A - c_{A,0}) \right] - \frac{UA_t}{\rho c_p V}(T - T_c) \tag{6.31}$$

We can proceed a step further to develop a more convenient form for eqs. (6.30) and (6.31) using the deviation variables. Assume that T_0 and $c_{A,0}$ are the steady-state conditions for the CSTR and for given input conditions $c_{A_i,0}$, $T_{i,0}$, and $T_{c,0}$. Then from eqs. (6.28) and (6.29), we take

$$0 = \frac{1}{\tau}(c_{A_i,0} - c_{A,0}) - k_0 e^{-E/RT_0} c_{A,0} \qquad (6.32)$$

and

$$0 = \frac{1}{\tau}(T_{i,0} - T_0) + J k_0 e^{-E/RT_0} c_{A,0} - \frac{UA_t}{\rho c_p V}(T_0 - T_{c,0}) \qquad (6.33)$$

Subtract eqs. (6.32) and (6.33) from (6.30) and (6.31), respectively, and take

$$\frac{dc_A}{dt} = \frac{1}{\tau}[(c_{A_i} - c_{A_i,0}) - (c_A - c_{A,0})] - k_0 \frac{E}{RT_0^2} e^{-E/RT_0} c_{A,0}(T - T_0)$$

$$- k_0 e^{-E/RT_0}(c_A - c_{A,0}) \qquad (6.34)$$

$$\frac{dT}{dt} = \frac{1}{\tau}[(T_i - T_{i,0}) - (T - T_0)] + J k_0 \left[\frac{E}{RT_0^2} e^{-E/RT_0} c_{A,0}(T - T_0) \right.$$

$$\left. + e^{-E/RT_0}(c_A - c_{A,0}) \right] - \frac{UA_t}{\rho c_p V}[(T - T_0) - (T_c - T_{c,0})] \qquad (6.35)$$

Define the following deviation variables:

$$c_A' = c_A - c_{A,0} \qquad c_{A_i}' = c_{A_i} - c_{A_i,0}$$

$$T' = T - T_0 \qquad T_i' = T_i - T_{i,0}$$

$$T_c' = T_c - T_{c,0}$$

Then eqs. (6.34) and (6.35) take the following form, in terms of the deviation variables:

$$\frac{dc_A'}{dt} = \frac{1}{\tau}(c_{A_i}' - c_A') - \left[\frac{k_0 E}{RT_0^2} e^{-E/RT_0} c_{A,0} \right] T' - [k_0 e^{-E/RT_0}]c_A' \qquad (6.36)$$

$$\frac{dT'}{dt} = \frac{1}{\tau}(T_i' - T') + J k_0 \left[\left(\frac{E}{RT_0^2} e^{-E/RT_0} c_{A,0} \right) T' + e^{-E/RT_0} c_A' \right]$$

$$- \frac{UA_t}{\rho c_p V}(T' - T_c') \qquad (6.37)$$

THINGS TO THINK ABOUT

1. What is computer simulation, and what is it used for?

2. Discuss the methods of interval halving, successive substitution, and Newton–Raphson for solving nonlinear algebraic equations. What are their relative advantages and disadvantages?

3. Do the same with Euler's and the fourth-order Runge–Kutta integration methods.

4. What is linearization?

5. Why are the linearized approximate models useful for process control purposes?

6. When is the linearized model more accurate—near or far from the point of linearization—and why?

7. What is the most attractive point of linearization for control purposes, and why?

8. Which linearization is more useful, the one around point A or the one around point B, and why? (See Figure Q6.1.)

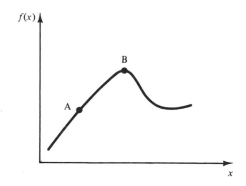

Figure Q6.1

9. What are the deviation variables? What is the point of linearization in order to define the deviation variables that will be useful for process control purposes?

10. Consider the tank system discussed in Example 6.1 (Figure 6.2a), where the flow rate of the outlet stream is proportional to the square root of the height of the liquid level. Show that we should *relinearize the balance equation every time we change the desired liquid level at steady state.*

11. The following differential equations provide the mathematical models for several processes. Which of them are linear and which nonlinear?

$$\text{Process I:} \quad x\frac{dx}{dt} + x = 10t + 5$$

$$\text{Process II:} \quad a_1\frac{dx_1}{dt} + a_2\frac{dx_2}{dt} = m_1(t) + d_1(t)$$

$$b_1\frac{dx_1}{dt} + b_2\frac{dx_2}{dt} = m_2(t) + d_2(t)$$

$$\text{Process III:} \quad a_1\frac{dx_1}{dt} + x_1\frac{dx_2}{dt} = m_1(t)$$

$$b_1\frac{dx_1}{dt} + b_2\frac{dx_2}{dt} = \cos \omega t$$

Laplace Transforms **7**

The use of Laplace transforms offers a very simple and elegant method of solving linear or linearized differential equations which result from the mathematical modeling of chemical processes.

The Laplace transforms also allow:

Simple development of input–output models which are very useful for control purposes (see Chapter 9)

Straightforward qualitative analysis of how chemical processes react to various external influences

It is for all the reasons cited above that the Laplace transforms have been included in a process control book, although they constitute a purely mathematical subject.

7.1 Definition of the Laplace Transform

Consider the function $f(t)$. The Laplace transform $\bar{f}(s)$ of the function $f(t)$ is defined as follows:

$$\mathcal{L}[f(t)] \equiv \bar{f}(s) = \int_0^\infty f(t)e^{-st}\, dt \qquad (7.1)$$

Note that a bar on top of a variable will signify the Laplace transform of that variable. This convention will be used throughout the text.

Remarks

1. A more rigorous definition of the Laplace transform is given by eq. (7.1a):

$$\mathcal{L}[f(t)] = \bar{f}(s) = \lim_{\substack{\epsilon \to 0^+ \\ T \to \infty}} \int_{\epsilon}^{T} f(t)e^{-st}\, dt \qquad (7.1a)$$

If the function $f(t)$ is piecewise continuous and defined for every value of time from $t = 0$ to $t = \infty$, the rigorous definition (7.1a) reduces to that of (7.1). For almost all the problems that we will be concerned with in this book, the simpler definition given by (7.1) will suffice.

2. From the definition (7.1) or (7.1a) we notice that Laplace transformation is a transformation of a function from the time domain (where time is the independent variable) to the s-domain (with s the independent variable). s is a variable defined in the complex plane (i.e., $s = a + jb$).

3. From the definition (7.1) or (7.1a), we notice that the Laplace transform of the function $f(t)$ exists if the integral $\int_{0}^{\infty} f(t)e^{-st}\, dt$ takes a finite value (i.e., remains bounded). Consider the function $f(t) = e^{at}$, where $a > 0$. Then

$$\mathcal{L}[e^{at}] = \int_{0}^{\infty} e^{at}e^{-st}\, dt = \int_{0}^{\infty} e^{(a-s)t}\, dt \qquad (7.2)$$

Now, if $a - s > 0$ or $s < a$, then the integral in (7.2) becomes unbounded. Consequently, the Laplace transform of e^{at} is defined only for $s > a$, which yields finite values for the integral in (7.2). All functions we are concerned with in this book will possess Laplace transforms so that we will not test their existence every time we need them.

4. The Laplace transformation is a linear operation:

$$\mathcal{L}[a_1 f_1(t) + a_2 f_2(t)] = a_1\mathcal{L}[f_1(t)] + a_2\mathcal{L}[f_2(t)] \qquad (7.3)$$

where a_1 and a_2 are constant parameters. The proof is straightforward:

$$\mathcal{L}[a_1 f_1(t) + a_2 f_2(t)] = \int_{0}^{\infty} [a_1 f_1(t) + a_2 f_2(t)]e^{-st}\, dt$$

$$= a_1 \int_{0}^{\infty} f_1(t)e^{-st}\, dt + a_2 \int_{0}^{\infty} f_2(t)e^{-st}\, dt$$

$$= a_1\mathcal{L}[f_1(t)] + a_2\mathcal{L}[f_2(t)]$$

7.2   Laplace Transforms of Some Basic Functions

Let us now apply the Laplace transformation on some basic functions that we will use repeatedly in the following chapters.

Exponential function

This function is defined as:

$$f(t) = e^{-at} \qquad \text{for } t \geq 0$$

Then

$$\mathcal{L}[e^{-at}] = \frac{1}{s + a} \tag{7.4}$$

Proof:

$$\mathcal{L}[e^{-at}] = \int_0^\infty e^{-at} e^{-st} \, dt = \int_0^\infty e^{-(s+a)t} \, dt = -\frac{1}{s+a}[e^{-(s+a)t}]_0^\infty = \frac{1}{s+a}$$

From (7.4) it is clear that

$$\mathcal{L}[e^{at}] = \frac{1}{s - a} \tag{7.5}$$

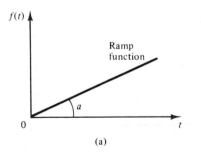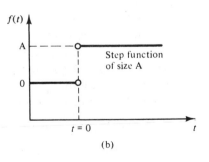

Figure 7.1 (a) Ramp function; (b) step function of size A.

Ramp function (figure 7.1a)

This function is defined as:

$$f(t) = at \qquad \text{for } t \geq 0 \text{ with } a = \text{constant}$$

Then

$$\mathcal{L}[at] = \frac{a}{s^2} \tag{7.6}$$

Proof:

$$\mathcal{L}[at] = \int_0^\infty ate^{-st} \, dt$$

Integrating by parts, where t and e^{-st} are the two functions, we take

$$\mathcal{L}[at] = \int_0^\infty ate^{-st} \, dt = \left[-\frac{at}{s} e^{-st} \right]_0^\infty + \int_0^\infty \frac{a}{s} e^{-st} \, dt$$

$$= (-0 + 0) + \frac{a}{s} \left[\frac{-1}{s} e^{-st} \right]_0^\infty = \frac{a}{s^2}$$

Trigonometric functions

Consider the sinusoidal function $f(t) = \sin \omega t$. Then

$$\mathcal{L}[\sin \omega t] = \frac{\omega}{s^2 + \omega^2} \tag{7.7}$$

Proof:

$$\mathcal{L}[\sin \omega t] = \int_0^\infty \sin \omega t \; e^{-st} \, dt = \int_0^\infty \frac{e^{j\omega t} - e^{-j\omega t}}{2j} e^{-st} \, dt$$

$$= \int_0^\infty \frac{1}{2j} [e^{-(s-j\omega)t} - e^{-(s+j\omega)t}] \, dt$$

$$= \frac{1}{2j} \left[-\frac{e^{-(s-j\omega)t}}{s - j\omega} + \frac{e^{-(s+j\omega)t}}{s + j\omega} \right]_0^\infty = \frac{1}{2j} \left(\frac{1}{s - j\omega} - \frac{1}{s + j\omega} \right)$$

$$= \frac{\omega}{s^2 + \omega^2}$$

Similarly, it can be proved that

$$\mathcal{L}[\cos \omega t] = \frac{s}{s^2 + \omega^2} \tag{7.8}$$

Note.   In the proof above we have used the Euler identity

$$\sin \alpha = \frac{e^{j\alpha} - e^{-j\alpha}}{2j}$$

For the proof of (7.8) use

$$\cos \alpha = \frac{e^{j\alpha} + e^{-j\alpha}}{2}$$

Step function (figure 7.1b)

This function is defined by

$$f(t) = \begin{cases} A & \text{for } t > 0 \\ 0 & \text{for } t < 0 \end{cases}$$

and its Laplace transform is

$$\mathcal{L}[\text{step function of size } A] = \frac{A}{s} \qquad (7.9)$$

Proof: We notice that a discontinuity in the value of the function exists at $t = 0$ such that $f(t = 0)$ is undefined. The definition of Laplace transform from eq. (7.1) requires the knowledge of the function at $t = 0$. The drawback is overcome if we consider the more precise mathematical definition of the Laplace transformation from eq. (7.1a):

$$\mathcal{L}[f(t)] = \lim_{\substack{\epsilon \to 0^+ \\ T \to \infty}} \int_{\epsilon}^{T} f(t) e^{-st} \, dt$$

Thus for the step function, the upper limit is $T = \infty$ but the lower limit is $t = 0+$ (i.e., a very small but finite positive time) instead of $t = 0$. Hence for step function we have

$$\mathcal{L}[A] = \int_{0+}^{\infty} A e^{-st} \, dt = \frac{A}{-s}[e^{-st}]_{0+}^{\infty} = \frac{A}{s}$$

Translated functions

Consider the function $f(t)$ shown in Figure 7.2a. If this function is delayed by t_o seconds, we take the function shown in Figure 7.2b, and if it is advanced by t_o seconds, then we have the curve of Figure 7.2c. The relationship among the three curves is

$$f(t + t_o) = \quad f(t) \quad = f(t - t_o)$$

curve in   curve in   curve in
Figure 7.2c   Figure 7.2a   Figure 7.2b

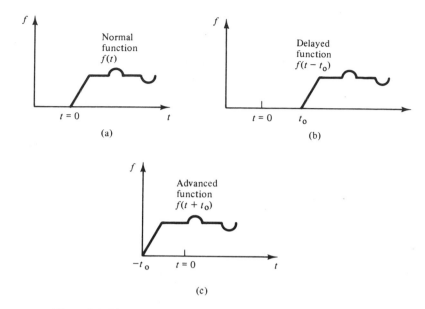

Figure 7.2 Time delayed and advanced forms of a given function.

Let $\mathcal{L}[f(t)] = \bar{f}(s)$ be the Laplace transform of $f(t)$. Then

$$\mathcal{L}[f(t - t_o)] = e^{-st_o}\bar{f}(s) \tag{7.10}$$

and

$$\mathcal{L}[f(t + t_o)] = e^{st_o}\bar{f}(s) \tag{7.11}$$

Proof:

$$\mathcal{L}[f(t - t_o)] = \int_0^\infty f(t - t_o)e^{-st}\,dt = e^{-st_o}\int_0^\infty f(t - t_o)e^{-s(t-t_o)}\,d(t - t_o)$$

since $dt = d(t - t_o)$. Let $t - t_o = \tau$; then

$$e^{-st_o}\int_0^\infty f(t - t_o)e^{-s(t-t_o)}\,d(t - t_o) = e^{-st_o}\int_{t_0}^\infty f(\tau)e^{-s\tau}\,d\tau$$

$$= e^{-st_o}\int_0^\infty f(\tau)e^{-s\tau}\,d\tau = e^{-st_o}\bar{f}(s)$$

Notice that in the last equality we replaced the lower bound $-t_o$ with 0. This will not change the value of the integral since $f(\tau) = 0$ for $\tau < 0$.

Equation (7.10) will be particularly useful in the computation of Laplace transforms of systems with dead time.

Example 7.1

Let us recall the flow of an incompressible liquid through a pipe (Example 4.9 and Figure 4.7a). From eq. (4.7) we have

$$T_{out}(t) = T_{in}(t - t_d) \qquad (4.7)$$

where T_{out} is the temperature of the liquid flowing out of the pipe and T_{in} is the temperature of the fluid flowing in the pipe. The temperature of the outlet is equal to the temperature of the inlet but dclayed by t_d, where t_d is the dead time (transportation lag), that is, the time required for a change in the inlet to reach the outlet of the pipe.

If $\mathcal{L}[T_{in}(t)] = \overline{T}_{in}(s)$, then, using (7.10), we have

$$\overline{T}_{out}(s) \equiv \mathcal{L}[T_{out}(t)] = \mathcal{L}[T_{in}(t - t_d)] = e^{-st_d}\overline{T}_{in}(s)$$

Unit pulse function

Consider the function in Figure 7.3a. The height is $1/A$ and the width A. Thus the area under the curve is

$$\text{area} = \frac{1}{A} A = 1$$

This function is called *unit pulse* function of duration A and is defined by

$$\delta_A(t) = \begin{cases} 0 & \text{for } t < 0 \\ \dfrac{1}{A} & \text{for } 0 < t < A \\ 0 & \text{for } t > A \end{cases}$$

It can also be described as the difference of two step functions of equal size $1/A$. The first step function occurs at time $t = 0$ while the second is delayed by A units of time. Thus if

$$\text{first step function: } f_1(t) = \begin{cases} 0 & t < 0 \\ \dfrac{1}{A} & t > 0 \end{cases}$$

$$\text{second step function: } f_2(t) = \left.\begin{cases} 0 & t < A \\ \dfrac{1}{A} & t > A \end{cases}\right\} = f_1(t - A)$$

then

$$\delta_A(t) = \text{unit pulse of duration } A = f_1(t) - f_2(t)$$

$$= f_1(t) - f_1(t - A)$$

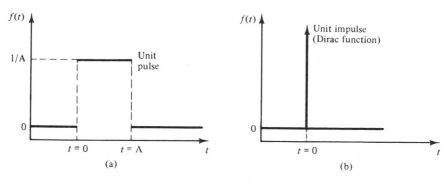

Figure 7.3 (a) Unit pulse; (b) unit impulse.

The Laplace transform of the unit pulse function of duration A is

$$\mathcal{L}[\delta_A(t)] = \frac{1}{A} \frac{1 - e^{-sA}}{s} \tag{7.12}$$

Proof:

$$\mathcal{L}[\delta_A(t)] = \mathcal{L}[f_1(t) - f_1(t - A)]$$

$$= \mathcal{L}[f_1(t)] - e^{-sA} \mathcal{L}[f_1(t)] = \frac{1}{As} - e^{-sA} \frac{1}{As} = \frac{1}{A} \frac{1 - e^{-sA}}{s}$$

Unit impulse function

Consider that the duration A of a unit pulse function is allowed to shrink, approaching zero, while the height $1/A$ approaches infinity. The area under the curve remains always equal to 1:

$$\lim_{A \to 0} \left(A \frac{1}{A} \right) = 1$$

As $A \to 0$ we take the function shown in Figure 7.3b. This function is called the *unit impulse* or *Dirac function* and it is usually represented by $\delta(t)$. It is defined as equal to zero for all times except for $t = 0$. Since the area under the unit pulse remains equal to 1, it is clear that this is true for the unit impulse:

$$\int_{-\infty}^{\infty} \delta(t) \, dt = 1$$

The Laplace transform of a unit impulse is

$$\mathcal{L}[\delta(t)] = 1 \tag{7.13}$$

Proof: Since $\delta(t) = \lim_{A \to 0} \delta_A(t)$,

$$\mathcal{L}[\delta(t)] = \mathcal{L}[\lim_{A \to 0} \delta_A(t)] = \int_0^\infty \lim_{A \to 0} \delta_A(t) e^{-st} \, dt$$

$$= \lim_{A \to 0} \int_0^\infty \delta_A(t) e^{-st} \, dt = \lim_{A \to 0} \left[\frac{1}{A} \frac{1 - e^{-sA}}{s} \right]$$

Using L'Hospital's rule, we have

$$\lim_{A \to 0} \left[\frac{1}{A} \frac{1 - e^{-sA}}{s} \right] = \lim_{A \to 0} \left[\frac{se^{-sA}}{s} \right] = 1$$

In Table 7.1 the Laplace transforms of some basic functions have been tabulated.

Remark. It is important to notice that the Laplace transforms of all the basic functions examined in this section and of additional functions shown in Table 7.1 are *ratios of two polynomials in s*. The only exceptions are the Laplace transforms of functions translated in time, which include the exponential terms $e^{-t_0 s}$. Therefore, for any function $f(t)$ (not including a time-translated term), we will have

$$\bar{f}(s) = \mathcal{L}[f(t)] = \frac{q_1(s)}{q_2(s)}$$

where $q_1(s)$ and $q_2(s)$ are two polynomials in s:

$$q_1(s) = k_m s^m + k_{m-1} s^{m-1} + \cdots + k_1 s + k_0$$
$$q_2(s) = \ell_n s^n + \ell_{n-1} s^{n-1} + \cdots + \ell_1 s + \ell_0$$

Example 7.2

If $f(t) = \cos \omega t$, then

$$\bar{f}(s) = \frac{s}{s^2 + \omega^2} = \frac{q_1(s)}{q_2(s)}$$

with $q_1(s) = 1 \cdot s + 0$ and $q_2(s) = 1 \cdot s^2 + \omega^2$. From Table 7.1 if $f(t) = e^{-at} \cos \omega t$, then

$$\bar{f}(s) = \frac{s + a}{(s + a)^2 + \omega^2} = \frac{q_1(s)}{q_2(s)}$$

with $q_1(s) = 1 \cdot s + a$ and $q_2(s) = 1 \cdot s^2 + 2a \cdot s + (a^2 + \omega^2)$.

TABLE 7.1
LAPLACE TRANSFORMS OF VARIOUS FUNCTIONS

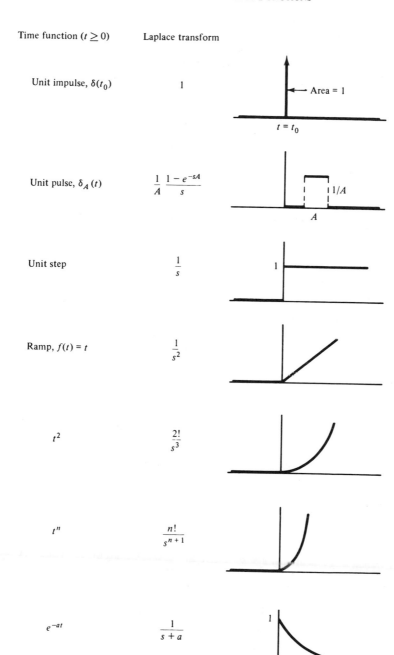

Time function ($t \geq 0$)	Laplace transform
Unit impulse, $\delta(t_0)$	1
Unit pulse, $\delta_A(t)$	$\dfrac{1}{A} \dfrac{1 - e^{-sA}}{s}$
Unit step	$\dfrac{1}{s}$
Ramp, $f(t) = t$	$\dfrac{1}{s^2}$
t^2	$\dfrac{2!}{s^3}$
t^n	$\dfrac{n!}{s^{n+1}}$
e^{-at}	$\dfrac{1}{s+a}$

TABLE 7.1 (cont.)

$t^n e^{-at}$ $\dfrac{n!}{(s+a)^{n+1}}$

$\sin(\omega t)$ $\dfrac{\omega}{s^2 + \omega^2}$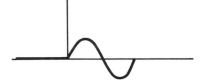

$\cos(\omega t)$ $\dfrac{s}{s^2 + \omega^2}$

$\sinh(\omega t)$ $\dfrac{\omega}{s^2 - \omega^2}$

$\cosh(\omega t)$ $\dfrac{s}{s^2 - \omega^2}$

$e^{-at}\sin(\omega t)$ $\dfrac{\omega}{(s+a)^2 + \omega^2}$

$e^{-at}\cos(\omega t)$ $\dfrac{(s+a)}{(s+a)^2 + \omega^2}$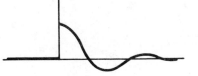

7.3   Laplace Transforms of Derivatives

$$\mathcal{L}\left[\frac{df(t)}{dt}\right] = s\bar{f}(s) - f(0) \qquad (7.14)$$

where $\bar{f}(s) = \mathcal{L}[f(t)]$.

Proof:

$$\mathcal{L}\left[\frac{df(t)}{dt}\right] = \int_0^\infty \frac{df(t)}{dt}\, e^{-st}\, dt = [e^{-st} f(t)]_0^\infty + \int_0^\infty se^{-st} f(t)\, dt$$

$$= [0 - f(0)] + s\int_0^\infty f(t)e^{-st}\, dt = s\bar{f}(s) - f(0)$$

Similarly, it can be proved that

$$\mathcal{L}\left[\frac{d^2f(t)}{dt^2}\right] = s^2\bar{f}(s) - sf(0) - f'(0) \qquad (7.15)$$

where $f'(0) = df(t)/dt$ evaluated at $t = 0$. In general,

$$\mathcal{L}\left[\frac{d^{(n)}f(t)}{dt^n}\right] = s^n\bar{f}(s) - s^{n-1}f(0) - s^{n-2}f'(0) - \cdots - sf^{(n-2)}(0) - f^{(n-1)}(0) \qquad (7.16)$$

From eqs. (7.14) through (7.16) we notice that in order to find the Laplace transform of any derivative, we need to have a number of initial conditions. To find the Laplace transform of an nth-order derivative, we need n initial conditions,

$$f(0), \quad f'(0), \quad f''(0), \quad \cdots, \quad f^{(n-1)}(0)$$

7.4   Laplace Transforms of Integrals

$$\mathcal{L}\left[\int_0^t f(t)\, dt\right] = \frac{1}{s}\bar{f}(s) \qquad (7.17)$$

where $\bar{f}(s) = \mathcal{L}[f(t)]$.

Proof:

$$\mathcal{L}\left[\int_0^t f(t)\, dt\right] = \int_0^\infty \left[\int_0^t f(t)\, dt\right]e^{-st}\, dt$$

Integrate by parts. Put $u = e^{-st}$ and $v = \int_0^t f(t)\, dt$. Then

$$du = -se^{-st}\, dt \qquad \text{and} \qquad dv = f(t)\, dt$$

Now

$$\int_0^\infty \left[\int_0^t f(t)\, dt \right] e^{-st}\, dt = -\frac{1}{s} \int_0^\infty v\, du = -\frac{1}{s} \left[(vu)\,|_0^\infty - \int_0^\infty u\, dv \right]$$

$$= -\frac{1}{s} \left[\int_0^t f(t)\, dt \cdot e^{-st} \right]_0^\infty + \frac{1}{s} \int_0^\infty f(t) e^{-st}\, dt$$

$$= -\frac{1}{s}(0 - 0) + \frac{1}{s}\bar{f}(s) = \frac{1}{s}\bar{f}(s)$$

7.5   Final-Value Theorem

$$\lim_{t \to \infty} f(t) = \lim_{s \to 0} [s\bar{f}(s)] \tag{7.18}$$

where $\bar{f}(s) = \mathcal{L}[f(t)]$.

Proof: Using the Laplace transform of a derivative [eq. (7.14)], we have

$$\int_0^\infty \frac{df(t)}{dt} e^{-st}\, dt = s\bar{f}(s) - f(0)$$

Take the limit of both sides as $s \to 0$:

$$\lim_{s \to 0} \int_0^\infty \frac{df(t)}{dt} e^{-st}\, dt = \lim_{s \to 0} [s\bar{f}(s) - f(0)]$$

Since variable s is independent of time t, we take

$$\int_0^\infty \lim_{s \to 0} \frac{df(t)}{dt} e^{-st}\, dt = \lim_{s \to 0} [s\bar{f}(s) - f(0)]$$

or

$$\int_0^\infty \frac{df(t)}{dt}\, dt = \lim_{t \to \infty} f(t) - f(0) = \lim_{s \to 0} [s\bar{f}(s)] - f(0)$$

Example 7.3

Let

$$\bar{f}(s) = \frac{s + 4}{s(s + 1)(s + 2)(s + 3)}$$

Find the value of $f(t)$ as $t \to \infty$. Using the final-value theorem, we have

$$\lim_{t \to \infty} f(t) = \lim_{s \to 0} [s\bar{f}(s)] = \lim_{s \to 0} s\left[\frac{s+4}{s(s+1)(s+2)(s+3)}\right]$$

$$= \lim_{s \to 0} \left[\frac{s+4}{(s+1)(s+2)(s+3)}\right] = -\frac{4}{6}$$

The final-value theorem allows us to compute the value that a function approaches as $t \to \infty$ when its Laplace transform is known.

7.6   Initial-Value Theorem

$$\lim_{t \to 0} f(t) = \lim_{s \to \infty} [s\bar{f}(s)] \qquad (7.19)$$

where

$$\bar{f}(s) = \int_0^\infty f(t)e^{-st} \, dt = \text{Laplace transform of } f(t)$$

The proof follows the same pattern as for the final-value theorem.

Example 7.4

Let

$$\bar{f}(s) = \frac{(s-1)(s+1)}{s(s+3)(s-4)}$$

Find $f(t = 0)$. Using the initial-value theorem, we have

$$\lim_{t \to 0} f(t) = \lim_{s \to \infty} [s\bar{f}(s)] = \lim_{s \to \infty} s\left[\frac{(s-1)(s+1)}{s(s+3)(s-4)}\right]$$

$$= \lim_{s \to \infty} \left[1 + \frac{s+11}{s^2 - s - 12}\right] = 1 + \lim_{s \to \infty} \frac{s+11}{s^2 - s - 12} = 1$$

THINGS TO THINK ABOUT

1. If $\bar{f}_1(s) = \mathcal{L}[f_1(t)]$ and $\bar{f}_2(s) = \mathcal{L}[f_2(t)]$, can we find the $\mathcal{L}[f_1(t)f_2(t)]$ for general functions $f_1(t)$ and $f_2(t)$?

2. Does the function $f_1 = 1/(t-1)$ possess a Laplace transform?

3. What is the Laplace transform of the function $f(t) = 5\cos 4t + e^{-t} + 5t$?

4. What is the Laplace transform of the vector function

$$\mathbf{f}(t) = \begin{bmatrix} a \sin t + b e^{-ct} \\ a + bt \\ \cos t + b \sin (t - t_d) \end{bmatrix}$$

5. Using Euler's identity,

$$\cos \alpha = \frac{e^{j\alpha} + e^{-j\alpha}}{2}$$

show that

$$\cos \omega t = \frac{s}{s^2 + \omega^2}$$

6. Show that

$$\mathcal{L}[f(t + t_d)] = e^{st_d} \bar{f}(s)$$

where $\bar{f}(s) = \mathcal{L}[f(t)]$.

7. Starting from the equation yielding the Laplace transform of a derivative,

$$\int_0^\infty \frac{df(t)}{dt} e^{-st} \, dt = s\bar{f}(s) - f(0)$$

prove the initial value theorem.

8. What functions have Laplace transforms which cannot be cast as ratios of two polynomials in s?

Solution of Linear **8**
Differential Equations
Using Laplace Transforms

As mentioned earlier, the primary use for the Laplace transforms is to solve linear differential equations or systems of linear (or linearized nonlinear) differential equations with constant coefficients. The procedure was developed by the English engineer Oliver Heaviside and it enables us to solve many problems without going through the trouble of finding the complementary and the particular solutions for linear differential equations. The same procedure can be extended to simple or systems of partial differential equations and to integral equations.

8.1   A Characteristic Example and the Solution Procedure

Assuming that $F_i = F$ (i.e., that the liquid level remains unchanged), the energy balance for a stirred tank heater (see Example 5.1) is

$$\frac{dT}{dt} + aT = \frac{1}{\tau} T_i + K T_{st} \tag{5.1}$$

Equation (5.1) can be expressed in terms of deviation variables,

$$\frac{dT'}{dt} + aT' = \frac{1}{\tau} T'_i + K T'_{st} \tag{5.3}$$

where $T' = T - T_s$, $T'_i = T_i - T_{i,s}$, and $T'_{st} = T_{st} - T_{st,s}$ are the deviation variables from the steady state defined by the values T_s, $T_{i,s}$, and $T_{st,s}$.

143

Assume that the heater is initially at steady state [i.e., $T'(0) = 0$]. At $t = 0$, the temperature of the inlet stream increases by a step of 10°F from its steady-state value and remains at this new level. Thus $T_i'(t) = 10$°F for $t > 0$. The temperature of the liquid in the tank will start increasing and we want to know how it changes with time. In other words, we must solve eq. (5.3).

Equation (5.3) is a linear equation with constant coefficients. We can use Laplace transforms to solve it. Let us examine the solution procedure.

Take the Laplace transforms of both sides of eq. (5.3):

$$\mathcal{L}\left[\frac{dT'}{dt}\right] + a\,\mathcal{L}[T'] = \frac{1}{\tau}\,\mathcal{L}[T_i'] + K\,\mathcal{L}[T_{st}']$$

or

$$[s\,\overline{T}'(s) - T'(0)] + a\,\overline{T}'(s) = \frac{1}{\tau}\,\overline{T}_i'(s) + K\overline{T}_{st}'(s) \tag{8.1}$$

Recall that $T'(0) = 0$, $\overline{T}_i'(s) = \mathcal{L}[\text{step function of } 10°F] = 10/s$ and $\overline{T}_{st}' = 0$. Then eq. (8.1) becomes

$$\overline{T}'(s) = \frac{1}{\tau}\,\frac{1}{s+a}\,\frac{10}{s} \tag{8.2}$$

The function $T'(t)$ whose Laplace transform is given by the right-hand side of eq. (8.2) is our solution. It is easy to show that

$$\overline{T}'(s) = \frac{1}{\tau}\,\frac{1}{s+a}\,\frac{10}{s} = \frac{10}{\tau a}\left[\frac{1}{s} - \frac{1}{s+a}\right] \tag{8.3}$$

From Table 7.1 we find easily that:

The function with Laplace transform 1/s is a unit step function.
The function with Laplace transform 1/(s + a) is e^{-at}.

Therefore, from eq. (8.3) we find that

$$T'(t) = \frac{10}{\tau a}\,(1 - e^{-at}) \tag{8.4}$$

$T'(t)$ given by eq. (8.4) is the solution to our initial differential equation (5.3). Indeed, taking the Laplace of eq. (8.4), it yields eq. (8.3). The procedure by which we find the time function when its Laplace transform is known is called the *inverse Laplace transformation* and is the most critical step while solving linear differential equations using Laplace transforms. To summarize the solution procedure described in the example above, we can identify the following steps:

1. Take the Laplace transform of both sides of the differential equation. Use eqs. (7.14), (7.15) and (7.16) to develop the Laplace transforms of the various derivatives. The initial conditions given for the differential equation are incorporated in this step with the transforms of the derivatives.
2. Solve the resulting algebraic equation in terms of the Laplace transform of the unknown function.
3. Find the time function that has as its Laplace transform the right-hand side of the equation obtained in step 2. This function is the desired solution since it satisfies the differential equation and the initial conditions.

Step 3 is the most tedious. Given a general expression such as

$$\overline{x}(s) = \frac{(s^2 + a_1 s + b_1)(s + c_1)}{s(s^3 + a_2 s + b_2 s + c_2)}$$

it is not obvious at all what the function $x(t)$ is, that has the foregoing Laplace transform. In Section 8.2 we will study a particular methodology for the inversion of Laplace transforms by *partial-fractions expansion*.

8.2 Inversion of Laplace Transforms: Heaviside Expansion

As pointed out above, the critical point in finding the solution to a differential equation using Laplace transforms is the inversion of the Laplace transforms. In this section we will study a method developed by Heaviside for the inversion of Laplace transforms known as *Heaviside or partial-fractions expansion*.

Assume that the Laplace transform of an unknown function $x(t)$ is given by

$$\overline{x}(s) = \frac{Q(s)}{P(s)} \tag{8.5}$$

where $Q(s)$ and $P(s)$ are polynomials in s of order m and n, respectively. The inversion of Laplace transforms using the expansion to partial fractions is composed of the following three steps:

1. Expand the $Q(s)/P(s)$ into a series of fractions,

$$\overline{x}(s) = \frac{Q(s)}{P(s)} = \frac{C_1}{r_1(s)} + \frac{C_2}{r_2(s)} + \cdots + \frac{C_n}{r_n(s)} \tag{8.6}$$

where $r_1(s), r_2(s), \ldots, r_n(s)$ are low-order polynomials such as first order, second order, and so on.

2. Compute the values of the constants $C_1, C_2, \ldots, C_n$ from eq. (8.6).
3. Find the inverse Laplace transform of every partial fraction. Then the unknown function $x(t)$ is given by

$$x(t) = \mathcal{L}^{-1}\left[\frac{C_1}{r_1(s)}\right] + \mathcal{L}^{-1}\left[\frac{C_2}{r_2(s)}\right] + \cdots + \mathcal{L}^{-1}\left[\frac{C_n}{r_n(s)}\right]$$

where $\mathcal{L}^{-1}$ symbolizes the inverse Laplace transform of the expression within the brackets. The inversion of each fraction can be done rather easily by inspection using tables of Laplace transforms for typical functions such as Tables 7.1 and 8.1.

TABLE 8.1
INVERSE LAPLACE TRANSFORMS OF SELECTED EXPRESSIONS

Laplace transform: $\bar{f}(s)$	Time function: $f(t)$
1. $\dfrac{1}{(s+a)(s+b)}$	$\dfrac{e^{-at} - e^{-bt}}{b - a}$
2. $\dfrac{1}{(s+a)(s+b)(s+c)}$	$\dfrac{e^{-at}}{(b-a)(c-a)} + \dfrac{e^{-bt}}{(c-b)(a-b)} + \dfrac{e^{-ct}}{(a-c)(b-c)}$
3. $\dfrac{s+a}{(s+b)(s+c)}$	$\dfrac{1}{c-b}[(a-b)e^{-bt} - (a-c)e^{-ct}]$
4. $\dfrac{a}{(s+b)^2}$	ate^{-bt}
5. $\dfrac{a}{(s+b)^3}$	$\dfrac{a}{2}t^2 e^{-bt}$
6. $\dfrac{a}{(s+b)^{n+1}}$	$\dfrac{a}{n!}t^n e^{-bt}$
7. $\dfrac{1}{s(as+1)}$	$1 - e^{-t/a}$
8. $\dfrac{1}{s(as+1)^2}$	$1 - \dfrac{a+t}{a}e^{-t/a}$
9. $\dfrac{\omega^2}{s(s^2 + 2\zeta\omega s + \omega^2)}$	$1 + \dfrac{e^{-\zeta\omega t}}{\sqrt{1-\zeta^2}}\sin(\omega\sqrt{1-\zeta^2}\,t - \phi)$ where $\cos\phi = -\zeta$
10. $\dfrac{s}{(1+as)(s^2+\omega^2)}$	$-\dfrac{1}{1+a^2\omega^2}e^{-t/a} + \dfrac{1}{\sqrt{1+a^2\omega^2}}\cos(\omega t - \phi)$ where $\phi = \tan^{-1} a\omega$
11. $\dfrac{s}{(s^2+\omega^2)^2}$	$\dfrac{1}{2\omega}t\sin\omega t$
12. $\dfrac{1}{(s+a)[(s+b)^2+\omega^2]}$	$\dfrac{e^{-at}}{(a-b)^2+\omega^2} + \dfrac{e^{-bt}\sin(\omega t - \phi)}{\omega[(a-b)^2+\omega^2]^{1/2}}$ where $\phi = \tan^{-1}\left(\dfrac{\omega}{a-b}\right)$

When $\bar{x}(s)$ is given as the ratio of two polynomials [eq. (8.5)], its expansion into a series of fractions is governed by the form and the roots of the polynomial in the denominator, $P(s)$. In general, we will distinguish two cases:

1. Polynomial $P(s)$ has n distinct (all different) roots, real or complex, or
2. Polynomial $P(s)$ has multiple roots.

We will examine each case separately using characteristic examples.

Distinct real roots of the polynomial $P(s)$

Consider the Laplace transform of the function $x(t)$ given by

$$\bar{x}(s) = \frac{s^2 - s - 6}{s^3 - 2s^2 - s + 2} = \frac{Q(s)}{P(s)} \tag{8.7}$$

The polynomial in the denominator is of third order,

$$P(s) = s^3 - 2s^2 - s + 2$$

and has three roots,

$$p_1 = 1 \qquad p_2 = -1 \qquad p_3 = 2$$

Therefore,

$$P(s) = s^3 - 2s^2 - s + 2 = (s - 1)(s + 1)(s - 2)$$

and eq. (8.7) becomes

$$\bar{x}(s) = \frac{s^2 - s - 6}{(s - 1)(s + 1)(s - 2)} \tag{8.8}$$

Expand (8.8) into partial fractions and take

$$\bar{x}(s) = \frac{s^2 - s - 6}{(s - 1)(s + 1)(s - 2)} = \frac{C_1}{s - 1} + \frac{C_2}{s + 1} + \frac{C_3}{s - 2} \tag{8.9}$$

where C_1, C_2, and C_3 are unknown constants to be evaluated. From eq. (8.9) it is clear that

$$x(t) = \mathcal{L}^{-1}\left[\frac{C_1}{s - 1}\right] + \mathcal{L}^{-1}\left[\frac{C_2}{s + 1}\right] + \mathcal{L}^{-1}\left[\frac{C_3}{s - 2}\right]$$

and using Table 7.1 we find that

$$x(t) = C_1 e^{1 \cdot t} + C_2 e^{-1 \cdot t} + C_3 e^{2 \cdot t} \tag{8.10}$$

which is the inverse Laplace transform of the expression in (8.7).

Let us see now how we can compute the constants C_1, C_2, and C_3.

Compute C_1. Multiply both sides of (8.9) by $(s - 1)$:

$$\frac{(s^2 - s - 6)(s - 1)}{(s - 1)(s + 1)(s - 2)} = C_1 + \frac{C_2(s - 1)}{s + 1} + \frac{C_3(s - 1)}{s - 2} \qquad (8.11)$$

Equation (8.11) holds for all values of s. Set $s - 1 = 0$ (i.e., $s = 1$).
    The last two terms in the right-hand side of (8.11) become zero and we take

$$C_1 = \left[\frac{s^2 - s - 6}{(s + 1)(s - 2)}\right]_{s=1} = 3$$

Compute C_2. Multiply both sides of (8.9) by $(s + 1)$:

$$\frac{(s^2 - s - 6)(s + 1)}{(s - 1)(s + 1)(s - 2)} = \frac{C_1(s + 1)}{s - 1} + C_2 + \frac{C_3(s + 1)}{s - 2}$$

Set $s + 1 = 0$ (i.e., $s = -1$):

$$C_2 = \left[\frac{s^2 - s - 6}{(s - 1)(s - 2)}\right]_{s=-1} = \frac{-2}{3}$$

Compute C_3. Multiply both sides of (8.9) by $(s - 2)$:

$$\frac{(s^2 - s - 6)(s - 2)}{(s - 1)(s + 1)(s - 2)} = \frac{C_1(s - 2)}{s - 1} + \frac{C_2(s - 2)}{s + 1} + C_3$$

Set $s - 2 = 0$ (i.e., $s = 2$):

$$C_3 = \left[\frac{s^2 - s - 6}{(s - 1)(s + 1)}\right]_{s=2} = \frac{-4}{3}$$

Therefore, eq. (8.10) yields

$$x(t) = 3e^t - \tfrac{2}{3} e^{-t} - \tfrac{4}{3} e^{2t}$$

Distinct complex roots of the polynomial $P(s)$

Consider the Laplace transform

$$\overline{x}(s) = \frac{s + 1}{s^2 - 2s + 5}$$

The polynomial $P(s)$ is of second order and has two distinct roots which *are not real* (as in the previous case) but complex conjugates:

$$p_1 = 1 + 2j \qquad \text{and} \qquad p_2 = 1 - 2j$$

Therefore,

$$P(s) = s^2 - 2s + 5 = [s - (1 + 2j)][s - (1 - 2j)]$$

Expansion into partial fractions yields

$$\overline{x}(s) = \frac{s + 1}{s^2 - 2s + 5} = \frac{s + 1}{[s - (1 + 2j)][s - (1 - 2j)]}$$

$$= \frac{C_1}{s - (1 + 2j)} + \frac{C_2}{s - (1 - 2j)}$$

(8.12)

and using the transforms of Table 7.1, we find that

$$x(t) = C_1 e^{(1+2j)t} + C_2 e^{(1-2j)t}$$

(8.13)

The constants C_1 and C_2 are computed as in the case with distinct real roots:

Compute C_1. Multiply both sides of (8.12) by $[s - (1 + 2j)]$:

$$\frac{(s + 1)[\cancel{s - (1 + 2j)}]}{[\cancel{s - (1 + 2j)}][s - (1 - 2j)]} = C_1 + \frac{C_2[s - (1 + 2j)]}{s - (1 - 2j)}$$

Set $s - (1 + 2j) = 0$ (i.e., $s = 1 + 2j$) and take

$$C_1 = \frac{1 - j}{2}$$

Compute C_2. Multiply both sides of (8.12) by $[s - (1 - 2j)]$ and then set $s - (1 - 2j) = 0$ (i.e. $s = 1 - 2j$) to find¹

$$C_2 = \frac{1 + j}{2}$$

Notice that the coefficients C_1 and C_2 are complex conjugates of each other. Put the values of C_1 and C_2 in (8.13) and find that

$$x(t) = \frac{1 - j}{2} e^{(1+2j)t} + \frac{1 + j}{2} e^{(1-2j)t}$$

or

$$x(t) = \frac{e^t}{2} [(1 - j)e^{2jt} + (1 + j)e^{-2jt}]$$

(8.14)

Let us recall Euler's identity

$$e^{j\alpha} = \cos \alpha + j \sin \alpha$$

(8.15)

Then we have

$$e^{2jt} = \cos 2t + j \sin 2t$$

and

$$e^{-2jt} = \cos (-2t) + j \sin (-2t)$$
$$= \cos 2t - j \sin 2t$$

In eq. (8.14), replace e^{2jt} and e^{-2jt} by their equal from the equations above and take

$$x(t) = \frac{e^t}{2} \{(1 - j)[\cos 2t + j \sin 2t] + (1 + j)[\cos 2t - j \sin 2t]\}$$

or

$$x(t) = e^t[\cos 2t + \sin 2t] \tag{8.16}$$

Recall the trigonometric identity

$$a_1 \cos b + a_2 \sin b = a_3 \sin (b + \phi) \tag{8.17}$$

where

$$a_3 = \sqrt{a_1^2 + a_2^2} \quad \text{and} \quad \phi = \tan^{-1} \left(\frac{a_1}{a_2}\right)$$

Apply (8.17) to eq. (8.16) and find

$$x(t) = e^t \sqrt{2} \sin (2t + \phi)$$

where $\phi = \tan^{-1} (1/1) = 45°$. Therefore, whenever the polynomial $P(s)$ has complex roots:

1. They will always be in complex pairs.
2. The coefficients of the corresponding terms in the partial fractions expansion will also be complex conjugates of each other.
3. They will give rise to a periodic term (e.g., sinusoidal wave).

Multiple roots of the polynomial $P(s)$

The expansion into partial fractions and the computation of the coefficients change when the polynomial $P(s)$ has multiple roots. Consider the Laplace transform

$$\overline{x}(s) = \frac{1}{(s + 1)^3(s + 2)} \tag{8.18}$$

The polynomial $P(s)$ has three roots equal and the fourth different:

$$p_1 = p_2 = p_3 = -1 \quad \text{and} \quad p_4 = -2$$

Expand (8.18) into partial fractions

$$\bar{x}(s) - \frac{1}{(s+1)^3(s+2)} = \frac{C_1}{s+1} + \frac{C_2}{(s+1)^2} + \frac{C_3}{(s+1)^3} + \frac{C_4}{s+2} \tag{8.19}$$

From Tables 7.1 and 8.1, we find that

$$\mathcal{L}^{-1}\left[\frac{C_2}{(s+1)^2}\right] = C_2 t e^{-t} \quad \text{and} \quad \mathcal{L}^{-1}\left[\frac{C_3}{(s+1)^3}\right] = \frac{C_3}{2} t^2 e^{-t}$$

Consequently, the inverse Laplace transform of (8.19) is easily found to be

$$x(t) = C_1 e^{-t} + C_2 t e^{-t} + \frac{C_3}{2} t^2 e^{-t} + C_4 e^{-2t} \tag{8.20}$$

Let us see then how can we compute the constants C_1, C_2, C_3, and C_4.

Compute C_4. This constant corresponds to the distinct root and can be computed using the procedure described earlier. Thus multiply both sides of (8.19) by $(s + 2)$ and then set $s + 2 = 0$ (i.e., $s = -2$) and find that $C_4 = -1$.

Compute C_3. Use the familiar procedure [i.e., multiply both sides of (8.19) by $(s + 1)^3$]

$$\frac{1}{s+2} = C_1(s+1)^2 + C_2(s+1) + C_3 + \frac{C_4(s+1)^3}{s+2} \tag{8.21}$$

Set $(s + 1)^3 = 0$ (i.e., $s = -1$) and find that $C_3 = +1$.

Compute C_2. The familiar procedure used above cannot be employed for the computation of C_2. Thus if we multiply both sides of (8.19) by $(s + 1)^2$, we take

$$\frac{1}{(s+1)(s+2)} = C_1(s+1) + C_2 + \frac{C_3}{s+1} + \frac{C_4(s+1)}{s+2}$$

Then, setting $s = -1$, the term involving C_3 becomes infinite. The same problem is encountered if we try to compute C_1. Therefore, an alternative procedure is needed to compute C_2 and C_1.

Differentiate both sides of (8.21) with respect to s and take

$$-\frac{1}{(s+2)^2} = 2C_1(s+1) + C_2 + C_4\frac{(s+1)^2(2s+5)}{(s+2)^2} \tag{8.22}$$

Set $s = -1$ and find that $C_2 = -1$.

Compute C_1. To obtain the value of C_1, differentiate (8.22) once more and take

$$\frac{2}{(s+2)^3} = 2C_1 + C_4 \cdot 2(s+1)\frac{s^2+5s+7}{(s+2)^3}$$

Set $s = -1$ and find $C_1 = +1$. Substitute the values of C_1, C_2, C_3, and C_4 in (8.20) and find that

$$x(t) = e^{-t}(1 - t + \tfrac{1}{2}t^2) - e^{-2t}$$

Remark.

If the polynomial $P(s)$ has multiple roots, the denominator of $\overline{x}(s)$ has a term $(s - p_i)^m$, where p_i is the multiple root which is repeated m times. In such case the partial-fractions expansion produces terms such as

$$\frac{C_1}{s - p_i} + \frac{C_2}{(s - p_i)^2} + \cdots + \frac{C_{m-1}}{(s - p_i)^{m-1}} + \frac{C_m}{(s - p_i)^m}$$

From Table 7.1 we know that

$$\mathcal{L}[t^n e^{-at}] = \frac{n!}{(s+a)^{n+1}}$$

Therefore, the terms of the expansion above lead to the following inverse Laplace transform:

$$\left[C_1 + C_2 t + \frac{C_3}{2!}t^2 + \cdots + \frac{C_{m-1}}{(m-2)!}t^{m-2} + \frac{C_m}{(m-1)!}t^{m-1} \right]e^{p_i t}$$

The constant C_m can be computed in the usual manner by multiplying both sides of the expansion with $(s - p_i)^m$ and setting $s = p_i$. The remaining constants $C_{m-1}, \cdots, C_2, C_1$ are computed by successive differentiations of the equation resulting from the multiplication of the expansion by $(s - p_i)^m$.

8.3 Examples of the Solution of Linear Differential Equations Using Laplace Transforms

In this section we will give two characteristic examples of solving linear differential equations using the Laplace transforms. The first example is the solution of a second-order differential equation, while in the second example we find the solution to a system of two differential equations.

The solution of any other linear differential equation or of a system of linear differential equations will follow the same general pattern outlined in the two examples.

Example 8.1: Solution of a Second-Order Differential Equation

Consider the following second-order differential equation

$$a_2 \frac{d^2x}{dt^2} + a_1 \frac{dx}{dt} + a_0x = f(t) \tag{8.23}$$

where $x(t)$ is considered to be in the form of a deviation variable with initial conditions

$$x(0) = \left(\frac{dx}{dt}\right)_{t=0} = 0 \tag{8.24}$$

Take the Laplace transform of (8.23),

$$a_2\left[s^2\overline{x}(s) - sx(0) - \left(\frac{dx}{dt}\right)_{t=0}\right] + a_1[s\overline{x}(s) - x(0)] + a_0\overline{x}(s) = \overline{f}(s)$$

or

$$\overline{x}(s) = \frac{\overline{f}(s)}{a_2s^2 + a_1s + a_0} + \frac{a_2sx(0) + a_2(dx/dt)_{t=0} + a_1x(0)}{a_2s^2 + a_1s + a_0} \tag{8.25}$$

Let us assume that $f(t)$ is a unit step function, giving $\overline{f}(s) = 1/s$. Using this expression for $\overline{f}(s)$ and the initial conditions given by (8.24), eq. (8.25) becomes

$$\overline{x}(s) = \frac{1}{s(a_2s^2 + a_1s + a_0)} \tag{8.26}$$

The polynomial $P^*(s) \equiv a_2s^2 + a_1s + a_0$ is called the *characteristic polynomial* of a second-order equation. To invert the right-hand side of (8.26) we need to know the roots of the polynomial $P^*(s)$. Depending on the values of the constants a_2, a_1, and a_0, we can distinguish three cases.

Case 1: $a_1^2 - 4a_2a_0 > 0$. Then we have *two distinct real roots*,

$$s_{1,2} = \frac{-a_1 \pm \sqrt{a_1^2 - 4a_2a_0}}{2a_2}$$

For example, let $a_1 = 4$, $a_2 = 1$, $a_0 = 3$; then $a_1^2 - 4a_2a_0 = 16 - 12$, $4 > 0$ and $s_1 = -1$ and $s_2 = -3$.

$$\frac{1}{s(a_2s^2 + a_1s + a_0)} = \frac{1}{s(s^2 + 4s + 3)} = \frac{1}{s(s + 3)(s + 1)}$$

$$= \frac{C_1}{s} + \frac{C_2}{s + 3} + \frac{C_3}{s + 1} \tag{8.27}$$

Multiply (8.27) by s and set $s = 0$. Find $C_1 = \frac{1}{3}$. Multiply (8.27) by $s + 3$

and set $s = -3$. Find $C_2 = \frac{1}{6}$. Multiply (8.27) by $s + 1$ and set $s = -1$. Find $C_3 = -\frac{1}{2}$. Then

$$x(t) = \mathcal{L}^{-1}\left[\frac{1/3}{s}\right] + \mathcal{L}^{-1}\left[\frac{1/6}{s+3}\right] + \mathcal{L}^{-1}\left[\frac{-1/2}{s+1}\right] = \frac{1}{3} + \frac{1}{6}e^{-3t} - \frac{1}{2}e^{-t}$$

Case 2: $a_1^2 - 4a_2a_0 = 0$. Then we have *two equal roots*:

$$s_1 = s_2 = \frac{-a_1}{2a_2}$$

Let $a_1 = 2$, $a_2 = 1$, $a_0 = 1$; then $a_1^2 - 4a_2a_0 = 4 - 4 \cdot 1 \cdot 1 = 0$:

$$s_1 = s_2 = -1$$

$$\frac{1}{s(a_2s^2 + a_1s + a_0)} = \frac{1}{s(s^2 + 2s + 1)} = \frac{1}{s(s+1)^2}$$

$$= \frac{C_1}{s} + \frac{C_2}{s+1} + \frac{C_3}{(s+1)^2} \tag{8.28}$$

Multiply (8.28) by s and set $s = 0$. Find $C_1 = 1$. Multiply (8.28) by $(s + 1)^2$ and take

$$\frac{1}{s} = \frac{C_1(s+1)^2}{s} + C_2(s+1) + C_3 \tag{8.29}$$

Set $s = -1$ and find $C_3 = -1$. Differentiate (8.29) with respect to s:

$$-\frac{1}{s^2} = 2C_1\frac{s+1}{s} - C_1\frac{(s+1)^2}{s^2} + C_2$$

Set $s = -1$ and find $C_2 = -1$. Then

$$x(t) = \mathcal{L}^{-1}\left[\frac{1}{s}\right] + \mathcal{L}^{-1}\left[-\frac{1}{s+1}\right] + \mathcal{L}^{-1}\left[-\frac{1}{(s+1)^2}\right]$$

$$= 1 - (1 + t)e^{-t}$$

Case 3: $a_1^2 - 4a_2a_0 < 0$. Then we have *two complex conjugate roots*.
    Let $a_1 = 2$, $a_2 = 2$, $a_0 = 1$; then $a_1^2 - 4a_2a_0 = 4 - 8 = -4 < 0$, and the two roots are:

$$s_1 = \frac{-1+j}{2} \quad \text{and} \quad s_2 = \frac{-1-j}{2}$$

$$\frac{1}{s(2s^2 + 2s + 1)} = \frac{1/2}{s\left[s - \dfrac{-1+j}{2}\right]\left[s - \dfrac{-1-j}{2}\right]} \tag{8.30}$$

$$= \frac{C_1}{s} + \frac{C_2}{s - \dfrac{-1+j}{2}} + \frac{C_3}{s - \dfrac{-1-j}{2}}$$

Multiply (8.30) by s, set $s = 0$, and find $C_1 = 1$.

Multiply (8.30) by $\left(s - \dfrac{-1+j}{2} \right)$, set $s = \dfrac{-1+j}{2}$, and find

$$C_2 = \frac{1}{-1-j} = \frac{-1+j}{(-1-j)(-1+j)} = (-1+j)/2$$

Multiply (8.30) by $\left(s - \dfrac{-1-j}{2} \right)$, set $s = \dfrac{-1-j}{2}$, and find

$$C_3 = \frac{1}{-1+j} = \frac{-1-j}{(-1+j)(-1-j)} = (-1+j)/2$$

Consequently,

$$x(t) = \mathcal{L}^{-1}\left[\frac{1}{s} \right] + (-1+j)\, \mathcal{L}^{-1}\left[\frac{1/2}{s - \dfrac{-1+j}{2}} \right] + (-1-j)\, \mathcal{L}^{-1}\left[\frac{1/2}{s - \dfrac{-1-j}{2}} \right]$$

or

$$x(t) = 1 + [\, (-1+j)e^{(-1+j)t/2} + (-1-j)e^{(-1-j)t/2} \,]/2 \qquad (8.31)$$

Recall Euler's identity:

$$e^{jat} = \cos at + j \sin at \qquad (8.15)$$

Then eq. (8.31) becomes

$$x(t) = 1 + e^{-t/2}\{(-1+j)[\cos (t/2) + j \sin (t/2)]$$
$$+ (-1-j)[\cos (t/2) - j \sin (t/2)]\}/2$$

or

$$x(t) = 1 - e^{-t/2}[\cos (t/2) + \sin (t/2)] \qquad (8.31a)$$

Use the trigonometric identity (8.17):

$$a_1 \cos b + a_2 \sin b = a_3 \sin (b + \phi) \qquad (8.17)$$

where

$$a_3 = \sqrt{a_1^2 + a_2^2} \quad \text{and} \quad \phi = \tan^{-1}\left(\frac{a_1}{a_2} \right)$$

Then eq. (8.31a) becomes

$$x(t) = 1 - \sqrt{2}\, e^{-t/2} \sin \left(\frac{t}{2} + \phi \right) \qquad (8.32)$$

where $\phi = \tan^{-1}(1/1) = \tan^{-1}(1) = 45°$.

The use of Laplace transforms is not limited to the solution of simple differential equations, like the second-order equation of Example 8.1. It extends to the solution of sets of differential equations. Consider, for example, the following system of linear differential equations:

$$\frac{dx_1}{dt} = a_{11}x_1 + a_{12}x_2 + b_{11}f_1(t) + b_{12}f_2(t) \tag{8.33}$$

$$\frac{dx_2}{dt} = a_{21}x_1 + a_{22}x_2 + b_{21}f_1(t) + b_{22}f_2(t) \tag{8.34}$$

with initial conditions $x_1(0) = x_2(0) = 0$. Taking the Laplace transforms of the equations above and after appropriate grouping, we find

$$(s - a_{11})\overline{x}_1(s) - a_{12}\overline{x}_2(s) = b_{11}\overline{f}_1(s) + b_{12}\overline{f}_2(s)$$

$$-a_{21}\overline{x}_1(s) + (s - a_{22})\overline{x}_2(s) = b_{21}\overline{f}_1(s) + b_{22}\overline{f}_2(s)$$

The last two equations form a *set of two linear algebraic equations* with $\overline{x}_1(s)$ and $\overline{x}_2(s)$ as the two unknown variables, and can be solved easily using, for example, Cramer's rule. Thus we find

$$\overline{x}_1(s) = \frac{[b_{11}\overline{f}_1(s) + b_{12}\overline{f}_2(s)](s - a_{22}) + a_{12}[b_{21}\overline{f}_1(s) + b_{22}\overline{f}_2(s)]}{s^2 - (a_{11} + a_{22})s - (a_{12}a_{21} - a_{11}a_{22})} \tag{8.35}$$

$$\overline{x}_2(s) = \frac{[b_{21}\overline{f}_1(s) + b_{22}\overline{f}_2(s)](s - a_{11}) + a_{21}[b_{11}\overline{f}_1(s) + b_{12}\overline{f}_2(s)]}{s^2 - (a_{11} + a_{22})s - (a_{12}a_{21} - a_{11}a_{22})} \tag{8.36}$$

The expressions above can now be inverted using the partial-fractions expansion, as it was described in Section 8.2, to find the unknown solution $x_1(t)$ and $x_2(t)$.

The solution procedure described above can be extended to larger systems of equations, but it is computationally more cumbersome. Let us now discuss the details of the solution procedure, in terms of an example.

Example 8.2: Solution of a Set of Linear Differential Equations

Find the solution of the following set of equations:

$$\frac{dx_1}{dt} = 2x_1 + 3x_2 + 1 \qquad \text{with} \qquad x_1(0) = 0$$

$$\frac{dx_2}{dt} = 2x_1 + x_2 + e^t \qquad \text{with} \qquad x_2(0) = 0$$

Take the Laplace transforms and after rearrangement find

$$(s - 2)\overline{x}_1(s) - 3\overline{x}_2(s) = \frac{1}{s}$$

$$-2\overline{x}_1(s) + (s - 1)\overline{x}_2(s) = \frac{1}{s - 1}$$

Using Cramer's rule to solve the foregoing system of linear algebraic equations in $\bar{x}_1(s)$, $\bar{x}_2(s)$, we find

$$\bar{x}_1(s) = \frac{\begin{vmatrix} 1/s & -3 \\ 1/(s-1) & s-1 \end{vmatrix}}{\begin{vmatrix} s-2 & -3 \\ -2 & s-1 \end{vmatrix}} \quad \text{and} \quad \bar{x}_2(s) = \frac{\begin{vmatrix} s-2 & 1/s \\ -2 & 1/(s-1) \end{vmatrix}}{\begin{vmatrix} s-2 & -3 \\ -2 & s-1 \end{vmatrix}}$$

or

$$\bar{x}_1(s) = \frac{s^2 + s + 1}{s(s-1)(s-4)(s+1)} \quad \text{and} \quad \bar{x}_2(s) = \frac{s^2 - 2}{s(s-1)(s-4)(s+1)}$$

Expand into partial fractions:

$$\bar{x}_1(s) = \frac{s^2 + s + 1}{s(s-1)(s-4)(s+1)} = \frac{C_1}{s} + \frac{C_2}{s-1} + \frac{C_3}{s-4} + \frac{C_4}{s+1}$$

$$\bar{x}_2(s) = \frac{s^2 - 2}{s(s-1)(s-4)(s+1)} = \frac{D_1}{s} + \frac{D_2}{s-1} + \frac{D_3}{s-4} + \frac{D_4}{s+1}$$

Compute the constants using the procedure described in Section 8.2 and find

$$\bar{x}_1(s) = \frac{1/4}{s} + \frac{-1/2}{s-1} + \frac{21/60}{s-4} - \frac{1/10}{s+1}$$

$$\bar{x}_2(s) = \frac{-1/2}{s} + \frac{1/6}{s-1} + \frac{14/60}{s-4} + \frac{1/10}{s+1}$$

Taking the inverses, we finally have

$$x_1(t) = \tfrac{1}{4} - \tfrac{1}{2} e^t + \tfrac{7}{20} e^{4t} - \tfrac{1}{10} e^{-t}$$

$$x_2(t) = -\tfrac{1}{2} + \tfrac{1}{6} e^t + \tfrac{7}{30} e^{4t} + \tfrac{1}{10} e^{-t}$$

THINGS TO THINK ABOUT

1. What is the characteristic polynomial for a first-order and a second-order system? Find its roots.
2. Why are we interested in the roots of the characteristic polynomial of an nth-order linear differential equation or a system of linear differential equations?
3. How does the procedure to compute the constants of the terms resulting from the partial fractions expansion vary in the presence of multiple roots?
4. What is the complementary solution, and what is the particular solution for (a) an nth-order linear differential equation, and (b) a 2 × 2 system of linear differential equations? What do these solutions mean? What factors determine them?

5. Consider the following system of simultaneous linear differential equations:

$$\frac{dx_1}{dt} = a_{11}x_1 + a_{12}x_2 + f_1(t) \qquad \text{with} \qquad x_1(0) = 0$$

$$\frac{dx_2}{dt} = a_{21}x_1 + a_{22}x_2 + f_2(t) \qquad \text{with} \qquad x_2(0) = 0$$

Show that this system can be converted to the following equivalent system:

$$\frac{d^2x_1}{dt^2} + b_1\frac{dx_1}{dt} + b_2x_1 = h(t)$$

and

$$\frac{dx_2}{dt} - a_{22}x_2 = a_{21}x_1 + f_2(t)$$

where b_1, and b_2 depend on a_{11}, a_{12}, a_{21}, a_{22} and $h(t)$ depends on $f_1(t)$, $f_2(t)$, and their derivatives. Notice that the modified system can be solved sequentially and thus more easily than the original system, which requires simultaneous solution.

Transfer Functions **9**
and the Input–Output
Models

The use of Laplace transforms allows us to form a very simple, convenient, and meaningful representation of chemical process dynamics. It is simple because it uses only algebraic equations (not differential equations, as we have seen in Part II). It is convenient because it allows a quick analysis of process dynamics and finally, it is meaningful because it provides directly the relationship between the *inputs* (disturbances, manipulated variables) and the *outputs* (controlled variables) of a process.

9.1 Transfer Function of a Process with a Single Output

Consider a simple processing system with a single input and a single output (Figure 9.1a). The dynamic behavior of the process is described by an nth-order linear (or linearized nonlinear) differential equation:

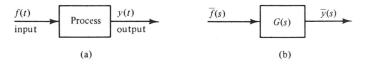

Figure 9.1 (a) Single-input, single-output process; (b) its block diagram.

159

$$a_n \frac{d^n y}{dt^n} + a_{n-1} \frac{d^{n-1} y}{dt^{n-1}} + \cdots + a_1 \frac{dy}{dt} + a_0 y = bf(t) \qquad (9.1)$$

where $f(t)$ and $y(t)$ are the input and output of the process, respectively. Both are expressed in terms of deviation variables.

Assume that the system is initially at steady state. Then

$$y(0) = \left[\frac{dy}{dt} \right]_{t=0} = \left[\frac{d^2 y}{dt^2} \right]_{t=0} = \cdots = \left[\frac{d^{n-1} y}{dt^{n-1}} \right]_{t=0} = 0 \qquad (9.2)$$

After taking the Laplace transform of both sides of (9.1) and using the initial conditions (9.2), we find that

$$\frac{\overline{y}(s)}{\overline{f}(s)} \equiv G(s) = \frac{b}{a_n s^n + a_{n-1} s^{n-1} + \cdots + a_1 s + a_0} \qquad (9.3)$$

$G(s)$ is called the *transfer function* of the system above, and in a simple algebraic form it relates the output of a process to its input (Figure 9.1b). The diagram of Figure 9.1b is also known as the *block diagram* for the system.

If the process has two inputs, $f_1(t)$ and $f_2(t)$, as shown in Figure 9.2a, its dynamic model is

$$a_n \frac{d^n y}{dt^n} + a_{n-1} \frac{d^{n-1} y}{dt^{n-1}} + \cdots + a_1 \frac{dy}{dt} + a_0 y = b_1 f_1(t) + b_2 f_2(t) \quad (9.4)$$

with the same initial conditions (9.2). From (9.4) we take

$$\overline{y}(s) = \frac{b_1}{a_n s^n + a_{n-1} s^{n-1} + \cdots + a_1 s + a_0} \overline{f}_1(s)$$

$$+ \frac{b_2}{a_n s^n + a_{n-1} s^{n-1} + \cdots + a_1 s + a_0} \overline{f}_2(s)$$

or, equivalently,

$$\overline{y}(s) = G_1(s) \overline{f}_1(s) + G_2(s) \overline{f}_2(s) \qquad (9.5)$$

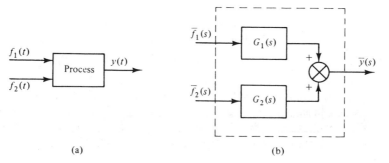

(a)   (b)

Figure 9.2 (a) Two-input, single-output process; (b) its block diagram.

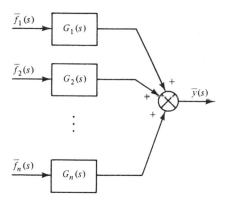

Figure 9.3 Block diagram of a process with several inputs and single output.

with

$$G_1(s) \equiv \frac{b_1}{a_n s^n + a_{n-1} s^{n-1} + \cdots + a_1 s + a_0} \quad \text{and}$$

$$\text{(9.6)}$$

$$G_2(s) \equiv \frac{b_2}{a_n s^n + a_{n-1} s^{n-1} + \cdots + a_1 s + a_0}$$

$G_1(s)$ and $G_2(s)$ are the two transfer functions which relate the output of the process to each one of its two inputs. Thus $G_1(s)$ relates the $\bar{y}(s)$ to the first input $\bar{f}_1(s)$, and $G_2(s)$ relates $\bar{y}(s)$ to the other input $\bar{f}_2(s)$. These relationships are shown by the *block diagram* of Figure 9.2b. A similar procedure can be applied to any system with one output and several inputs. Figure 9.3 shows the block diagram for such a system.

Summarizing all the above, we can define the transfer function between an input and an output as follows:

transfer function $\equiv G(s)$

$$= \frac{\text{Laplace transform of the output, in deviation form}}{\text{Laplace transform of the input, in deviation form}} \quad \text{(9.7)}$$

Remarks

1. The transfer function allows the development of a simpler *input output model* than that discussed in Section 5.1.
2. It describes completely the dynamic behavior of the output when the corresponding input changes are given. Thus, for a particular variation of the input $f(t)$, we can find its transform $\bar{f}(s)$, and from (9.7) we see that the response of the system is

$$\bar{y}(s) = G(s)\bar{f}(s)$$

Take the inverse Laplace transform of $G(s)\bar{f}(s)$ and you have the response $y(t)$ in the time domain.

3. To find the transfer function for a nonlinear system, it must first be linearized around a steady state and be expressed in terms of deviation variables.

Example 9.1: Transfer Functions of a Stirred Tank Heater

The mathematical model of the stirred tank heater in terms of deviation variables was developed in Example 5.1 and is given by eq. (5.3):

$$\frac{dT'}{dt} + aT' = \frac{1}{\tau} T'_i + KT'_{st} \qquad (5.3)$$

where T', T'_i, and T'_{st} are deviation variables, and

$$a = \frac{1}{\tau} + K \qquad \frac{1}{\tau} = \frac{F_i}{V} \qquad K = \frac{UA_t}{V\rho c_p}$$

Take the Laplace transforms of both sides of (5.3):

$$(s + a)\bar{T}'(s) = \frac{1}{\tau} \bar{T}'_i(s) + K\bar{T}'_{st}(s)$$

or

$$\bar{T}'(s) = \frac{1/\tau}{s + a} \bar{T}'_i(s) + \frac{K}{s + a} \bar{T}'_{st}(s) \qquad (9.8)$$

Define the two transfer functions

$$G_1(s) = \frac{\bar{T}'(s)}{\bar{T}'_i(s)} \qquad \text{and} \qquad G_2(s) = \frac{\bar{T}'(s)}{\bar{T}'_{st}(s)}$$

Then

$$\bar{T}'(s) = G_1(s)\bar{T}'_i(s) + G_2(s)\bar{T}'_{st}(s) \qquad (9.8a)$$

and Figure 9.4 shows the block diagram for the tank heater. $G_1(s)$ relates

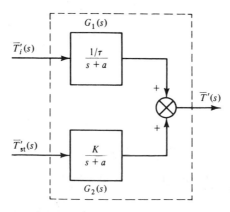

Figure 9.4 Block diagram of a tank heater.

the temperature of the liquid in the tank to that of the inlet stream, while $G_2(s)$ relates the temperature of the liquid in the tank to that of the steam.

Note. Compare the input–output model given by eq. (9.8) and Figure 9.4 to the more complex model developed in Example 5.1 [eq. (5.5) and Figure 5.2].

9.2 Transfer Function Matrix of a Process with Multiple Outputs

Consider a process (Figure 9.5a) with two inputs, $f_1(t)$ and $f_2(t)$, and two outputs, $y_1(t)$ and $y_2(t)$. Let its mathematical model be given by the following two linear differential equations, with all the variables in deviation form:

$$\frac{dy_1}{dt} = a_{11}y_1 + a_{12}y_2 + b_{11}f_1(t) + b_{12}f_2(t) \qquad (9.9a)$$

$$\frac{dy_2}{dt} = a_{21}y_1 + a_{22}y_2 + b_{21}f_1(t) + b_{22}f_2(t) \qquad (9.9b)$$

The initial conditions are

$$y_1(0) = y_2(0) = 0$$

Take the Laplace transforms of both sides of eqs. (9.9a) and (9.9b) and solve with respect to $\bar{y}_1(s)$ and $\bar{y}_2(s)$. (For the details of this procedure, see Section 8.3 and Example 8.2.) Then

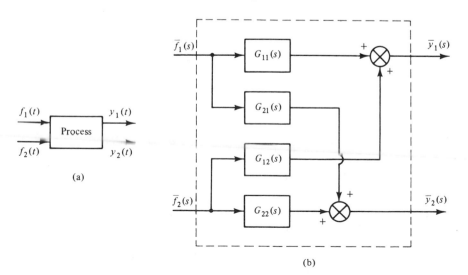

(b)

Figure 9.5 (a) Two-input, two-output process; (b) its block diagram.

$$\overline{y}_1(s) = \frac{[(s - a_{22})b_{11} + a_{12}b_{21}]}{P(s)}\overline{f}_1(s) + \frac{[(s - a_{22})b_{12} + a_{12}b_{22}]}{P(s)}\overline{f}_2(s) \quad (9.10a)$$

$$\overline{y}_2(s) = \frac{[(s - a_{11})b_{21} + a_{21}b_{11}]}{P(s)}\overline{f}_1(s) + \frac{[(s - a_{11})b_{22} + a_{21}b_{12}]}{P(s)}\overline{f}_2(s) \quad (9.10b)$$

$P(s)$ is the characteristic polynomial defined by $P(s) \equiv s^2 - (a_{11} + a_{22})s - (a_{12}a_{21} - a_{11}a_{22})$. Equations (9.10a) and (9.10b) can be written as follows:

$$\overline{y}_1(s) = G_{11}(s)\overline{f}_1(s) + G_{12}(s)\overline{f}_2(s) \qquad (9.11a)$$

$$\overline{y}_2(s) = G_{21}(s)\overline{f}_1(s) + G_{22}(s)\overline{f}_2(s) \qquad (9.11b)$$

where the transfer functions G_{11}, G_{12}, G_{21}, and G_{22} are defined as follows [from eqs. (9.10a) and (9.10b)]:

$$G_{11}(s) \equiv \frac{b_{11}s + (a_{12}b_{21} - a_{22}b_{11})}{P(s)} \qquad G_{12}(s) \equiv \frac{b_{12}s + (a_{12}b_{22} - a_{22}b_{12})}{P(s)}$$

$$G_{21}(s) \equiv \frac{b_{21}s + (a_{21}b_{11} - a_{11}b_{21})}{P(s)} \qquad G_{22}(s) \equiv \frac{b_{22}s + (a_{21}b_{12} - a_{11}b_{22})}{P(s)}$$

The block diagram of the system is shown in Figure 9.5b.

Remarks

1. Equations (9.11a) and (9.11b) can be written as follows in a matrix notation:

$$\begin{bmatrix} \overline{y}_1(s) \\ \overline{y}_2(s) \end{bmatrix} = \begin{bmatrix} G_{11}(s) & G_{12}(s) \\ G_{21}(s) & G_{22}(s) \end{bmatrix} \begin{bmatrix} \overline{f}_1(s) \\ \overline{f}_2(s) \end{bmatrix} \qquad (9.12)$$

The matrix of the transfer functions is called the *transfer function matrix*.

2. For a system with two inputs and two outputs, such as the one discussed above, we have $2 \times 2 = 4$ transfer functions to relate all outputs to all inputs. For a general process with M inputs and N outputs, we will have $N \times M$ transfer functions or a transfer function matrix with N rows (number of outputs) and M columns (number of inputs).

Example 9.2: Transfer Function Matrix of a CSTR

In Example 6.4 we developed the linearized model of a continuous stirred tank reactor in terms of deviation variables, given by eqs. (6.36) and (6.37). After rearranging the terms in these equations, we take

$$\frac{dc'_A}{dt} + \left[\frac{1}{\tau} + k_0 e^{-E/RT_0}\right]c'_A + \left[\frac{k_0 E}{RT_0^2} e^{-E/RT_0} c_{A,0}\right]T' = \frac{1}{\tau}c'_{A_i} \quad (9.13a)$$

$$\frac{dT'}{dt} + \left[\frac{1}{\tau} - \frac{Jk_0E}{RT_0^2}\, e^{-E/RT_0} c_{A,0} + \frac{UA_t}{\rho c_p V}\right] T' - [Jk_0 e^{-E/RT_0}] c_A'$$

$$= \frac{1}{\tau}\, T_i' + \frac{UA_t}{\rho c_p V}\, T_c' \qquad (9.13b)$$

Simplify the notation by defining

$$a_{11} = \frac{1}{\tau} + k_0 e^{-E/RT_0} \qquad a_{12} = \frac{k_0 E}{RT_0^2}\, e^{-E/RT_0} c_{A,0}$$

$$a_{21} = -Jk_0 e^{-E/RT_0} \qquad a_{22} = \frac{1}{\tau} - \frac{Jk_0 E}{RT_0^2}\, e^{-E/RT_0} c_{A,0} + \frac{UA_t}{\rho c_p V}$$

and

$$b_1 = \frac{1}{\tau} \qquad b_2 = \frac{UA_t}{\rho c_p V}$$

Then eqs. (9.13a) and (9.13b) become

$$\frac{dc_A'}{dt} + a_{11} c_A' + a_{12} T' = b_1 c_{A_i}' \qquad (9.14a)$$

$$\frac{dT'}{dt} + a_{21} c_A' + a_{22} T' = b_1 T_i' + b_2 T_c' \qquad (9.14b)$$

The initial conditions are

$$c_A'(0) = T'(0) = 0$$

Take the Laplace transforms of (9.14a) and (9.14b):

$$(s + a_{11})\overline{c}_A'(s) + a_{12}\overline{T}'(s) = b_1 \overline{c}_{A_i}'(s)$$

$$a_{21}\overline{c}_A'(s) + (s + a_{22})\overline{T}'(s) = b_1 \overline{T}_i'(s) + b_2 \overline{T}_c'(s)$$

Solve for $\overline{c}_A'(s)$ and $\overline{T}'(s)$ and take

$$\overline{c}_A'(s) = \frac{b_1(s + a_{22})}{P(s)}\, \overline{c}_{A_i}'(s) - \frac{a_{12} b_1}{P(s)}\, \overline{T}_i'(s) - \frac{a_{12} b_2}{P(s)}\, \overline{T}_c'(s) \qquad (9.15a)$$

$$\overline{T}'(s) = -\frac{a_{21} b_1}{P(s)}\, \overline{c}_{A_i}'(s) + \frac{b_1(s + a_{11})}{P(s)}\, \overline{T}_i'(s) + \frac{b_2(s + a_{11})}{P(s)}\, \overline{T}_c'(s) \qquad (9.15b)$$

where $P(s) \equiv s^2 + (a_{11} + a_{22})s + (a_{11}a_{22} - a_{12}a_{21})$.

In a matrix form, eqs. (9.15a) and (9.15b) are written as follows:

$$\begin{bmatrix} \overline{c}_A'(s) \\ \overline{T}'(s) \end{bmatrix} = \begin{bmatrix} G_{11}(s) & G_{12}(s) & G_{13}(s) \\ G_{21}(s) & G_{22}(s) & G_{23}(s) \end{bmatrix} \begin{bmatrix} \overline{c}_{A_i}'(s) \\ \overline{T}_i'(s) \\ \overline{T}_c'(s) \end{bmatrix}$$

In Table 9.1 we see the six transfer functions corresponding to the CSTR. These can be derived easily from eqs. (9.15a) and (9.15b). The transfer function matrix is nonsquare since the number of inputs is not equal to the number of outputs:

$$\text{transfer function matrix} = \mathbf{G}(s) = \begin{bmatrix} G_{11} & G_{12} & G_{13} \\ G_{21} & G_{22} & G_{23} \end{bmatrix}$$

Figure 9.6 shows the input–output model for the CSTR in block diagram form.

TABLE 9.1
COMPONENTS OF THE TRANSFER FUNCTION MATRIX FOR THE CSTR

Output	Input	G_{ij} Element	Transfer function
$\overline{c}_A'(s)$	$\overline{c}_{A_i}'(s)$	G_{11}	$b_1(s + a_{22})/P(s)$
	$\overline{T}_i'(s)$	G_{12}	$-a_{12}b_1/P(s)$
	$\overline{T}_c'(s)$	G_{13}	$-a_{12}b_2/P(s)$
$\overline{T}'(s)$	$\overline{c}_{A_i}'(s)$	G_{21}	$-a_{21}b_1/P(s)$
	$\overline{T}_i'(s)$	G_{22}	$b_1(s + a_{11})/P(s)$
	$\overline{T}_c'(s)$	G_{23}	$b_2(s + a_{11})/P(s)$

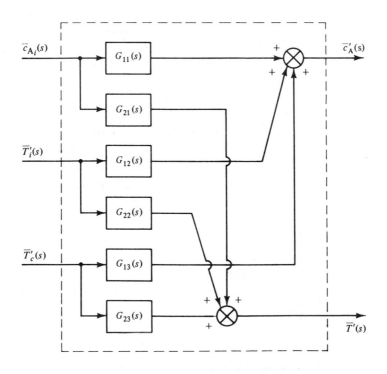

Figure 9.6 Block diagram of a CSTR.

9.3   Poles and Zeros of a Transfer Function

According to the definition of a transfer function, we have

$$\frac{\overline{y}(s)}{\overline{f}(s)} = G(s)$$

In general, the transfer function $G(s)$ will be the ratio of two polynomials,

$$G(s) = \frac{Q(s)}{P(s)}$$

The only exception are systems with time delays which introduce exponential terms (see Section 7.2). For physically realizable systems, the polynomial $Q(s)$ will always be of lower order than the polynomial $P(s)$. The reasons will become clear in subsequent chapters. For the time being, all the examples we have covered satisfy this restriction.

The roots of the polynomial $Q(s)$ are called the *zeros of the transfer function*, or the *zeros of the system* whose dynamics are described by the transfer function $G(s)$. When the variable s takes on as values the zeros of $G(s)$, the transfer function becomes zero.

The roots of the polynomial $P(s)$ are called the *poles of the transfer function*, or equivalently, the *poles of the system*. At the poles of a system the transfer function becomes infinity.

The poles and the zeros of a system play an important role in the dynamic analysis of processing systems and the design of effective controllers. As we proceed, their usefulness will become clearer.

Example 9.3: Poles and Zeros of the Stirred Tank Heater

The input–output model of the tank heater was developed in Example 9.1 and it is given by

$$\overline{T}'(s) = G_1(s)\overline{T}'_i(s) + G_2(s)\overline{T}'_{st}(s) \qquad (9.8a)$$

The transfer function $G_1(s)$ is

$$G_1(s) = \frac{1/\tau}{s + a}$$

and has *no zeros* and *one pole* at $s = -a$. Similarly, the transfer function $G_2(s)$, which is given by

$$G_2 = \frac{K}{s + a}$$

has *no zeros* and *one pole* at $s = -a$. Notice that the two transfer functions have a common pole.

Example 9.4: Poles and Zeros in a CSTR

The transfer functions corresponding to the CSTR were developed in Example 9.2 and are summarized in Table 9.1. All six transfer functions have a common denominator,

$$P(s) = s^2 + (a_{11} + a_{22})s + (a_{11}a_{22} - a_{12}a_{21})$$

and therefore common poles. Since $P(s)$ is a second-order polynomial, the system has two poles, which are given by

$$p_{1,2} = \frac{-(a_{11} + a_{22}) \pm \sqrt{(a_{11} - a_{22})^2 + 4a_{12}a_{21}}}{2}$$

With respect to the zeros, the six transfer functions differ.

$G_{12}(s)$, $G_{13}(s)$, *and* $G_{21}(s)$ *have no zeros*.
$G_{22}(s)$ *and* $G_{23}(s)$ *have one common zero at* $s = -a_{11}$.
$G_{11}(s)$ *has one zero at* $s = -a_{22}$.

9.4   Qualitative Analysis of the Response of a System

The dynamic response of an output y is given by

$$\bar{y}(s) = G(s)\bar{f}(s)$$

For given input $f(t)$ we can find easily its Laplace transform $\bar{f}(s)$, while the transfer function $G(s)$ is known for the particular system. Therefore, the response $y(t)$ in the time domain can be found if we invert the term $G(s)\bar{f}(s)$.

Furthermore, in general,

$$G(s) = \frac{Q(s)}{P(s)}$$

while the Laplace transform of all common inputs can also be expressed as the ratio of two polynomials (see examples in Chapters 7 and 8 as well as Tables 7.1 and 8.1):

$$\bar{f}(s) = \frac{r(s)}{q(s)}$$

Consequently,

$$\bar{y}(s) = \frac{Q(s)}{P(s)} \frac{r(s)}{q(s)} \tag{9.16}$$

To invert the right-hand side of (9.16) using the method of partial fractions we need to know the roots of the polynomial $P(s)$ [i.e., the poles of the system] and the roots of the polynomial $q(s)$. The terms resulting from the inversion by partial fractions are uniquely characterized by the poles of the system and the roots of $q(s)$. Therefore, *if we know where the poles of a system are located, we can determine the qualitative characteristics of the system's response to a particular input, without additional computations.*

Let us use the following general example to clarify the statement above. Suppose that the transfer function of a system is given by

$$G(s) = \frac{Q(s)}{P(s)} = \frac{Q(s)}{(s - p_1)(s - p_2)(s - p_3)^m(s - p_4)(s - p_4^*)(s - p_5)} \quad (9.17)$$

where p_1, p_2, p_3, p_4, p_4^*, and p_5 are the roots of $P(s)$ [i.e., the poles of the system located at various points of the complex plane (see Figure 9.7)]. The partial-fractions expansion of $G(s)$ will yield the following terms:

$$G(s) = \frac{C_1}{s - p_1} + \frac{C_2}{s - p_2} + \left[\frac{C_{31}}{s - p_3} + \frac{C_{32}}{(s - p_3)^2} + \cdots + \frac{C_{3m}}{(s - p_3)^m} \right]$$

$$+ \frac{C_4}{s - p_4} + \frac{C_4^*}{s - p_4^*} + \frac{C_5}{s - p_5}$$

The following observations can be made for the location of the poles:

1. *Real, distinct poles*, such as p_1 and p_2, are located on the real axis (Figure 9.7). During the inversion, they give rise to exponential

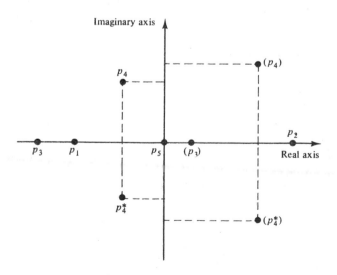

Figure 9.7 Location of poles in the complex plane.

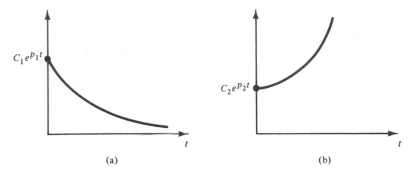

Figure 9.8 (a) Exponential decay; (b) exponential growth.

terms such as $C_1 e^{p_1 t}$ and $C_2 e^{p_2 t}$. Since $p_1 < 0$, $C_1 e^{p_1 t}$ decays exponentially to zero as $t \rightarrow \infty$ (Figure 9.8a). Also, because $p_2 > 0$, $C_2 e^{p_2 t}$ grows exponentially to infinity with time (Figure 9.8b). Therefore, *distinct poles on the negative real axis produce terms that decay to zero with time, while real positive poles make the response of the system grow toward infinity with time.*

2. *Multiple, real poles*, such as p_3, which is repeated m times. Such poles give rise to terms such as

$$\left[C_{31} + \frac{C_{32}}{1!} t + \frac{C_{33}}{2!} t^2 + \cdots + \frac{C_{3m}}{(m-1)!} t^{m-1} \right] e^{p_3 t} \qquad (9.18)$$

The term within the brackets grows toward infinity with time. The behavior of the exponential term depends on the value of the pole p_3:

If $p_3 > 0$ then $e^{p_3 t} \rightarrow \infty$ as $t \rightarrow \infty$.
If $p_3 < 0$ then $e^{p_3 t} \rightarrow 0$ as $t \rightarrow \infty$.
If $p_3 = 0$ then $e^{p_3 t} = 1$ for all times.

Therefore, *a real, multiple pole gives rise to terms which either grow to infinity, if the pole is positive or zero, or decay to zero if the pole is negative.*

3. *Complex conjugate poles*, such as p_4 and p_4^*. We should emphasize that *complex poles always appear in conjugate pairs* and never alone. Let $p_4 = \alpha + j\beta$ and $p_4^* = \alpha - j\beta$. In Section 8.2 we have seen that conjugate pairs of complex roots give rise to terms such as $e^{\alpha t} \sin(\beta t + \phi)$. The $\sin(\beta t + \phi)$ is a periodic, oscillating function, while the behavior of $e^{\alpha t}$ depends on the value of the real part α. Thus:

If $\alpha > 0$, then $e^{\alpha t} \rightarrow \infty$ as $t \rightarrow \infty$, and $e^{\alpha t} \sin(\beta t + \phi)$ grows to infinity in an oscillating manner (Figure 9.9a).
If $\alpha < 0$, then $e^{\alpha t} \rightarrow 0$ as $t \rightarrow \infty$, and $e^{\alpha t} \sin(\beta t + \phi)$ decays to

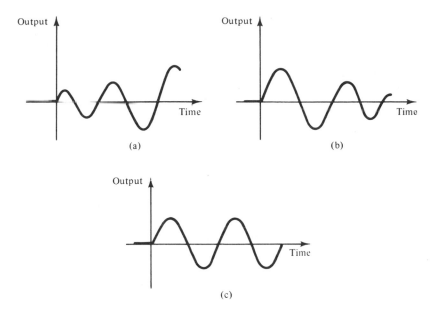

Figure 9.9 Oscillations with (a) growing, (b) decaying, and (c) sustained amplitude.

zero in an oscillating manner with ever-decreasing amplitude (Figure 9.9b).

If $\alpha = 0$, then $e^{\alpha t} = 1$ for all times, and $e^{\alpha t} \sin(\beta t + \phi) = \sin(\beta t + \phi)$, which oscillates continuously (Figure 9.9c) with constant amplitude.

Therefore, *a pair of complex conjugate poles gives rise to oscillatory behavior, whose amplitude may grow continuously if the real part of the complex poles is positive, decay to zero if it is negative, or remain unchanged, if the real part of the poles is zero.*

4. *Poles at the origin:* Pole p_5 is located at the origin of the complex plane (i.e., $p_5 = 0 + j \cdot 0$). Therefore, $C_5/(s - p_5) = C_5/s$ and after inversion it gives a constant term C_5.

Remarks

1. The observations above are general and can be applied to any system. Thus we can find the qualitative characteristics of a system's response if we know where the poles of the corresponding transfer function are located. It is obvious that for a particular input $f(t)$ we should consider the additional roots introduced by the denominator of $\bar{f}(s)$, before we can have the complete picture of the qualitative response of a system.

2. Poles to the right of the imaginary axis give rise to terms which grow to infinity with time. Such systems with unbounded behav-

ior are called *unstable*. Therefore, a system will be *stable* (i.e., with bounded behavior) if all the poles of its transfer function are located to the left of the imaginary axis (Figure 9.7). In subsequent chapters we will define more precisely the stability of a system.

THINGS TO THINK ABOUT

1. Define the transfer function. Why is it useful?

2. For a process with four inputs (disturbances and manipulated variables) and three measured outputs, how many transfer functions should you formulate, and why? What is the corresponding transfer function matrix?

3. In Section 5.1 we developed a different type of input–output model. Would you prefer that over the input–output model based on the transfer function concept? Elaborate on your answer.

4. What is the block diagram of a process? What type of information does it convey?

5. Equations (4.4a) and (4.5b) constitute the complete mathematical model of a stirred tank heater. Develop the input–output model for the process by formulating the necessary transfer functions. Draw the corresponding block diagram. Analyze the interactions among inputs and outputs. What do you observe? (*Hint*: Start by linearizing the modeling equations and expressing the variables in deviation form.)

6. Draw the block diagram of the distillation column shown in Figure 4.10. Can you develop analytically the transfer functions among the various inputs and outputs? If yes, explain how, but do not do it.

7. Consider the stirred tank heater of Example 9.1. Is it a stable system or not, and why? For what values of the parameters a, τ, and K is it stable? Can it become unstable?

8. Does the location of the zeros of a system affect its response to external inputs? Elaborate on your answer.

9. Repeat question 8, but take the location of the poles of a system into account.

10. Show that the poles of a 2×2 system are also the eigenvalues of the matrix of constant coefficients in the dynamic model of the system.

11. Under what conditions can the CSTR of Example 9.2 become unstable?

12. A multiple pole p_3 which is repeated m times gives rise to terms such as those given in (9.18). The terms within the brackets grow toward infinity with time, independently of where the pole p_3 is located. Explain then, why the overall term of (9.18) decays to zero when p_3 is located on the negative real axis?

Dynamic Behavior **10**
of First-Order Systems

The previous chapters of Part III have provided us with all the tools we need to analyze the dynamic behavior of typical processing systems when their inputs change in some fashion (e.g., step, ramp, impulse, sinusoid, etc.). In this section we examine the so-called *first-order systems*. In particular, we will study:

1. What a first-order system is and what physical phenomena give rise to first-order systems.
2. What its characteristic parameters are.
3. How it responds to the various changes in the input variables (disturbances and/or manipulated variables).

10.1 What is a First-Order System?

A first-order system is one whose output $y(t)$ is modeled by a first-order differential equation. Thus in the case of linear (or linearized) system, we have

$$a_1 \frac{dy}{dt} + a_0 y = bf(t) \tag{10.1}$$

where $f(t)$ is the input (forcing function). If $a_0 \neq 0$, then eq. (10.1) yields

$$\frac{a_1}{a_0} \frac{dy}{dt} + y = \frac{b}{a_0} f(t)$$

Define

$$\frac{a_1}{a_0} = \tau_p \qquad \text{and} \qquad \frac{b}{a_0} = K_p$$

and take

$$\tau_p \frac{dy}{dt} + y = K_p f(t) \tag{10.2}$$

τ_p is known as the *time constant* of the process and K_p is called the *steady-state gain* or *static gain* or simply the *gain* of the process. Their physical meaning will become clear in the next three sections.

If $y(t)$ and $f(t)$ are in terms of deviation variables around a steady state, the initial conditions are

$$y(0) = 0 \qquad \text{and} \qquad f(0) = 0$$

From eq. (10.2) it is easily found that the transfer function of a first-order process is given by

$$G(s) = \frac{\bar{y}(s)}{\bar{f}(s)} = \frac{K_p}{\tau_p s + 1} \tag{10.3}$$

A first-order process with a transfer function given by eq. (10.3) is also known as *first-order lag, linear lag*, or *exponential transfer lag*.

If, on the other hand, $a_0 = 0$, then from eq. (10.1) we take

$$\frac{dy}{dt} = \frac{b}{a_1} f(t) = K_p' f(t)$$

which gives a transfer function

$$G(s) = \frac{\bar{y}(s)}{\bar{f}(s)} = \frac{K_p'}{s} \tag{10.4}$$

In such case the process is called *purely capacitive* or *pure integrator*.

10.2　Processes Modeled as First-Order Systems

The first-order processes are characterized by:

1. Their capacity to store material, energy, or momentum
2. The resistance associated with the flow of mass, energy, or momentum in reaching the capacity.

Thus the dynamic response of tanks that have the capacity to store liquids or gases can be modeled as first-order. The resistance is associated with the pumps, valves, weirs, and pipes which are attached to the inflowing or outflowing liquids or gases. Similarly, the temperature response of solid, liquid, or gaseous systems which can store thermal energy (thermal capacity, c_p) is modeled as first-order. For such systems the resistance is associated with the transfer of heat through walls, liquids, or gases. In other words, a process that possesses a capacity to store mass or energy and thus act as a buffer between inflowing and outflowing streams will be modeled as a first-order system. The stirred tank heater of Example 4.4 and the mixing processes of Example 4.11 are typical examples of first-order processes.

It is clear from the above that the first-order lags should be the most common class of dynamic components in a chemical plant, with the capacity to store primarily mass and energy.

Let us examine now some typical capacity processes modeled as first-order systems.

Example 10.1: First-Order System with a Capacity for Mass Storage

Consider the tank shown in Figure 10.1a. The volumetric (volume/time) flow in is F_i and the outlet volumetric flow rate is F_o. In the outlet stream there is a resistance to flow, such as a pipe, valve, or weir. Assume that the effluent flow rate F_o is related linearly to the hydrostatic pressure of the liquid level h, through the resistance R:

$$F_o = \frac{h}{R} = \frac{\text{driving force for flow}}{\text{resistance to flow}} \tag{10.5}$$

At any time point, the tank has the capacity to store mass. The total mass balance gives

$$A\frac{dh}{dt} = F_i - F_o = F_i - \frac{h}{R}$$

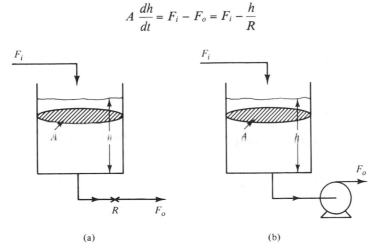

(a)   (b)

Figure 10.1 Systems with capacity for mass storage: (a) first-order lag; (b) pure capacitive.

or

$$AR \frac{dh}{dt} + h = RF_i \tag{10.6}$$

where A is the cross-sectional area of the tank. At steady state

$$h_s = RF_{i,s} \tag{10.6a}$$

and from eqs. (10.6) and (10.6a), we take the following equation in terms of deviation variables:

$$AR \frac{dh'}{dt} + h' = RF'_i \tag{10.7}$$

where $h' = h - h_s$ and $F'_i = F_i - F_{i,s}$. Let

$$\tau_p = AR = \text{time constant of the process}$$

$$K_p = R \quad = \text{steady-state gain of the process}$$

Then the transfer function is

$$G(s) = \frac{\overline{h}'(s)}{\overline{F}'_i(s)} = \frac{K_p}{\tau_p s + 1} \tag{10.8}$$

Certain notes are in order.

1. The cross-sectional area of the tank, A, is a measure of its capacitance to store mass. Thus the larger the value of A, the larger the storage capacity of the tank.
2. Since $\tau_p = AR$, we can say that for the tank we have

(time constant) = (storage capacitance) × (resistance to flow) (10.9)

Example 10.2: First-Order System with a Capacity for Energy Storage

The liquid of a tank is heated with saturated steam, which flows through a coil immersed in the liquid (Figure 10.2). The energy balance for the system yields

$$V \rho c_p \frac{dT}{dt} = Q = UA_t(T_{st} - T) \tag{10.10}$$

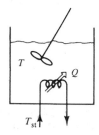

Figure 10.2 System with capacity for energy storage.

where V = volume of liquid in the tank
ρ, c_p = liquid's density and heat capacity
    U = overall heat transfer coefficient between steam and liquid
    A_t = total heat transfer area
    T_{st} = temperature of the saturated steam

The steady state is given by

$$0 = UA_t(T_{st,s} - T_s) \tag{10.11}$$

Subtract (10.11) from (10.10) and take the following equation in terms of deviation variables:

$$V\rho c_p \frac{dT'}{dt} = UA_t(T'_{st} - T') \tag{10.12}$$

where $T' = T - T_s$ and $T'_{st} = T_{st} - T_{st,s}$. The Laplace transform of (10.12) will yield the following transfer function:

$$G(s) \equiv \frac{\overline{T'}(s)}{\overline{T'_{st}}(s)} = \frac{1}{\dfrac{V\rho c_p}{UA_t}s + 1} = \frac{K_p}{\tau_p s + 1} \tag{10.13}$$

where τ_p = time constant of the process = $V\rho c_p / UA_t$
    K_p = steady-state gain = 1

Remarks.

1. Eq. (10.13) demonstrates clearly that this is a first-order lag system.
2. The system possesses capacity to store thermal energy and a resistance to the flow of heat characterized by U.
3. The capacity to store thermal energy is measured by the value of the term $V\rho c_p$. The resistance to the flow of heat from the steam to the liquid is expressed by the term $1/(UA_t)$. Therefore, we notice that the time constant of this system is given by the same equation as that of the tank system in Example 10.1:

$$\text{time constant} = \tau_p = \frac{V\rho c_p}{UA_t}$$

$$= (\text{storage capacitance}) \times (\text{resistance to flow})$$

Example 10.3: Pure Capacitive System

Consider the tank discussed in Example 10.1 with the following difference:

The effluent flow rate F_o is determined by a constant-displacement pump and not by the hydrostatic pressure of the liquid level h (Figure 10.1b)

In such case the total mass balance around the tank yields

$$A \frac{dh}{dt} = F_i - F_o \qquad (10.14)$$

At steady state

$$0 = F_{i,s} - F_o \qquad (10.15)$$

Subtract eq. (10.15) from (10.14) and take the following equation in terms of deviation variables:

$$A \frac{dh'}{dt} = F'_i$$

which yields the following transfer function:

$$G(s) = \frac{\overline{h}'(s)}{\overline{F}'_i(s)} = \frac{1/A}{s} \qquad (10.16)$$

10.3   Dynamic Response of a Pure Capacitive Process

The transfer function for such process is given by eq. (10.4):

$$G(s) = \frac{\overline{y}(s)}{\overline{f}(s)} = \frac{K'_p}{s} \qquad (10.4)$$

Let us examine how $y(t)$ changes with time, when $f(t)$ undergoes a unit step change:

$$f(t) = 1 \qquad \text{for } t > 0$$

We know that for a unit step change,

$$\overline{f}(s) = \frac{1}{s}$$

Therefore, eq. (10.4) yields

$$\overline{y}(s) = \frac{K'_p}{s^2}$$

and after inversion we find (see Table 7.1) that

$$y(t) = K'_p t$$

We notice that the output grows linearly with time in an unbounded fashion (Figure 10.3). Thus

$$y(t) \longrightarrow \infty \qquad \text{as} \qquad t \longrightarrow \infty$$

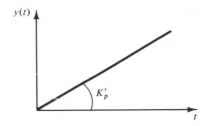

Figure 10.3 Unbounded response of pure capacitive process.

Such response, characteristic of a pure capacitive process, lends the name *pure integrator* because it behaves as if there were an integrator between its input and output.

A pure capacitive process will cause serious control problems, because it cannot balance itself. In the tank of Example 10.3, we can adjust manually the speed of the constant-displacement pump, so as to balance the flow coming in and thus keep the level constant. But any small change in the flow rate of the inlet stream will make the tank flood or run dry (empty). This attribute is known as *non-self-regulation*.

Processes with integrating action most commonly encountered in a chemical process are tanks with liquids, vessels with gases, inventory systems for raw materials or products, and so on.

10.4 Dynamic Response of a First-Order Lag System

The transfer function for such systems is given by eq. (10.3):

$$G(s) = \frac{\overline{y}(s)}{\overline{f}(s)} = \frac{K_p}{\tau_p s + 1} \qquad (10.3)$$

Let us examine how it responds to a unit step change in $f(t)$. Since $\overline{f}(s) = 1/s$, from eq. (10.3) we take

$$\overline{y}(s) = \frac{K_p}{s(\tau_p s + 1)} = \frac{K_p}{s} - \frac{K_p \tau_p}{\tau_p s + 1} \qquad (10.17)$$

Inverting eq. (10.17), we take

$$y(t) = K_p(1 - e^{-t/\tau_p}) \qquad (10.18)$$

If the step change in $f(t)$ were of magnitude A, the response would be

$$y(t) = AK_p(1 - e^{-t/\tau_p}) \qquad (10.19)$$

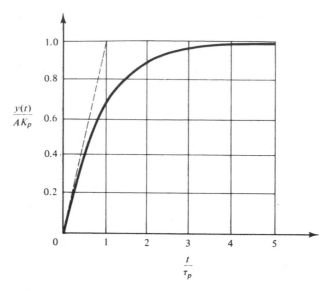

Figure 10.4 Dimensionless response of first-order lag to step input change.

Figure 10.4 shows how $y(t)$ changes with time. The plot is in terms of the dimensionless coordinates

$$\frac{y(t)}{AK_p} \quad \text{versus} \quad \frac{t}{\tau_p}$$

and as such can be used to determine the response of any typical first-order system, independently of the particular values of A, K_p, and τ_p.

Several features of the plot of Figure 10.4 are characteristic of the response of first-order systems and thus worth remembering. These features are:

1. A first-order lag process is *self-regulating*. Unlike a purely capacitive process, it reaches a new steady state. In terms of the tank system in the Example 10.1, when the inlet flow rate increases by unit step, the liquid level goes up. As the liquid level goes up, the hydrostatic pressure increases, which in turn increases the flow rate F_o of the effluent stream [see eq. (10.5)]. This action works toward the restoration of an equilibrium state (steady state).

2. The slope of the response at $t = 0$ is equal to 1.

$$\left. \frac{d[y(t)/AK_p]}{d(t/\tau_p)} \right|_{t=0} = (e^{-t/\tau_p})_{t=0} = 1$$

This implies that if the initial rate of change of $y(t)$ were to be maintained, the response would reach its final value in one time constant (see the dashed line of Figure 10.4). The corollary conclusions are:

The smaller the value of the time constant τ_p, the steeper the initial response of the system.

Equivalently,

The time constant τ_p of a process is a measure of the time necessary for the process to adjust to a change in its input.

3. The value of the response $y(t)$ reaches 63.2% of its final value when the time elapsed is equal to one time constant, τ_p. Subsequently, we have:

Time elapsed	$2\tau_p$	$3\tau_p$	$4\tau_p$
$y(t)$ as percentage of its ultimate value	86.5	95	98

Thus, after four time constants, the response has essentially reached its ultimate value.

4. The ultimate value of the response (i.e., its value at the new steady state) is equal to K_p for a unit step change in the input, or AK_p for a step of size A. This is easily seen from eq. (10.19), which yields $y \longrightarrow AK_p$ as $t \longrightarrow \infty$. This characteristic explains the name *steady state* or *static gain* given to the parameter K_p, since for any step change Δ(input), in the input, the resulting change in the output steady state is given by

$$\Delta(\text{output}) = K_p\,\Delta(\text{input}) \qquad (10.20)$$

Equation (10.20) also tells us by how much we should change the value of the input in order to achieve a desired change in the output, for a process with given gain, K_p. Thus, to effect the same change in the output, we need:

A small change in the input if K_p is large (very sensitive systems)

A large change in the input if K_p is small

Example 10.4: Effect of Parameters on the Response of a First-Order System

Consider the tank system of Example 10.1. It possesses two parameters:

The cross-sectional area of the tank, A
The resistance to the flow of the liquid, R

or from another but equivalent point of view:

The time constant of the process, τ_p
The static gain, K_p

Consider two tanks with different cross-sectional areas A_1 and A_2, where $A_1 > A_2$, and the same resistance, R. From eq. (10.9) we find that $\tau_{p_1} > \tau_{p_2}$ (i.e., *the tank with the larger capacity has a larger time constant*) while the static gains remain the same. When we subject the two tanks to the same unit step changes in the inlet flow rates, the liquid level in each tank responds according to eq. (10.19) and its behavior is shown in Figure 10.5a. We notice that the level of the tank with the smaller cross-sectional area responds faster at the beginning, but ultimately, both levels reach the same steady-state values. This is in agreement with our physical experience. Suppose now that both tanks have different cross-sectional areas A_1 and A_2 and different flow resistances R_1 and R_2, such that

$$\frac{A_1}{A_2} = \frac{R_2}{R_1} \tag{10.21}$$

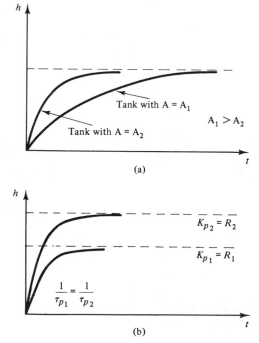

Figure 10.5 Effect of (a) time constant and (b) static gain, in the response of first-order lag systems.

Equation (10.21) yields

$$\tau_{p_1} = A_1 R_1 = A_2 R_2 = \tau_{p_2}$$

But since $A_1 > A_2$, then from eq. (10.21) $R_2 > R_1$, which implies that $K_{p_2} > K_{p_1}$. Figure 10.5b shows the responses of the two tanks to a unit step change in the input. Since both tanks have the same time constant, they have the same initial speed of response. But as time goes on, the tank with the larger resistance R_2 allows less liquid out of the tank. Thus the liquid level grows more in this tank and its ultimate value is larger than the value of the level in the tank with resistance R_1. This again agrees with our physical experience and also demonstrates the fact that *the larger the static gain of a process, the larger the steady-state value of its output for the same input change*.

10.5   First-Order Systems with Variable Time Constant and Gain

In previous sections we assumed that the coefficients of the first-order differential equation [see eq. (10.1)] were constant. This led to the conclusion that the time constant τ_p and steady-state gain K_p of the process were constant. But this is not true for a large number of components in a chemical process. As a matter of fact, in a chemical plant, we will more often encounter processes with variable time constants and gains than not.

Let us examine two characteristic examples.

Example 10.5: Tank System with Variable Time Constant and Gain

For the tank system discussed in Example 10.1, assume that the effluent flow rate, F_o, is not a linear function of the liquid level, but is given by the following relationship (which holds for turbulent flow);

$$F_o = \beta\sqrt{h} \qquad \beta = \text{constant}$$

Then the material balance yields the following nonlinear equation:

$$A\frac{dh}{dt} + \beta\sqrt{h} = F_i$$

Linearize this equation around a steady state and put it in terms of deviation variables (this problem was solved in Examples 6.1 and 6.2):

$$A\frac{dh'}{dt} + \frac{\beta}{2\sqrt{h_s}}h' = F_i' \tag{6.15}$$

or

$$\tau_p\frac{dh'}{dt} + h' = K_p F_i'$$

where $\tau_p = 2A\sqrt{h_s}/\beta$ and $K_p = 2\sqrt{h_s}/\beta$. We notice that both the time constant τ_p and the steady-state gain K_p depend on the steady-state value of the liquid level, h_s. Since we can vary the value of h_s by varying the steady-state value of the inlet flow rate $F_{i,s}$, we conclude that the system has variable time constant and static gain.

Example 10.6: Heater with Variable Time Constant

Let us return to the heater system discussed in Example 10.2. The time constant and the static gain for the heater were found to be

$$\tau_p = \frac{V\rho c_p}{UA_t} \qquad \text{and} \qquad K_p = 1$$

The overall heat transfer coefficient, U, does not remain the same for a long period of operation. Corrosion, dirt, or various other solids deposited on the internal or external surfaces of the heating coil result in a gradual decrease of the heat transfer coefficient. This, in turn, will cause the time constant of the system to vary. This example is characteristic of what can happen to even simple first-order systems.

The question then arises as to how one handles first-order systems with variable time constants and static gains in order to find the dynamic response of such systems. There are two possible solutions:

1. We can use the analytical solutions that are available for first-order differential equations with variable coefficients. Such solutions are quite complicated and of very little value to us for process control purposes.
2. We can assume that such systems possess constant time constants and static gains for a certain limited period of time only. At the end of such a period we will change the values of τ_p and K_p and consider that we have *a new first-order system* with new but constant τ_p and K_p, which will be changed again at the end of the next period. Such an *adaptive procedure* can be used successfully if the time constant and the static gain of a process change slowly, in which case the time period of relatively constant values is rather long.

THINGS TO THINK ABOUT

1. What is a first-order system, and how do you derive the transfer functions of a first-order lag or of a purely capacitive process?
2. What is the principal characteristic of the first-order processes, and what causes the appearance of a purely capacitive process?

3. In Examples 10.1 and 10.2 it was found that for a first-order process

   (time constant) = (storage capacity) × (resistance to flow)

   Is this appropriate for an isothermal, constant-volume CSTR, where a simple, irreversible reaction, A ⟶ B, takes place?

4. Show that a tank with variable cross-sectional area along its height also has variable time constant and static gain.

5. Discuss a system that stores momentum and exhibits first-order dynamics.

6. How would you regulate the purely capacitive process of the tank in Example 10.3 so that it does not flood or run dry?

7. Consider a closed vessel with air flowing in it. Is this a pure capacitive or a first-order lag system? Answer the same question if the vessel is also supplied with an exit for the air.

8. Study the response of a first-order lag to a unit impulse input. [Recall that for a unit impulse, $\bar{f}(s) = 1$.]

9. Study the response of a first-order lag to a sinusoidal input. What do you observe in its behavior after a long time (i.e., as $t \longrightarrow \infty$)?

Dynamic Behavior 11

of Second-Order Systems

Systems with first-order dynamic behavior are not the only ones encountered in a chemical process. An output may change, under the influence of an input, in a drastically different way from that of a first-order system, following higher-order dynamics. In this chapter we analyze (1) the physical origin of systems with second-order dynamics, and (2) their dynamic characteristics. The analysis of systems with higher than second-order dynamics is left for Chapter 12.

11.1 What Is a Second-Order System?

A second-order system is one whose output, $y(t)$, is described by the solution of a second-order differential equation. For example, the following equation describes a second-order linear system:

$$a_2 \frac{d^2y}{dt^2} + a_1 \frac{dy}{dt} + a_0 y = b f(t) \tag{11.1}$$

If $a_0 \neq 0$, then eq. (11.1) yields

$$\tau^2 \frac{d^2y}{dt^2} + 2\zeta\tau \frac{dy}{dt} + y = K_p f(t) \tag{11.2}$$

where $\tau^2 = a_2/a_0$, $2\zeta\tau = a_1/a_0$, and $K_p = b/a_0$. Equation (11.2) is in the standard form of a second-order system, where

$\tau = $ *natural period* of oscillation of the system

$\zeta =$ *damping factor*

$K_p =$ *steady state*, or *static*, or simply *gain* of the system

The physical meaning of the parameters τ and ζ will become clear in the next two sections, while K_p has the same significance as for first-order systems.

If eq. (11.2) is in terms of deviation variables, the initial conditions are zero and its Laplace transformation yields the following standard transfer function for a second-order system;

$$G(s) = \frac{\overline{y}(s)}{\overline{f}(s)} = \frac{K_p}{\tau^2 s^2 + 2\zeta\tau s + 1} \tag{11.3}$$

Systems with second- or higher-order dynamics can arise from several physical situations. These can be classified into three categories:

1. *Multicapacity processes*; processes that consist of two or more capacities (first-order systems) in series, through which material or energy must flow. In Section 11.3 we discuss the characteristics of such systems.
2. *Inherently second-order systems*, such as the fluid or mechanical solid components of a process that possess inertia and are subjected to acceleration. Such systems are rare in chemical processes. They will be discussed briefly in Section 11.4 and three examples are given in Appendix 11A at the end of this chapter.
3. *A processing system with its controller*, may exhibit second- or higher-order dynamics. In such cases, the controller which has been installed on a processing unit introduces additional dynamics which, when coupled with the dynamics of the unit, give rise to second- or higher-order behavior. An example in Section 11.5 will demonstrate this point.

The very large majority of the second- or higher-order systems encountered in a chemical plant come from multicapacity processes or the effect of process control systems. Very rarely we will find systems with appreciable inherent second- or higher-order dynamics.

11.2   Dynamic Response of a Second-Order System

Before we proceed to examine the physical origin of second- and higher-order systems, let us analyze the dynamic response of a second-order system to a unit step input. Such analysis will provide us with all the fundamental dynamic features of a second-order system.

For a unit step change in the input $f(t)$, eq. (11.3) yields

$$\overline{y}(s) = \frac{K_p}{s(\tau^2 s^2 + 2\zeta\tau s + 1)} \tag{11.4}$$

The two poles of the second-order transfer function are given by the roots of the characteristic polynomial,

$$\tau^2 s^2 + 2\zeta\tau s + 1 = 0$$

and they are

$$p_1 = -\frac{\zeta}{\tau} + \frac{\sqrt{\zeta^2 - 1}}{\tau} \quad \text{and} \quad p_2 = -\frac{\zeta}{\tau} - \frac{\sqrt{\zeta^2 - 1}}{\tau} \tag{11.5}$$

Therefore, eq. (11.4) becomes

$$\overline{y}(s) = \frac{K_p/\tau^2}{s(s - p_1)(s - p_2)} \tag{11.6}$$

and the form of the response $y(t)$ will depend on the location of the two poles, p_1 and p_2, in the complex plane (see Section 9.4). Thus we can distiguish three cases:

Case A: When $\zeta > 1$, we have two distinct and real poles.
Case B: When $\zeta = 1$, we have two equal poles (multiple pole).
Case C: When $\zeta < 1$, we have two complex conjugate poles.

Let us examine each case separately.

Case A: Overdamped response, when $\zeta > 1$.
   In this case the inversion of eq. (11.6) by partial-fractions expansion yields

$$y(t) = K_p\left[1 - e^{-\zeta t/\tau}\left(\cosh \sqrt{\zeta^2 - 1}\frac{t}{\tau} + \frac{\zeta}{\sqrt{\zeta^2 - 1}}\sinh \sqrt{\zeta^2 - 1}\frac{t}{\tau}\right)\right] \tag{11.7}$$

where cosh $(\cdot)$ and sinh $(\cdot)$ are the hyberbolic trigonometric functions defined by

$$\sinh \alpha = \frac{e^\alpha - e^{-\alpha}}{2} \quad \text{and} \quad \cosh \alpha = \frac{e^\alpha + e^{-\alpha}}{2}$$

The response has been plotted in Figure 11.1a for various values of ζ, $\zeta > 1$. It is known as *overdamped response* and resembles a little the response of a first-order system to a unit step input. But when compared to a first-order response we notice that the system initially delays to respond and then its response is rather *sluggish*. It becomes more sluggish as ζ increases (i.e., as the system becomes more heavily over-

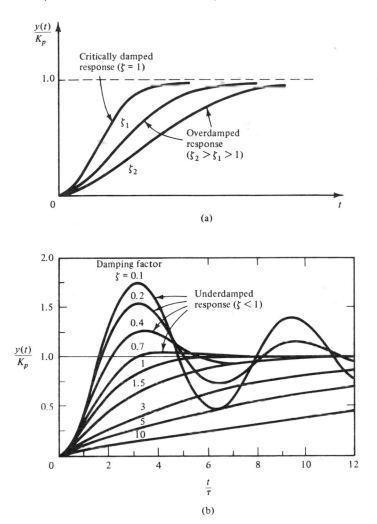

Figure 11.1 Dimensionless response of second-order system to input step change.

damped). Finally, we notice that as time goes on, the response approaches its ultimate value asymptotically. As it was the case with first-order system, the gain is given by

$$K_p = \frac{\Delta(\text{output steady state})}{\Delta(\text{input steady state})}$$

Overdamped are the responses of multicapacity processes, which result from the combination of first-order systems in series, as we will see in Section 11.3.

Case B: Critically damped response, when $\zeta = 1$.
   In this case, the inversion of eq. (11.4) gives the result

$$y(t) = K_p \left[1 - \left(1 + \frac{t}{\tau} \right) e^{-t/\tau} \right] \tag{11.8}$$

The response is also shown in Figure 11.1a. We notice that a second-order system with *critical damping* approaches its ultimate value faster than does an overdamped system.

Case C: Underdamped response, when $\zeta < 1$.
   The inversion of eq. (11.4) in this case yields

$$y(t) = K_p \left[1 - \frac{1}{\sqrt{1 - \zeta^2}} e^{-\zeta t/\tau} \sin (\omega t + \phi) \right] \tag{11.9}$$

where

$$\omega = \frac{\sqrt{1 - \zeta^2}}{\tau} \tag{11.10}$$

and

$$\phi = \tan^{-1} \left[\frac{\sqrt{1 - \zeta^2}}{\zeta} \right]$$

The response has been plotted in Figure 11.1b for various values of the damping factor, ζ. From the plots we can observe the following:

1. The underdamped response is initially faster than the critically damped or overdamped responses, which are characterized as *sluggish*.
2. Although the underdamped response is initially faster and reaches its ultimate value quickly, it does not stay there, but it starts oscillating with progressively decreasing amplitude. This oscillatory behavior makes an underdamped response quite distinct from all previous ones.
3. The oscillatory behavior becomes more pronounced with smaller values of the damping factor, ζ.

It must be emphasized that almost all the underdamped responses in a chemical plant are caused by the interaction of the controllers with the process units they control. Therefore, it is a type of response that we will encounter very often, and it is wise to become well acquainted with its characteristics.

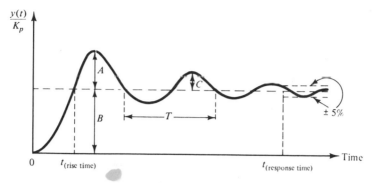

Figure 11.2 Characteristics of an underdamped response.

Characteristics of an underdamped response

Let us use as reference the underdamped response shown in Figure 11.2, in order to define the terms employed to describe an underdamped response.

1. *Overshoot:* Is the ratio A/B, where B is the ultimate value of the response and A is the maximum amount by which the response exceeds its ultimate value. The overshoot is a function of ζ, and it can be shown that it is given by the following expression:

$$\text{overshoot} = \exp\left(\frac{-\pi\zeta}{\sqrt{1 - \zeta^2}}\right) \qquad (11.11)$$

Figure 11.3 shows the plot of overshoot versus ζ given by eq. (11.11). We notice that the overshoot increases with decreasing ζ, while as ζ approaches 1 the overshoot approaches zero (critically damped response).

2. *Decay ratio:* Is the ratio C/A (i.e., the ratio of the amounts above the ultimate value of two successive peaks). The decay ratio can be shown to be related to the damping factor ζ through the equation

$$\text{decay ratio} = \exp\left(\frac{-2\pi\zeta}{\sqrt{1 - \zeta^2}}\right) = (\text{overshoot})^2 \qquad (11.12)$$

Equation (11.12) has been also plotted in Figure 11.3.

3. *Period of oscillation:* From eq. (11.10) we see that the radian frequency (rad/time) of the oscillations of an underdamped response is given by

$$\omega = \text{radian frequency} = \frac{\sqrt{1 - \zeta^2}}{\tau} \qquad (11.10)$$

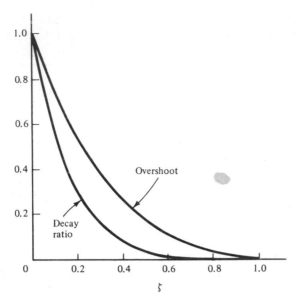

Figure 11.3 Effect of damping factor on overshoot and decay ratio.

To find the period of the oscillation T (i.e., the time elapsed between two successive peaks), use the well-known relationships $\omega = 2\pi f$ and $f = 1/T$ where f = cyclical frequency. Thus

$$T = \frac{2\pi\tau}{\sqrt{1-\zeta^2}} \tag{11.13}$$

4. *Natural period of oscillation*: A second-order system with $\zeta = 0$ is a system free of any damping. Its transfer function is

$$G(s) = \frac{K_p}{\tau^2 s^2 + 1} = \frac{K_p/\tau^2}{\left(s - j\dfrac{1}{\tau}\right)\left(s + j\dfrac{1}{\tau}\right)} \tag{11.14}$$

that is, it has two purely imaginary poles (on the imaginary axis) and according to the analysis of Section 9.4, it will oscillate continuously with a constant amplitude and a natural frequency [see eq. (11.14)]

$$\omega_n = \frac{1}{\tau} \tag{11.15}$$

The corresponding cyclical period T_n is given by

$$T_n = 2\pi\tau \tag{11.16}$$

It is this property of the parameter τ that gave it its name.

5. *Response time:* The response of an underdamped system will reach its ultimate value in an oscillatory manner as $t \rightarrow \infty$. For practical purposes, it has been agreed to consider that the response reached its final value when it came within ±5% of its final value and stayed there. The time needed for the response to reach this situation is known as the *response time*, and it is also shown in Figure 11.2.

6. *Rise time:* This term is used to characterize the speed with which an underdamped system responds. It is defined as the time required for the response to reach its final value for the first time (see Figure 11.2). From Figure 11.1b we notice that the smaller the value of ζ, the shorter the rise time (i.e., the faster the response of the system), but at the same time the larger the value of the overshoot.

Remark.   In subsequent chapters (Part IV), our objective during the design of a controller will be proper selection of the corresponding ζ and τ values, so that the overshoot is small, the rise time short, the decay ratio small, and the response time short. We will realize that it will not be possible to achieve all these objectives for the same values of ζ and τ, and that an acceptable compromise should be defined. Good understanding of the underdamped behavior of a second-order system will help tremendously in the design of efficient controllers.

11.3   Multicapacity Processes as Second-Order Systems

When material or energy flows through a single capacity, we get a first-order system. If on the other hand, mass or energy flows through a series of two capacities, the behavior of the system is described by second-order dynamics. Two multicapacity systems are shown in Figure 11.4, each with two mass capacities (the two tanks).

Examine the two systems of Figure 11.4 more closely to identify a significant qualitative difference between them. In system 1 (Figure 11.4a), tank 1 feeds tank 2 and thus it affects its dynamic behavior, whereas the opposite is not true. Such a system is characteristic of a large class of *noninteracting capacities* or *noninteracting first-order systems in series*. On the contrary, in system 2, tank 1 affects the dynamic behavior of tank 2, and vice versa, because the flow rate F_1 depends on the difference between liquid levels h_1 and h_2. This system represents *interacting capacities* or *interacting first-order systems in series*.

Multicapacity processes do not have to involve more than one physical processing unit. It is quite possible that all capacities are associated

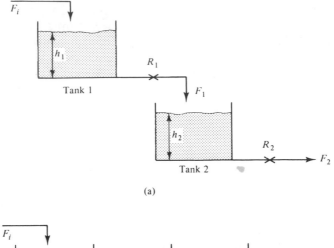

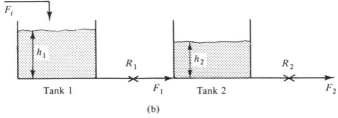

Figure 11.4 (a) Noninteracting and (b) interacting tanks.

with the same processing unit. For example, the stirred tank heater is a multicapacity process with capacity to store mass and energy. A distillation column is another example of a multicapacity process. Every tray has a mass storage capacity (liquid holdup), which, in turn, also allows for thermal energy storage capacity.

Let us now see how multicapacity processes result in second-order systems. We start with noninteracting capacities.

Noninteracting capacities

When a system is composed of two noninteracting capacities, it is described by a set of two differential equations of the general form:

$$\tau_{p1}\frac{dy_1}{dt} + y_1 = K_{p1}f_1(t) \qquad \text{first capacity} \qquad (11.17a)$$

$$\tau_{p2}\frac{dy_2}{dt} + y_2 = K_{p2}y_1(t) \qquad \text{second capacity} \qquad (11.17b)$$

In other words, the first system affects the second by its output, but it is not affected by it (Figure 11.5a). Equation (11.17a) can be solved first

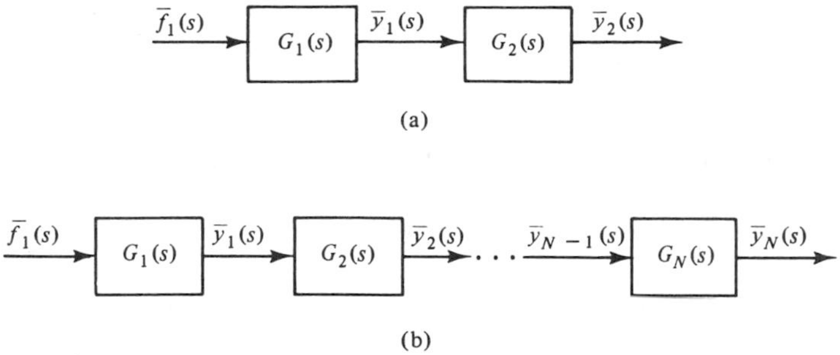

(a)

(b)

Figure 11.5 Noninteracting capacities in series.

and then we can solve eq. (11.17b). This *sequential solution is characteristic of noninteracting capacities in series.* The corresponding transfer functions are

$$G_1(s) = \frac{\overline{y}_1(s)}{\overline{f}_1(s)} = \frac{K_{p1}}{\tau_{p1}s + 1} \qquad G_2(s) = \frac{\overline{y}_2(s)}{\overline{y}_1(s)} = \frac{K_{p2}}{\tau_{p2}s + 1}$$

The *overall transfer function* between the external input $f_1(t)$ and $y_2(t)$ is

$$G_o(s) = \frac{\overline{y}_2(s)}{\overline{f}_1(s)} = \frac{\overline{y}_2(s)}{\overline{y}_1(s)}\frac{\overline{y}_1(s)}{\overline{f}_1(s)} = G_1(s)G_2(s) = \frac{K_{p1}}{\tau_{p1}s + 1}\frac{K_{p2}}{\tau_{p2}s + 1} \qquad (11.18)$$

or

$$G_o(s) = \frac{K'_p}{(\tau')^2 s^2 + 2\zeta'\tau' s + 1} \qquad (11.19)$$

where

$$(\tau')^2 = \tau_{p1}\tau_{p2} \qquad 2\zeta'\tau' = \tau_{p1} + \tau_{p2} \qquad K'_p = K_{p1}K_{p2}$$

Equation (11.19) indicates very clearly that the overall response of the system is second-order. From eq. (11.18) we also notice that the two poles of the overall transfer function are real and distinct:

$$p_1 = -\frac{1}{\tau_{p1}} \quad \text{and} \quad p_2 = -\frac{1}{\tau_{p2}}$$

If the time constants τ_{p1} and τ_{p2} are equal, we have two equal poles. Therefore, *noninteracting capacities always result in an overdamped or critically damped second-order system* and never in an underdamped system. The response of two noninteracting capacities to a unit step change in the input will be given by eq. (11.7) for the overdamped case, or eq. (11.8) for the critically damped. Instead of eq. (11.7), we can use the following equivalent form for the response:

$$y(t) = K'_p \left[1 + \frac{1}{\tau_{p2} - \tau_{p1}} (\tau_{p1} e^{-t/\tau_{p1}} - \tau_{p2} e^{-t/\tau_{p2}}) \right] \qquad (11.20)$$

where $K'_p = K_{p1} K_{p2}$. Equation (11.20) can be derived easily by simple inversion of eq. (11.18), where $\bar{f}_1(s) = 1/s$.

For the case of N noninteracting capacities (Figure 11.5b), it is easy to show that the overall transfer function is given by

$$G_o(s) = G_1(s) G_2(s) \cdots G_N(s) = \frac{K_{p1} K_{p2} \cdots K_{pN}}{(\tau_{p1} s + 1)(\tau_{p2} s + 1) \cdots (\tau_{pN} s + 1)} \qquad (11.21)$$

Example 11.1: Two Noninteracting Material Capacities in Series

System 1 in Figure 11.4a is such a system. The transfer functions for the two tanks are

$$G_1(s) = \frac{\bar{h}'_1(s)}{\bar{F}'_i(s)} = \frac{K_{p1}}{\tau_{p1} s + 1} \qquad \text{and} \qquad G_2(s) = \frac{\bar{h}'_2(s)}{\bar{F}'_i(s)} = \frac{K_{p2}}{\tau_{p2} s + 1}$$

where, according to Example 10.1, we have

$$K_{p1} = R_1 \qquad K_{p2} = R_2 \qquad \tau_{p1} = A_1 R_1 \qquad \tau_{p2} = A_2 R_2$$

and variables h'_1, h'_2, F'_i, and F'_1 are in deviation form. Since

$$F'_1 = \frac{h'_1}{R_1}$$

we can easily find that the overall transfer function is

$$G_o(s) = \frac{\bar{h}'_2(s)}{\bar{F}'_i(s)} = \frac{K_{p2}}{(\tau_{p1} s + 1)(\tau_{p2} s + 1)} \qquad (11.22)$$

Equation (11.22) indicates that the relationship between the external input, $F_i(t)$, and the final output, $h_2(t)$, is that of an overdamped second-order system. Using eq. (11.20) for the response of two noninteracting capacities with $\tau_{p1} \neq \tau_{p2}$ we find that

$$h'_2(t) = K_{p2} \left[1 + \frac{1}{\tau_{p2} - \tau_{p1}} (\tau_{p1} e^{-t/\tau_{p1}} - \tau_{p2} e^{-t/\tau_{p2}}) \right]$$

Figure 11.6 shows the qualitative features of the response, which are the same as those of an overdamped system. A comparison with the first-order response would be instructive. Thus from Figure 11.6 we notice that:

1. The response of the overdamped multicapacity system to step input change is S-shaped (i.e., initially changes slowly and then it picks up speed). This is in contrast to a first-order response which has the largest rate of change at the beginning. This *sluggishness* or *delay* is also known as *transfer lag* and is characteristic of multicapacity systems.

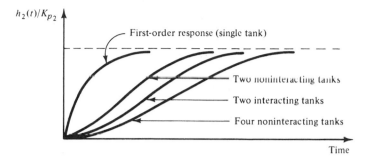

Figure 11.6 Effects of interaction on response to input step changes.

2. As the number of capacities in series increases, the delay in the initial response (sluggishness) becomes more pronounced.

Interacting capacities

In order to analyze the characteristics of such a system, we will use the two-capacity system 2 of Figure 11.4b. The mass balances yield

$$A_1 \frac{dh_1}{dt} = F_i - F_1 \qquad \text{tank 1} \qquad (11.23a)$$

$$A_2 \frac{dh_2}{dt} = F_1 - F_2 \qquad \text{tank 2} \qquad (11.23b)$$

Assume linear resistances to flow:

$$F_1 = \frac{h_1 - h_2}{R_1} \qquad \text{and} \qquad F_2 = \frac{h_2}{R_2}$$

Then eqs. (11.23a) and (11.23b) become

$$A_1 R_1 \frac{dh_1}{dt} + h_1 - h_2 = R_1 F_i \qquad (11.24a)$$

$$A_2 R_2 \frac{dh_2}{dt} + \left(1 + \frac{R_2}{R_1}\right)h_2 - \frac{R_2}{R_1}h_1 = 0 \qquad (11.24b)$$

We notice that eqs. (11.24a) and (11.24b) *must be solved simultaneously.* This is the distinguishing characteristic of interacting capacities and indicates the mutual effect of the two capacities.

The steady-state equivalents of eqs. (11.24a) and (11.24b) are

$$h_{1,s} - h_{2,s} = R_1 F_{i,s} \qquad (11.25a)$$

$$\left(1 + \frac{R_2}{R_1}\right)h_{2,s} - \frac{R_2}{R_1}h_{1,s} = 0 \qquad (11.25b)$$

Subtract (11.25a) from (11.24a) and (11.25b) from (11.24b) and after introducing the deviation variables, take

$$A_1 R_1 \frac{dh_1'}{dt} + h_1' - h_2' = R_1 F_i' \qquad (11.26a)$$

$$A_2 R_2 \frac{dh_2'}{dt} + \left(1 + \frac{R_2}{R_1}\right) h_2' - \frac{R_2}{R_1} h_1' = 0 \qquad (11.26b)$$

where $h_1' = h_1 - h_{1,s}$, $h_2' = h_2 - h_{2,s}$, and $F_i' = F_i - F_{i,s}$. Take the Laplace transforms of eqs. (11.26a) and (11.26b) and find

$$(A_1 R_1 s + 1)\overline{h}_1'(s) - \overline{h}_2'(s) = R_1 \overline{F}_i'(s)$$

$$-\frac{R_2}{R_1}\overline{h}_1'(s) + \left[A_2 R_2 s + \left(1 + \frac{R_2}{R_1}\right)\right]\overline{h}_2'(s) = 0$$

Solve these algebraic equations with respect to $\overline{h}_1'(s)$ and $\overline{h}_2'(s)$ and find

$$\overline{h}_1'(s) = \frac{(\tau_{p2} R_1)s + (R_1 + R_2)}{\tau_{p1}\tau_{p2}s^2 + (\tau_{p1} + \tau_{p2} + A_1 R_2)s + 1} \overline{F}_i'(s) \qquad (11.27a)$$

$$\overline{h}_2'(s) = \frac{R_2}{\tau_{p1}\tau_{p2}s^2 + (\tau_{p1} + \tau_{p2} + A_1 R_2)s + 1} \overline{F}_i'(s) \qquad (11.27b)$$

where $\tau_{p1} = A_1 R_1$ and $\tau_{p2} = A_2 R_2$ are the time constants of the two tanks. Equations (11.27a) and (11.27b) indicate that the responses of both tanks follow second-order dynamics. Compare eq. (11.27b) for the interacting tanks with eq. (11.22), which corresponds to the noninteracting tanks. We notice that they differ only in the coefficient of s in the denominator by the term, $A_1 R_2$. This term may be thought of as the *interaction factor* and indicates the degree of interaction between the two tanks. The larger the value of $A_1 R_2$, the larger the interaction between the two tanks.

Remarks

1. From eq. (11.27b) it is easily found that the two poles of the transfer function are given by

$$p_{1,2} = \frac{-(\tau_{p1} + \tau_{p2} + A_1 R_2) \pm \sqrt{(\tau_{p1} + \tau_{p2} + A_1 R_2)^2 - 4\tau_{p1}\tau_{p2}}}{2\tau_{p1}\tau_{p2}} \qquad (11.28)$$

But

$$(\tau_{p1} + \tau_{p2} + A_1 R_2)^2 - 4\tau_{p1}\tau_{p2} > 0$$

Therefore, p_1 and p_2 are distinct and real poles. Consequently, the *response of interacting capacities is always overdamped*.

2. Since the response is overdamped with poles p_1 and p_2 given by eq. (11.28), then eq. (11.27b) can be written as follows:

$$\frac{\overline{h}_2'(s)}{\overline{F}_i'(s)} = \frac{R_2/\tau_{p_1}\tau_{p_2}}{(s - p_1)(s - p_2)} = \frac{(\tau_1\tau_2)R_2/\tau_{p_1}\tau_{p_2}}{(\tau_1 s + 1)(\tau_2 s + 1)} \qquad (11.29)$$

where $\tau_1 = -1/p_1$ and $\tau_2 = -1/p_2$. Equation (11.29) implies that two *interacting capacities can be viewed as noninteracting capacities but with modified effective time constants*. Thus, whereas initially the two interacting tanks had effective time constants τ_{p_1} and τ_{p_2}, when they are viewed as noninteracting, they have different time constants τ_1 and τ_2.

3. Assume that the two tanks have the same time constants, $\tau_{p_1} = \tau_{p_2} = \tau$. Then, from eq. (11.28), we take

$$\frac{\tau_1}{\tau_2} = \frac{p_2}{p_1} = \frac{-(2\tau + A_1R_2) + \sqrt{A_1^2 R_2^2 + 4\tau A_1 R}}{-(2\tau + A_1R_2) - \sqrt{A_1^2 R_2^2 + 4\tau A_1 R}} \neq 1$$

Thus we see that the effect of interaction is to *change the ratio of the effective time constants for the two tanks* (i.e., one tank becomes faster in its response and the other slower). Since the overall response of $h_2(t)$ is affected by both tanks, the slower tank becomes the controlling and the overall response becomes more sluggish due to the interaction. Therefore, *interacting capacities are more sluggish than the noninteracting*.

Example 11.2: Dynamics of Two Interacting Tanks

Consider two interacting tanks such as those of Figure 11.4b. Let $A_1 = A_2$ and $R_1 = R_2/2$. Then $\tau_{p_1} = \tau_{p_2}/2 = \tau$. From eq. (11.27b) we take

$$\overline{h}_2'(s) = \frac{R_2}{2\tau_*^2 s^2 + 5\tau s + 1}\overline{F}_i'(s) = \frac{R_2}{(0.44\tau s + 1)(4.56\tau s + 1)}\overline{F}_i'(s) \ (11.30)$$

For a unit step change in $F_i'(t)$ [i.e., for $\overline{F}_i'(s) = 1/s$], eq. (11.30) after inversion yields

$$h_2'(t) = R_2(1 - 1.11 e^{-t/4.56\tau} + 0.11 e^{-t/0.44\tau})$$

or

$$h_2'(t) = 1 - 1.11 e^{-t/4.56\tau} + 0.11 e^{-t/0.44\tau}$$

If the two tanks were noninteracting, the transfer function of the system would be given by eq. (11.22):

$$\frac{\overline{h}_2'(s)}{\overline{F}_2'(s)} = \frac{R_2}{(\tau_{p_1}s + 1)(\tau_{p_2}s + 1)} = \frac{R_2}{(\tau s + 1)(2\tau s + 1)}$$

which after inversion yields

$$h_2'(t) = R_2(1 + e^{-t/\tau} - 2e^{-t/2\tau})$$

or

$$F'_2(t) = 1 + e^{-t/\tau} - 2e^{-t/2\tau}$$

Let us compare the responses of the two systems:

1. They are both overdamped. As such they have the characteristics discussed in Section 11.2 (i.e., they are S-shaped with no oscillations).
2. For the system of the two noninteracting tanks the time constants are τ and 2τ. For the case of the interacting tanks, the effective time constants have become 0.44τ and 4.56τ (i.e., one was decreased and the other was increased). Their ratio from $1/2$ changed to $0.44/4.56 = 0.10$.
3. As a result of the change in the effective time constants, the response of the interacting tanks is more sluggish, or more damped, than the response of the noninteracting tanks. Figure 11.6 dramatizes this result.

Example 11.3: The Stirred Tank Heater as a System with Two Interacting Capacities

The stirred tank heater of Example 4.4 is characterized by its capacity to store mass and energy. It is easy to show that these two capacities interact when the inlet flowrate changes. Thus, a change in the inlet flowrate affects the liquid level in the tank, which in turn affects the temperature of the liquid. Consequently, the temperature response to an inlet flowrate change exhibits second-order overdamped characteristics. The reader should note that the two capacities do not interact when the inlet temperature changes. Therefore, the temperature response to inlet temperature changes exhibits first-order characteristics.

This example demonstrates that multiple capacities need not correspond to physically different units (as in the case of interacting tanks, Example 11.2), but could be present within the same processing unit.

Remark. Consider the linearized mass and energy balances for a constant holdup CSTR [see eqs. (9.14a) and (9.14b) in Example 9.2]. The reader should note that these two equations could be perceived as characterizing two interacting capacities. Therefore, he or she could erroneously conclude that the response of c_A or T to inlet changes is always second-order overdamped (as is the case with the stirred tank heater, above). But, this is not true. A CSTR is not only characterized by its capacity to store material A and energy. Its distinguishing characteristic is the kinetic rate term, which denotes disappearance of component A and which is not present in the stirred tank heater. Such reaction terms may produce not only overdamped but also underdamped or inverse responses (for inverse response see Section 12.3). Therefore, interacting capacities will always yield overdamped response, unless they also include mass or/and energy generation (or disappearance) terms, in which case they may yield a variety of responses.

11.4   Inherently Second-Order Processes

Such a process can exhibit underdamped behavior, and consequently it cannot be decomposed into two first-order systems in series (interacting or noninteracting) with physical significance, like the systems we examined in previous sections. They occur rather rarely in a chemical process, and they are associated with the motion of liquid masses or the mechanical translation of solid parts, possessing: (1) inertia to motion, (2) resistance to motion, and (3) capacitance to store mechanical energy. Since resistance and capacitance are characteristic of the first-order systems, we conclude that the inherently second-order systems are characterized by their inertia to motion. The three examples in Appendix 11A clearly demonstrate this feature.

Newton's law applied on a given system yields

$$\left(\begin{array}{c} \text{balance of forces} \\ \text{on the system} \end{array}\right) = (\text{mass of system}) \times (\text{acceleration}) \quad (11.31)$$

Since

$$\text{acceleration} = \frac{d(\text{velocity})}{dt}$$

and

$$\text{velocity} = \frac{d}{dt}(\text{spatial displacement})$$

we conclude that

$$\left(\begin{array}{c} \text{balance of forces} \\ \text{on the system} \end{array}\right) \qquad\qquad (11.32)$$
$$= (\text{mass of system}) \times \frac{d^2}{dt^2}(\text{spatial displacement})$$

The second term of the right-hand side gives rise to the second-order behavior of the system. Equation (11.31) or its equivalent (11.32) is the starting point for the examples of Appendix 11A.

11.5   Second-Order Systems Caused by the Presence of Controllers

The presence of a control system in a chemical process can change the order of the process and produce a dynamic behavior which the process cannot exhibit without the controller. In the chapters of Part IV we will have the opportunity to examine many such situations. For the time being, let us consider a simple example.

Example 11.4: First-Order Process with Second-Order Dynamics Due to the Presence of a Control System

Consider the tank shown in Figure 10.1a (Example 10.1). This is a simple first-order system with a transfer function given by eq. (10.8). We would like to control the liquid level at a desired value when the inlet flow rate F_i undergoes step changes. To do that we use the feedback control system shown in Figure 11.7a. This control system measures the liquid level and compares it with the desired steady-state value. If the level is higher than the desired value, it increases the effluent flow rate F_o by opening the control valve V, while it closes the valve when the level is lower than the desired value. Let us now see how the presence of this controller changes the order of the dynamic behavior of the tank from first- to second-order.

The dynamic mass balance around the tank gives

$$A \frac{dh}{dt} = F_i - F_o \tag{11.33}$$

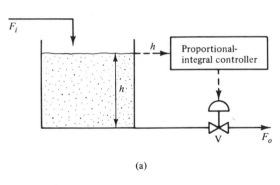

(a)

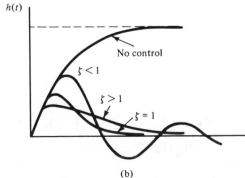

(b)

Figure 11.7 (a) Feedback control; (b) the resulting second-order behavior of the liquid level.

while at the desired steady state we have

$$0 = F_{i,s} - F_{o,s} \tag{11.34}$$

Subtract (11.34) from (11.33) and take

$$A\frac{dh'}{dt} = F_i' - F_o' \tag{11.35}$$

where the deviation variables are defined by $h' = h - h_s$, $F_i' = F_i - F_{i,s}$, and $F_o' = F_o - F_{o,s}$. When the liquid level is not at the desired value, then $h' \neq 0$. The measuring device measures h and this value is compared to the desired value h_s. The deviation (error) h' is used by the controller to increase or decrease the effluent flow rate according to the relationship

$$F_o = F_{o,s} + K_c h' + \frac{K_c}{\tau_I}\int_0^t h'(t)\,dt \tag{11.36}$$

where K_c and τ_I are constant parameters with positive values. According to (11.36):

1. When $h' = 0$, then $F_o = F_{o,s}$ and the valve V stays where it is.
2. When $h' < 0$ (i.e., the level goes down), then from eq. (11.36) we have $F_o < F_{o,s}$ (i.e., the controller reduces the effluent rate and the level starts increasing).
3. When $h' > 0$ (i.e., the level goes up), then from eq. (11.36) we find that $F_o > F_{o,s}$ (i.e., the controller increases the effluent rate and the level decreases).

The control action described by eq. (11.36) is called *proportional-integral control*, because the value of the manipulated variable is determined by two terms, one of which is *proportional to the error* h', and the other *proportional to the time integral of the error*.
    In eq. (11.35) replace F_o' with its equal given by (11.36) and take

$$A\frac{dh'}{dt} + K_c h' + \frac{K_c}{\tau_I}\int_0^t h'\,dt = F_i' \tag{11.37}$$

The Laplace transform of (11.37) gives

$$As\overline{h}'(s) + K_c \overline{h}'(s) + \frac{K_c}{\tau_I}\frac{1}{s}\overline{h}'(s) = \overline{F}_i'(s)$$

01

$$\left[\frac{A\tau_I}{K_c}s^2 + \tau_I s + 1\right]\overline{h}'(s) = \frac{\tau_I s}{K_c}\overline{F}_i'(s). \tag{11.38}$$

From eq. (11.38) we find that the transfer function between the external input $\overline{F}_i'(s)$ and the output $\overline{h}'(s)$ is that of a second-order system and given by

$$\frac{\overline{h}'(s)}{\overline{F}_i'(s)} = \frac{K_p s}{\tau^2 s^2 + 2\zeta\tau s + 1}$$

where $\tau^2 = A\tau_I/K_c$, $2\zeta\tau = \tau_I$, and $K_p = \tau_I/K_c$. From the equations above we find that

$$\tau = \sqrt{\frac{A\tau_I}{K_c}} \qquad \text{and} \qquad \zeta = \frac{1}{2}\sqrt{\frac{K_c\tau_I}{A}}$$

Depending on the values of the control parameters K_c and τ_I, we may have the following cases:

1. $\sqrt{K_c\tau_I/A} < 2$. Then, $\zeta < 1$ and the response $\bar{h}'(s)$ to a step input in $\bar{F}_i'(s)$ is that of an underdamped system.
2. $\sqrt{K_c\tau_I/A} = 2$. Then $\zeta = 1$ and the response is critically damped.
3. Finally, $\sqrt{K_c\tau_I/A} > 2$. Then $\zeta > 1$ and we have an overdamped response. In Figure 11.7b we can see the dynamic response of the liquid level to a step change in the inlet flow rate, with and without control.

Example 11.4 demonstrates very clearly how the simple first-order dynamic behavior of a tank can change to that of a second-order when a proportional-integral controller is added to the process. Also, it indicates that the control parameters K_c and τ_I can have a very profound effect on the dynamic behavior of the system, which can range from an underdamped to an overdamped response.

THINGS TO THINK ABOUT

1. What is a second-order system? Write the differential equation describing its behavior in the time domain and give its transfer function.

2. Explain the physical significance of the two parameters τ and ζ of a second-order system. Consult Refs. 11 (Section 10.II) and 12 (Chapter 8).

3. Identify the three classes of second-order systems and give one representative example for each class. What is the origin of the most second-order systems in chemical processes?

4. Discuss the overdamped, critically damped, and underdamped responses of a second-order system. Identify their distinguishing characteristics.

5. Describe the characteristics of an underdamped response.

6. Develop the expressions for the overshoot and the decay ratio [eqs. (11.11) and (11.12)].

7. How do you understand the interaction or noninteraction of several capacities in multicapacity processes? Give the general set of two differential equations describing (a) two noninteracting capacities, and (b) two interacting capacities.

8. Explain why two interacting capacities have more sluggish response than two equivalent but noninteracting capacities.

9. Show that as the number of noninteracting first-order systems in series increases, the response of the system becomes more sluggish.

10. Develop the equations giving the response of a second-order system to a unit impulse input for $\zeta > 1$, $\zeta = 1$, and $\zeta < 1$.

11. Prove eq. (11.20) for two noninteracting capacities.

12. A drum boiler (Figure PII.10) has a capacity to store material and thermal energy. Are these capacities interacting or not?

13. What is the origin of the most common systems with inherent second-order dynamics? Describe an example. You can use Refs. 11 and 12.

14. In Example 11.4, if you use proportional control only, would you change the order of the tank's dynamic behavior?

APPENDIX 11A

Examples of Physical Systems with Inherent Second-Order Dynamics

Systems with inherent second-order dynamics can exhibit oscillatory (underdamped) behavior but are rather rare in chemical processes. In this appendix we present three simple units which can be encountered in chemical plants and which possess second-order dynamics.

Simple manometers and externally mounted level indicators

Consider the simple U-tube manometer shown in Figure 11A.1a. When the pressures at the top of the two legs are equal, the two liquid

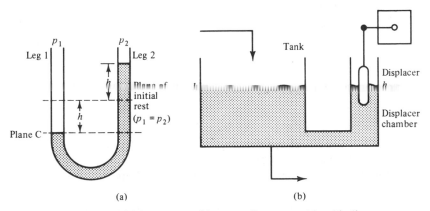

(a)   (b)

Figure 11A.1 (a) Manometer; (b) externally mounted level indicator.

levels are at rest at the same horizontal plane. Let us assume that suddenly a pressure difference $\Delta p = p_1 - p_2$ is imposed on the two legs of the manometer. We like to know the dynamic response of the levels in the two legs.

Let us apply Newton's law given by eq. (11.31) on the plane C of the manometer. We take

$$\begin{pmatrix} \text{force due to pressure} \\ p_1 \text{ on leg 1} \end{pmatrix} - \begin{pmatrix} \text{force due to pressure} \\ p_2 \text{ on leg 2} \end{pmatrix}$$

$$- \begin{pmatrix} \text{force due to liquid} \\ \text{level difference} \\ \text{in the two legs} \end{pmatrix} - \begin{pmatrix} \text{force due to} \\ \text{fluid friction} \end{pmatrix}$$

$$= \begin{pmatrix} \text{mass of liquid} \\ \text{in the tube} \end{pmatrix} \times (\text{acceleration})$$

or

$$p_1 A_1 - p_2 A_2 - \rho \frac{g}{g_c} A_2 (2h) - \begin{pmatrix} \text{force due to} \\ \text{fluid friction} \end{pmatrix} = \frac{m}{g_c} \frac{dv}{dt} \qquad (11A.1)$$

where p_1, p_2 = pressures at the top of legs 1 and 2, respectively

$\quad\quad A_1, A_2$ = cross-sectional areas of legs 1 and 2, respectively; typically $A_1 = A_2 = A$

$\quad\quad \rho$ = density of liquid in manometer

$\quad\quad g$ = acceleration gravity

$\quad\quad g_c$ = conversion constant

$\quad\quad m$ = mass of liquid in the manometer = $\rho A L$

$\quad\quad v$ = average velocity of the liquid in the tube

$\quad\quad h$ = deviation of liquid level from the initial plane of rest

$\quad\quad L$ = length of liquid in the manometer tubes

Poiseuille's equation for laminar flow in a pipe can be used to relate the force due to fluid friction with the flow velocity. Thus we have (Poiseuille's equation)

$$\text{volumetric flow rate} = A \frac{dh}{dt} = \frac{\pi R^4}{8\mu} \frac{\Delta P}{L} \qquad (11A.2)$$

where R = radius of the pipe through which liquid flows

$\quad\quad \mu$ = viscosity of the flowing liquid

$\quad\quad L$ = length of the pipe

$\quad\quad \Delta P$ = pressure drop due to fluid friction along the tube of length L

Therefore, applying Poiseuille's equation to the flow of liquid in the manometer we take

$$\left(\begin{array}{c}\text{force due to}\\\text{fluid friction}\end{array}\right) = \frac{\Delta p\,\pi R^2}{g_c} = A\,\frac{8\mu L}{R^2 g_c}\frac{dh}{dt} \tag{11A.3}$$

where $\Delta p = p_1 - p_2$. Recall also that the fluid velocity and acceleration are given by

$$v = \frac{dh}{dt} \quad \text{and} \quad \frac{dv}{dt} = \frac{d^2 h}{dt^2} \tag{11A.4}$$

Put eq. (11A.3) and (11A.4) in equation (11A.1) and take

$$\Delta p\,A - \frac{2\rho g A}{g_c}h - \frac{8\mu L A}{R^2 g_c}\frac{dh}{dt} = \frac{\rho A L}{g_c}\frac{d^2 h}{dt^2}$$

Finally, after dividing both sides by $2\rho g A/g_c$, we take

$$\left(\frac{L}{2g}\right)\frac{d^2 h}{dt^2} + \frac{4\mu L}{\rho g R^2}\frac{dh}{dt} + h = \frac{g_c}{2\rho g}\Delta p \tag{11A.5}$$

Define $\tau^2 = L/2g$, $2\zeta\tau = 4\mu L/\rho g R^2$, and $K_p = g_c/2\rho g$ and take

$$\tau^2 \frac{d^2 h}{dt^2} + 2\zeta\tau\,\frac{dh}{dt} + h = K_p\,\Delta p \tag{11A.6}$$

Therefore, the transfer function between h and Δp is

$$\frac{\overline{h}(s)}{\overline{\Delta p}(s)} = \frac{K_p}{\tau^2 s^2 + 2\zeta\tau s + 1} \tag{11A.7}$$

Both eq. (11A.6) and (11A.7) indicate the inherent second-order dynamics of the manometer.

For the measurement of liquid levels quite often we use the *externally mounted displacement-type transmitter*, which is shown in Figure 11A.1b. We notice that the system of the tank-displacer chamber has many similarities with the manometer. The cross-sectional areas of the two legs are unequal and the Δp (external) pressure difference is caused by a change in the liquid level of the main tank. Therefore, we expect that the response of the level in the displacer chamber, h_m, will follow second-order dynamics with respect to a change in the liquid level of the tank, h:

$$\frac{\overline{h}_m(s)}{\overline{h}(s)} = \frac{K_{p_m}}{\tau_m^2 s^2 + 2\zeta_m\tau_m s + 1} \tag{11A.8}$$

Variable capacitance differential pressure transducer

The variable capacitance differential pressure transducer is a very popular device which is used to sense and transmit pressure differences. Figure 11A.2 shows a schematic of such a device. A pressure signal is

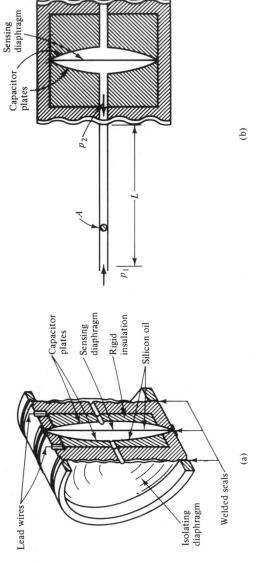

Figure 11A.2 Variable capacitance differential pressure transducer. (a) Differential pressure sensing element. (b) Pressure signal transmission system.

transferred through an isolating diaphragm and fill liquid in a sealed capillary system with a differential-pressure sensing element (Figure 11A.2a) attached at the other end of the capillary (Figure 11A.2b). Here, the pressure is transmitted through a second isolating diaphragm and fill liquid (silicone oil), to a sensing diaphragm. A reference pressure will balance the sensing diaphragm on the other side of this diaphragm. The position of the sensing diaphragm is detected by capacitor plates on both sides of the diaphragm. A change in pressure p_1 of a processing unit (e.g., a change in the pressure of a vessel, or a change in the liquid level in a tank, etc.) will make the pressure p_2 change at the end of the capillary tube.

A force balance around the capillary will yield

$$\begin{pmatrix} \text{force due to the} \\ \text{pressure } p_1 \text{ of the process} \\ \text{exercised at the end 1} \\ \text{of the capillary} \end{pmatrix} - \begin{pmatrix} \text{force due to the} \\ \text{pressure } p_2 \text{ exercised} \\ \text{at the end 2 of the} \\ \text{capillary} \end{pmatrix}$$

$$= (\text{mass}) \times (\text{acceleration})$$

or

$$p_1 A - p_2 A = \left(\frac{AL\rho}{g_c}\right)\frac{d^2x}{dt^2} \tag{11A.9}$$

where A = cross-sectional area of the capillary
L = length of the capillary tube
ρ = density of the liquid in the capillary tube
x = fluid displacement in the capillary tube
= displacement of diaphragm

The force p_2A at the end of the capillary is balanced by two forces:

$$p_2 A = \begin{pmatrix} \text{resistance exerted by} \\ \text{the diaphragm} \\ \text{which acts like a spring} \end{pmatrix} + \begin{pmatrix} \text{viscous friction force} \\ \text{exercised by the fluid} \end{pmatrix} \tag{11A.10}$$

$$= Kx + C\frac{dx}{dt}$$

where K = Hooke's constant for the diaphragm
C = damping coefficient of the viscous liquid in front of the diaphragm

Substitute p_2A in eq. (11A.9) by its equal given by equation (11A.10) and take

$$\left(\frac{AL\rho}{Kg_c}\right)\frac{d^2x}{dt^2} + \frac{C}{K}\frac{dx}{dt} + x = \frac{A}{K}p_1 \tag{11A.11}$$

Equation (11A.11) clearly indicates that the response of the device (i.e., the diaphragm displacement, x) follows second-order dynamics to any changes in process pressure p_1. If we define $\tau^2 = AL\rho/Kg_c$, $2\zeta\tau = C/K$, and $K_p = A/K$, we take the transfer function

$$\frac{\overline{x}(s)}{\overline{p}_1(s)} = \frac{K_p}{\tau^2 s^2 + 2\zeta\tau s + 1}$$

Pneumatic valve

The pneumatic valve is the most commonly used final control element. It is a system that exhibits inherent second-order dynamics.

Consider a typical pneumatic valve like that of Figure 11A.3. The position of the stem (or, equivalently, of the plug at the end of the stem) will determine the size of the opening for flow and consequently the size of the flow (flow rate). The position of the stem is determined by the balance of all forces acting on it. These forces are:

pA = force exerted by the compressed air at the top of the diaphragm; pressure p is the signal that opens or closes the valve and A is the area of the diaphragm; this force acts downward.

Kx = force exerted by the spring attached to the stem and the diaphragm; K is the Hooke's constant for the spring and x is the displacement; it acts upward.

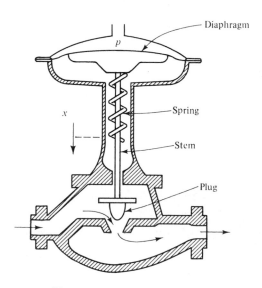

Figure 11A.3 Pneumatic valve.

$C \dfrac{dx}{dt}$ = frictional force exerted upward and resulting from the close contact of the stem with valve packing; C is the friction coefficient between stem and packing.

Apply Newton's law and take

$$pA - Kx - C\frac{dx}{dt} = \left(\frac{M}{g_c}\right)\frac{d^2x}{dt^2}$$

or

$$\left(\frac{M}{Kg_c}\right)\frac{d^2x}{dt^2} + \frac{C}{K}\frac{dx}{dt} + x = \frac{A}{K}p$$

Let $\tau^2 = M/Kg_c$, $2\zeta\tau = C/K$, and $K_p = A/K$ and take

$$\tau^2 \frac{d^2x}{dt^2} + 2\zeta\tau\frac{dx}{dt} + x = K_p p$$

The last equation indicates that the stem position x follows inherent second-order dynamics. The transfer function is

$$\frac{\overline{x}(s)}{\overline{p}(s)} = \frac{A/K}{(M/Kg_c)s^2 + \dfrac{C}{K}s + 1} \tag{11A.12}$$

Usually, $M \ll Kg_c$ and as a result, the dynamics of a pneumatic valve can be approximated by that of first-order system.

Dynamic Behavior **12**
of Higher-Order Systems

Systems with higher than second-order dynamics are not uncommon in chemical processes. Three classes of higher-order systems are most often encountered:

1. N first-order processes in series (multicapacity processes)
2. Processes with dead time
3. Processes with inverse response

In this chapter we analyze their typical dynamic characteristics.

12.1 *N* Capacities in Series

In Section 11.3 we found that two capacities in series, interacting or noninteracting, give rise to a second-order system. If we extend the same procedure to N capacities (first-order systems) in series, we find that the overall response is of Nth order; that is, the denominator of the overall transfer function is an Nth-order polynomial,

$$a_N s^N + a_{N-1} s^{N-1} + \cdots + a_1 s + a_0$$

If the N capacities are noninteracting, the overall transfer function is given by eq. (11.21):

$$G_o(s) = G_1(s)G_2(s)\cdots G_N(s) = \frac{K_1 K_2 \cdots K_N}{(\tau_1 s + 1)(\tau_2 s + 1)\cdots(\tau_N s + 1)} \quad (11.21)$$

where $G_1(s)$, $G_2(s)$, ..., $G_N(s)$ are the transfer functions of the N capacities. For interacting capacities the overall transfer function is more complex.

In Section 11.3 we studied the basic dynamic characteristics of two capacities in series when the input is changed by a step. Similar analysis is possible for N capacities in series. The following general conclusions can easily be drawn from the discussion in Section 11.3.

1. N noninteracting capacities in series:
   (a) The response has the characteristics of an overdamped system; that is, it is S-shaped and sluggish.
   (b) Increasing the number of capacities in series increases the sluggishness of the response.
2. N interacting capacities in series:
   Interaction increases the sluggishness of the overall response.

It is clear, therefore, that a process with N capacities in series will necessitate a controller that will not only keep the final output at a desired value but will also try to improve the speed of the system's response.

Let us now examine some typical examples of processes with N capacities in series.

Example 12.1: Jacketed Coolers as Multicapacity Processes

Consider the batch cooler shown in Figure 12.1a. The content of the tank is a mixture of components A and B and is being cooled by constant flow of cold water circulating through the jacket. We can identify the following three capacities in series:

Heat capacity of the mixture in the tank
Heat capacity of the tank's wall
Heat capacity of the coolant in the jacket.

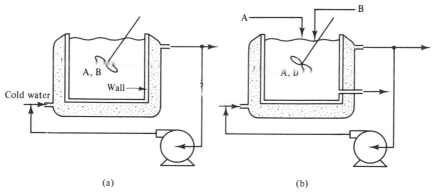

(a)   (b)

Figure 12.1 Jacketed coolers: (a) batch; (b) continuous flow

It is easy to show that the three capacities interact.

For the jacketed continuous flow cooler of Figure 12.1b we have more interacting capacities:

Total material capacity of the tank

Tank's capacity for component A

Heat capacity of the tank's content

Heat capacity of the tank's wall

Heat capacity of the cold water in the jacket

Again, all five capacities are interacting.

According to what we have said above, we expect that the response of the coolers to input changes will be overdamped and rather sluggish.

Example 12.2: Staged Processes as Multicapacity Systems

Distillation and gas absorption columns are very often encountered in chemical processes for the separation of a mixture into its components. Both systems have a number of trays. Each tray has material and heat capacities. Therefore, each column with N trays can be considered as a system with $2N$ capacities in series. From the physics of distillation and absorption it is easy to see that the $2N$ capacities interact.

Therefore, a step change in the liquid flow rate of the solvent at the top of the absorption column produces a very delayed, sluggish response for the content of solvent in the valuable component A (see Figure PII.13). This is because the input change has to travel through a large number of interacting capacities in series.

Similarly, a step change in the reflux ratio of a distillation column (see Figure 4.10) will quickly have an effect on the composition of the overhead product while the composition of the bottoms stream will respond very sluggishly (delayed and slow).

Finally, a step change in the steam flow rate of the reboiler will have almost an immediate effect on the composition of the bottoms stream. On the contrary, the effect on the composition of the overhead product will be delayed and slow.

12.2   Dynamic Systems with Dead Time

For all the systems we examined in Chapters 10 and 11 and Section 12.1, we have assumed that there is no dead time between an input and the output; that is, whenever a change took place in the input variable, its effect was instantaneously observed in the behavior of the output variable. This is not true and contrary to our physical experience. Virtually all physical processes will involve some time delay between the input and the output.

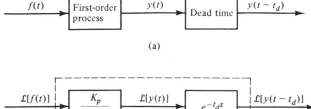

(a)

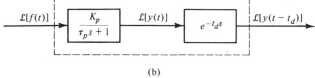

(b)

Figure 12.2 (a) Process with dead time; (b) its block diagram.

Consider a first-order system with a dead time t_d between the input $f(t)$ and the output $y(t)$. We can represent such system by a series of two systems as shown in Figure 12.2a (i.e., a first-order system in series with a dead time). For the first-order system we have the following transfer function:

$$\frac{\mathcal{L}[y(t)]}{\mathcal{L}[f(t)]} = \frac{\overline{y}(s)}{\overline{f}(s)} = \frac{K_p}{\tau_p s + 1}$$

while for the dead time we have [see Section 7.2, eq. (7.10)]

$$\frac{\mathcal{L}[y(t - t_d)]}{\mathcal{L}[y(t)]} = e^{-t_d s}$$

Therefore, the transfer function between the input $f(t)$ and the delayed output $y(t - t_d)$ is given by (see also Figure 12.2b)

$$\frac{\mathcal{L}[y(t - t_d)]}{\mathcal{L}[f(t)]} = \frac{K_p e^{-t_d s}}{\tau_p s + 1} \tag{12.1}$$

Similarly, the transfer function for a second-order system with delay is given by

$$\frac{\mathcal{L}[y(t - t_d)]}{\mathcal{L}[f(t)]} = \frac{K_p e^{-t_d s}}{\tau^2 s^2 + 2\zeta\tau s + 1} \tag{12.2}$$

Remarks

1. Figure 12.3 shows the response of first- and second-order systems with dead time to a step change in the input.
2. Quite often the exponential term is approximated by the first- or second-order Padé approximations

$$e^{-t_d s} \approx \frac{1 - \dfrac{t_d}{2} s}{1 + \dfrac{t_d}{2} s} \quad \text{first-order approximation} \tag{12.3a}$$

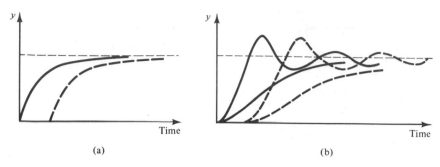

Figure 12.3 Response of time-delayed systems to step input change: (a) first order; (b) second-order.

$$e^{-t_d s} \approx \frac{(t_d)^2 s^2 - 6 t_d s + 12}{(t_d)^2 s^2 + 6 t_d s + 12} \quad \text{second-order approximation} \quad (12.3b)$$

3. Processes with dead time are difficult to control because the output does not contain information about current events.

12.3 Dynamic Systems with Inverse Response

The dynamic behavior of certain processes deviates drastically from what we have seen so far. Figures 12.4b and 12.5b show the response of such systems to a step change in the input. We notice that initially the response is in the opposite direction to where it eventually ends up. Such behavior is called *inverse response* or *nonminimum phase response* and it is exhibited by a small number of processing units.

> ### *Example 12.3: Inverse Response of the Liquid Level in a Boiler System*
>
>   Consider the simple drum boiler shown in Figure PII.10. If the flow rate of the cold feedwater *is increased* by a step, the total volume of the boiling water and consequently the liquid level *will be decreased* for a short period and then it will start increasing, as shown by the response in Figure 12.4b. Such behavior is the *net result of two opposing effects* and can be explained as follows:
>
> 1. The cold feedwater causes a temperature drop which decreases the volume of the entrained vapor bubbles. This leads to a decrease of the liquid level of the boiling water, following first-order behavior (curve 1 in Figure 12.4b), that is, $-K_1/(\tau_1 s + 1)$.
> 2. With constant heat supply, the steam production remains constant and consequently the liquid level of the boiling water will start

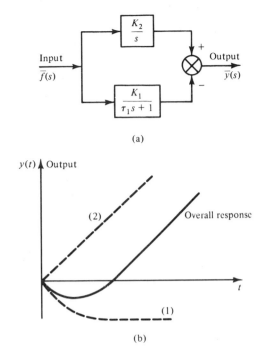

(a)

(b)

Figure 12.4 (a) Block diagram of liquid level in a boiler system; (b) its inverse response.

increasing in an integral form (pure capacity), leading to a pure capacitive response, K_2/s (curve 2 in Figure 12.4b).

3. The result of the two opposing effects is given by (see also Figure 12.4a)

$$\frac{K_2}{s} - \frac{K_1}{\tau_1 s + 1} = \frac{(K_2\tau_1 - K_1)s + K_2}{s(\tau_1 s + 1)} \tag{12.4}$$

and for

$$K_2\tau_1 < K_1$$

the second term $[-K_1/(\tau_1 s + 1)]$ dominates initially and we take the inverse response. If the condition above is not satisfied, we do not have inverse response.

Note.   When $K_2\tau_1 < K_1$, then from eq. (12.4) we notice that *the transfer function has a positive zero*, at the point $s = -K_2/(K_2\tau_1 - K_1) > 0$.

This example demonstrates that the inverse response is the result of two opposing effects. Table 12.1 shows several such opposing effects between first- or second-order systems. In all cases we notice that *when the system possesses an inverse response, its transfer function has a positive zero*. In general, the transfer function of a system with inverse response

TABLE 12.1
SYSTEMS WITH INVERSE RESPONSE

1. Pure capacitive minus first-order response (Figure 12.4):

$$G(s) = \frac{K_2}{s} - \frac{K_1}{\tau_1 s + 1} = \frac{(K_2 \tau_1 - K_1)s + K_2}{s(\tau_1 s + 1)}$$

for $K_2 \tau_1 < K_1$ zero $= -K_2/(K_2 \tau_1 - K_1) > 0$.

2. Difference between two first-order responses (Figure 12.5):

$$G(s) = \frac{K_1}{\tau_1 s + 1} - \frac{K_2}{\tau_2 s + 1} = \frac{(K_1 \tau_2 - K_2 \tau_1)s + (K_1 - K_2)}{(\tau_1 s + 1)(\tau_2 s + 1)}$$

for $\frac{\tau_1}{\tau_2} > \frac{K_1}{K_2} > 1$ zero $= -(K_1 - K_2)/(K_1 \tau_2 - K_2 \tau_1) > 0$

3. Difference between two first-order responses with dead time:

$$G(s) = \frac{K_1 e^{-t_1 s}}{\tau_1 s + 1} - \frac{K_2 e^{-t_2 s}}{\tau_2 s + 1}$$

for $K_1 > K_2$ and $t_1 > t_2 \geq 0$.

4. Second-order minus first-order response:

$$G(s) = \frac{K_1}{\tau^2 s^2 + 2\zeta \tau s + 1} - \frac{K_2}{\tau_2 s + 1}$$

for $K_1 > K_2$.

5. Difference between two second-order responses:

$$G(s) = \frac{K_1}{\tau_1^2 s^2 + 2\zeta_1 \tau_1 s + 1} - \frac{K_2}{\tau_2^2 s^2 + 2\zeta_2 \tau_2 s + 1}$$

for $\tau_1^2/\tau_2^2 > K_1/K_2 > 1$.

6. Difference between two second-order responses with dead time:

$$G(s) = \frac{K_1 e^{-t_1 s}}{\tau_1^2 s^2 + 2\zeta_1 \tau_1 s + 1} - \frac{K_2 e^{-t_2 s}}{\tau_2^2 s^2 + 2\zeta_2 \tau_2 s + 1}$$

for $K_1 > K_2$ and $t_1 > t_2 \geq 0$.

is given by

$$G(s) = \frac{b_m s^m + b_{m-1} s^{m-1} + \cdots + b_1 s + b_0}{a_n s^n + a_{n-1} s^{n-1} + \cdots + a_1 s + a_0}$$

where one of the roots of the numerator (i.e., one of the zeros of the transfer function) has positive real part.

Systems with inverse response are particularly difficult to control and require special attention.

Example 12.4: Inverse Response from Two Opposing First-Order Systems

Figure 12.5a shows another possibility of inverse response. Two opposing effects result from two different first-order processes, yielding an overall response equal to

$$\bar{y}(s) = \left(\frac{K_1}{\tau_1 s + 1} - \frac{K_2}{\tau_2 s + 1}\right)\bar{f}(s)$$

or

$$\bar{y}(s) = \frac{(K_1\tau_2 - K_2\tau_1)s + (K_1 - K_2)}{(\tau_1 s + 1)(\tau_2 s + 1)}\bar{f}(s)$$

We have inverse response when:

Initially (at $t = 0+$) Process 2, which reacts faster than Process 1 (i.e. $K_2/\tau_2 > K_1/\tau_1$, see Section 10.4), dominates the response of the overall system, but

Ultimately Process 1 reaches a higher steady-state value than Process 2 (i.e., $K_1 > K_2$), and forces the response of the overall system in the opposite direction.

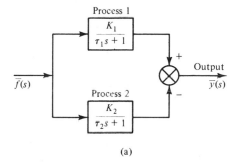

(a)

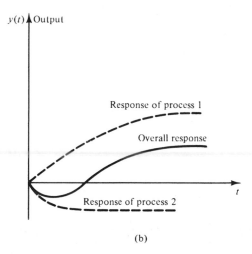

(b)

Figure 12.5 (a) Block diagram of two opposing first-order systems; (b) the resulting inverse response.

Figure 12.5b shows the inverse response of the overall system.

Note.   When $(\tau_1/\tau_2) > (K_1/K_2) > 1$ the process exhibits inverse response, and we find that *the system's transfer function has a positive zero*:

$$z = -\frac{K_1 - K_2}{K_1\tau_2 - K_2\tau_1} > 0$$

THINGS TO THINK ABOUT

1. How would you define a higher-order system?

2. Using the definition above, why is a system with dead time a higher-order system? (*Hint*: Consider the Taylor series expansion of an exponential term. See also item 4.)

3. Show that as the number of noninteracting or interacting capacities in series increases, the response of the system becomes more sluggish.

4. Consider N identical noninteracting capacities in series, with gain K_p and time constant τ_p for each capacity. Show that as $N \to \infty$, the response of the system approaches the response of a system with dead time τ_p and overall gain K_p.

5. In an ideal binary distillation column the dynamics of each tray can be described by first-order systems. Are these capacities interacting or not? What general types of responses would you expect for the overhead and bottoms compositions to a step change in the feed composition?

6. How many capacities can you identify in the mixing process of Example 4.11? Are they interacting or not?

7. What is the most common transfer function encountered in chemical processes? Why?

8. What is an inverse response, and what causes it?

9. Show qualitatively that the response of the bottoms composition of a distillation column to a step change in the vapor boilup, V, can exhibit inverse behavior. Consult Refs. 1 and 16.

10. Why do you think a system with inverse response is difficult to control?

REFERENCES FOR PART III

Chapter 6.   Two very good references on computer simulation (digital or analog) are the following books by Luyben and Franks:

1. *Process Modeling, Simulation, and Control for Chemical Engineers*, by W. L. Luyben, McGraw-Hill Book Company, New York (1973).

2. *Modeling and Simulation in Chemical Engineering*, by R. G. E. Franks, John Wiley & Sons, Inc., New York (1972).

Both books provide a series of examples drawn from the area of chemical engineering and demonstrate how digital computer simulation of chemical processes can enhance our ability to understand the dynamics and develop better controllers for such systems. Computer programs in FORTRAN for typical systems are also included. For more details on the numerical techniques for the solution of algebraic or differential equations, the reader is encouraged to consult the following two classic books:

3. *Digital Computation for Chemical Engineers*, by L. Lapidus, McGraw-Hill Book Company, New York (1962).
4. *Applied Numerical Methods*, by B. Carnahan, H. R. Luther, and J. D. Wilkes, John Wiley & Sons, Inc., New York (1969).

The notion and the characteristics of the Taylor series expansion as well as the linear approximation of nonlinear systems can be found in all the standard texts on calculus.

In Section 6.1 of the following book, Douglas discusses a procedure that allows us to ascertain the range of values around the point of linearization for which the linearized model is acceptable:

5. *Process Dynamics and Control*, Vol. 1, by J. M. Douglas, Prentice-Hall, Inc., Englewood Cliffs, N.J. (1972).

Example 6.1 was motivated by the physical system analyzed in Section 2.3 of the following work, where the reader can find more information:

6. *Introduction to Chemical Engineering Analysis*, by T. W. F. Russell and M. M. Denn, John Wiley & Sons, Inc., New York (1972).

Chapters 7 and 8. The Laplace transformation has been the object of a large body of mathematical research. For more details on the theoretical aspects of Laplace transforms, the reader will find useful the following book:

7. *Operational Mathematics*, 2nd ed., by R. V. Churchill, McGraw-Hill Book Company, New York (1950).

For the use of Laplace transforms to the solution of differential equations (ordinary, partial, or sets of), the book by Jenson and Jeffreys can be very valuable:

8. *Mathematical Methods in Chemical Engineering*, by V. G. Jenson and G. V. Jeffreys, Academic Press Ltd., London (1963).

In the following two references the reader can find tables with the Laplace transformation of a large number of functions:

9. *Feedback and Control Systems*, by J. J. DiStefano III, A. R. Stubberud, and J. J. Williams, Schaums Outline Series, McGraw-Hill Book Company, New York (1967).

10. *Handbook of Mathematical Functions*, by M. Abramowitz and I. A. Stegun (eds.), Dover Publications, New York (1972).

Chapter 10. The following book, by Weber, is an excellent reference for the dynamics of first-order systems. The interested reader will find (Chapters 8 and 9) an extensive coverage of first-order systems based on mass, energy, and momentum balances, with a large number of examples. It also provides a valuable physical interpretation of the notion of capacity for various processing systems.

11. *An Introduction to Process Dynamics and Control*, by T. W. Weber, John Wiley & Sons, Inc., New York (1973).

In the books by Douglas [Ref. 5] and Coughanowr and Koppell [Ref. 12] the reader can study the response of first-order systems to impulse or sinusoidal inputs. The response of a capacity process to a sinusoidal input is also given in Chapter 17 of the present text.

12. *Process Systems Analysis and Control*, by D. R. Coughanowr and L. B. Koppell, McGraw-Hill Book Company, New York (1965).

Chapter 11. The book by Weber [Ref. 11] is also an excellent reference for the development and physical interpretation of second-order systems (Chapter 10). It contains examples of inherently second-order systems which the reader will find quite useful. In the books by Coughanowr and Koppell [Ref. 12] and Douglas [Ref. 5] the reader can find discussions of the response of second-order systems to impulse and sinusoidal inputs. For more information on externally mounted level measuring systems, manometers, and their dynamic second-order characteristics, the reader can consult Refs. 14 (Chapter 18), 11 (Chapter 10), or the book by Shinskey [Ref. 13, Chapter 3].

13. *Process Control Systems*, 2nd ed., by F. G. Shinskey, McGraw-Hill Book Company, New York (1979).

14. *Techniques of Process Control*, by P. S. Buckley, John Wiley & Sons, Inc., New York (1964).

The following two references can be consulted for further details on the variable capacitance differential pressure transducer and the pneumatic control valve:

15. "Process Dynamics: Part 2. Process Control Loops," by J. C. Guy, *Chem. Eng.*, p. 111 (Aug. 24, 1981).

16. *Measurements and Control Applications for Practicing Engineers*, by J. O. Hougen, Cahners Books, Boston (1972).

Chapter 12. Luyben [Ref. 1, Section 11.5] has a good discussion on the inverse response of the bottoms composition of a distillation column to a change in the vapor boilup. Iinoya and Altpeter [Ref. 17] discuss the characteristics of systems that exhibit inverse response and give a table of the most common physical situations (transfer functions) that give rise to inverse response.

17. "Inverse Response in Process Control," by K. Iinoya and R. J. Altpeter, *Ind. Eng. Chem.*, *54*(7), 39 (1962).

In the book by Shinskey [Ref. 13] the reader can find further discussion on the inverse response of a drum boiler.

PROBLEMS FOR PART III

Chapter 6

III.1 Equations (4.8a), (4.9a), and (4.10b) in Example 4.10 describe the dynamic behavior of a continuous stirred tank reactor with a simple, exothermic and irreversible reaction, A → B. Develop a numerical procedure that solves these equations and can be implemented on a digital computer. Also, describe a numerical procedure for solving the algebraic steady-state equations of the reactor above. (*Note*: For this problem you need to be familiar with numerical techniques for the solution of differential and algebraic equations on a computer.)

III.2 Do the same as in Problem III.1 for the equations describing the dynamic and steady-state behavior of the binary distillation column modeled in Example 4.13.

III.3 Linearize the following single-input, single-output nonlinear dynamic models.

(a) $\dfrac{dy}{dt} = \alpha y + \beta y^2 + \gamma \ln y \qquad \alpha, \beta, \gamma : \text{constants}$

(b) $\dfrac{dy}{dt} = \dfrac{1-y}{y} m + ym^2 + \sin \alpha m \qquad \alpha : \text{constant}$

(c) $\dfrac{dy}{dt} = ym - 2y + m$

III.4 Linearize the following multiple-input, multiple-output nonlinear dynamic models.

(a) $\dfrac{dy_1}{dt} = 2y_1^2 + 3y_1 y_2 + m_1 - m_2^3$

$\dfrac{dy_2}{dt} = \sqrt{y_2} - 4\, m_1 m_2$

(b) $2\dfrac{dy_1}{dt} - 3\dfrac{dy_2}{dt} = y_1^2 y_2 - 0.5 y_2 m_1 + m_2 y_1^3$

$\dfrac{dy_1}{dt} + \dfrac{dy_2}{dt} = \ln y_2 + y_1 \cos 2m_1 - \sqrt{m_2}$

III.5 Linearize the nonlinear models of the following processes and place them in deviation variables form.
(a) Stirred tank heater (Example 4.4).
(b) Mixing process (Example 4.11).
(c) Tubular heat exchanger (Example 4.12; be careful here).

III.6 Linearize the nonlinear models of the following processes and place them in deviation variables form:
(a) Batch reactor where the following reactions with first-order kinetics take place:

$$A \longrightarrow B \longrightarrow C$$

Can we use steady-state operation as the point of linearization for this system?
(b) Batch mixing system of Problem II.6 (Figure PII.6). Identify the point of linearization for this system.
(c) Continuous stirred tank reactor system of Problem II.8 (Figure PII.7).
(d) Drum boiler of Problem II.11 (Figure PII.10).
(e) Flash drum of Problem II.14 (Figure 4.6).

III.7 Develop the linearized models for the following processes.
(a) Packed absorption column with N ideal plates. The mass balance of component i at the nth ideal plate is given by

$$H \frac{dx_n}{dt} = Lx_{n+1} - (L + mV)x_n + mVx_{n-1}$$

where H = holdup of liquid solvent at the nth plate
$\quad\quad L$ = flow rate of the liquid solvent
$\quad\quad V$ = flow rate of the gaseous stream
$\quad\quad x_n$ = mole fraction of component i at the nth plate
$\quad\quad m$ = slope of the equilibrium curve for component i

(b) Nonisothermal tubular reactor with a simple reaction $A \rightarrow B$ (see also Problem II.10 and Figure PII.9). The mass and heat balances are given by

$$\frac{\partial x}{\partial t} + v \frac{\partial x}{\partial z} = -kx$$

$$c_p \rho A \frac{\partial T}{\partial t} + c_p \rho v A \frac{\partial T}{\partial z} = hA_t(T_c - T) + (-\Delta H_r)kAx$$

where x and T are the composition and temperature of the reacting mixture within the reactor; ρ, c_p, A, v, h, A_t, T_c, and $(-\Delta H_r)$ are constant-value parameters; and k is the kinetic rate parameter given by

$$k = k_0 \exp\left(\frac{-E}{RT}\right)$$

with k_0, E, and R being constants.

Chapter 7

III.8 Find analytically the Laplace transforms of the following functions.
(a) $f(t) = t^n$
(b) $f(t) = e^{-at} \sin \omega t$
(c) $f(t) = \cosh \omega t$
(d) $f(t) = t^n e^{-at}$
(e) $f(t) = 5 + t + e^{-2t} - \cos 0.5t + t^2 e^{-t}$

III.9 Prove the initial value theorem.

III.10 Using the final value theorem find the value that each of the following functions reaches as $t \to \infty$
(a) $y(t) = 2e^{-0.1t}$
(b) $y(t) = 1 - 5e^{-0.5t} \sin (6t + 2)$
(c) $y(t) = 2 - e^{2t}[\cosh 2t + 4 \sinh 2t]$

Chapter 8

III.11 Solve the following linear differential equations using Laplace transforms.

(a) $\dfrac{d^2 y}{dt^2} + 3 \dfrac{dy}{dt} - y = 5t \qquad$ with $\left(\dfrac{dy}{dt}\right)_{t=0} = 0$ and $y(0) = 2$

(b) $2 \dfrac{d^3 y}{dt^3} - \dfrac{d^2 y}{dt^2} + 3y = 1 + \sin 2t \qquad$ with $y''(0) = y'(0) = y(0) = 0$

(c) $\dfrac{d^2 y}{dt^2} - 2 = 0 \qquad$ with $y'(0) = y(0) = 0$

III.12 Using Laplace transforms, solve the following sets of linear differential equations.

(a) $\dfrac{dy_1}{dt} - 2 \dfrac{dy_2}{dt} = 1 - y_1 + y_2 \qquad\qquad y_1(0) = 1$

$\quad\ 3 \dfrac{dy_1}{dt} + \dfrac{dy_2}{dt} = y_2 - e^{-t} \qquad\qquad\quad y_2(0) = 2$

(b) $\dfrac{dy_1}{dt} - y_1 + 2y_2 = 2 \sin t \qquad\qquad\quad y_1(0) = 0$

$\quad\ \dfrac{dy_2}{dt} + 2 \dfrac{dy_1}{dt} - 5y_1 - 3y_2 = 1 - e^{-0.1t} \qquad y_2(0) = 0$

III.13 Using Laplace transforms, find the solution of the following linear partial differential equations.

(a) $\dfrac{\partial y}{\partial t} + 2 \dfrac{\partial y}{\partial z} = 2 \qquad$ with $y(0, z) = 0$ and $y(t, 0) = 5$

(b) $\dfrac{\partial y}{\partial t} - y - \dfrac{\partial y}{\partial z} = 1 - e^{-0.5t} \qquad$ with $y(0, z) = 2$ and $y(t, 0) = 0$

III.14 Find the inverse Laplace transforms of the following expressions.

(a) $\bar{y}(s) = \dfrac{1}{\tau^2 s^2 + 2\zeta \tau s + 1} \qquad$ where $\zeta \lessgtr 1$

(b) $\bar{y}(s) = \dfrac{s - 1}{s(s + 1)(s + 2)}$

(c) $\bar{y}(s) = \dfrac{2s + 1}{s^2(s + 1)(s - 2)}$

(d) $\bar{y}(s) = \dfrac{s^2 + 4s + 3}{(s^2 + 1)(s^2 - 7s + 12)}$

(e) $\bar{y}(s) = \dfrac{(s - 1)(s + 2)}{(s + 1)^2(s + 2)(s^2 - 2s + 5)}$

(f) $\bar{y}(s) = \dfrac{s^3 + 2s^2 - s - 2}{s^3 + 6s^2 + 11s + 30}$

III.15 Using Laplace transforms, solve the following partial differential equation which describes the dynamic behavior of a tubular heat exchanger (see Example 4.12):

$$\rho c_p A \frac{\partial T}{\partial t} + \rho c_p v A \frac{\partial T}{\partial z} = \pi D U (T_{st} - T)$$

where ρ, c_p, A, v, U, and D are constant parameters. Assume that the system is at steady state when the steam temperature T_{st} experiences a unit step change.

III.16 Using Laplace transforms find the linearized dynamic behavior of:
(a) A stirred tank heater (Example 4.4)
(b) A mixing process (Example 4.11)
to unit step changes in their inputs. Assume that the systems are initially at steady state.

III.17 Use Laplace transforms and find the dynamic response of the following linearized systems to unit step input changes.
(a) Storage tank systems of Problem II.1 (Figure PII.1).
(b) Storage tank system of Problem II.2 (Figure PII.2).
(c) Continuous stirred tank reactor (Example 6.4).
Assume that each system is initially at steady state.

III.18 Using Laplace transforms find the dynamic behavior of an isothermal batch reactor where the following reactions take place:

$$A \xrightarrow{k_1} B \xrightarrow{k_2} C$$

Assume first-order kinetics for the two reactions. Also, plot the concentration of A, B, and C versus time.

Chapter 9

III.19 Derive the transfer function between effluent F_3 and inlet F_1 for:
(a) The two storage tank systems of Problem II.1 (Figure PII.1).
(b) The storage tank system of Problem II.2 (Figure PII.2).
Identify the poles and zeros of each system and sketch the dynamic response of F_3 to unit step changes in F_1.

III.20 Derive the transfer functions relating the outputs to the inputs of the mixing process described in Example 4.11. Draw the corresponding block diagram. Is the process stable? Determine its new steady-state composition when c_{A_1} changes by a unit step. (*Note*: Assume that the heat of solution is not a function of the concentration.)

III.21 Let $G(s)$ be the transfer function between an input m and an output y. Show that the new steady state resulting from a step change in the input m, is given by $A[G(s)]_{s=0}$, where A is the size of the step change in m.

III.22 Find the transfer function between the effluent temperature T_3 and the inlet temperature T_1 for the system of two stirred tank heaters described in Problem II.3 (Figure PII.3). Draw the corresponding block diagram. Sketch the response of T_3 to a unit impulse change in T_1. Is the process stable?

III.23 Do the same work as in Problem III.22 for the heaters' system of Problem II.4 (Figure PII.4).

III.24 Without computing analytically the transfer functions, develop the block diagrams for the following processes:
(a) The two-tank mixing process of Problem II.5 (Figure PII.5).
(b) The two-tank mixing process of Problem II.5, assuming that the heat of solution is independent of the concentrations.
(c) The drum boiler of Problem II.11 (Figure PII.10).
(d) The flash drum of Problem II.14 (Figure 4.6).

III.25 Without computing analytically the transfer functions between the various external inputs and outputs develop the block diagrams and the corresponding input–output models for the following systems.
(a) An ideal binary distillation column (see Example 4.13).
(b) A gas absorption column (see Problem II.23 and Figure PII.13).

III.26 Develop the block diagram for the plant of Problem II.20 (Figure PII.11) without computing explicitly the transfer functions between the various inputs and outputs.

III.27 In Example 9.2 we developed the transfer function matrix for a continuous stirred tank reactor. Determine:
(a) The location of the two poles of the reactor.
(b) The two conditions that parameters a_{11}, a_{12}, a_{21}, and a_{22} should satisfy in order to have stable response to external disturbances.

III.28 Sketch, qualitatively, the response of systems with the following transfer functions. Assume unit step input changes.

(a) $G(s) = \dfrac{s+1}{(s+2)(s+3)}$

(b) $G(s) = \dfrac{s+1}{(s+2)(s+3)} e^{-0.5s}$

(c) $G(s) = \dfrac{s+1}{s(s+2)(s+3)}$

(d) $G(s) = \dfrac{s^2 + 4s + 3}{(s^2 + 1)(s^2 - 7s + 12)}$

(e) $G(s) = \dfrac{(s - 1)(s + 3)}{(s + 1)^2(s + 2)(s^2 - 2s + 5)}$

Chapter 10

III.29 Find the dynamic response of a first-order lag system with time constant $\tau_p = 0.5$ and static gain $K_p = 1$ to (a) a unit impulse input change, (b) a unit pulse input change of duration 5, and (c) a sinusoidal input change, sin $0.5t$. Determine the behavior of the output after long time (as $t \to \infty$) for each of the input changes above.

III.30 Repeat Problem III.29 but for a pure capacity process with gain $K'_p = 1$.

III.31 Consider a process whose output exhibits first-order behavior to input changes:

$$\tau_p \frac{dy}{dt} + y = K_p m(t)$$

The values of the time constant τ_p and static gain K_p are not well known. Develop an experiment whereby changing $m(t)$ in a certain way and recording the values of $y(t)$ with time you can compute the values of the unknown parameters τ_p and K_p.

III.32 Consider a first-order system with $\tau_p = 0.5$ min and $K_p = 1$. Initially, the system is at steady state. Then the input changes linearly with time:

$$m(t) = t$$

(a) Develop an expression that shows how the output changes with time in response to the input above.

(b) What is the minimum and what is the maximum difference between the output $y(t)$ and input $m(t)$? When do these extreme points occur?

(c) Plot the input $m(t)$ and output $y(t)$ in the same graph as functions of time.

III.33 Consider the three storage tanks in Figure PIII.1. For each of these systems, (a) develop the transfer functions between the liquid levels and the inlet streams, (b) determine the time constants and process static gains, and (c) determine which of the three systems have constant and which variable time constants and process gains. Assume that the flow rates of all free effluent streams are linear functions of the corresponding liquid levels: [Flowrates in figure are steady state values]

$$F = 2\left(\frac{ft^3}{ft \cdot min}\right) h(ft)$$

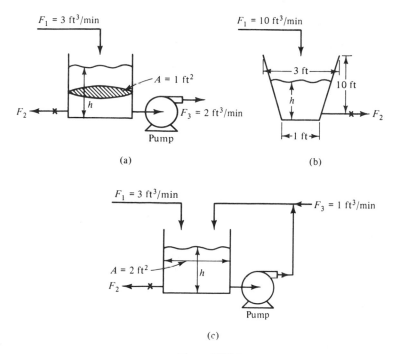

Figure PIII.1

III.34 Repeat Problem III.33 assuming nonlinear dependence of an effluent flow rate on the liquid level:

$$F = 0.5 \sqrt{h}$$

III.35 Consider the storage tank of Figure PIII.1a. Suppose that we want to control the liquid level in the tank at the height of 5 ft, by manipulating the effluent flow rate F_2, according to the following *proportional control* law:

$$F_2 = 10(5 - h) + 1$$

(a) Develop the transfer function between h and F_1.
(b) Determine the time constant and static gain of the tank, under control.
(c) Compute the dynamic response of the liquid level to a step change in F_1 by 1 ft³/min [i.e., find how $h(t)$ changes with time].
(d) Compute the new steady-state value for the liquid level.

III.36 Repeat Problem III.35 for the storage tanks shown in Figure PIII.1b and c.

III.37 Consider the gas storage tank shown in Figure PIII.2a.

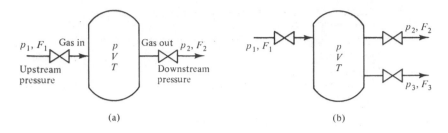

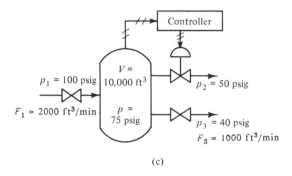

Figure PIII.2

(a) Write the material balance for this tank if the gas flow rates are determined by the pressure difference between an upstream and a downstream point:

$$F = \alpha\sqrt{\Delta p} \qquad \alpha = \text{constant}$$

(b) Derive the transfer functions between the tank pressure p and the inlet and outlet pressures, p_1 and p_2, respectively. Draw the corresponding block diagram.
(c) How do you define the mass storage capacity for this tank?
(d) Determine the time constant and static gain for the system.
(e) Find the dynamic response of tank's pressure to a unit step change in the inlet or outlet pressure. Flowrates are per minute.

III.38 Repeat the work of Problem III.37 but for the gas storage tank of Figure PIII.2b. Notice the presence of an additional effluent stream.

III.39 Consider the gas storage tank of Figure PIII.2c. Suppose that we want to control the tank pressure at the desired value of 75 psig, by manipulating the downstream pressure p_2. A proportional controller is used for this purpose:

$$p_2 = 10(75 - p) + 50$$

Assume that the gases are ideal.
(a) Derive the transfer functions between the tank pressure p and the possible disturbances p_1 (inlet) and p_3 (outlet).

(b) Determine the time constant when the temperature of the gas in the system is $T = 100°F$ or $T = 200°F$.

(c) Find the dynamic response of the pressure in the tank to a 10-psig step change in the inlet stream, for both temperatures in the tank.

(d) Compare the dynamic behavior for the two cases in part (c) and indicate how the gas temperature affects the response of the tank's pressure.

III.40 Figure PIII.3a shows a cross-sectional view of the bulb of a mercury thermometer. Let T be the temperature of the surrounding fluid and T_m the thermometer reading. Assume that the film coefficient of heat transfer, h_f (Btu/hr, ft^2, °F), determines the amount of heat transferred from the surrounding fluid to the mercury of the bulb (i.e., the resistance of the glass wall to heat transfer is negligible). Also, assume that the heat capacity of the mercury is much larger than that of the glass wall. Finally, the wall is assumed not to expand or contract.

(a) Show that the thermometer reading follows a first-order behavior to any changes in the surrounding temperature.

(b) Derive the transfer function between T_m and T.

(c) Develop the equations that define the time constant and static gain and compare them to those derived in Example 10.2.

(d) Sketch the qualitative response of T_m to a step change in T for various values of the film coefficient h_f. Using physical arguments, explain the observed behavior.

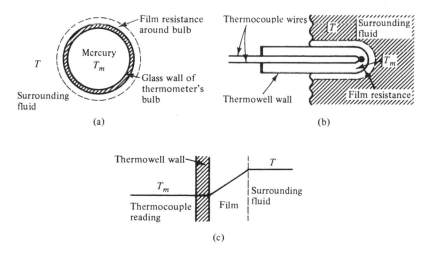

Figure PIII.3

III.41 Thermocouples are commonly used to measure the temperature of a process fluid. Figure PIII.3b shows a thermocouple within its thermowell. Assume that all resistance to heat transfer resides with the external film around the thermowell wall (see Figure PIII.3c). In other words, there is no resistance to the heat transfer between the thermowell wall and the thermocouple.

(a) Show that the thermocouple reading T_m follows first-order behavior to any changes in the surrounding temperature T.

(b) Derive the transfer function between T_m and T, and develop the equations defining the time constant and static gain of the thermocouple's response.

(c) The external film heat transfer coefficient, h_f, depends on the physical properties of the surrounding fluid, as well as the flow conditions. Sketch the qualitative response of T_m to a step change in T for various values of h_f. Using physical arguments, explain the observed behavior.

III.42 Consider the mixing tank shown in Figure PIII.4a. Assume that all inlet and outlet flow rates remain constant but that the compositions of the inlets can change.

(a) Show that the effluent composition follows first-order behavior to any changes in the composition of the inlet streams.

(b) Derive the transfer functions between the composition of the effluent and the compositions of the inlet streams. Draw the corresponding block diagram.

(c) Describe what is the characteristic capacity of the mixing tank and define the time constant and static gains for this system.

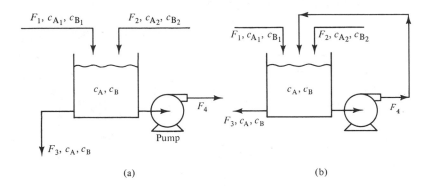

Figure PIII.4

III.43 Repeat the work of Problem III.42 but for the mixing tank system of Figure PIII.4b.

III.44 Show that the concentration c_A of reactant A in an isothermal continuous stirred tank reactor exhibits first-order dynamics to changes in the inlet composition, c_{A_i}. The reaction is irreversible, A → B, and has first-order kinetics (i.e., $r = kc_A$). Furthermore: (a) identify the time constant and static gain for the system, (b) derive the transfer function between c_A and c_{A_i}, (c) draw the corresponding block diagram, and (d) sketch the qualitative response of c_A to a unit pulse change in c_{A_i}. The reactor has a volume V, and the inlet and outlet flow rates are equal to F.

III.45 Consider an isothermal batch reactor, where the following reactions with first-order kinetics take place:

$$A \longrightarrow B \longrightarrow C$$

Show that the concentration c_A of reactant A in the reacting mixture exhibits first-order dynamic behavior with respect to the initial concentration, $c_A(t = 0)$. Identify the time constant and explain what it means in physical terms.

Chapter 11

III.46 Show that the following systems exhibit second-order overdamped response.
   (a) Effluent temperature T_3 to changes in inlet temperature T_1 (see Figure PII.3 for the two tank heaters of Problem II.3).
   (b) Effluent temperature T_3 to changes in inlet temperature T_1 (see Problem II.4 and Figure PII.4).
   (c) Effluent concentration c_{A_2} to changes in inlet composition c_{A_0} (see the two isothermal CSTRs of Problem II.9, Figure PII.8).

III.47 In Appendix 11A we found that simple manometers, externally mounted level indicators, variable capacitance transducers, and pneumatic valves exhibit inherent second-order dynamic behavior. Design all the systems above so that they exhibit underdamped behavior with a decay ratio equal to $\frac{1}{4}$. In other words, find the conditions that the values of the physical parameters of these systems should satisfy in order for the device to exhibit underdamped behavior with decay ratio equal to 1/4.

III.48 Determine the dynamic response of an overdamped second-order system to the following changes of input: (a) unit impulse, (b) unit pulse of duration 5 min., and (c) sinusoid, sin $2t$. Determine the resulting steady-state behaviour.

III.49 Repeat Problem III.48 but for an underdamped second-order system.

III.50 Consider an underdamped second-order system whose τ and ζ are not well known. Develop an experiment whereby changing the input $m(t)$ in a certain way and recording the output $y(t)$ with time you can determine the values of τ and ζ.

III.51 Consider a second-order system with the following transfer function:

$$G(s) = \frac{\overline{y}(s)}{\overline{m}(s)} = \frac{1}{s^2 + s + 1}$$

Introduce a step change of magnitude 5 into the system and find (a) per cent overshoot, (b) decay ratio, (c) maximum value of $y(t)$, (d) ultimate value of $y(t)$, (e) rise time, and (f) period of oscillation.

III.52 Let the input of a second-order system change in a sinusoidal manner,

$$m(t) = 1 \sin 2t$$

Show that the ultimate response of the system (a) is also a sustained sinusoidal wave, (b) has an amplitude equal to

$$\frac{1}{\sqrt{(1 - 4\tau^2)^2 + (4\zeta\tau)^2}}$$

and (c) lags behind the input wave by an angle ϕ equal to

$$\phi = \tan^{-1}\left(\frac{-4\zeta\tau}{1 - 4\tau^2}\right)$$

III.53 Use the results of Problem III.52 and describe an experimental procedure that allows us to compute the unknown values of τ and ζ for a second-order system.

III.54 Which of the following second-order systems are equivalent to two first-order systems in series and which are not?
(a) $G(s) = 1/(s^2 + 3s + 2)$
(b) $G(s) = 1/(s^2 + 1.9s + 0.7)$
(c) $G(s) = 1/(s^2 + 5)$
(d) $G(s) = 1/(s^2 + s + 2)$

III.55 Consider a thermocouple inside a thermowell as shown in Figure PIII.3b. Assume that the resistance to heat transfer does not come only from the external film between the surrounding fluid and the thermowell wall but also from the internal film between the thermowell wall and the thermocouple (Figure PIII.5). Let h_{ext} and h_{int} be the heat transfer coefficients for these two films.
(a) Show that the thermocouple reading T_m follows a second-order dynamic behavior with respect to any changes in the surrounding fluid temperature.
(b) Is it an overdamped or underdamped response and why?
(c) Design the thermocouple in such a way that it exhibits a slightly overdamped behavior (e.g., $\zeta = 1.2$).

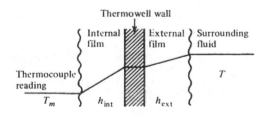

Figure PIII.5

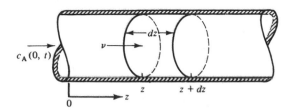

Figure PIII.6

Chapter 12

III.56 Consider a liquid mixture of A and B which flows through a cylindrical tube (Figure PIII.6). Assume that the flow is turbulent and consequently that the plug-flow approximation is valid. Plug flow implies that the concentration of A or B is a function of time and the position along the length of the tube only (i.e., there are no radial concentration gradients).

(a) Show that the material balance around a cylindrical element of length dz (see Figure PIII.6) leads to the following equation:

$$\frac{\partial (v c_A)}{\partial z} + \frac{\partial c_A}{\partial t} = 0$$

where $c_A \equiv c_A(z, t)$ is the concentration of A at time t and a cross-sectional area at distance z from the inlet of the tube. Also, v is the linear velocity of the liquid through the tube, which is assumed to be constant (it is a poor assumption for long tube lengths).

(b) Show that the transfer function between $c_A(z, t)$ and the inlet concentration $c_A(0, t)$ is given by

$$G(s) = \frac{\overline{c}_A(z, s)}{\overline{c}_A(0, s)} = e^{-t_d s}$$

where $t_d = z/v$ (i.e., the time required by the liquid to travel a distance z).

(c) Find the transient response of the concentration at $z = 20$ when a unit impulse in the inlet concentration is applied at time $t = 0$.

(d) Find the transient response of the concentration along the tube length when the inlet concentration changes by a unit step at $t = 0$.

III.57 Problem III.56 implies that a dead-time element is basically a distributed system, described by a partial differential equation. Suppose that the tube of Problem III.56 (Figure PIII.6) is approximated by a series of N identical well-stirred tanks (Figure PIII.7).

(a) Find the transfer function between the concentration of the effluent from tank N and the inlet concentration to the first tank.

(b) Assuming that the total volume of the N tanks remains the same, show that the transfer function between the effluent concentration

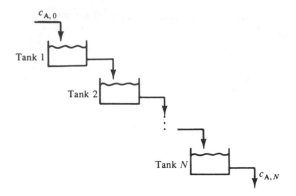

Figure PIII.7

from tank N and the inlet concentration to the first tank, as $N \to \infty$, is given by

$$G(s) = \frac{\overline{c}_{A,N}(s)}{\overline{c}_{A,0}(s)} = e^{-t_d s}$$

t_d is the time required by the liquid to travel from the inlet of the first tank to the outlet of the Nth tank.

III.58 Consider the mercury thermometer examined in Problem III.40. In that problem, assuming that the glass bulb (see Figure PIII.3a) does not expand or contract, we found that the thermometer reading exhibits first-order dynamics to changes in the surrounding temperature. In the present problem assume that the glass bulb does expand and contract significantly with changes in surrounding temperature.

   (a) Show qualitatively that the reading of the mercury thermometer may exhibit inverse response.

   (b) We have two mercury thermometers, whose mercury bulbs are constructed from two different materials with different thermal expansion or contraction coefficients. Which one is more likely to exhibit inverse response?

   (c) Suppose that the volume of a glass bulb exhibits first-order dynamics to changes in the surrounding temperature with a time constant $\tau_p = \alpha$ and gain $K_p = \beta$. If the volume occupied by the mercury of the thermometer's bulb exhibits first-order dynamics with $\tau_p = 0.1$ and $K_p = 1$ to any change in the temperature of the surrounding fluid, find the conditions that α and β should satisfy so that the thermometer reading exhibits inverse response.

   (d) Draw the corresponding block diagram for part (c).

III.59 Consider a system with the following transfer function:

$$G(s) = \frac{K_1 e^{-t_1 s}}{\tau_1 s + 1} - \frac{K_2 e^{-t_2 s}}{\tau_2 s + 1}$$

   (a) Draw the block diagram for this system.

(b) Find the conditions that must be satisfied by the six parameters K_1, K_2, τ_1, τ_2, t_1, and t_2 so that the system exhibits inverse response.

(c) If the system exhibits inverse response, plot the response of the system to a unit step input change.

III.60 Find the conditions that parameters K_1, K_2, τ, ζ, and τ_2 must satisfy so that a system with the following transfer function exhibits inverse response:

$$G(s) = \frac{K_1}{\tau^2 s^2 + 2\zeta\tau s + 1} - \frac{K_2}{\tau_2 s + 1}$$

Draw its response to a unit step change when $\zeta > 1$ and when $\zeta < 1$.

III.61 For each of the systems with transfer functions given below, (a) draw the corresponding block diagram, (b) identify the poles and zeros of the transfer function, (c) plot the response to a unit step input change, and determine the ultimate response to a sinusoidal input $\sin 2t$.

(a) $G(s) = \dfrac{10}{0.1s + 1} - \dfrac{5}{0.04s + 1}$

(b) $G(s) = \dfrac{10}{0.2s + 1} - \dfrac{5}{0.3s + 1}$

(c) $G(s) = \dfrac{10 \cdot e^{-s}}{0.1s + 1} - \dfrac{5}{0.04s + 1}$

(d) $G(s) = \dfrac{10 \cdot e^{-s}}{0.2s + 1} - \dfrac{5}{0.3s + 1}$

Analysis and Design of Feedback Control Systems

IV

An important feature of a good control system design algorithm is that it provides the practicing engineer with a framework within which to cast his problem and provides a systematic design procedure which can be applied to a larger number of similar problems.

A. Kestenbaum, R. Shinnar, and F. E. Thau*

In Part III we studied the dynamic behavior of various typical processing systems under the influence of changes in the input variables (disturbances or manipulated variables). In doing so, we were not concerned about having the system respond in a specific manner. In other words, we were not interested in controlling the behavior of the process.

Starting with Part IV, our main concern will be: How can we control a process in order to exhibit a certain desired response in the presence of input changes? First, we will study the most common control configuration, known as feedback, which we touched upon very briefly in Chapter 2. Then, in Parts V and VI we will discuss additional control configurations such as feedforward, cascade, ratio, override, split range, and multivariable.

Thus in the subsequent six chapters of Part IV, we will do the following:

1. Discuss the notion of the feedback loop and describe the hardware elements needed for its implementation.

*"Design Concepts for Process Control," Ind. Eng. Chem. Proc. Des. Dev., 15 (1), 2 (1976).

2. Identify the types of feedback controllers which are available for process control and examine their effect on the response of a chemical process.
3. Analyze the stability characteristics of a feedback control system and learn how to design the appropriate feedback system to control a given process.
4. Solve some special problems that are encountered during the design of feedback controllers.

Introduction **13**
to Feedback Control

In Chapter 1 we introduced the notion of a feedback control system. In this chapter we expand the discussion by introducing the hardware elements of a feedback system and the types of the available controllers.

13.1 Concept of Feedback Control

Consider the generalized process shown in Figure 13.1a. It has an output y, a potential disturbance d, and an available manipulated variable m.

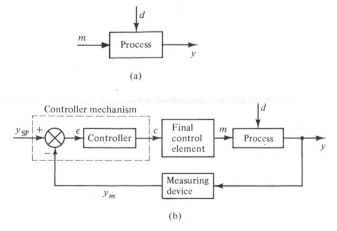

Figure 13.1 (a) process and (b) corresponding feedback loop.

241

The disturbance d (also known as *load* or *process load*) changes in an unpredictable manner and our control objective is to keep the value of the output y at desired levels. A feedback control action takes the following steps:

1. Measures the value of the output (flow, pressure, liquid level, temperature, composition) using the appropriate measuring device. Let y_m be the value indicated by the measuring sensor.
2. Compares the indicated value y_m to the desired value y_{SP} (*set point*) of the output. Let the deviation (*error*) be $\epsilon = y_{SP} - y_m$.
3. The value of the deviation ϵ is supplied to the main controller. The controller in turn changes the value of the manipulated variable m in such a way as to reduce the magnitude of the deviation ϵ. Usually, the controller does not affect the manipulated variable directly but through another device (usually a control valve), known as the *final control element*.

Figure 13.1b summarizes pictorially the foregoing three steps.

The system in Figure 13.1a is known as *open loop*, in contrast to the feedback-controlled system of Figure 13.1b, which is called *closed loop*. Also, when the value of d or m changes, the response of the first is called *open-loop response* while that of the second is the *closed-loop response*. The origin of the term closed-loop is evident from Figure 13.1b.

Example 13.1: Feedback Control Systems

The following represent some typical feedback control systems which are often encountered in chemical processes.

1. *Flow control:* Two feedback systems are shown in Figure 13.2a and b, controlling the flow rate F at the desired value F_{SP}.
2. *Pressure control:* The feedback system in Figure 13.2c controls the pressure of the gases in the tank, at the desired pressure p_{SP}.
3. *Liquid-level control:* Figure 13.2d and e show two feedback systems used for the control of the liquid levels at the bottom of a distillation column and its condenser accumulation tank.
4. *Temperature control:* The system in Figure 13.2f controls the temperature of the exiting hot stream at the desired value T_{SP}.
5. *Composition control:* Composition is the controlled variable in the blending system of Figure 13.2g. The desired value is c_{SP}.

Remark. To simplify the presentation of a feedback control system, we will usually replace the diagrammatic details of a controller mechanism with a simple circle carrying one of the following characterizations:

FC for flow control
PC for pressure control
LC for liquid-level control
TC for temperature control
CC for composition control

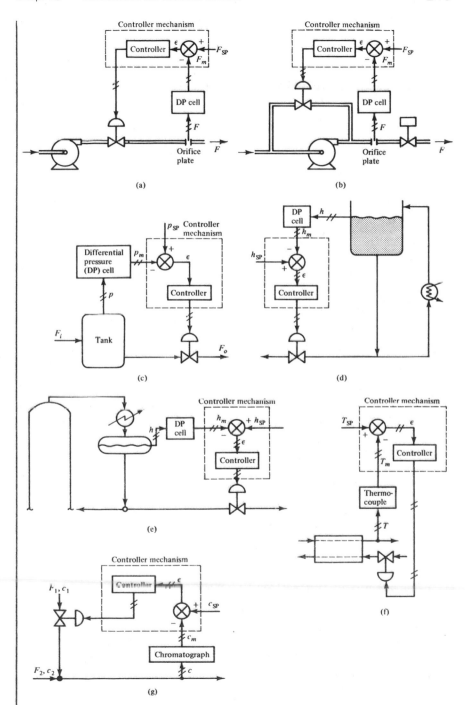

Figure 13.2 Examples of feedback systems: (a) and (b) flow control; (c) pressure control; (d) and (e) liquid-level control; (f) temperature control; (g) composition control.

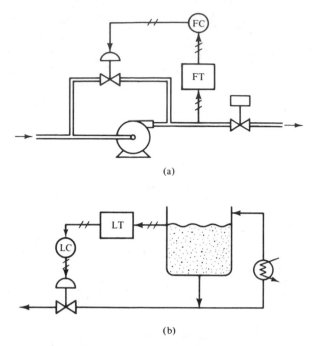

(a)

(b)

Figure 13.3 Simplified representations of feedback loops.

Also, little squares with the characterizations LT, TT, PT, FT, and CT are used to indicate level, temperature, pressure, flow, and concentration measurement and transmission. Figure 13.3a and b is equivalent to Figure 13.2b and d, respectively.

All the examples above indicate that the basic hardware components of a feedback control loop are the following:

1. *Process:* the material equipment along with the physical or chemical operations which take place (tanks, heat exchangers, reactors, separators, etc.).
2. *Measuring instruments or sensors:* for example, thermocouples (for temperature), bellows, or diaphragms (for pressure or liquid level), orifice plates (for flow), gas chromatographs or various types of spectroscopic analyzers (for composition), and so on.
3. *Transmission lines:* used to carry the measurement signal from the sensor to the controller and the control signal from the controller to the final control element. These lines can be either pneumatic (compressed air or liquid) or electrical.
4. *Controller:* also includes the function of the comparator. This is the unit with logic that decides by how much to change the value of the manipulated variable. It requires the specification of the desired value (set point).

5. *Final control element:* usually, a control valve or a variable-speed metering pump. This is the device that receives the control signal from the controller and implements it by physically adjusting the value of the manipulated variable.

Each of the elements above should be viewed as a physical system with an input and an output. Consequently, their behavior can be described by a differential equation or equivalently by a transfer function. In the following sections of this chapter we take a closer look at the dynamics of these hardware elements.

13.2   Types of Feedback Controllers

Between the measuring device and the final control element comes the controller (Figure 13.1b). Its function is to receive the measured output signal $y_m(t)$ and after comparing it with the set point y_{SP} to produce the actuating signal $c(t)$ in such a way as to return the output to the desired value y_{SP}. Therefore, the input to the controller is the error $\epsilon(t) = y_{SP} - y_m(t)$, while its output is $c(t)$. The various types of continuous feedback controllers differ in the way they relate $\epsilon(t)$ to $c(t)$.

The output signal of a feedback controller depends on its construction and may be a pneumatic signal (compressed air) for pneumatic controllers or an electrical one for electronic controllers.

There are three basic types of feedback controllers: (1) proportional, (2) proportional-integral, and (3) proportional-integral-derivative. The details of construction may differ among the various manufacturers, but their functions are essentially the same. Let us study each one separately.

Proportional controller (or P controller)

Its actuating output is proportional to the error:

$$c(t) \quad K_c \epsilon(t) \mid c_s \qquad\qquad (13.1)$$

where K_c = proportional gain of the controller and c_s = controller's bias signal (i.e., its actuating signal when $\epsilon = 0$).

A proportional controller is described by the value of its *proportional gain* K_c or equivalently by its *proportional band* PB, where PB = $100/K_c$. The proportional band characterizes the range over which the error must change in order to drive the actuating signal of the controller over its full range. Usually,

$$1 \leq PB \leq 500$$

It is clear that

The larger the gain K_c, or equivalently, the smaller the proportional band, the higher the sensitivity of controller's actuating signal to deviations ϵ will be.

Define the deviation $c'(t)$ of the actuating signal by

$$c'(t) = c(t) - c_s$$

and take

$$c'(t) = K_c \epsilon(t) \tag{13.2}$$

The last equation yields the following transfer function for a proportional controller

$$G_c(s) = K_c \tag{13.3}$$

Proportional-integral controller (or PI controller)

Most commonly it is known as *proportional-plus-reset* controller. Its actuating signal is related to the error by the equation

$$c(t) = K_c \epsilon(t) + \frac{K_c}{\tau_I} \int_0^t \epsilon(t)\, dt + c_s \tag{13.4}$$

where τ_I is the *integral time constant* or *reset time* in minutes. The reset time is an adjustable parameter and is sometimes referred to as *minutes per repeat*. Usually it varies in the range

$$0.1 \leq \tau_I \leq 50 \text{ min}$$

Some manufacturers do not calibrate their controllers in terms of τ_I but in terms of its reciprocal, $1/\tau_I$ (repeats per minute), which is known as the *reset rate*.

At this point it is instructive to examine the origin of the term "reset." Consider that the error changes by a step of magnitude ϵ. Figure 13.4 shows the response of the output of a controller as it is computed from eq. (13.4). We observe that initially the controller output is $K_c \epsilon$ (the contribution of the integral term is zero). After a period of τ_I minutes the contribution of the integral term is

$$\frac{K_c}{\tau_I} \int_0^{\tau_I} \epsilon(t)\, dt = \frac{K_c}{\tau_I} \epsilon \tau_I = K_c \epsilon$$

that is, the integral control action has "repeated" the response of the

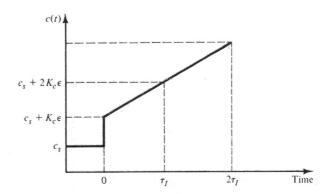

Figure 13.4 Response of PI controller to step change in error.

proportional action. This repetition takes place every τ_I minutes and has lent its name to the reset time. Therefore:

Reset time is the time needed by the controller to repeat the initial proportional action change in its output.

The integral action causes the controller output $c(t)$ to change as long as an error exists in the process output. Therefore, *such a controller can eliminate even small errors*.

From eq. (13.4) it is easy to show that the transfer function of a proportional-integral controller is given by

$$G_c(s) = K_c\left(1 + \frac{1}{\tau_I s}\right) \tag{13.5}$$

Remark

The integral term of a PI controller causes its output to continue changing as long as there is a non-zero error. Often the errors cannot be eliminated quickly, and given enough time they produce larger and larger values for the integral term, which in turn keeps increasing the control action until it is saturated (e.g. the valve completely open or closed). This condition is called *integral windup* and occurs during manual operational changes like; shut-down, change-over, etc. When the process is returned to automatic operation, the control action will remain saturated leading to large overshoots. A PI controller needs special provisions to cope with integral windup.

Proportional-integral-derivative controller (or PID controller)

In the industrial practice it is commonly known as *proportional-plus-reset-plus-rate* controller.

The output of this controller is given by

$$c(t) = K_c \epsilon(t) + \frac{K_c}{\tau_I} \int_0^t \epsilon(t)\, dt + K_c \tau_D \frac{d\epsilon}{dt} + c_s \qquad (13.6)$$

where τ_D is the *derivative time constant* in minutes.

With the presence of the derivative term, $(d\epsilon/dt)$, the PID controller *anticipates what the error will be in the immediate future* and applies a control action which is proportional to the current rate of change in the error. Due to this property, the derivative control action is sometimes referred to as *anticipatory control*.

The major drawbacks of the derivative control action are the following:

1. For a response with constant nonzero error it gives no control action since $d\epsilon/dt = 0$.
2. For a noisy response with almost zero error it can compute large derivatives and thus yield large control action, although it is not needed.

From eq. (13.6) we can easily derive the transfer function of a PID controller,

$$G_c(s) = K_c \left(1 + \frac{1}{\tau_I s} + \tau_D s \right) \qquad (13.7)$$

13.3 Measuring Devices (Sensors)

The successful operation of any feedback control system depends, in a very critical manner, on the good measurement of the controlled output and the uncorrupted transmission of the measurement to the controller. The first requirement implies the need for an accurate measuring device while the second necessitates good and effective transmission lines.

There are a large number of commercial sensors. They differ either in the basic measuring principle they employ or in their construction characteristics. For more details the reader can consult the various references at the end of Part IV or the technical booklets circulated by the various manufacturers. Table 13.1 lists typical measuring devices encountered in various applications of process control.

Let us look more closely at the various typical sensors used to measure the most common process outputs.

TABLE 13.1
TYPICAL MEASURING DEVICES FOR PROCESS CONTROL

Measured process variable	Measuring device	Comments
Temperature	Thermocouples	} Most common for relatively
	Resistance thermometers	} low temperatures
	Filled-system thermometers	
	Bimetal thermometers	
	Radiation pyrometers	Used for high temperatures
	Oscillating quartz crystal	
Pressure	Manometers	With floats or displacers
	Bourdon-tube elements	}
	Bellows elements	} Based on the elastic
	Diaphragm elements	} deformation of materials
	Strain gages	}
	Piezoresistivity elements	} Used to convert pressure
	Piezoelectric elements	} to electrical signal
Flow	Orifice plates	}
	Venturi flow nozzle	} Measuring pressure drop
	Dahl flow tube	} across a flow constriction
	Kennison flow nozzle	}
	Turbine flow meters	
	Ultrasound	
	Hot-wire anemometry	For high precision
Liquid level	Float-actuated devices	} Coupled with various types
	Displacer devices	} of indicators and signal
		} converters
	Liquid head pressure devices	
	Conductivity measurement	} Good for systems with
	Dielectric measurement	} two phases
	Sonic resonance	
Composition	Chromatographic analyzers	Long times required for analysis
	Infrared analyzers	}
	Ultraviolet analyzers	} Convenient for one or
	Visible-radiation analyzers	} two chemicals
	Turbidimetry analyzers	}
	Paramagnetism analyzers	} Not very convenient for
	Nephelometry analyzers	} process control
	Potentiometry	
	Conductimetry	
	Oscillometric analyzers	
	pH Meters	
	Polarographic analyzers	
	Coulometers	
	Spectrometers (x-ray, electron, ion, Mössbauer, Raman, etc.)	} Expensive for low-cost control loops
	Differential thermal analyzers	}

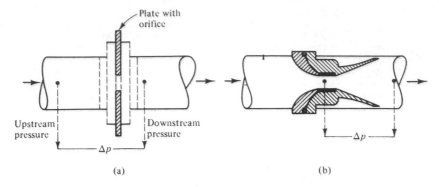

Figure 13.5 Flow sensors: (a) orifice plate; (b) venturi tube.

Flow sensors

The flow sensors most commonly employed in the industrial practice are those that measure the pressure gradient developed across a flow constriction. Then, using the well-known (from fluid mechanics) equation of Bernoulli, we can compute the flow rate. Such devices can be used for both gases and liquids. The *orifice plate* (Figure 13.5a), *venturi tube* (Figure 13.5b), and *Dall flow tube* are typical examples of sensors based on the foregoing principle. The first is more popular due to its simplicity and low cost. The last two are more expensive but also more accurate.

A different sensor is the *turbine flow meter*, which uses the number of turbine revolutions to compute the flow rate of liquids quite accurately.

Flow sensors have very fast dynamics and they are usually modeled by simple algebraic equations:

$$\text{flow} = \alpha\sqrt{\Delta p} \tag{13.8}$$

where α is a constant determined by the construction characteristics of the flow sensor, and Δp is the pressure difference between a point at the flow constriction and a point with fully developed flow.

Pressure or pressure-actuated sensors

Such sensors are used to measure the pressure of a process or the pressure difference which is employed to compute a liquid level or a flow rate (orifice plate, venturi tube). The *variable capacitance differential pressure transducer* has become very popular. Figure 11A.2 shows a schematic of such a device. Pressure differences cause small displacements of the sensing diaphragm. The position of the sensing diaphragm

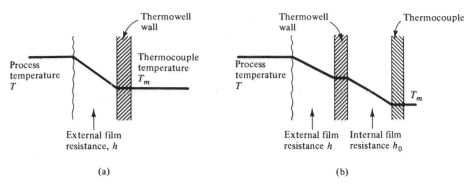

Figure 13.6 Thermocouples with: (a) external film resistance only; (b) external and internal film resistances.

is detected by capacitor plates on both sides of the diaphragm. The differential capacitance between the sensing diaphragm and the capacitor plates is converted into dc voltage. A force balance around the sensing diaphragm leads to the following second-order model:

$$\tau^2 \frac{d^2z}{dt^2} + 2\zeta\tau \frac{dz}{dt} + z = K_p\,\Delta p \qquad (13.9)$$

where z = displacement of the sensing diaphragm

  Δp = actuating pressure difference

  $\tau,\ \zeta,\ K_p$ = three parameters of a second-order system, defined in this case by the constructional characteristics of the device

For details on the development of eq. (13.9), see Appendix 11A. Various other types of sensors, all of them measuring the displacement of a mechanical part under the influence of Δp, are also in use.

Temperature sensors

The most common are *thermocouples, resistance bulb thermometers*, and *thermistors*. All provide measurement in terms of electrical signals. Independently of their constructional differences, their basic dynamic behavior can be examined in terms of the temperature profiles in Figure 13.6a and b. The temperature-sensing element is always inside a thermowell (Figure 13.7). In the first case (Figure 13.6a) we assume

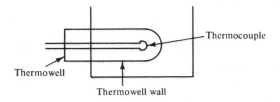

Figure 13.7 Typical thermocouple arrangement.

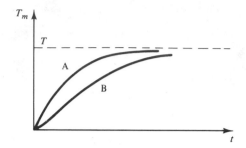

Figure 13.8 Response of thermocouples with single film resistance (curve a) and two film resistances (curve b).

that the major resistance to heat transfer is located outside the thermowell casing. In such a case we have a single capacity with resistance and as we know from Chapter 10, it is modeled by a first-order system:

$$\tau_p \frac{dT_m}{dt} + T_m = T \tag{13.10}$$

In the second case (Figure 13.6b) we have major heat transfer film resistances inside and outside the thermowell casing. This is equivalent to two capacities in series and as we know from Chapter 11, the thermocouple reading will exhibit second-order (overdamped) behavior:

$$\tau^2 \frac{d^2 T_m}{dt^2} + 2\zeta\tau \frac{dT_m}{dt} + T_m = T \tag{13.11}$$

The parameters τ and ζ depend on the constructional and material characteristics of the temperature-sensing device (i.e., thermocouple, casing, materials of construction). It is clear that the response of a thermocouple modeled by eq. (13.11) is slower than that of a thermocouple modeled by eq. (13.10) (see Figure 13.8).

Composition analyzers

Typical examples of such sensors are *gas chromatographs* and various types of *spectroscopic analyzers*. They are used to measure the composition of liquids or gases in terms of one or two key components or in terms of all components present in a process stream.

The dominant dynamic feature of composition analyzers is the time delay (dead time) in their response, which can be quite large. Thus, for a chromatographic column, the time required by the sample to travel from the process stream to the column, plus the time required to travel through the column, plus the time needed by the detector at the end of the column to respond, can be quite large. Such long time delays result in ineffective control.

Other features characteristic of composition analyzers are: (1) their low operational reliability (easy decalibration or breakdown), and (2) their relatively high cost.

13.4   Transmission Lines

These are used to carry the measurement signal to the controller and the control signal to the final control element. There are two types of transmission lines: the pneumatic (compressed air, liquids) and the electrical.

Unless the process changes very fast or the transmission lines are very long, the dynamic behavior of a pneumatic transmission line can be neglected from consideration. When the assumptions above do not hold, it has been found that the following transfer function correlates successfully the pressure at the outlet (P_o) to the pressure at the inlet (P_i) of a pneumatic transmission line:

$$\frac{\overline{P}_o(s)}{\overline{P}_i(s)} = \frac{e^{-\tau_d s}}{\tau_p s + 1}$$

with $\tau_d/\tau_p \simeq 0.25$.

Note.   In the subsequent chapters, as a rule, we will neglect the dynamics of pneumatic transmission lines.

13.5   Final Control Elements

These are the hardware components of the control loops that implement the control action. They receive the output of a controller (actuating signal) and adjust accordingly the value of the manipulated variable.

The most common final control element is the *pneumatic valve* (Figure 11A.3). This is an air-operated valve which controls the flow through an orifice by positioning appropriately a plug. The plug is attached at the end of a stem which is supported on a diaphragm at the other end. As the air pressure (controller output) above the diaphragm increases, the stem moves down and consequently the plug restricts the flow through the orifice. Such a valve is known as an "*air-to-close*" valve (Figure 13.9a). If the air supply above the diaphragm is lost, the valve will "*fail open*" since the spring would push the stem and the plug upward. There are pneumatic valves with opposite actions, (i.e. "*air-to-open*" which "*fail closed*") (Figure 13.9b). The most commercial valves

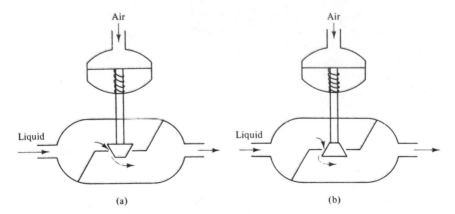

Figure 13.9 Pneumatic valves: (a) fail open; (b) fail closed.

move from fully open to fully closed as the air pressure at the top of the diaphragm changes from 3 to 15 psig.

  In Appendix 11A we developed the mathematical model that describes the dynamic behavior of a pneumatic control valve. This was shown to be of second-order. But the response to changes, of most small

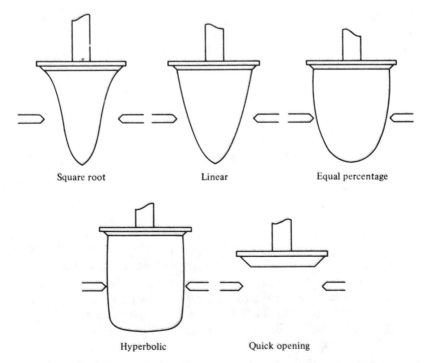

Figure 13.10 Types of plugs for pneumatic valves. Reproduced from Buckley P.S., *Techniques of Process Control*, by permission.

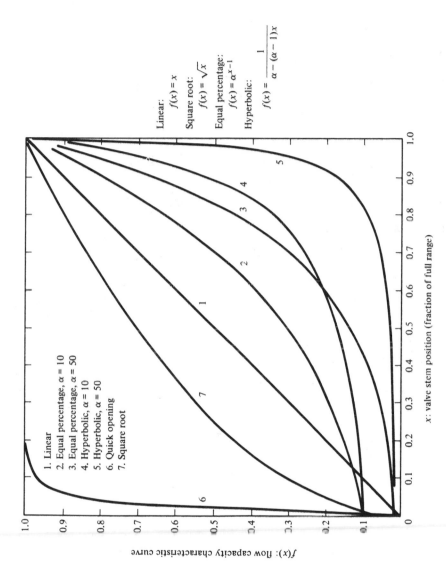

Figure 13.11 Flow capacity characteristics for various valves. Reproduced from Buckley P.S., *Techniques of Process Control*, by permission.

255

or medium-size valves, is so fast that the dynamics can be neglected. In such a case only a constant gain term will remain which relates the output from the controller (air pressure signal) to the fluid flow through the valve.

For nonflashing liquids the flow through the valve is given by

$$F = Kf(x) \sqrt{\frac{\Delta p}{\rho}}$$

where Δp = pressure drop across the valve

K = constant which depends on the valve size

ρ = specific gravity of the flowing liquid (relative to water)

$f(x)$ = valve flow characteristic curve

The valve flow characteristic curve, $f(x)$, depends on the geometrical shape of the plug's surface. Figure 13.10 shows the most common types of plugs while Figure 13.11 shows the flow capacity characteristics for the various valves.

Other final control elements include relays to start or stop various equipment, variable-drive motors for fans or pumps, heavy load electrohydraulic actuators, and so on.

THINGS TO THINK ABOUT

1. From all that you know so far, what are the strengths and weaknesses of a feedback control system?

2. Describe one example of (a) flow control, (b) pressure control, (c) liquid level control, (d) temperature control, and (e) composition control, which are not the same as the examples covered in this chapter. Draw the appropriate diagrams.

3. Define an open-loop and a closed-loop system. Why do we use the terms open-loop and closed-loop? Also define open-loop or closed-loop response.

4. What are the basic hardware components of a feedback control loop? Identify the hardware elements present in a feedback loop for the temperature control of a stirred tank heater.

5. Write Bernoulli's equation for two points of a Venturi tube and show how you can compute the flow rate through the tube by measuring the pressure difference between the two points [i.e., prove the essence of eq. (13.8)].

6. The model for a variable capacitance pressure transducer was developed in Appendix 11A and is given by eq. (13.9). It shows that the system is inherently second-order and can exhibit underdamped response. What does this mean for the applicability of such device?

7. Is it possible to have an oscillatory behavior by the indicated temperature (T_m) of a thermocouple if the measured temperature (T) changes by a step? Elaborate on your answer.

8. Discuss some of the factors you should take into account before deciding whether to use an air-to-close or air-to-open pneumatic control valve.

9. Compute the response of a PD (proportional-derivative) controller to a ramp change in the error ϵ (i.e., $\epsilon = \alpha t$ with α = constant). Sketch the contributions of the proportional and derivative actions separately. On the basis of this example discuss the anticipatory nature of the derivative control term.

10. Consult Refs. 6 (Chapter 15) and 7 (Chapter 10) and discuss the factors that affect the selection of the valve type (i.e., linear, square root, equal percentage, and hyperbolic).

11. When an error $\epsilon(t)$ persists for a long time, the value of the integral $\int \epsilon(t) \, dt$ increases significantly and may lead the output of a PI controller to its maximum allowable value. We say that the controller has been saturated and in physical terms it means that the valve is fully open or closed before the control action has been completed (i.e., before the error has been driven to zero). This situation is also known as *reset windup*. How would you handle such a situation? You can consult Refs. 7 and 15.

Dynamic Behavior **14**
of Feedback-Controlled
Processes

In Chapter 13 we defined the basic notion of a feedback control system and we discussed its hardware components. In this chapter we examine the dynamic behavior of a process that is controlled by a feedback control system, when; (1) the value of the disturbance (load) d, or (2) the desired value of the set point y_{SP} change.

14.1 Block Diagram and the Closed-Loop Response

Consider the generalized closed-loop system shown in Figure 13.1b. For each of its four components (process, measuring device, controller mechanism, and final control element) we can write the corresponding transfer function relating its output to its inputs. In particular, if we neglect the dynamics of the transmission lines, we have:

Process:

$$\bar{y}(s) = G_p(s)\bar{m}(s) + G_d(s)\bar{d}(s) \tag{14.1}$$

Measuring device:

$$\bar{y}_m(s) = G_m(s)\bar{y}(s) \tag{14.2}$$

Controller mechanism:

$$\bar{\epsilon}(s) = \bar{y}_{SP}(s) - \bar{y}_m(s) \qquad \text{comparator} \tag{14.3a}$$

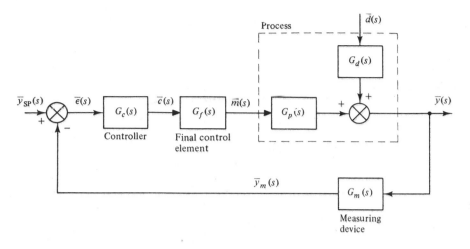

Figure 14.1 Block diagram of generalized closed-loop system.

$$\overline{c}(s) = G_c(s)\overline{\epsilon}(s) \qquad \text{control action} \qquad (14.3b)$$

Final control element:

$$\overline{m}(s) = G_f(s)\overline{c}(s) \qquad\qquad (14.4)$$

where G_p, G_d, G_m, G_c, and G_f are the transfer functions between the corresponding inputs and outputs.

Figure 14.1 shows the *block diagram for the generalized closed-loop system* and it is nothing more than a pictorial representation of eqs. (14.1), (14.2), (14.3a), (14.3b), and (14.4). Notice the direct correspondence between the block diagram of Figure 14.1 and the schematic of Figure 13.1b.

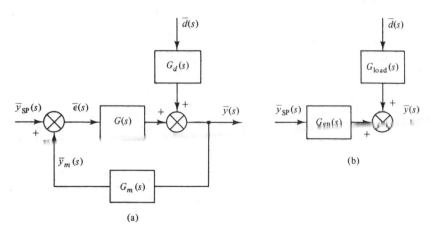

Figure 14.2 Simplified block diagrams.

The series of blocks between the comparator and the controlled output (i.e., G_c, G_f, and G_p) constitutes the *forward path*, while the block G_m is on the *feedback path* between the controlled output and the comparator. If $G = G_c G_f G_p$, then Figure 14.2a shows a simplified but equivalent version of the block diagram.

Algebraic manipulation of the equations above yields

$$\overline{m}(s) = G_f(s)\overline{c}(s) = G_f(s)G_c(s)\overline{e}(s) \qquad \text{using eq. (14.3b)}$$

$$= G_f(s)G_c(s)[\overline{y}_{SP}(s) - \overline{y}_m(s)] \qquad \text{using eq. (14.3a)}$$

$$= G_f(s)G_c(s)[\overline{y}_{SP}(s) - G_m(s)\overline{y}(s)] \qquad \text{using eq. (14.2)}$$

Put the last expression in eq. (14.1):

$$\overline{y}(s) = G_p(s)\{G_f(s)G_c(s)[\overline{y}_{SP}(s) - G_m(s)\overline{y}(s)]\} + G_d(s)\overline{d}(s)$$

and after readjustment take

$$\overline{y}(s) = \frac{G_p(s)G_f(s)G_c(s)}{1 + G_p(s)G_f(s)G_c(s)G_m(s)}\overline{y}_{SP}(s)$$

$$+ \frac{G_d(s)}{1 + G_p(s)G_f(s)G_c(s)G_m(s)}\overline{d}(s) \tag{14.5}$$

Equation (14.5) gives the *closed-loop response* of the process. We notice that it is composed of two terms. The first term shows the effect on the output of a change in the set point, while the second constitutes the effect on the output of a change in the load (disturbance). The corresponding transfer functions are known as *closed-loop transfer functions*. In particular,

$$\frac{G_p G_f G_c}{1 + G_p G_f G_c G_m} \equiv \frac{G}{1 + GG_m} = G_{SP} \tag{14.6}$$

is the closed-loop transfer function for a change in the set point and

$$\frac{G_d}{1 + G_p G_f G_c G_m} \equiv \frac{G_d}{1 + GG_m} = G_{load} \tag{14.7}$$

is the closed-loop transfer function for a change in the load. Figure 14.2b shows a block diagram equivalent to that of Figure 14.2a but further simplified.

For every feedback control system we can distinguish two types of control problems: the *servo* and the *regulator* problem.

Servo problem: The disturbance does not change [i.e., $\overline{d}(s) = 0$] while the set point undergoes a change. The feedback controller acts in such a way as to keep y close to the changing y_{SP}. In such a case,

$$\overline{y}(s) = G_{SP}(s)\overline{y}_{SP}(s) \tag{14.8}$$

Regulator problem: The set point remains the same [i.e., $\overline{y}_{SP}(s) = 0$] while the load changes. Then

$$y(s) = G_{load}(s)\overline{d}(s) \qquad (14.9)$$

and the feedback controller tries to eliminate the impact of the load changes and keep y at the desired set point.

From eqs. (14.6) and (14.7) it can be easily seen that *the closed-loop overall transfer functions G_{SP} and G_{load} depend not only on the process dynamics but also on the dynamics of the measuring sensor, controller and final control element.*

Example 14.1: Closed-Loop Response of the Liquid Level in a Tank

Consider the liquid-level control system for the tank of Figure 14.3a.

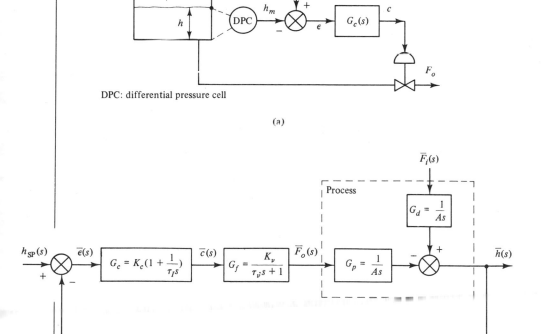

(a)

(b)

Figure 14.3 (a) Closed-loop of liquid-level control; (b) corresponding block diagram.

The level h is the controlled output while F_i is the load (disturbance) and F_o the manipulated variable. The transfer functions for each component of the feedback loop are as follows:

Process. The material balance around the tank gives

$$A \frac{dh}{dt} = F_i - F_o$$

and we find easily that

$$\bar{h}(s) = \frac{1}{As} \bar{F}_i(s) - \frac{1}{As} \bar{F}_o(s) \tag{14.10}$$

Measuring device. This can be a variable capacitance differential pressure transducer (Section 13.3), measuring the pressure of a liquid column of height h. The dynamic response of the sensor is given by eq. (13.9). Let $\Delta p = \alpha h$, where α is a constant. Then take

$$\tau^2 \frac{d^2 z}{dt^2} + 2\zeta\tau \frac{dz}{dt} + z = K_p \, \Delta p = K_p \alpha h$$

where $z \equiv h_m$ (i.e., the value indicated by the measuring device). Therefore, the transfer function for the sensor is

$$\bar{h}_m(s) = \frac{K_p \alpha}{\tau^2 s^2 + 2\zeta\tau s + 1} \bar{h}(s) \tag{14.11}$$

Controller. Let h_{SP} be the set point. Then

$$\bar{\epsilon}(s) = \bar{h}_{SP}(s) - \bar{h}_m(s)$$

and for a PI controller [eq. (13.5)]

$$\bar{c}(s) = K_c \left(1 + \frac{1}{\tau_I s} \right) \bar{\epsilon}(s) \tag{14.12}$$

Control valve. Let us assume that for the control valve of this system, the response is that of a first-order system:

$$\bar{F}_o(s) = \frac{K_v}{\tau_v s + 1} \bar{c}(s) \tag{14.13}$$

Figure 14.3b shows the block diagram for the closed-loop system with the transfer functions for each component of the loop. The closed-loop response of the liquid level will be given by eq. (14.5), where the transfer functions G_p, G_d, G_m, G_c, and G_f are shown in Figure 14.3b. The servo problem arises when the inlet flow rate F_i remains constant and we change the desired set point. In this case the controller acts in such a way as to keep the liquid level h close to the changing desired value h_{SP}. On the other hand, for the regulator problem the set point h_{SP} remains the same and the feedback controller acts in such a way as to eliminate the impact of the changing load and keep h at the desired value h_{SP}.

Example 14.2: Closed-Loop Temperature Response of a Tank Heater

Consider the temperature control system for the heater of Figure 14.4a. The temperature T is the controlled output while the inlet temperature T_i is the load and the steam temperature is the manipulated variable. The transfer functions for each component of the feedback loop are:

Process. If T, T_i, and T_{st} are deviation variables, then from eq. (9.8) the response of the process is given by

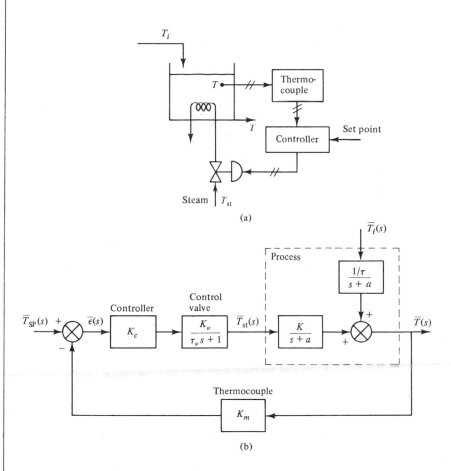

Figure 14.4 (a) Closed-loop of temperature control; (b) corresponding block diagram.

$$\overline{T}(s) = \frac{1/\tau}{s+a}\,\overline{T}_i(s) + \frac{K}{s+a}\,\overline{T}_{st}(s) \qquad (14.14)$$

The parameters τ, a, and K have been defined in Examples 5.1 and 9.1.

Temperature sensor (thermocouple). Assume that the response of the thermocouple is very fast and its dynamics can be neglected. Thus

$$\overline{T}_m(s) = K_m\overline{T}(s) \qquad (14.15)$$

Controller. Let T_{SP} be the set point. Then

$$\overline{\epsilon}(s) = \overline{T}_{SP}(s) - \overline{T}_m(s) \qquad (14.16a)$$

and for a proportional controller the actuating output is given by

$$\overline{c}(s) = K_c\overline{\epsilon}(s) \qquad (14.16b)$$

Control valve. Assume first-order dynamics:

$$\overline{T}_{st}(s) = \frac{K_v}{\tau_v s + 1}\,\overline{c}(s) \qquad (14.17)$$

Figure 14.4b shows the block diagram for the closed-loop system with the transfer functions for each component of the loop. The closed-loop response is easily found to be

$$\overline{T}(s) = G_{SP}(s)\overline{T}_{SP}(s) + G_{load}(s)\,\overline{T}_i(s)$$

where the closed-loop transfer functions G_{SP} and G_{load} are defined as follows:

$$G_{SP}(s) = \frac{\left[\dfrac{K}{s+a}\right][K_c]\left[\dfrac{K_v}{\tau_v s + 1}\right]}{1 + \left[\dfrac{K}{s+a}\right][K_m][K_c]\left[\dfrac{K_v}{\tau_v s + 1}\right]}$$

and

$$G_{load}(s) = \frac{\left[\dfrac{1/\tau}{s+a}\right]}{1 + \left[\dfrac{K}{s+a}\right][K_m][K_c]\left[\dfrac{K_v}{\tau_v s + 1}\right]}$$

Remark. To expedite the construction of the overall closed-loop transfer functions for any feedback control loop, use the following rules:

1. The denominator of the overall transfer functions for both the load and the set point changes is the same. It is given by:

     1 + product of the transfer functions in the loop

or

$$1 + G_p G_m G_c G_f$$

2. The numerator of an overall closed-loop transfer function is the product of the transfer functions on the forward path between the set point or the load and the controlled output. Thus:
   (a) The transfer functions on the forward path between the set point y_{SP} and output y are: G_c, G_f, and G_p. Therefore, the numerator is $G_c\ G_f\ G_p$.
   (b) The transfer functions on the forward path between the load d and the output is only G_d. Thus the corresponding numerator is G_d.

Verify these two rules with the overall closed-loop transfer functions G_{SP} and G_{load} [eqs. (14.6) and (14.7)]. Also, these rules can be used to formulate the closed-loop transfer function between an input anywhere in the loop and the output.

14.2 Effect of Proportional Control on the Response of a Controlled Process

Let us now examine how the response of a normal, uncontrolled process is changed when a simple proportional, integral, or derivative feedback controller is incorporated. In this section we consider only the proportional controller and its effect on the most commonly encountered first- and second-order systems. The effects of integral and derivative control actions will be studied in the following two sections.

The closed-loop response of a process is given by eq. (14.5). To simplify the analysis assume that

$$G_m(s) = 1 \quad \text{and} \quad G_f(s) = 1$$

Also, for a proportional controller,

$$G_c(s) = K_c$$

and eq. (14.5) yields

$$\bar{y}(s) = \frac{G_p(s)K_c}{1 + G_p(s)K_c}\,\bar{y}_{SP}(s) + \frac{G_d(s)}{1 + G_p(s)K_c}\,\bar{d}(s) \qquad (14.18)$$

First-order systems

For first-order systems

$$\tau_p \frac{dy}{dt} + y = K_p m + K_d d \qquad \text{with } y(0) = m(0) = d(0) = 0$$

which gives

$$\bar{y}(s) = \frac{K_p}{\tau_p s + 1} \, \bar{m}(s) + \frac{K_d}{\tau_p s + 1} \, \bar{d}(s)$$

Thus for the uncontrolled system we have:

Time constant: τ_p
Static gains: K_p for the manipulation   and K_d for the load

Put

$$G_p(s) = \frac{K_p}{\tau_p s + 1} \quad \text{and} \quad G_d(s) = \frac{K_d}{\tau_p s + 1}$$

in eq. (14.18) and take the closed-loop response:

$$\bar{y}(s) = \frac{K_p K_c}{\tau_p s + 1 + K_p K_c} \, \bar{y}_{SP}(s) + \frac{K_d}{\tau_p s + 1 + K_p K_c} \, \bar{d}(s)$$

Rearrange the last equation and take

$$\bar{y}(s) = \frac{K'_p}{\tau'_p s + 1} \, \bar{y}_{SP}(s) + \frac{K'_d}{\tau'_p s + 1} \, \bar{d}(s) \tag{14.19}$$

where

$$\tau'_p = \frac{\tau_p}{1 + K_p K_c} \tag{14.20a}$$

$$K'_p = \frac{K_p K_c}{1 + K_p K_c} \tag{14.20b}$$

and

$$K'_d = \frac{K_d}{1 + K_p K_c} \tag{14.20c}$$

The parameters K'_p and K'_d are known as *closed-loop static gains*.
From eq. (14.19) we conclude that the closed-loop response of a first-order system has the following characteristics:

1. It *remains first-order* with respect to load and set point changes.
2. The time constant has been reduced (i.e., $\tau'_p < \tau_p$), which means that the *closed-loop response has become faster*, than the open-loop response, to changes in set point or load.
3. The static gains have been decreased.

To gain a better insight into the effect of the proportional controller, consider unit step changes in the set point (servo problem) and the load (regulator problem) and examine the resulting closed-loop responses. For the servo problem, $\bar{y}_{SP}(s) = 1/s$ and $\bar{d}(s) = 0$. Then eq. (14.19) yields

$$\bar{y}(s) = \frac{K'_p}{\tau'_p s + 1} \frac{1}{s}$$

and after inversion we find that

$$y(t) = K'_p(1 - e^{-t/\tau'_p}) \tag{14.21}$$

Figure 14.5a shows the response of the closed-loop system to a unit step change in the set point. We notice that:

The ultimate response, after $t \to \infty$, never reaches the desired new set point. There is always a discrepancy called *offset* which is equal to

offset = (new set point) − (ultimate value of the response)

$$= 1 - K'_p = 1 - \frac{K_p K_c}{1 + K_p K_c} = \frac{1}{1 + K_p K_c}$$

The offset is characteristic effect of proportional control. It decreases as K_c becomes larger and theoretically

$$\text{offset} \longrightarrow 0 \quad \text{when} \quad K_c \longrightarrow \infty$$

For the regulator problem, $\bar{y}_{SP}(s) = 0$. Consider a unit step change in the load, [i.e., $\bar{d}(s) = 1/s$]. Then eq. (14.19) yields

$$\bar{y}(s) = \frac{K'_d}{\tau'_p s + 1} \frac{1}{s}$$

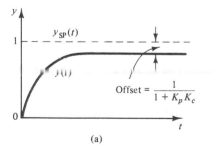

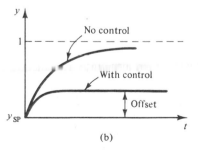

(a)   (b)

Figure 14.5 Closed-loop responses of first-order systems with P control, to: (a) unit step change in set point; (b) unit step change in load.

and after inversion

$$y(t) = K'_d(1 - e^{-t/\tau'_p})$$

Figure 14.5b shows this response to a unit step change in the load. We notice again that the proportional controller cannot keep the response at the desired set point but instead it exhibits an offset:

$$\text{offset} = (\text{set point}) - (\text{ultimate value of response})$$

$$= 0 - K'_d = -\frac{K_d}{1 + K_p K_c}$$

The benefit of the proportional control in the presence of load changes can be seen from Figure 14.5b. Although it cannot keep the process response at the desired set point and introduces an offset, the response is much closer to the desired set point than would have been with no control at all. Furthermore, as we increase the gain K_c the offset decreases and theoretically,

$$\text{offset} \longrightarrow 0 \qquad \text{when} \qquad K_c \longrightarrow \infty$$

Remarks

1. Although the offset tends to zero as $K_c \to \infty$, we will never use extremely large values of K_c for proportional control. The reason will become very clear in the next chapter, where we will study the stability of closed-loop systems.
2. If $G_m = K_m$ and $G_f = K_f$, it is easy to show that the offsets become:
   For set point unit step changes,

$$\text{offset} = 1 - \frac{K_p K_c K_f}{1 + K_p K_c K_f K_m}$$

   For load unit step changes,

$$\text{offset} = -\frac{K_d}{1 + K_p K_c K_f K_m}$$

   Remark (1) still holds.
3. In subsequent sections we will examine only the response for the servo problem assuming that the reader has gained the facility to repeat a similar analysis for the regulator problem.
4. Processes having the term $1/s$ in their transfer function, when they are controlled with proportional controller, do not exhibit offset for set point changes but they do for sustained load changes (e.g., step changes). Let us demonstrate this important feature for the liquid-level control system shown in Figure 14.6a. The output

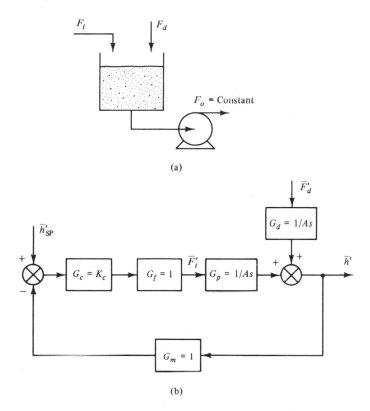

Figure 14.6 (a) Pure integrator; (b) corresponding closed-loop block diagram.

F_o is constant and the level is controlled by manipulating the inlet flow rate F_i. The load (disturbance) is the flow rate F_d. In terms of deviation variables, the mass balance around the tank yields

$$A \frac{dh'}{dt} = F'_i + F'_d$$

and in the Laplace domain,

$$\bar{h}'(s) = \frac{1}{As} \bar{F}'_i(s) + \frac{1}{As} \bar{F}'_d(s)$$

Therefore,

$$G_p(s) = \frac{1}{As}$$

Consider proportional control and for simplicity, $G_m = G_f = 1$.

The closed-loop block diagram is shown in Figure 14.6b and gives

$$\overline{h}'(s) = \frac{1}{\dfrac{A}{K_c} s + 1}\, \overline{h}'_{SP}(s) + \frac{1/K_c}{\dfrac{A}{K_c} s + 1}\, \overline{F}'_d(s)$$

For a unit step change in the set point we have $\overline{h}'_{SP}(s) = 1/s$ and $\overline{F}'_d(s) = 0$. Then

$$\overline{h}'(s) = \frac{1}{\dfrac{A}{K_c} s + 1}\, \frac{1}{s}$$

From the final-value theorem,

$$h'(t \to \infty) = \lim_{s \to 0} [s\overline{h}'(s)] = 1$$

Therefore,

$$\text{offset} = h'_{SP} - h'(t \to \infty) = 1 - 1 = 0$$

For a load unit step change,

$$\overline{h}'(s) = \frac{1/K_c}{\dfrac{A}{K_c} s + 1}\, \frac{1}{s}$$

and

$$h'(t \to \infty) = \lim_{s \to 0} [s\overline{h}'(s)] = \frac{1}{K_c}$$

Therefore,

$$\text{offset} = 0 - \frac{1}{K_c} = \frac{-1}{K_c} \neq 0$$

For liquid-level control systems such as the one of Figure 14.6a, usually we are not interested in maintaining the liquid level exactly at the desired value but within a certain range. In such case the value of the offset $1/K_c$ may be acceptable for reasonably large K_c. Therefore, the foregoing two conclusions lead to the following statement:

Liquid level can be controlled effectively with proportional control.

A similar conclusion can be reached for gas pressure systems whose transfer function also includes the term $1/s$.

Second-order systems (servo problem)

The transfer function for a second-order process is

$$G_p(s) = \frac{\overline{y}(s)}{\overline{m}(s)} = \frac{K_p}{\tau^2 s^2 + 2\zeta\tau s + 1}$$

Put this expression in eq. (14.18) and recalling that for the servo problem $\overline{d}(s) = 0$, we take

$$\overline{y}(s) = \frac{K'_p}{(\tau')^2 s^2 + 2\zeta'\tau' s + 1}\, \overline{y}_{SP}(s) \tag{14.22}$$

where

$$\tau' = \frac{\tau}{\sqrt{1 + K_p K_c}} \tag{14.23a}$$

$$\zeta' = \frac{\zeta}{\sqrt{1 + K_p K_c}} \tag{14.23b}$$

$$K'_p = \frac{K_p K_c}{1 + K_p K_c} \tag{14.23c}$$

From the above we notice that the closed-loop response of a second-order system with proportional control has the following characteristics:

It remains second-order.

The static gain decreases.

Both the natural period and damping factor decrease. This implies that an overdamped process, with proportional control and appropriate value of K_c, may become underdamped (oscillatory).

Consider a unit step change in the set point [i.e., $\overline{y}_{SP}(s) = 1/s$]. Then

$$\overline{y}(s) = \frac{K'_p}{(\tau')^2 s^2 + 2\zeta'\tau' s + 1}\, \frac{1}{s}$$

Depending on the value of ζ', the inverse of the expression above may be given by

Equation (11.7) for the overdamped case ($\zeta' > 1$), or
Equation (11.8) for the critically damped case ($\zeta' = 1$), or
Equation (11.9) for the underdamped case ($\zeta' < 1$)

Independently of the particular value of ζ', the ultimate value of $y(t)$ is given by the final-value theorem. Thus

$$y(t \to \infty) = \lim_{s \to 0} [s\overline{y}(s)] = K_p' = \frac{K_p K_c}{1 + K_p K_c}$$

Consequently, we again notice the presence of offset:

offset = (new set point) − (ultimate value of response)

$$= 1 - \frac{K_p K_c}{1 + K_p K_c} = \frac{1}{1 + K_p K_c}$$

Again, offset → 0 for $K_c \to \infty$.

Remarks

1. Depending on the value of the damping factor ζ for the uncontrolled second-order system, eq. (14.23b) shows that $\zeta' \lessgtr 1$. If $\zeta' > 1$, the overdamped response of the closed-loop system is very sluggish. Therefore, we prefer to increase the value of K_c and make $\zeta' < 1$. Then the closed-loop response reacts faster but it becomes oscillatory. Also, by increasing K_c, the offset decreases.
2. The increase in the speed of system's response and the decrease in the offset, both very desirable features, come at the expense of higher overshoots (maximum errors) and longer oscillating responses. Thus, as K_c increases, causing ζ' to decrease:
   (a) From eq. (11.11) we see that the overshoot increases while
   (b) Equation (11.12) shows that the decay ratio also increases.
   (c) Finally, eq. (11.13) shows that the period of oscillation for the closed-loop response decreases as ζ' decreases.

All the features above are demonstrated in Figure 14.7.

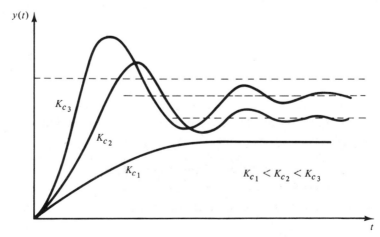

Figure 14.7 Effect of gain on the closed-loop response of second-order systems with proportional control.

14.3   Effect of Integral Control Action

In this section we repeat an analysis similar to that of the preceding section but using an integral instead of a proportional controller. Not to overwhelm the reader with the repetition of algebraic manipulations, we will limit our attention to first-order systems and for the servo problem only.

Recall that for the servo problem, $\overline{d}(s) = 0$, and eq. (14.5) yields

$$\overline{y}(s) = \frac{G_p G_f G_c}{1 + G_p G_f G_c G_m} \overline{y}_{SP}(s) \tag{14.24}$$

For simplicity let

$$G_m = G_f = 1$$

For a first-order process we have

$$G_p = \frac{K_p}{\tau_p s + 1}$$

and for a simple integral control action,

$$G_c = K_c \frac{1}{\tau_I s}$$

Substitute G_m, G_p, G_c, G_f in eq. (14.24), by their equals, and take

$$\overline{y}(s) = \frac{\left(\dfrac{K_p}{\tau_p s + 1}\right)\left(K_c \dfrac{1}{\tau_I s}\right)}{1 + \left(\dfrac{K_p}{\tau_p s + 1}\right)\left(K_c \dfrac{1}{\tau_I s}\right)} \overline{y}_{SP}(s)$$

or

$$\overline{y}(s) = \frac{1}{\tau^2 s^2 + 2\zeta\tau s + 1} \overline{y}_{SP}(s) \tag{14.25}$$

where

$$\tau = \sqrt{\frac{\tau_I \tau_p}{K_p K_c}} \tag{14.26a}$$

$$\zeta = \frac{1}{2}\sqrt{\frac{\tau_I}{\tau_p K_p K_c}} \tag{14.26b}$$

Equation (14.25) indicates an important effect of the integral control action,

It increases the order of dynamics for the closed-loop response.

Thus for a first-order uncontrolled process, the response of the closed-loop becomes second-order and consequently it may have drastically different dynamic characteristics. Furthermore, as we have seen in Sections 11.3 and 12.1, by increasing the order of a system, its response becomes more sluggish. Thus:

Integral control action alone is expected to make the response of the closed-loop system more sluggish.

Let us examine the dynamic behavior of the closed-loop system when the set point changes by a unit step. From eq. (14.25) we take

$$\overline{y}(s) = \frac{1}{\tau^2 s^2 + 2\zeta\tau s + 1} \frac{1}{s}$$

The shape of the response $y(t)$ depends on the value of ζ (overdamped, critically damped, or underdamped), but the ultimate value of the response can be found from the final-value theorem (Section 7.5):

$$y(t \to \infty) = \lim_{s \to 0} [s\overline{y}(s)] = \lim_{s \to 0} \left[\frac{1}{\tau^2 s^2 + 2\zeta\tau s + 1} \right] = 1$$

Therefore,

$$\text{offset} = 1 - 1 = 0$$

This indicates the most characteristic effect of integral action:

Integral control action eliminates any offset.

The reader can verify easily that for the regulator problem the integral control action produces a second-order closed-loop response and leads again to zero offset.

Remarks

1. Equation (14.26b) indicates that the form of the closed-loop response (i.e., overdamped, critically damped, underdamped) depends on the values of the controller gain K_c and reset time τ_I. Therefore, tuning the integral control action for the appropriate values of K_c and τ_I is an important question and will be discussed in Chapters 16 and 18.
2. From eq. (14.26b) we observe that as K_c increases, the damping factor ζ decreases. The consequences of decreasing ζ are:
   (a) The response moves in general from sluggish overdamped to faster but oscillatory underdamped behavior.

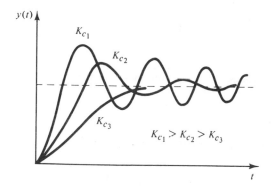

Figure 14.8 Effect of gain on the closed-loop response of first-order systems with integral control only.

(b) The overshoot and the decay ratio of the closed-loop response both increase [see eqs. (11.11) and (11.12) and Figure 11.3].
Therefore, we conclude that we can improve the speed of the closed-loop response at the expense of higher deviations and longer oscillations. Figure 14.8 summarizes the foregoing characteristics for set point changes.
3. From eq. (14.26b) we also observe that as τ_I decreases, ζ decreases too. Therefore, the consequences of decreasing τ_I on the closed-loop response will be as above in Remark 2 (i.e., increased speed comes at the expense of higher overshoots and long oscillations). Figure 14.9 demonstrates these effects very clearly.
4. The conclusions drawn by Remarks 1 and 2 can be restated as follows:

Increasing the integral control action (i.e., increasing K_c and

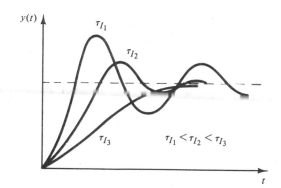

Figure 14.9 Effect of reset time on the closed-loop response of first-order systems with integral action only.

decreasing τ_I) makes the response of the closed-loop system more sensitive.

In Chapter 15 we will see that such trends lead to instability of the closed-loop response.

14.4   Effect of Derivative Control Action

For derivative control action alone, we have

$$G_c = K_c \tau_D s$$

Assuming again for simplicity that $G_m = G_f = 1$, the closed-loop response of a first-order system with derivative control action is given by

$$\overline{y}(s) = \frac{\dfrac{K_p}{\tau_p s + 1} K_c \tau_D s}{1 + \dfrac{K_p}{\tau_p s + 1} K_c \tau_D s} \overline{y}_{SP}(s)$$

or

$$\overline{y}(s) = \frac{K_p K_c \tau_D s}{(\tau_p + K_p K_c \tau_D)s + 1} \overline{y}_{SP}(s) \qquad (14.27)$$

Equation (14.27) leads to the following observations on the effects that the derivative control action has on the closed-loop response of a system:

1. The derivative control *does not change the order of the response*. In the example above it has remained first order.
2. From eq. (14.27) it is clear that the effective time constant of the closed-loop response is $(\tau_p + K_p K_c \tau_D)$, i.e., larger than τ_p. This means that the *response of the controlled process is slower than that of the original first-order process*. Furthermore, as K_c increases, the effective time constant increases and the response becomes progressively slower.

Remarks

1. It is very instructive to examine the effect of the derivative control action on the response of a second-order system. Assuming

again that $G_m = G_f = 1$, the closed-loop response for the servo problem is

$$\overline{y}(s) = \frac{\dfrac{K_p}{\tau^2 s^2 + 2\zeta\tau s + 1} K_c \tau_D s}{1 + \dfrac{K_p}{\tau^2 s^2 + 2\zeta\tau s + 1} K_c \tau_D s}\, \overline{y}_{SP}(s)$$

or

$$\overline{y}(s) = \frac{K_p K_c \tau_D s}{\tau^2 s^2 + (2\zeta\tau + K_p K_c \tau_D)s + 1}\, \overline{y}_{SP}(s)$$

From the last equation we observe that:
(a) The natural period of the closed-loop response remains the same while
(b) The new damping factor ζ' can be found from the equation

$$2\zeta'\tau = 2\zeta\tau + K_p K_c \tau_D$$

(i.e., $\zeta' > \zeta$). Therefore, the closed-loop response is more damped and the damping increases as K_c or τ_D increase. This characteristic produces more robust behavior by the controlled process.
2. The decrease in the speed of the response and the increase in the damping demonstrate that

the derivative control action produces more robust behavior by the controlled process.

14.5   Effect of Composite Control Actions

Although proportional control can be used alone, this is almost never the case for integral or derivative control actions. Instead, proportional-integral (PI) and proportional-integral-derivative (PID) are the usual controllers employing integral and derivative modes of control.

Effect of PI control

Combination of proportional and integral control modes leads to the following effects on the response of a closed-loop system:

1. The order of the response increases (effect of integral mode).
2. The offset is eliminated (effect of integral mode).

3. As K_c increases, the response becomes faster [effect of proportional and integral modes] and more oscillatory to set point changes [i.e., the overshoot and decay ratio increase (effect of integral mode)]. Large values of K_c create a very sensitive response and may lead to instability (see Chapter 15).

4. As τ_I decreases, for constant K_c, the response becomes faster but more oscillatory with higher overshoots and decay ratios (effect of integral mode).

Effect of PID control

Combination of the three control modes leads to a closed-loop response which has in general the same qualitative dynamic characteristics as those resulting from PI control alone. Let us now describe the main benefit introduced by the derivative control action.

We have seen that the presence of integral control slows down the closed-loop response of a process. To increase the speed of the closed loop response we can increase the value of the controller gain K_c. But increasing enough K_c in order to have acceptable speeds, the response becomes more oscillatory and may lead to instability. The introduction of the derivative mode brings a stabilizing effect to the system. Thus we can achieve acceptable response speed by selecting an appropriate value for the gain K_c while maintaining moderate overshoots and decay ratios.

Figure 14.10 summarizes the effect of a PID controller on the response of a controlled process. Notice that although increasing K_c leads to faster responses, the overshoot remains almost the same and the settling time is shorter. Both are results of the derivative control action.

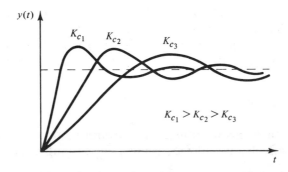

Figure 14.10 Effect of gain on the closed-loop response of first-order systems with PID control.

THINGS TO THINK ABOUT

1. Develop the block diagram of a generalized feedback control system with one disturbance, incorporating in each block the appropriate transfer function and on each stream the appropriate variable.

2. Develop the closed-loop responses for set point and load changes.

3. Repeat items 1 and 2 for a process with two disturbances. Can the feedback controller handle simultaneous changes in both loads?

4. Define in physical terms the servo and regulator control problems.

5. The following block diagram (Figure Q14.1) corresponds to a control system with two loops. Reduce the block diagram to a simpler one, such as that shown in Figure Q14.2, by identifying the appropriate transfer functions G_1, G_2, and G_3.

6. What are the relative advantages and disadvantages of the proportional, integral, and derivative control actions? What are their characteristic effects on the closed-loop response of a process?

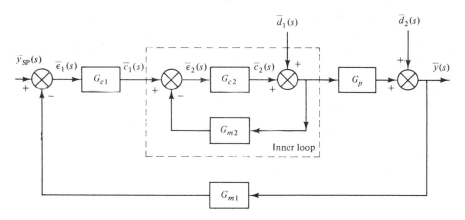

Figure Q14.1

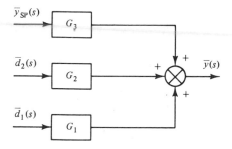

Figure Q14.2

7. The proportional control leads to a lower static gain for the closed-loop response compared to the gain of the uncontrolled process [see eqs. (14.20b) and (14.20c)]. Is a lower gain more or less favorable for the controlled process? Recall the definition of the static gain from Section 10.4.

8. What is the order of the closed-loop dynamic response for a second-order process with PI control? Can the PI control destabilize such a process?

9. Discuss the effects of K_c and τ_I on the closed-loop response of a process controlled with PI.

10. Discuss the effects of K_c, τ_I, and τ_D on the closed-loop response of a process controlled with PID.

11. Consider a first-order process. Could you have almost the same closed-loop responses with PI and PID controllers and appropriate values of their adjustable parameters?

12. Repeat item 11 but for a second-order process.

13. Which one the three controllers, P, PI, PID, would give more robust closed-loop response to an underdamped second-order system?

14. Integral control action makes a process (a) faster or slower; (b) more oscillatory or less; (c) with larger deviations from the set point or smaller? Explain your answers.

15. Repeat item 14 but for derivative control action.

Stability Analysis 15
of Feedback Systems

In Chapter 14 we examined the dynamic characteristics of the response of closed-loop systems, and developed the closed-loop transfer functions that determine the dynamics of such systems. It is important to emphasize again that the *presence of measuring devices, controllers, and final control elements changes the dynamic characteristics of an uncontrolled process*. Thus nonoscillatory first-order processes may acquire oscillatory behavior with PI control. Oscillatory second-order processes may become unstable with a PI controller and an unfortunate selection of K_c and τ_I.

While designing a feedback control system (i.e., selecting its components and tuning its controller), we are seriously concerned about its stability characteristics. Therefore, before we proceed with the particular details of designing a feedback control loop, we will study the notion of stability and analyze the stability characteristics of closed-loop systems.

15.1 The Notion of Stability

In Section 1.2 we introduced a simpleminded notion of stability. A system was considered unstable if, after it had been disturbed by an input change, its output "took off" and did not return to the initial state of rest. Figure 1.6 shows typical outputs for unstable processes. Example 1.2 also described the unstable operation of a CSTR.

How do we define a stable or unstable system? There are different ways, depending on the mathematical rigorousness of the definition and its practical utility for realistic applications. In this text we employ the following definition, which is often known as *bounded input, bounded output stability*:

> A dynamic system is considered to be stable if for every bounded input it produces a bounded output, regardless of its initial state.

Every system that is not stable according to the definition above will be called unstable. To complete the definition, consider that:

> *"Bounded" is an input that always remains between an upper and a lower limit (e.g., sinusoidal, step, but not the ramp).*
>
> *Unbounded outputs exist only in theory and not in practice because all physical quantities are limited. Therefore, the term "unbounded" means very large.*

According to the definition above, a system with response like those of Figure 15.1a is stable, while Figure 15.1b shows the responses of unstable systems.

Let us consider a dynamic system with input m and output y. Its dynamic behavior can be described by a transfer function $G(s)$,

$$\overline{y}(s) = G(s)\overline{m}(s)$$

In Section 9.4 we concluded that if $G(s)$ has a pole with positive real part, it gives rise to a term $C_1 e^{pt}$ which grows continuously with time, thus producing an unstable system. The transfer function $G(s)$ can correspond to an uncontrolled process or it can be the closed-loop transfer function of a controlled system (e.g., G_{SP} or G_{load}). Therefore,

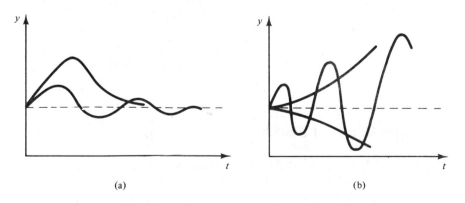

(a)   (b)

Figure 15.1 (a) Stable and (b) unstable responses.

the stability analysis of a system can be treated in a unified way independently if it is controlled or uncontrolled.

The location of the poles of a transfer function gives us the first criterion for checking the stability of a system:

> If the transfer function of a dynamic system has even one pole with positive real part, the system is unstable.

Therefore, all poles of a transfer function must be in the left-hand part of a complex plane, for the system to be stable.

Example 15.1: Stabilization of an Unstable Process with P Control

Consider a process with the following response:

$$\overline{y}(s) = \frac{10}{s-1}\,\overline{m}(s) + \frac{5}{s-1}\,\overline{d}(s)$$

Clearly, this process is unstable because its transfer function possesses a

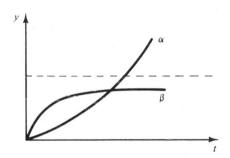

Figure 15.2 Curve α, open-loop unstable response; curve β, closed-loop stable response with P control.

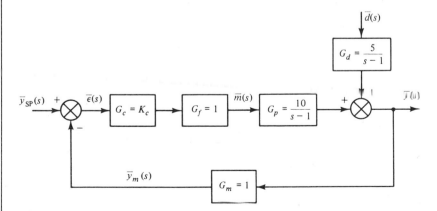

Figure 15.3 Block diagram for the system of Example 15.1.

pole at $s = 1 > 0$. Figure 15.2 (curve α) shows the response of the uncontrolled system to a unit step change in the load d which verifies its unstable character. Let us introduce a feedback control system with proportional control only. Assume that for the measuring sensor and the final control element

$$G_m = G_f = 1$$

Figure 15.3 shows the block diagram of the closed-loop system.

The closed-loop response of the system is given by eq. (14.5), which for the present system becomes

$$\overline{y}(s) = \frac{10K_c}{s - (1 - 10K_c)} \overline{y}_{SP}(s) + \frac{5}{s - (1 - 10K_c)} \overline{d}(s)$$

From the last equation we conclude that the closed-loop transfer functions

$$G_{SP} = \frac{10K_c}{s - (1 - 10K_c)} \qquad G_{load} = \frac{5}{s - (1 - 10K_c)}$$

have negative the common pole if $K_c > \frac{1}{10}$. Therefore, the original system can be stabilized with simple proportional control. Figure 15.2 (curve β) shows the dynamic response of the controlled system to a unit step change in the load for $K_c = 1$. Compare it to the behavior of the uncontrolled system and realize the stabilizing effect of the controller.

Example 15.2: Destabilization of a Stable Process with PI Control

Consider a second-order process with the following transfer function:

$$G_p(s) = \frac{1}{s^2 + 2s + 2}$$

The system has two complex poles with *negative real parts*:

$$p_1 = -1 + j \qquad \text{and} \qquad p_2 = -1 - j$$

Therefore, according to our criterion the system is stable. Indeed, if we make a unit step change in the input, the response of the system is as shown in Figure 15.4a. Introduce a PI controller. Let the measuring ele-

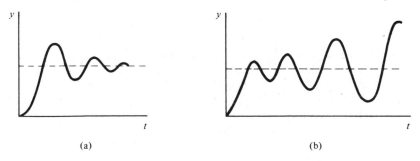

(a)　　　　　　　　　　　　(b)

Figure 15.4 (a) Open-loop stable response; (b) destabilized response under PI control.

ment and the final control element have the following transfer functions:

$$G_m(s) = G_f(s) = 1$$

The closed-loop response to set point changes is given by

$$\bar{y}(s) = \frac{G_p G_c}{1 + G_p G_c} \, \bar{y}_{SP}(s) = G_{SP} \bar{y}_{SP}(s)$$

To examine the stability of the closed-loop response, we have to find where the poles of G_{SP} are located.

$$G_{SP} = \frac{G_p G_c}{1 + G_p G_c} = \frac{\dfrac{1}{s^2 + 2s + 2} K_c \dfrac{\tau_I s + 1}{\tau_I s}}{1 + \dfrac{1}{s^2 + 2s + 2} K_c \dfrac{\tau_I s + 1}{\tau_I s}} = \frac{K_c(\tau_I s + 1)/\tau_I}{s^3 + 2s^2 + (2 + K_c)s + \dfrac{K_c}{\tau_I}}$$

Let

$$K_c = 100 \qquad \text{and} \qquad \tau_I = 0.1$$

Then the poles of G_{SP} are determined by the roots of the polynomial

$$s^3 + 2s^2 + (2 + 100)s + \frac{100}{0.1}$$

and are found to be

$$p_1 = -7.185 \qquad p_2 = 2.59 + j(11.5) \qquad p_3 = 2.59 - j(11.5)$$

We notice that p_2 and p_3 have *positive real parts*. Therefore, according to our criterion the closed-loop response is unstable. Figure 15.4b shows the response of the system to a unit step change of the set point. Compare it to the response of the uncontrolled system and notice the destabilizing effect of the PI controller. For different values of K_c and τ_I the response becomes stable. Indeed, lowering the gain to $K_c = 10$ and increasing $\tau_I = 0.5$, we find that all the poles of G_{SP} have negative real parts (i.e., the closed-loop system is stable).

15.2 The Characteristic Equation

Examples 15.1 and 15.2 dramatized the effect that a feedback control loop may have on the stability characteristics of a process. In this section we organize and systematize our analysis, introducing and defining some appropriate terms.

Consider the generalized feedback control system shown in Figure 14.1. The closed-loop response for such system is given by eq. (14.5):

$$\bar{y}(s) = \frac{G_p G_f G_c}{1 + G_p G_f G_c G_m} \, \bar{y}_{SP}(s) + \frac{G_d}{1 + G_p G_f G_c G_m} \, \bar{d}(s) \quad (14.5)$$

or equivalently

$$\bar{y}(s) = G_{SP} \bar{y}_{SP}(s) + G_{load} \bar{d}(s)$$

The stability characteristics of the closed-loop response will be determined by the poles of the transfer functions G_{SP} and G_{load}. *These poles are common for both transfer functions* (because they have common denominator) and are given by the solution of the equation

$$1 + G_p G_f G_c G_m = 0 \tag{15.1}$$

Equation (15.1) is called the *characteristic equation* for the generalized feedback system of Figure 14.1.

Let $p_1, p_2, \ldots, p_n$ be the n roots of the characteristic equation (15.1):

$$1 + G_p G_f G_c G_m = (s - p_1)(s - p_2) \cdots (s - p_n)$$

Then we can state the following criterion for the stability of a closed-loop system:

A feedback control system is stable if all the roots of its characteristic equation have negative real parts (i.e., are to the left of the imaginary axis).

If any root of the characteristic equation is on or to the right of the imaginary axis (i.e., it has real part zero or positive), the feedback system is unstable.

Remarks

1. The stability criterion stated above secures stable response of a feedback system independently if the input changes are in the set point or the load. The reason is that the roots of the characteristic equation are the common poles of the two transfer functions, G_{SP} and G_{load}, which determine the stability of the closed loop with respect to changes in the set point and the load, respectively.
2. The product

$$G_{OL} = G_p G_f G_c G_m$$

will be called *open-loop transfer function* because it relates the measurement indication y_m to the set point y_{SP} if the feedback loop is broken just before the comparator:

$$\overline{y}_m(s) = G_{OL}(s)\overline{y}_{SP}(s)$$

Therefore, the characteristic equation can be written as follows:

$$1 + G_{OL} = 0$$

and we notice that *it depends only on the transfer functions of the elements in the loop* [i.e., it does not depend on G_d which is outside the loop].

3. The roots of the characteristic equation are also the poles of the closed-loop transfer functions, G_{SP} and G_{load}. For this reason they are often called *closed-loop poles*.

Example 15.3: Stability Analysis of Two Feedback Loops

In Example 15.1 we have

$$G_p = \frac{10}{s-1} \qquad G_f = 1 \qquad G_m = 1 \qquad G_c = K_c$$

Therefore, the corresponding characteristic equation is

$$1 + G_p G_f G_c G_m = 1 + \frac{10}{s-1} \cdot 1 \cdot K_c \cdot 1 = 0$$

which has the root

$$p = 1 - 10K_c$$

and the system is stable if $p < 0$ (i.e., $K_c > 1/10$).

For the system of Example 15.2, we have

$$G_p = \frac{1}{s^2 + 2s + 2} \qquad G_f = 1 \qquad G_m = 1 \qquad G_c = K_c\left(1 + \frac{1}{\tau_I s}\right)$$

The corresponding characteristic equation is

$$1 + G_p G_f G_c G_m = 1 + \frac{1}{s^2 + 2s + 2} \cdot 1 \cdot K_c\left(1 + \frac{1}{\tau_I s}\right) \cdot 1 = 0$$

For $K_c = 100$ and $\tau_I = 0.1$, the equation above yields

$$s^3 + 2s^2 + 102s + 1000 = 0$$

with roots -7.185, $2.59 + j(11.5)$ and $2.59 - j(11.5)$. The closed-loop system is unstable because two roots of the characteristic equation have positive real parts.

15.3 Routh–Hurwitz Criterion for Stability

The criterion of stability for closed-loop systems does not require calculation of the actual values of the roots of the characteristic polynomial. It only requires that we know if any root is to the right of the imaginary axis. The Routh–Hurwitz procedure allows us to test if any root is to the right of the imaginary axis and thus reach quickly a conclusion as to the stability of the closed-loop system without computing the actual values of the roots.

Expand the characteristic equation into the following polynomial form:

$$1 + G_p G_f G_c G_m \equiv a_0 s^n + a_1 s^{n-1} + \cdots + a_{n-1} s + a_n = 0$$

Let a_0 be positive. If it is negative, multiply both sides of the equation above by -1.

First test. If any of the coefficients $a_1, a_2, \ldots, a_{n-1}, a_n$ is negative, there is at least one root of the characteristic equation which has positive real part and the corresponding system is unstable. No further analysis is needed.

Second test. If all coefficients $a_0, a_1, a_2, \ldots, a_{n-1}, a_n$ are positive, then from the first test we cannot conclude anything about the location of the roots. Form the following array (known as the Routh array):

$$
\begin{array}{cccccc}
Row\ 1 & a_0 & a_2 & a_4 & a_6 & \cdots \\
2 & a_1 & a_3 & a_5 & a_7 & \cdots \\
3 & A_1 & A_2 & A_3 & \cdot\cdot & \cdots \\
4 & B_1 & B_2 & B_3 & \cdot & \cdots \\
5 & C_1 & C_2 & C_3 & \cdot & \cdots \\
& \cdot & \cdot\ \cdot\ \cdot\ \cdot\ \cdot\ \cdot\ \cdot \\
n+1 & W_1 & W_2 & \cdot & \cdot & \cdots
\end{array}
$$

where

$$
A_1 = \frac{a_1 a_2 - a_0 a_3}{a_1} \qquad A_2 = \frac{a_1 a_4 - a_0 a_5}{a_1} \qquad A_3 = \frac{a_1 a_6 - a_0 a_7}{a_1} \cdots
$$

$$
B_1 = \frac{A_1 a_3 - a_1 A_2}{A_1} \qquad B_2 = \frac{A_1 a_5 - a_1 A_3}{A_1} \cdots
$$

$$
C_1 = \frac{B_1 A_2 - A_1 B_2}{B_1} \qquad C_2 = \frac{B_1 A_3 - A_1 B_3}{B_1} \cdots
$$

etc.

Examine the elements of the first column of the array above:

$$
a_0 \quad a_1 \quad A_1 \quad B_1 \quad C_1 \quad \cdots \quad W_1
$$

(a) If any of these elements is negative, we have at least one root to the right of the imaginary axis and the system is unstable.

(b) The number of sign changes in the elements of the first column is equal to the number of roots to the right of the imaginary axis.

Therefore, a system is stable if all the elements in the first column of the Routh array are positive.

Example 15.4: Stability Analysis with the Routh–Hurwitz Criterion

Consider the feedback control system of Example 15.2. The characteristic equation is

$$s^3 + 2s^2 + (2 + K_c)s + \frac{K_c}{\tau_I} = 0$$

The corresponding Routh array can now be formed:

Row 1	1	$2 + K_c$
2	2	$\dfrac{K_c}{\tau_I}$
3	$\dfrac{2(2 + K_c) - K_c/\tau_I}{2}$	0
4	$\dfrac{K_c}{\tau_I}$	

The elements of the first column are

$$\left[1, \quad 2, \quad \frac{2(2 + K_c) - K_c/\tau_I}{2}, \quad \frac{K_c}{\tau_I} \right]$$

All are always positive except the third, which can be positive or negative depending on the values of K_c and τ_I.

1. If $K_c = 100$ and $\tau_I = 0.1$, the third element becomes $-398 < 0$, which means that the system is unstable. We have two sign changes in the elements of the first column. Therefore, we have two roots with positive real parts (see Example 15.2).
2. If $K_c = 10$ and $\tau_I = 0.5$, the third element is equal to $+2 > 0$, and the system is stable since all the elements of the first column are positive.
3. In general, the system is stable if K_c and τ_I satisfy the condition

$$2(2 + K_c) > \frac{K_c}{\tau_I}$$

Example 15.5: Critical Stability Conditions for a Feedback Loop

Return to Example 15.4 and let $\tau_I = 0.1$. Then the third element of the first column in the Routh array becomes

$$\frac{2(2 + K_c) - 10K_c}{2}$$

The value of K_c that makes the third element zero is

$$K_c = 0.5$$

and constitutes the critical condition for stability of the PI feedback control system. Therefore, according to the Routh–Hurwitz test, we have:

P Pı PıD

1. If $K_c < 0.5$, all the elements of the first column in the Routh array are positive and the system is stable (i.e., all the roots of the characteristic equation are located to the left of the imaginary axis).
2. If $K_c > 0.5$, the third element of the first column of the Routh array becomes negative. We have two sign changes in the elements of the first column; therefore, we have two roots of the characteristic equation located to the right of the imaginary axis.

It is clear therefore that as K_c increases, two roots of the characteristic equation move toward the imaginary axis and when $K_c = 0.5$, we have two roots on the imaginary axis (pure imaginary) which give rise to sustained sinusoidal term.

Remark.   The two purely imaginary roots can be found from the equation

$$2s^2 + \frac{K_c}{\tau_I} = 0$$

that is,

$$2s^2 + \frac{0.5}{0.1} = 0$$

and they are

$$\pm j(1.58)$$

The coefficients 2 and K_c/τ_I are the elements of the Routh array in the row which is located just before the row whose first column element is zero (i.e., the elements of the second row).

15.4 Root-Locus Analysis

The preceding examples have demonstrated very vividly that the stability characteristics of a closed-loop system depend on the value of gain K_c. Thus in Example 15.1 we notice that the closed-loop system becomes stable when $K_c > 1/10$. Also, in Example 15.4, the system is stable when

$$2(2 + K_c) > \frac{K_c}{\tau_I}$$

which for $\tau_I = 0.1$ yields

$$0 < K_c < 0.5$$

The *root loci* are merely the plots, in the complex plane, of the roots of the characteristic equation as the gain K_c is varied from zero to infinity. As such they are very useful in determining the stability characteristics of a closed-loop system as the gain K_c changes. Let us examine the construction of the root locus using a specific example.

Example 15.6: Root Locus of Two Capacities in Series with P Control

The two capacities in series may be two stirred tanks, two heaters, and so on, and have a transfer function

$$G_p(s) = \frac{K_p}{(\tau_1 s + 1)(\tau_2 s + 1)}$$

Let

$$G_m = G_f = 1 \quad \text{and} \quad G_c = K_c$$

Then the characteristic equation is

$$1 + \frac{K_p}{(\tau_1 s + 1)(\tau_2 s + 1)} K_c = 0$$

or

$$(\tau_1 s + 1)(\tau_2 s + 1) + K = 0 \quad \text{where} \quad K = K_p K_c$$

Consider K as the changing parameter instead of the gain K_c, and make the following observations:

1. When $K = 0$ (i.e., $K_c - 0$) the characteristic equation has as its roots the poles of the process:

$$p_1 = \frac{-1}{\tau_1} \quad \text{and} \quad p_2 = \frac{-1}{\tau_2}$$

2. As K increases from the zero value, the roots of the characteristic equation are given by

$$p_{1,2} = \frac{-(\tau_1 + \tau_2) \pm \sqrt{(\tau_1 + \tau_2)^2 - 4\tau_1\tau_2(1 + K)}}{2\tau_1\tau_2}$$

They are distinct real and negative as long as $(\tau_1 + \tau_2)^2 - 4\tau_1\tau_2(1 + K) > 0$, or

$$K < \frac{(\tau_1 + \tau_2)^2}{4\tau_1\tau_2} \quad 1$$

that is, as long as K_c satisfies the inequality

$$K_c < \frac{1}{K_p}\left[\frac{(\tau_1 + \tau_2)^2}{4\tau_1\tau_2} - 1\right] \tag{15.2}$$

3. When

$$K_c = \frac{1}{K_p}\left[\frac{(\tau_1 + \tau_2)^2}{4\tau_1\tau_2} - 1\right] \tag{15.3}$$

we have two equal roots,

$$p_1 = p_2 = -\frac{\tau_1 + \tau_2}{2\tau_1\tau_2}$$

4. For

$$K_c > \frac{1}{K_p}\left[\frac{(\tau_1 + \tau_2)^2}{4\tau_1\tau_2} - 1\right] \tag{15.4}$$

we have again two distinct roots which are complex conjugates of each other:

$$p_{1,2} = \frac{-(\tau_1 + \tau_2) \pm j\ \sqrt{4\tau_1\tau_2(1 + K) - (\tau_1 + \tau_2)^2}}{2\tau_1\tau_2} \tag{15.5}$$

Notice that the real part is equal to

$$-\frac{\tau_1 + \tau_2}{2\tau_1\tau_2}$$

and independent of K, while the imaginary part tends to infinity as $K \to \infty$.

Using the information above, we can construct the root locus of the system as follows:

1. The beginning of the root locus corresponds to $K_c = 0$ and is given by the points $A(-1/\tau_1, 0)$ and $B(-1/\tau_2, 0)$ (see Figure 15.5).
2. As long as K_c satisfies inequality (15.2), we have two distinct real and negative roots. Therefore, the root locus is given by two distinct curves which emanate from points A and B and remain on the real axis. Furthermore, the two curves move toward each other and meet at point C (Figure 15.5). At this point, K_c has the value given by eq. (15.3) and we have a double root.
3. For larger values of K_c satisfying inequality (15.4), we have again two distinct curves of the root locus because we have distinct, complex conjugate roots. Since the real part of the complex roots is constant (see eq. 15.5), the two branches of the root locus are perpendicular to the real axis and extend to infinity as $K_c \to \infty$.

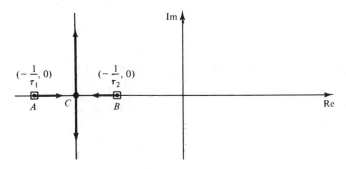

Figure 15.5 Root locus of the system in Example 15.6.

The complete root locus is given in Figure 15.5 and since all its branches are located to the left of the imaginary axis, we conclude that *the closed-loop system is stable for any value of K_c*. Furthermore, we conclude that for K_c satisfying inequality (15.2) the response of the system to a step input is not oscillatory because the imaginary part of the two roots is zero. It becomes oscillatory for K_c satisfying inequality (15.4).

Example 15.6 demonstrated that the root locus of a system not only provides information about the stability of a closed-loop system but also informs us about its general dynamic response characteristics as K_c changes. Therefore, the root locus analysis can be the basis of a feedback control loop design methodology, whereby the movement of the closed-loop poles (i.e., the roots of the characteristic equation) due to the change of the proportional controller gain can be clearly displayed.

Construction of the root locus for the system of Example 15.6 was rather simple. For higher-order systems, to find the exact location of the root locus branches we need a computer program that can find the roots of a high-order polynomial. Such programs are available in any computer system and the interested reader can find one in Ref. 7.

Quite often, though, we are not interested in the exact location of the root-locus branches and simple, but qualitatively correct graphs, will suffice to draw the general conclusions about the dynamic behavior of a closed-loop system. References 12, 13 and 14 give a set of general rules which can be used to draw the approximate root locus of any given system.

Let us close this chapter with one more example of the construction of the root locus for a reactor system and its use for the analysis of the system's dynamic response.

Example 15.7: Root Locus for a Reactor with Proportional Control

Douglas [Ref. 12] has developed the model for the reactor shown in Figure 15.6. The control objective is to keep the concentration of the desired product C as close as possible to a given steady-state value despite the upsets in the inputs of the reactor. He attempts to achieve this control

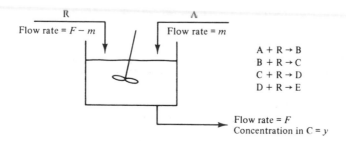

Figure 15.6 Reactor system of Example 15.7.

objective with a proportional controller which measures the concentration of C and manipulates the flow rate of the reactant A.

The transfer function for the process is

$$G_p(s) = \frac{\bar{y}(s)}{\bar{m}(s)} = \frac{2.98(s + 2.25)}{(s + 1.45)(s + 2.85)^2(s + 4.35)}$$

Assuming instantaneous responses with gain unity for the measuring device and the valve that controls the flow rate of A,

$$G_m = G_f = 1$$

we have the following characteristic equation for the closed-loop system:

$$1 + \frac{2.98(s + 2.25)}{(s + 1.45)(s + 2.85)^2(s + 4.35)} K_c = 0 \qquad (15.6)$$

When $K_c = 0$, it is easy to find that the roots of eq. (15.6) are

$$p_1 = -1.45 \qquad p_2 = p_3 = -2.85 \qquad p_4 = -4.35$$

As K_c increases, we need an iterative, trial-and-error, numerical procedure to find the roots of the characteristic equation. Such a solution is feasible through the use of a digital computer. Table 15.1 shows how the locations of the four roots change with the value of K_c. These results have been transferred in Figure 15.7, which displays the four branches of the root locus for the closed-loop reactor system.

TABLE 15.1
ROOTS OF THE CHARACTERISTIC EQUATION FOR THE SYSTEM OF EXAMPLE 15.7

K_c	p_1	p_2	p_3	p_4
0	−1.45	−2.85	−2.85	−4.35
1	−1.71	−2.30+ j(0.9)	−2.30− j(0.9)	−4.74
5	−1.98	−1.71+ j(1.83)	−1.71− j(1.83)	−5.87
20	−2.15	−1.09+ j(3.12)	−1.09− j(3.12)	−7.20
50	−2.20	−0.48+ j(4.35)	−0.48− j(4.35)	−8.61
100	−2.24	+0.35+ j(5.40)	+0.35− j(5.40)	−9.75

Let us examine the root-locus branches of Figure 15.7 and draw some conclusions on the dynamic response of the closed-loop reactor system as the proportional gain K_c changes from zero to infinity.

1. The system is stable for gain values up to 50 because all the roots are located to the left of the imaginary axis. For a gain value between 50 and 100 the root locus crosses the imaginary axis and moves to the right of the imaginary axis. Therefore, there is a critical value between 50 and 100 for which the closed-loop response of the reactor becomes unstable.
2. For any value of $K_c > 0$ until the critical value there are two complex conjugate roots with negative real parts. They imply that the response of the reactor to an input change will be a decaying oscillation.

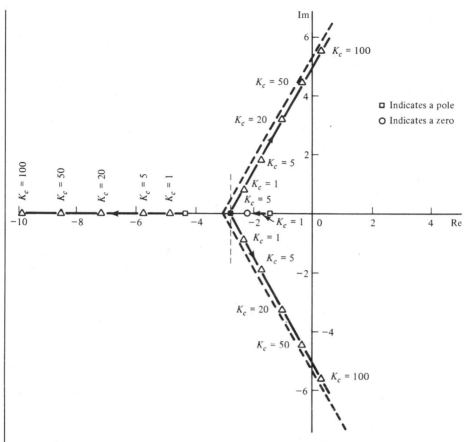

Figure 15.7 Root locus of the reactor in Example 15.7. Reproduced from J.M. Douglas, *Process Dynamics and Control: Control Systems Synthesis*, Vol. 2, (c) 1972, p. 112. Reprinted by permission of Prentice-Hall, Inc., Englewood Cliffs, N.J.

3. For K_c larger than the critical value (where the system becomes unstable) the roots that cause the instability are complex conjugates with positive real parts. Consequently, the unstable response of the closed-loop system to an input change will be oscillatory with growing amplitude.

THINGS TO THINK ABOUT

1. Define what is known as bounded input, bounded output stability.

2. Based on the definition above, examine if a system with a pole at $s = 0$ is stable.

3. Define the following terms: open-loop transfer function, characteristic equation, closed-loop poles.

4. If a closed-loop response is stable with respect to changes in the set point, is it stable to changes in the load? If yes, why?

5. How does the pole location determine the stability of an uncontrolled or controlled process?

6. Does the location of the zeros of a transfer function affect the response of an uncontrolled process?

7. What is the major advantage of the Routh–Hurwitz criterion for examining the stability of a system?

8. What conclusions can be drawn if one element in the first column of the Routh array is zero? Consult Refs. 13, 14, and Example 15.5.

9. The root-locus analysis cannot handle easily systems with dead time. Why? Show how systems with dead time could be handled with root-locus analysis.

10. Examples 15.6 and 15.7 indicate that the root locus has as many branches as the number of poles of the open-loop transfer function. Thus, in Example 15.6 the open-loop transfer function has two poles and the root locus two branches, whereas for Example 15.7 we have four poles and four branches. Is this true for any closed-loop system? Explain.

Design of Feedback **16**
Controllers

In this chapter we confront the following two critical questions: How do we select the type of the feedback controller (i.e., P, PI, or PID), and how do we adjust the parameters of the selected controller (i.e., K_c, τ_I, τ_D) in order to achieve an "optimum" response for the controlled process? These two questions, as well as the methods that lead to their resolution, define the content of the present chapter.

16.1 Outline of the Design Problems

Consider the block diagram of a general closed-loop system shown in Figure 14.1

When the load or the set point change, the response of the process deviates and the controller tries to bring the output again close to the desired set point. Figure 16.1 shows the response of a controlled process to a unit step change in the load when different types of controllers have been used. We notice that different controllers have different effects on the response of the controlled process. Thus the first design question arises:

Question 1: *What type of feedback controller should be used to control a given process?*

Given that we have decided somehow to use PI control, we still need to

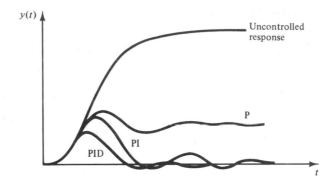

Figure 16.1 Response of a system to unit step change in load with no control, and various types of feedback controllers.

select the value of the gain K_c and the reset time τ_I. Figures 14.8 and 14.9 demonstrate very clearly that these two parameters have an important effect on the response of the controlled process. Thus the second design question:

> **Question 2:** *How do we select the best values for the adjustable parameters of a feedback controller?*

This is known as the *controller tuning* problem.

To answer these two design questions we need to have a quantitative measure in order to compare the alternatives and select the best type of controller and the best values of its parameters. Thus the third design question arises:

> **Question 3:** *What performance criterion should we use for the selection and the tuning of the controller?*

There are a variety of performance criteria we could use, such as:

> *Keep the maximum deviation (error) as small as possible.*
> *Achieve short settling times.*
> *Minimize the integral of the errors until the process has settled to its desired set point, and so on.*

As we will see in the following sections, different performance criteria lead to different control designs.

Let us now study the questions above in more detail and provide the initial guidelines for the design of a feedback controller.

16.2   Simple Performance Criteria

We start with the performance criteria since we need to establish some basis for the comparison of alternative controller designs, and because its selection constitutes the principal difficulty during the design of a feedback system.

Consider two different feedback control systems producing the two closed-loop responses shown in Figure 16.2. Response A has reached the desired level of operation faster than response B. If our criterion for the design of the controller had been

Return to the desired level of operation as soon as possible

then, clearly, we would select the controller which gives the closed-loop response of type A. But, if our criterion had been

Keep the maximum deviation as small as possible

or

Return to the desired level of operation and stay close to it in the shortest time

we would have selected the other controller, yielding the closed-loop response of type B. Similar dilemmas will be encountered quite often during the design of a controller.

For every process control application, we can distinguish

Steady-state performance criteria
Dynamic response performance criteria

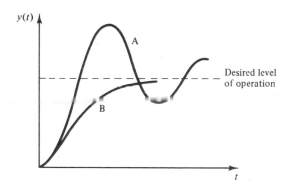

Figure 16.2 Alternative closed-loop responses.

The principal steady-state performance criterion usually is *zero error at steady state*. We have seen already that in most situations the proportional controller cannot achieve zero steady-state error, while a PI controller can. Also, we know that for proportional control the steady-state error (offset) tends to zero as $K_c \to \infty$. No further discussion is needed on the steady-state performance criteria.

The evaluation of the dynamic performance of a closed-loop system is based on two types of commonly used criteria:

1. Criteria that use only a few points of the response. They are simpler, but only approximate.
2. Criteria that use the entire closed-loop response from time $t = 0$ until t = very large. These are more precise but also more cumbersome to use.

In the remainder of this section we deal with the first category of simple performance criteria and we leave the more complicated criteria for Section 16.3.

The simple performance criteria are based on some characteristic features of the closed-loop response of a system. The most often quoted are (see Figure 11.2):

Overshoot

Rise time (i.e., time needed for the response to reach the desired value for the first time)

Settling time (i.e., time needed for the response to settle within ±5% of the desired value)

Decay ratio

Frequency of oscillation of the transient

Every one of the characertistics above could be used by the designer as the basic criterion for selecting the controller and the values of its adjusted parameters. Thus we could design the controller in order to have minimum overshoot, or minimum settling time, and so on. It must be emphasized, though, that one simple characteristic does not suffice to describe the desired dynamic response. Usually, we require that more objectives be satisfied (i.e., minimize overshoot, minimize settling time, etc.). Unfortunately, controller designs based on multiple criteria lead to conflicting response characteristics. For example, Figure 14.10 shows that for a PID controller, by decreasing the value of the overshoot (through a decrease in the value of gain K_c) we increase the settling time. Such conflicts will always arise while using simple design criteria such as those above. The control designer must intervene and subjectively balance the conflicting characteristics.

From all the performance criteria above, the decay ratio has been the most popular by the practicing engineers. Specifically, experience has shown that a decay ratio (see Figure 11.2)

$$\frac{C}{A} = \frac{1}{4}$$

is a reasonable trade-off between a fast rise time and a reasonable settling time. This criterion is usually known as the *one-quarter decay ratio* criterion.

Example 16.1: Controller Tuning with the One-Quarter Decay Ratio Criterion

Consider the servo control problem of a first-order process with PI controller. It can be easily shown that the closed-loop response is given by the following equation, when $G_m = G_f = 1$:

$$\overline{y}(s) = \frac{\tau_I s + 1}{\tau^2 s^2 + 2\zeta\tau s + 1} \overline{y}_{SP}(s)$$

where

$$\tau = \sqrt{\frac{\tau_I \tau_p}{K_p K_c}}$$

and

$$\zeta = \frac{1}{2}\sqrt{\frac{\tau_I}{\tau_p K_p K_c}}(1 + K_p K_c)$$

We notice that the closed-loop response is second-order.

For the selection of the "best" values for K_c and τ_I we will use simple criteria stemming from the underdamped response of a second-order system. Select the one-quarter decay ratio criterion. From eq. (11.12) we know that

$$\text{decay ratio} = \exp\left(\frac{-2\pi\zeta}{\sqrt{1 - \zeta^2}}\right)$$

Therefore, for our problem we have

$$\exp\left[\frac{-2\pi \cdot \frac{1}{2}\sqrt{\dfrac{\tau_I}{\tau_p K_p K_c}}(1 + K_p K_c)}{\sqrt{1 - \dfrac{1}{4}\dfrac{\tau_I}{\tau_p K_p K_c}(1 + K_p K_c)^2}}\right] = \frac{1}{4}$$

After algebraic simplifications we take

$$-2\pi\sqrt{\frac{\tau_I}{4\tau_p K_p K_c - \tau_I(1 + K_p K_c)^2}}(1 + K_p K_c) = \ln\left(\frac{1}{4}\right) \qquad (16.1)$$

Equation (16.1) has two unknowns: K_c and τ_I. Therefore, we will have several controller settings which satisfy the one-quarter decay ratio criterion. Let $K_p = 0.1$ and $\tau_p = 10$. Then, we find the following solutions:

$K_c = 1$	$K_c = 10$	$K_c = 30$	$K_c = 50$	$K_c = 100$
$\tau_I = 0.153$	$\tau_I = 0.464$	$\tau_I = 0.348$	$\tau_I = 0.258$	$\tau_I = 0.153$

and so on. The question is which one to select. Usually, we select first the proportional gain K_c so that the controller has the necessary "strength" to push the response back to the desired set point and then we choose the corresponding τ_I value so that the one-quarter decay ratio is satisfied.

16.3 Time-Integral Performance Criteria

The shape of the complete closed-loop response, from time $t = 0$ until steady state has been reached, could be used for the formulation of a dynamic performance criterion. Unlike the simple criteria that use only isolated characteristics of the dynamic response (e.g., decay ratio, settling time), the criteria of this category are based on the entire response of the process. The most often used are:

1. *Integral of the square error* (ISE), where

$$ISE = \int_0^\infty \epsilon^2(t) \, dt \qquad (16.2a)$$

2. *Integral of the absolute value of the error* (IAE), where

$$IAE = \int_0^\infty |\epsilon(t)| \, dt \qquad (16.2b)$$

3. *Integral of the time-weighted absolute error* (ITAE), where

$$ITAE = \int_0^\infty t \, |\epsilon(t)| \, dt \qquad (16.2c)$$

Note that $\epsilon(t) = y_{SP}(t) - y(t)$ is the deviation (error) of the response from the desired set point.

The problem of designing the "best" controller can now be formulated as follows:

Select the type of the controller and the values of its adjusted parameters in such a way as to minimize the ISE, IAE, or ITAE of the system's response.

Which one of the three criteria above we will use depends on the characteristics of the system we want to control and some additional requirements we impose on the controlled response of the process. The following are some general guidelines:

> *If we want to strongly suppress large errors, ISE is better than IAE because the errors are squared and thus contribute more to the value of the integral.*
>
> *For the suppression of small errors, IAE is better than ISE because when we square small numbers (smaller than one) they become even smaller.*
>
> *To suppress errors that persist for long times, the ITAE criterion will tune the controllers better because the presence of large t amplifies the effect of even small errors in the value of the integral.*

Figure 16.3 demonstrates, in a qualitative manner, the shape of the expected closed-loop responses. When we tune the controller parameters using ISE, IAE, or ITAE performance criteria, we should remember the following two points:

1. Different criteria lead to different controller designs.
2. For the same time integral criterion, different input changes lead to different designs.

Let us analyze these two statements on the basis of the following example.

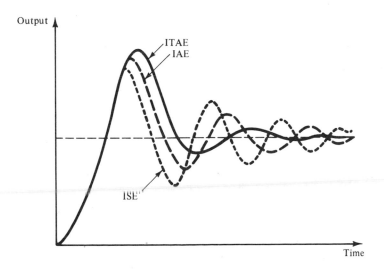

Figure 16.3 Closed-loop responses using various time integral criteria.

Example 16.2: Controller Tuning Using Time-Integral Criteria

Consider the feedback system shown in Figure 16.4. The closed-loop response is

$$\bar{y}(s) = \frac{\tau_I s + 1}{\dfrac{\tau_I}{20K_c}s^2 + \tau_I\left(1 + \dfrac{1}{20K_c}\right)s + 1}\bar{y}_{SP}(s) + \frac{(\tau_I/20K_c)s}{\dfrac{\tau_I}{20K_c}s^2 + \tau_I\left(1 + \dfrac{1}{20K_c}\right)s + 1}\bar{d}(s)$$

or

$$\bar{y}(s) = \frac{\tau_I s + 1}{\tau^2 s^2 + 2\zeta\tau s + 1}\bar{y}_{SP}(s) + \frac{(\tau_I/20\ K_c)s}{\tau^2 s^2 + 2\zeta\tau s + 1}\bar{d}(s) \tag{16.3}$$

where

$$\tau = \sqrt{\frac{\tau_I}{20K_c}} \tag{16.4a}$$

and

$$\zeta = \frac{1}{2}\sqrt{\frac{\tau_I}{20K_c}}(1 + 20K_c) \tag{16.4b}$$

To select the best values for K_c and τ_I, we can use one of the three criteria ISE, IAE, or ITAE. Furthermore, we can consider changes either in the load or the set point. Finally, even if we select set point changes, we still need to decide what kind of changes we will consider (i.e., step, sinusoidal, impulse, etc.). Let us say that we select ISE as the criterion and unit step changes in the set point. From eq. (16.3) we have

$$\bar{y}(s) = \frac{\tau_I s + 1}{\tau^2 s^2 + 2\zeta\tau s + 1}\frac{1}{s}$$

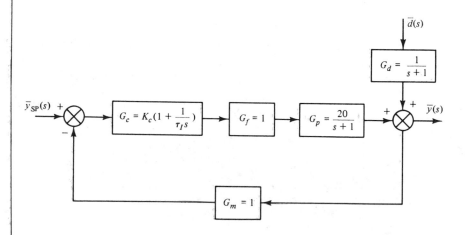

Figure 16.4 Closed-loop system of Example 16.2.

Invert the last equation and find (if $\zeta < 1$)

$$
y(t) = 1 + \frac{e^{-\zeta t/\tau}}{\sqrt{1 - \zeta^2}} \left[\frac{\tau_I}{\tau} \sin\left(\sqrt{1 - \zeta^2}\, \frac{t}{\tau} \right) \right.
$$

$$
\left. - \sin\left(\sqrt{1 - \zeta^2}\, \frac{t}{\tau} + \tan^{-1} \frac{\sqrt{1 - \zeta^2}}{\zeta} \right) \right]
$$

(16.5)

Then solve the following optimization problem:

Minimize ISE $= \displaystyle\int_0^\infty [y_{SP} - y(t)]^2 dt$ by selecting the values of τ and ζ, where $y(t)$ is given by eq. (16.5).

The optimal values of τ and ζ are given by the solution of the following equations (conditions for optimality):

$$
\frac{\partial(\text{ISE})}{\partial \tau} = \frac{\partial(\text{ISE})}{\partial \zeta} = 0
$$

Let τ^* and ζ^* be the optimal values. Then, from eqs. (16.4a) and (16.4b), we can find the corresponding optimal values for the controller parameters τ_I and K_c.

If the criterion was the ITAE, we would have to solve the following problem:

Minimize ITAE $= \displaystyle\int_0^\infty t\,|\,y_{SP} - y(t)\,|\,dt$ by selecting the values of τ and ζ where $y(t)$ is given by eq. (16.5).

The solution τ^* and ζ^* is given by the equations

$$
\frac{\partial(\text{ITAE})}{\partial \tau} = \frac{\partial(\text{ITAE})}{\partial \zeta} = 0
$$

and in turn, from eqs. (16.4a) and (16.4b) we can find the optimal K_c and τ_I.

It is clear that the solutions of the foregoing two problems with different criteria will be, in general, different.

Let us consider now unit step changes in the load. Equation (16.3) yields

$$
\bar{y}(s) = \frac{(\tau_I/20\, K_c)s}{\tau^2 s^2 + 2\zeta\tau s + 1} \frac{1}{s}
$$

and after inversion,

$$
y(t) = \frac{(\tau_I/20\, K_c)e^{-\zeta t/\tau}}{\tau \sqrt{1 - \zeta^2}} \sin\left(\sqrt{1 - \zeta^2}\, \frac{t}{\tau} \right)
$$

(16.6)

We can find the optimal values of K_c and τ_I, following a similar procedure as previously. Since the response $y(t)$ is now different than it was for a unit step change in the set point [compare eqs. (16.6) and (16.5)], we expect that the optimal settings of K_c and τ_I will be different even if we use the same criterion (i.e., ISE or ITAE).

Remark.   A proportional controller leads to a non-zero offset. Therefore, the value of the time-integral criteria, ISE, IAE, or ITAE is infinite, and the use of such criteria for tuning proportional controllers is analytically "difficult". In such cases, it is equivalent to tune the proportional controller for minimum offset, within the range of allowable values for the proportional gain.

16.4   Select the Type of Feedback Controller

Which one of the three popular feedback controllers should be used to control a given process? The question can be answered in a very systematic manner as follows:

1. Define an appropriate performance criterion (e.g., ISE, IAE, or ITAE).
2. Compute the value of the performance criterion using a P, or PI, or PID controller with the best settings for the adjusted parameters K_c, τ_I, and τ_D.
3. Select that controller which gives the "best" value for the performance criterion.

This procedure, although mathematically rigorous, has several serious practical drawbacks:

It is very tedious.

It relies on models (transfer functions) for the process, sensor, and final control element which may not be known exactly.

It incorporates certain ambiguities as to which is the most appropriate criterion and what input changes to consider.

Fortunately, we can select the most appropriate type of feedback controller using only general qualitative considerations stemming from the analysis in Chapter 14. There we had examined the effect of the proportional, integral, and derivative control modes on the response of a system. In summary, the conclusions were as follows:

1. Proportional control
   (a) Accelerates the response of a controlled process.
   (b) Produces an offset (i.e., nonzero steady-state error) for all processes except those with terms $1/s$ (integrators) in their

transfer function, such as the liquid level in a tank or the gas pressure in a vessel (see Remark 4 in Section 14.2).

2. Integral control
   (a) Eliminates any offset.
   (b) The elimination of the offset usually comes at the expense of higher maximum deviations.
   (c) Produces sluggish, long oscillating responses.
   (d) If we increase the gain K_c to produce faster response, the system becomes more oscillatory and may be led to instability.
3. Derivative control
   (a) Anticipates future errors and introduces appropriate action
   (b) Introduces a stabilizing effect on the closed-loop response of a process

Figure 16.1 reflects in a very simple way all the characteristics noted above.

It is clear from the above that a three-mode PID controller should be the best. This is true in the sense that it offers the highest flexibility to achieve the desired controlled response by having three adjustable parameters. At the same time, it introduces a more complex tuning problem because we have to adjust three parameters. To balance the quality of the desired response against the tuning difficulty we can adopt the following rules in selecting the most appropriate controller.

1. *If possible, use simple proportional controller.* Simple proportional controller can be used if (a) we can achieve acceptable offset with moderate values of K_c or (b) the process has an integrating action (i.e., a term $1/s$ in its transfer function) for which the P control does not exhibit offset. Therefore, for gas pressure or liquid-level control we can use only P controller.
2. *If a simple P controller is unacceptable, use a PI.* A PI controller should be used when proportional control alone cannot provide sufficiently small steady-state errors (offsets). Therefore, PI will seldom be used in liquid-level or gas presure control systems but very often (almost always) for flow control. The response of a flow system is rather fast. Consequently, the speed of the closed-loop system remains satisfactory despite the slowdown caused by the integral control mode.
3. *Use a PID controller to increase the speed of the closed-loop response and retain robustness.* The PI eliminates the offset but reduces the speed of the closed-loop response. For a multicapacity process whose response is very sluggish, the addition of a PI controller makes it even more sluggish. In such cases the addition of the derivative control action with its stabilizing effect allows

the use of higher gains which produce faster responses without excessive oscillations. Therefore, derivative action is recommended for temperature and composition control where we have sluggish, multicapacity processes.

Example 16.3. Selecting the Type of Controller for Various Processes

Let us discuss various processes that are to be controlled by feedback control systems. We will address primarily the question of selecting the appropriate type of feedback controller.

1. *Liquid-level control:* Consider the two liquid-level control systems for the bottom of a distillation column and its condenser's accumulation drum (Figure 13.2d and e). Our control objective is to keep each liquid level within a certain range around the desired set point. Consequently, proportional control alone is satisfactory.
2. *Gas pressure control:* Our objective is to regulate the pressure p in the tank of Figure PIII.2c, when the inlet pressure p_1 or the pressure p_3 in a downstream process change. Usually, we want to maintain p within a certain range around a desired value, thus making a proportional controller satisfactory for our purpose.
3. *Vapor pressure control:* Here we can have loops that react quite fast or are relatively slow. Consider, for example, the two configurations shown in Figure 16.5. The loop in Figure 16.5a measures the pressure and manipulates the flow of vapor, thus affecting directly and

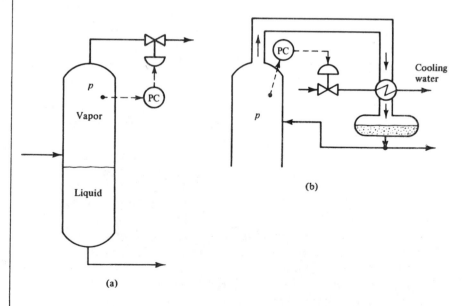

Figure 16.5 Pressure control loops: (a) direct effect, fast response; (b) indirect effect, slow response.

quickly the vapor pressure in the process. For such systems with fast response, a PI controller is satisfactory. It eliminates any undesirable offset while maintaining acceptable speed of response (despite some slowdown caused by the integral mode of control). For the system in Figure 16.5b the situation is different. Here, the vapor pressure is controlled indirectly by the flow of cooling water which affects the amount of vapor condensed. Such systems may be used for controlling the pressure in a distillation column. The slow dynamics of the heat transfer process are introduced in the control loop. We expect that the response of this sytem will be rather slow. A PI controller will make it even slower and if we attempted to use high gains to speed up the response, we may get a system with undesirable, highly oscillatory response. Therefore, a PID controller should be selected because it will provide enough speed and robustness.

4. *Flow control:* Consider the two flow control systems shown in Figure 13.2a and b. Both respond quite fast. Therefore, a PI controller is satisfactory because it eliminates offsets and retains acceptable speed of response.

5. *Temperature control:* Consider the temperature control system shown in Figure 16.6. Our objective is to keep the temperature of the reacting mixture at a desired value. Since the reaction is endothermic, this is accomplished by manipulating the flow of steam in the jacket around the reactor. Between the manipulated variable and the measured temperature we have two rather slow processes: (a) heat transfer between the reacting mixture and the temperature sensor (see Section 13.3), and (b) heat transfer from steam to the reacting mixture. We expect, therefore, that the overall response will be rather sluggish and a PI controller will make it even more so. Consequently, for such systems a PID controller would be the most appropriate, because it can allow high gains for faster response without undermining the stability of the system.

6. *Composition control:* Here we have a similar situation to that of temperature control (i.e., very slow response caused by slow com-

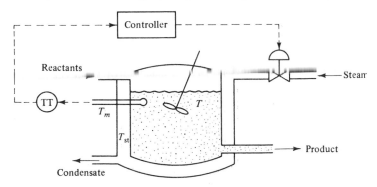

Figure 16.6 Temperature control in a jacketed CSTR.

position sensors). Therefore, a PID controller should be the most appropriate.

16.5 Controller Tuning

After the type of feedback controller has been selected, we still have the problem of deciding what values to use for its adjusted parameters. This is known as the *controller tuning* problem. There are three general approaches we can use for tuning a controller:

1. Use simple criteria such as the one-quarter decay ratio (see Example 16.1), minimum settling time, minimum largest error, and so on. Such an approach is simple and easily implementable on an actual process. Usually, it provides multiple solutions (see Example 16.1). Additional specifications on the closed-loop performance will then be needed to break the multiplicity and select a single set of values for the adjusted parameters.
2. Use time integral performance criteria such as ISE, IAE, or ITAE (see Example 16.2). This approach is rather cumbersome and relies heavily on the mathematical model (transfer function) of the process. Applied experimentally on an actual process, it is time consuming.
3. Use semiempirical rules which have been proven in practice.

In this section we discuss the most popular of the empirical tuning methods, known as the *process reaction curve method*, developed by Cohen and Coon.

Consider the control system of Figure 16.7, which has been "opened" by disconnecting the controller from the final control element. Introduce a step change of magnitude A in the variable c which

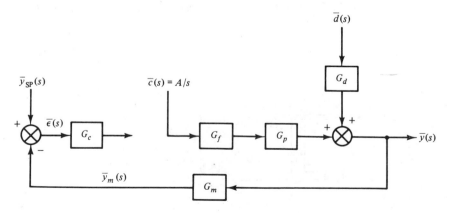

Figure 16.7 "Opened" control loop.

actuates the final control element. In the case of a valve, c is the stem position. Record the value of the output with respect to time. The curve $y_m(t)$ is called the *process reaction curve*. Between y_m and c we have the following transfer function (see Figure 16.7):

$$G_{PRC}(s) = \frac{\overline{y}_m(s)}{\overline{c}(s)} = G_f(s)G_p(s)G_m(s) \qquad (16.7)$$

The last equation shows that the process reaction curve is affected not only by the dynamics of the main process but also by the dynamics of the measuring sensor and final control element.

Cohen and Coon observed that the response of most processing units to an input change, such as the above, had a sigmoidal shape (see Figure 16.8a), which can be adequately approximated by the response of a first-order system with dead time (see the dashed curve in Figure 16.8b):

$$G_{PRC}(s) = \frac{\overline{y}_m(s)}{\overline{c}(s)} \simeq \frac{Ke^{-t_d s}}{\tau s + 1} \qquad (16.8)$$

which has three parameters: static gain K, dead time t_d, and time constant τ. From the approximate response of Figure 16.8b, it is easy to estimate the values of the three parameters. Thus

$$K = \frac{output\ (at\ steady\ state)}{input\ (at\ steady\ state)} = \frac{B}{A}$$

$\tau = B/S$, where S is the slope of the sigmoidal response at the point of inflection

t_d = time elapsed until the system responded

Cohen and Coon used the approximate model of eq. (16.8) and estimated the values of the parameters K, t_d, and τ as indicated above.

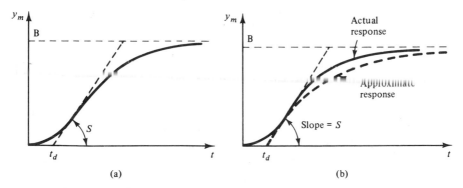

Figure 16.8 (a) Process reaction curve; (b) its approximation with a first-order plus dead-time system.

Then they derived expressions for the "best" controller settings using load changes and various performance criteria, such as:

One-quarter decay ratio
Minimum offset
Minimum integral square error (ISE)

The results of their analysis are summarized below.

1. For proportional controllers, use

$$K_c = \frac{1}{K} \frac{\tau}{t_d} \left(1 + \frac{t_d}{3\tau} \right) \tag{16.9}$$

2. For proportional-integral controllers, use

$$K_c = \frac{1}{K} \frac{\tau}{t_d} \left(0.9 + \frac{t_d}{12\tau} \right) \tag{16.10a}$$

$$\tau_I = t_d \frac{30 + 3t_d/\tau}{9 + 20t_d/\tau} \tag{16.10b}$$

3. For proportional-integral-derivative controllers, use

$$K_c = \frac{1}{K} \frac{\tau}{t_d} \left(\frac{4}{3} + \frac{t_d}{4\tau} \right) \tag{16.11a}$$

$$\tau_I = t_d \frac{32 + 6t_d/\tau}{13 + 8t_d/\tau} \tag{16.11b}$$

$$\tau_D = t_d \frac{4}{11 + 2t_d/\tau} \tag{16.11c}$$

Remarks

1. The controller settings given by eqs. (16.9), (16.10), and (16.11) are based on the assumption that the first-order plus dead-time system is a good approximation for the sigmoidal response of the open-loop real process. It is possible, though, that the approximation may be poor. In such a case the Cohen–Coon settings should be viewed only as first guesses needing certain on-line correction.
2. Why do most of the "opened" loops have a sigmoidal response like that of Figure 16.8a? The answer is rather clear using the analysis of Chapters 10 through 12. There we noticed that almost all physical processes encountered in a chemical plant are simple first-order or multicapacity processes whose response has the general overdamped shape of Figures 11.1a and 11.6. The oscillatory underdamped behavior is produced mainly by the presence of

feedback controllers. Therefore, when we "open" the loop (Figure 16.7) and thus disconnect the controller, the response takes the sigmoidal shape of an overdamped system.

3. From eqs. (16.9), (16.10a), and (16.11a), which give the value of the proportional gain K_c for the three controllers, we notice the following:

(a) The gain of the PI controller is lower than that of the P controller. This is due to the fact that the integral control mode makes the system more sensitive (may even lead to instability) and thus the gain value needs to be more conservative.

(b) The stabilizing effect of the derivative control mode allows the use of higher gains in the PID controller (higher than the gain for P or PI controllers).

Example 16.4: Tuning Feedback Controllers through Process Reaction Curves

In this example we examine how the dynamics of various typical processes influence the tuning results recommended by Cohen and Coon.

1. *Processes with very short time delay (dead time):* When t_d is very small (almost zero) the process reaction curve (Figure 16.8a) reminds us of the response of a simple first-order system. The Cohen–Coon settings dictate an extremely large value for the proportional gain K_c [see eqs. (16.9), (16.10a), and (16.11a)]. In real practice we will use the largest possible gain to reduce the offset if a proportional controller is employed. If a PI controller is used, the value of gain will be determined by the desired response characteristics.

2. *Multicapacity processes:* These constitute the large majority of real processes. Consider two first-order systems in series with

$$G_p = \frac{K_p}{(\tau_1 s + 1)(\tau_2 s + 1)}$$

Let the measuring device and the control valve (final control element) have first-order dynamics:

$$G_m = \frac{K_m}{\tau_m s + 1} \quad \text{and} \quad G_f = \frac{K_f}{\tau_f s + 1}$$

Then the transfer function between the control actuating variable c and the recorded measurement of the output y_m is given by [see eq. (16.7)]

$$G_{PRC} = G_f G_p G_m = \frac{K_f K_p K_m}{(\tau_f s + 1)(\tau_1 s + 1)(\tau_2 s + 1)(\tau_m s + 1)} \quad (16.12)$$

Equation (16.12) indicates that the process reaction curve has the same dynamic characteristics as the response of a system composed of four first-order systems in series (i.e., it is a sigmoidal curve).

Figure 16.9 shows the process reaction curve for the following values:

$$K_p = 1.0 \qquad K_m = 1.0 \qquad K_f = 1.0$$
$$\tau_1 = 5 \qquad \tau_2 = 2 \qquad \tau_f = 0.00 \qquad \tau_m = 10.0$$

Draw the tangent at the inflection point and find

$$S = \text{slope at the inflection point} = 0.05$$

$$B = \text{ultimate response} = 1.0$$

$$\tau = \text{effective time constant} = B/S = 1.0/0.05 = 20$$

$$t_d = \text{dead time} = 2.5$$

$$K = \text{gain} = B/A = 1.0/1.0 = 1.0$$

Therefore, the process reaction curve can be approximated by the response of the following first-order system with dead time:

$$G_{\text{PRC}} = \frac{1.0e^{-2.5s}}{20s + 1}$$

The approximate response is also shown in Figure 16.9. We notice that the approximation is satisfactory until the response has reached 40% of its final value.

Using the Cohen–Coon suggested settings, we find:
For the proportional controller:

$$K_c = 8.3$$

For the proportional-integral controller:

$$K_c = 7.3 \qquad \text{and} \qquad \tau_I = 6.6$$

For the proportional-integral-derivative controller:

$$K_c = 10.9 \qquad \tau_I = 5.85 \qquad \tau_D = 0.89$$

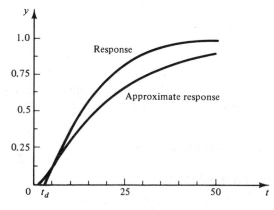

Figure 16.9 True and approximate process reaction curves for the multi-capacity process of Example 16.4.

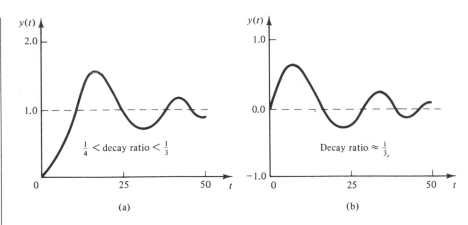

Figure 16.10 Closed-loop responses of multicapacity process in Example 16.4 for: (a) set-point and (b) load unit step changes.

Figure 16.10 shows the closed-loop responses with the foregoing settings for set point (Figure 16.10a) and load changes (Figure 16.10b). We notice that *the Cohen–Coon settings produce underdamped behavior with rather good decay ratio.*

Example 16.5: Controller Tuning for Poorly Known Processes

The methodology of controller tuning using process reaction curves is particularly appealing if the dynamics of the main process, or measuring sensor or final control element are poorly known (i.e., we do not know exactly the order of dynamics or the values of the parameters). In such case the process reaction curve reveals the effects of all the dynamic components (i.e., process, sensor, and final control element) and provides an experimental, approximate model for the overall process.

Take as example the temperature control system for the reactor of Figure 16.6. It is quite a complex system and we may not know with satisfactory precision all or a few of the following:

The reaction kinetics
The heat of reaction
The mixing characteristics in the tank
The heat capacity of the reacting mixture
The overall heat transfer coefficient between steam and reacting mixture
The effective order of the thermocouple's dynamics
The gain and time constant of the thermocouple
The characteristics of the steam valve

The process reaction curve for this system provides us with an experimental model of the overall process which we can use to tune the controller without requiring detailed knowledge of the dynamics for the reactor, heating jacket, thermocouple, and control valve.

THINGS TO THINK ABOUT

1. What are the principal questions that arise during the design of a feedback controller? Discuss them on the basis of a physical example.

2. What is meant by controller tuning?

3. Discuss the two classes of dynamic performance criteria. Give physical examples and demonstrate how different criteria lead to different controller designs.

4. Can you design a controller that minimizes the rise and settling times simultaneously? Explain.

5. Can you design a controller that minimizes the overshoot and settling time simultaneously? Explain.

6. What are the relative advantages and disadvantages of the three time-integral criteria, ISE, IAE, and ITAE? How would you select the most appropriate for a particular application?

7. Why do simple criteria such as minimum overshoot, minimum settling time, and one-quarter decay ratio lead to multiple solutions? How do you break the multiplicity and come up with a single solution?

8. Why do the time-integral criteria lead to unique solutions?

9. Discuss a set of simple heuristic rules you could use to select the most appropriate type of feedback controller for a particular system.

10. Discuss the philosophy of the methodology that leads to the Cohen–Coon settings for feedback controllers.

11. How do you understand the "opening" of the control loop shown in Figure 16.7? Explain in practical terms how one tunes a feedback controller for an existing process in a chemical plant.

12. Why do most of the process reaction curves have an overdamped, sigmoidal shape? Can you develop a physically meaningful system which has a reaction curve with an underdamped, oscillatory shape?

13. Are the Cohen–Coon settings reliable for all processes? Explain.

14. What is the value of the proportional gain K_c for a pure dead-time system according to the Cohen–Coon settings? Is it reasonable? Explain.

15. If the dynamics of the process or measuring sensor are not well known, what tuning techniques would you use? Discuss your answer.

Frequency Response Analysis of Linear Processes **17**

In Chapters 17 and 18 we will study a new technique which is often used to design feedback controllers. Quite different from everything we have seen so far, it is called *frequency response analysis*.

When a linear system is subjected to a sinusoidal input, its ultimate response (after a long time) is also a sustained sinusoidal wave. This characteristic, which will be proved in Section 17.2, constitutes the basis of frequency response analysis.

With frequency response analysis we are interested primarily in determining how the features of the output sinusoidal wave (amplitude, phase shift) change with the frequency of the input sinusoid. In this chapter we deal only with the basic premises of frequency response analysis, leaving its use in controller design for Chapter 18.

17.1 Response of a First-Order System to a Sinusoidal Input

Consider a simple first-order system with the transfer function

$$G(s) = \frac{\overline{y}(s)}{\overline{f}(s)} = \frac{K_p}{\tau_p s + 1} \tag{17.1}$$

Let $f(t)$ be a sinusoidal input with amplitude A and frequency ω:

$$f(t) = A \sin \omega t$$

Then

$$\bar{f}(s) = \frac{A\omega}{s^2 + \omega^2} \tag{17.2}$$

Substitute $\bar{f}(s)$ from eq. (17.2) into eq. (17.1) and take

$$\bar{y}(s) = \frac{K_p}{\tau_p s + 1} \frac{A\omega}{s^2 + \omega^2}$$

Expand into partial fractions and find

$$\bar{y}(s) = \frac{C_1}{s + 1/\tau_p} + \frac{C_2}{s + j\omega} + \frac{C_3}{s - j\omega}$$

Compute the constants C_1, C_2, and C_3 and find the inverse Laplace transform

$$y(t) = \frac{K_p A \omega \tau_p}{\tau_p^2 \omega^2 + 1} e^{-t/\tau_p} - \frac{K_p A \omega \tau_p}{\tau_p^2 \omega^2 + 1} \cos \omega t + \frac{K_p A}{\tau_p^2 \omega^2 + 1} \sin \omega t$$

As $t \to \infty$, $e^{-t/\tau_p} \to 0$ and the first term disappears. Thus, after a long time, the response of a first-order system to a sinusoidal input is given by

$$y_{ss}(t) = -\frac{K_p A \omega \tau_p}{\tau_p^2 \omega^2 + 1} \cos \omega t + \frac{K_p A}{\tau_p^2 \omega^2 + 1} \sin \omega t \tag{17.3}$$

Use the following trigonometric identity:

$$a_1 \cos b + a_2 \sin b = a_3 \sin (b + \phi)$$

where

$$a_3 = \sqrt{a_1^2 + a_2^2} \quad \text{and} \quad \phi = \tan^{-1} \left(\frac{a_1}{a_2} \right)$$

Then, eq. (17.3) yields

$$y_{ss}(t) = \frac{K_p A}{\sqrt{\tau_p^2 \omega^2 + 1}} \sin (\omega t + \phi) \tag{17.4}$$

where

$$\phi = \tan^{-1} -\omega \tau_p \tag{17.5}$$

From eqs. (17.4) and (17.5), we observe that:

1. The ultimate response (also referred to as steady state) of a first-order system to a sinusoidal input is also a sinusoidal wave with the same frequency ω.

2. The ratio of the output amplitude to the input amplitude is called the *amplitude ratio* and is a function of the frequency:

$$AR = \text{amplitude ratio} = \frac{K_p}{\sqrt{\tau_p^2 \omega^2 + 1}} \qquad (17.6)$$

3. The output wave lags behind (*phase lag*) the input wave by an angle $|\phi|$, which is also a function of the frequency ω [see eq. (17.5)]. Figure 17.1 shows the ultimate response of the system and its relationship to the input wave.

The three observations above hold not only for first-order systems but are true for any order linear system. Before we proceed with the generalization of the results above, let us make the following remarks related to the algebra of complex numbers.

Remarks

1. Consider a complex number W defined by

$$W = a + jb$$

where $a = \text{Re}(W) = $ real part of W, and $b = \text{Im}(W) - $ imaginary part of W. Define the following terms:

The *modulus* or *absolute value* or *magnitude* of W is represented by $|W|$ and defined by

$$|W| = \sqrt{[\text{Re}(W)]^2 + [\text{Im}(W)]^2} \qquad (17.7)$$

The *phase angle* or *argument* of W is represented by $\star W$ or $arg(W)$ and defined by

$$\star W = \tan^{-1}\left[\frac{\text{Im}(W)}{\text{Re}(W)}\right] = \theta \qquad (17.8)$$

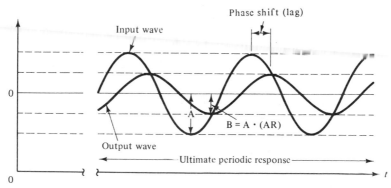

Figure 17.1 Ultimate response of first-order system to sinusoidal input.

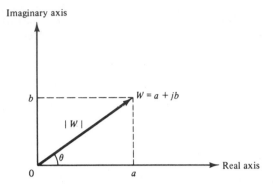

Figure 17.2 Complex plane and complex numbers.

From Figure 17.2 it is clear that

$$a = |W|\cos\theta \quad\text{and}\quad b = |W|\sin\theta$$

and

$$W = |W|\cos\theta + j|W|\sin\theta$$

Recall also that

$$\cos\theta = \frac{e^{j\theta} + e^{-j\theta}}{2} \quad\text{and}\quad \sin\theta = \frac{e^{j\theta} - e^{-j\theta}}{2j}$$

Then

$$W = |W|\frac{e^{j\theta} + e^{-j\theta}}{2} + j|W|\frac{e^{j\theta} - e^{-j\theta}}{2j} = |W|e^{j\theta} \quad (17.9)$$

2. Let, $Z = a - jb$. Then it is easily shown that

$$|W| = |Z| \quad\text{and}\quad \arg Z = -\arg W \quad (17.10)$$

3. Put $s = j\omega$ in eq. (17.1) and take

$$G(j\omega) = \frac{K_p}{j\omega\tau_p + 1} = \frac{K_p}{j\omega\tau_p + 1}\frac{-j\omega\tau_p + 1}{-j\omega\tau_p + 1}$$

or

$$G(j\omega) = \frac{K_p}{\tau_p^2\omega^2 + 1} - j\frac{K_p\omega\tau_p}{\tau_p^2\omega^2 + 1}$$

$G(j\omega)$ is a complex number. Therefore, according to eqs. (17.7) and (17.8), modulus of $G(j\omega) = \dfrac{K_p}{\sqrt{\tau_p^2\omega^2 + 1}}$ = amplitude ratio [see eq. (17.6)]

and

argument of $G(j\omega) = \tan^{-1} -\omega\tau_p$ = phase lag [see eq. (17.5)]

The last two relationships indicate that *the amplitude ratio and phase lag for the ultimate response of a first-order system are equal to the modulus and argument, respectively, of its transfer function when $s = j\omega$.*

This is an important result which we will generalize in Section 17.2 for any linear system.

17.2   Frequency Response Characteristics of a General Linear System

Consider a general linear system with the transfer function

$$G(s) = \frac{\overline{y}(s)}{\overline{f}(s)} = \frac{Q(s)}{P(s)} \tag{17.11}$$

where $Q(s)$ and $P(s)$ are polynomials of orders m and n, respectively, with $m < n$. We will prove that:

1. The ultimate response of a system to a sinusoidal input is also a sinusoidal wave.
2. The ratio of the output amplitude to the input amplitude is a function of the frequency ω and is given by the modulus of $G(s)$ if we put $s = j\omega$:

$$AR = \text{modulus of } G(j\omega)$$

3. The output wave is shifted with respect to the input wave by an angle ϕ which is a function of the frequency ω given by

$$\phi = \text{argument of } G(j\omega)$$

Proof. For a sinusoidal input $f(t) = A \sin \omega t$ we have $\overline{f}(s) = A\omega/(s^2 + \omega^2)$ and eq. (17.11) yields

$$\overline{y}(s) = G(s)\frac{A\omega}{s^2 + \omega^2}$$

Expand the last equation into partial fractions:

$$\bar{y}(s) = G(s)\frac{A\omega}{s^2 + \omega^2} = G(s)\frac{A\omega}{(s + j\omega)(s - j\omega)}$$

$$= \underbrace{\left[\frac{C_1}{s - p_1} + \frac{C_2}{s - p_2} + \cdots + \frac{C_n}{s - p_n}\right]}_{\substack{\text{Terms arising from expansion} \\ \text{of } G(s) \text{ into partial fractions}}} + \frac{a}{s + j\omega} + \frac{b}{s - j\omega} \tag{17.12}$$

where $p_1, p_2, \ldots, p_n$ are the poles of $G(s)$. The terms

$$\frac{C_1}{s - p_1}, \frac{C_2}{s - p_2}, \ldots, \frac{C_n}{s - p_n}$$

give rise to exponential terms

$$e^{p_1 t}, e^{p_2 t}, \ldots, e^{p_n t}$$

If the poles $p_1, p_2, \ldots, p_n$ have negative real parts, all the terms above decay to zero as $t \to \infty$ (see Section 9.4). Therefore, inverting eq. (17.12), we find that the ultimate response is given by

$$y_{ss}(t) = ae^{-j\omega t} + be^{j\omega t}$$

From eq. (17.12) compute constants a and b (as discussed in Section 8.2) and find

$$a = \frac{AG(-j\omega)}{-2j} \quad \text{and} \quad b = \frac{AG(j\omega)}{2j}$$

Therefore,

$$y_{ss}(t) = -\frac{AG(-j\omega)}{2j}e^{-j\omega t} + \frac{AG(j\omega)}{2j}e^{j\omega t} \tag{17.13}$$

Use eqs. (17.9) and (17.10) to express the complex numbers $G(-j\omega)$ and $G(j\omega)$ in polar form:

$$G(-j\omega) = |G(-j\omega)|e^{-j\phi} = |G(j\omega)|e^{-j\phi}$$

and

$$G(j\omega) = |G(j\omega)|e^{j\phi}$$

where $\phi = $ argument of $G(j\omega)$. Substitute the values of $G(-j\omega)$ and $G(j\omega)$ in eq. (17.13):

$$y_{ss}(t) = -\frac{A|G(j\omega)|}{2j}e^{-j(\omega t + \phi)} + \frac{A|G(j\omega)|}{2j}e^{j(\omega t + \phi)}$$

$$= A \, | G(j\omega) | \, \frac{e^{j(\omega t + \phi)} - e^{-j(\omega t + \phi)}}{2j}$$

or

$$y_{ss}(t) = A \, | G(j\omega) | \, \sin(\omega t + \phi)$$

The last equation proves what we set out to prove:

1. The ultimate response as $t \to \infty$ is sinusoidal with frequency ω.
2. The amplitude ratio is

$$AR = \frac{A \, | G(j\omega) |}{A} = | G(j\omega) | \qquad (17.14a)$$

3. The output sinusoidal wave has been shifted by the angle

$$\phi = \text{argument of } G(j\omega) \qquad (17.14b)$$

Example 17.1: Frequency Response of a Pure Capacitive Process

The transfer function is

$$G(s) = \frac{K_p}{s}$$

Put $s = j\omega$ and take

$$G(j\omega) = \frac{K_p}{j\omega} = \frac{K_p}{j\omega} \frac{j\omega}{j\omega} = 0 - j \frac{K_p}{\omega}$$

Consequently, for the ultimate response:

1. The amplitude ratio is

$$AR = | G(j\omega) | = \frac{K_p}{\omega} \qquad (17.15)$$

2. The phase shift is

$$\phi = \tan^{-1} -\infty = -90° \qquad (17.16)$$

that is, the ultimate sinusoidal response of the system *lags behind* the input wave by 90°.

Example 17.2: Frequency Response of N Noninteracting Capacities in Series

The transfer function is [see eq. (11.21)]

$$G(s) = G_1(s)G_2(s) \cdots G_N(s) = \frac{K_1}{\tau_1 s + 1} \frac{K_2}{\tau_2 s + 1} \cdots \frac{K_N}{\tau_N s + 1}$$

Put $s = j\omega$ and take

$$G(j\omega) = G_1(j\omega)G_2(j\omega) \cdots G_N(j\omega) \tag{17.17}$$

But, according to eq. (17.9),

$$G_1(j\omega) = |G_1(j\omega)|e^{j\phi_1}, \quad G_2(j\omega) = |G_2(j\omega)|e^{j\phi_2}, \quad \ldots,$$

$$G_N(j\omega) = |G_N(j\omega)|e^{j\phi_N}$$

where $\phi_1, \phi_2, \ldots, \phi_N$ are the arguments of $G_1(j\omega)$, $G_2(j\omega), \ldots, G_N(j\omega)$. Then eq. (17.17) becomes

$$G(j\omega) = [|G_1(j\omega)||G_2(j\omega)| \cdots |G_N(j\omega)|]e^{j(\phi_1+\phi_2+\cdots+\phi_N)}$$

Consequently, the response has the following characteristics:

1. Amplitude ratio:

$$AR = |G(j\omega)| = |G_1(j\omega)||G_2(j\omega)| \cdots |G_N(j\omega)| \tag{17.18}$$

or

$$AR = \frac{K_1 K_2 \cdots K_N}{\sqrt{1 + \tau_1^2\omega^2}\sqrt{1 + \tau_2^2\omega^2} \cdots \sqrt{1 + \tau_N^2\omega^2}} \tag{17.18a}$$

2. Phase shift:

$$\phi = \phi_1 + \phi_2 + \cdots + \phi_N \tag{17.19}$$

or

$$\phi = \tan^{-1}-\omega\tau_1 + \tan^{-1}-\omega\tau_2 + \cdots + \tan^{-1}-\omega\tau_N \tag{17.19a}$$

Since $\phi < 0$, the response *lags* behind the input.

Example 17.3: Frequency Response of a Second-Order System

For a second-order system the transfer function is

$$G(s) = \frac{K_p}{\tau^2 s^2 + 2\zeta\tau s + 1}$$

Put $s = j\omega$ and take

$$G(j\omega) = \frac{K_p}{(-\tau^2\omega^2 + 1) + j2\zeta\tau\omega} = \frac{K_p}{(-\tau^2\omega^2 + 1) + j2\zeta\tau\omega}\frac{(-\tau^2\omega^2 + 1) - j2\zeta\tau\omega}{(-\tau^2\omega^2 + 1) - j2\zeta\tau\omega}$$

or

$$G(j\omega) = \frac{K_p(1 - \tau^2\omega^2)}{(1 - \tau^2\omega^2)^2 + (2\zeta\tau\omega)^2} - j\frac{K_p \cdot 2\zeta\tau\omega}{(1 - \tau^2\omega^2)^2 + (2\zeta\tau\omega)^2}$$

Therefore, the ultimate response has the following characteristics:

1. Amplitude ratio:

$$AR = |G(j\omega)| = \frac{K_p}{\sqrt{(1 - \tau^2\omega^2)^2 + (2\zeta\tau\omega)^2}} \tag{17.20}$$

2. Phase shift:

$$\phi = \text{argument of } G(j\omega) = \tan^{-1}\left(-\frac{2\zeta\tau\omega}{1 - \tau^2\omega^2}\right) \qquad (17.21)$$

which is a *phase lag* since $\phi < 0$.

Example 17.4: Frequency Response of a Pure Dead-Time Process

The transfer function is

$$G(s) = e^{-\tau_d s}$$

Put $s = j\omega$ and take

$$G(j\omega) = e^{-j\tau_d\omega}$$

Clearly:

$$\text{amplitude ratio} = |G(j\omega)| = 1 \qquad (17.22)$$

$$\phi = \text{phase shift} = \text{argument of } G(j\omega) = -\tau_d\omega \qquad (17.23)$$

that is, a phase lag, since $\phi < 0$.

Example 17.5: Frequency Response of Feedback Controllers

Let us now shift our attention to the various types of feedback controllers.

1. *Proportional controller:* The transfer function is

$$G_c(s) = K_c$$

Therefore,

$$\text{AR} = K_c \qquad \text{and} \qquad \phi = 0$$

2. *Proportional-integral controller:* The transfer function is

$$G_c(s) = K_c\left(1 + \frac{1}{\tau_I s}\right)$$

Therefore,

$$\text{AR} = |G_c(j\omega)| = K_c\sqrt{1 + \frac{1}{(\omega\tau_I)^2}} \qquad (17.24)$$

$$\phi = \arg G_c(j\omega) = \tan^{-1}\left(\frac{-1}{\omega\tau_I}\right) < 0 \qquad (17.25)$$

3. *Proportional-derivative controller:* The transfer function is

$$G_c(s) = K_c(1 + \tau_D s)$$

Therefore,

$$\text{AR} = |G_c(j\omega)| = K_c\sqrt{1 + \tau_D^2\omega^2} \qquad (17.26)$$

$$\phi = \arg G_c(j\omega) = \tan^{-1} \tau_D \omega > 0 \qquad (17.27)$$

The positive phase shift is called *phase lead* and implies that the controller's output leads the input(!). This is another manifestation of the derivative control anticipating future developments (see also Section 13.2).

4. *Proportional-integral-derivative controller:* The transfer function is

$$G_c(s) = K_c\left(1 + \frac{1}{\tau_I s} + \tau_D s\right)$$

and it is easy to show that

$$\text{AR} = |G_c(j\omega)| = K_c\sqrt{\left(\tau_D\omega - \frac{1}{\tau_I\omega}\right)^2 + 1} \qquad (17.28)$$

$$\phi = \tan^{-1}\left(\tau_D\omega - \frac{1}{\tau_I\omega}\right) \qquad (17.29)$$

Notice that ϕ can take positive (phase lead) or negative (phase lag) values, depending on the values of τ_D, τ_I, and ω.

17.3   Bode Diagrams

The Bode diagrams (in honour of H. W. Bode) constitute a convenient way to represent the frequency response characteristics of a system. As we can see from Eqs. (17.14a) and (17.14b), the amplitude ratio and the phase shift of the ultimate response of a system are functions of the frequency ω. The Bode diagrams consist of a pair of plots showing;

How the logarithm of the amplitude ratio varies with frequency
How the phase shift varies with frequency

In order to cover a large range of frequencies, we use a logarithmic scale for the frequencies.

Let us now examine the Bode diagrams of some simple dynamic systems that we have encountered in previous chapters.

First-order system

For a first-order system we have seen that

$$\text{amplitude ratio} = \text{AR} = \frac{K_p}{\sqrt{1 + \tau_p^2\omega^2}} \qquad (17.6)$$

$$\text{phase lag} = \phi = \tan^{-1} -\tau_p\omega \qquad (17.5)$$

For simplification, let $K_p = 1$. Then, from eq. (17.6) we find that

$$\log AR = -\tfrac{1}{2}\log(1 + \tau_p^2\omega^2) \qquad (17.30)$$

For convenience, since τ_p is constant, regard $\tau_p\omega$ as the independent variable instead of ω. The plot of $\log AR$ versus $\log \tau_p\omega$ is shown in Figure 17.3a (solid line) and can be constructed from eq. (17.30) for various values of the frequency ω. Instead of the very elaborate numerical work needed to plot this graph, we can give an approximate sketch by considering its asymptotic behavior as $\omega \to 0$ and as $\omega \to \infty$. Thus we have:

1. As $\omega \to 0$, then $\tau_p\omega \to 0$ and from eq. (17.30) $\log AR \to 0$ or $AR \to 1$. This is the *low-frequency asymptote* shown by a dashed line in Figure 17.3a. It is a horizontal line passing through the point $AR = 1$.
2. As $\omega \to \infty$, then $\tau_p\omega \to \infty$ and from eq. (17.30) $\log AR \simeq -\log \tau_p\omega$. This is the *high-frequency asymptote* shown also by a dashed line in Figure 17.3a. It is a line with a slope of -1 passing through the point $AR = 1$ for $\tau_p\omega = 1$. The frequency $\omega = 1/\tau_p$ is known as *the corner frequency*. At the corner frequency, as can be seen from Figure 17.3a, the deviation of the true value of AR from the asymptotes is maximum.

The plot of phase shift versus $\tau_p\omega$ is shown in Figure 17.3b. It can be

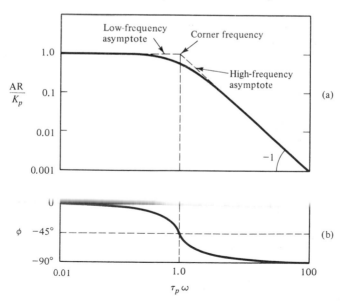

Figure 17.3 Bode diagrams for first-order system.

constructed from eq. (17.5) and we can easily verify the following characteristics of the plot:

As $\omega \to 0$, then $\phi \to 0$.

As $\omega \to \infty$, then $\phi \to tan^{-1}(-\infty) = -90°$.

At $\omega = 1/\tau_p$ (corner frequency), $\phi = tan^{-1}(-1) = -45°$.

Note. If $K_p \neq 1$, then as can be seen from eq. (17.6), the low-frequency asymptote shifts vertically by the value log K_p. Equation (17.5) shows that K_p has no effect on the phase shift.

Pure capacitive process

For such processes we know that (see Example 17.1)

$$AR = \frac{K_p}{\omega} \quad \text{and} \quad \phi = -90°$$

The Bode plots are easily constructed and shown in Figure 17.4.

Second-order system

In Example 17.3 we found that

$$AR = \frac{K_p}{\sqrt{(1 - \tau^2\omega^2)^2 + (2\zeta\tau\omega)^2}} \quad \text{and} \quad \phi = tan^{-1}\left(\frac{-2\zeta\tau\omega}{1 - \tau^2\omega^2}\right)$$

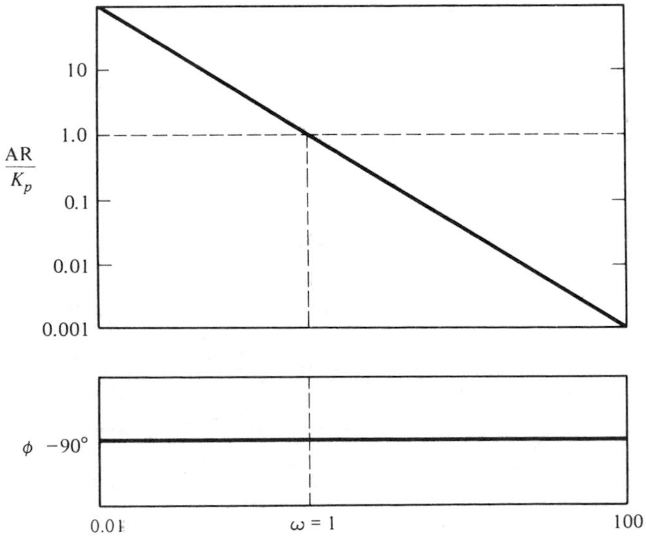

Figure 17.4 Bode plots for pure capacitive process.

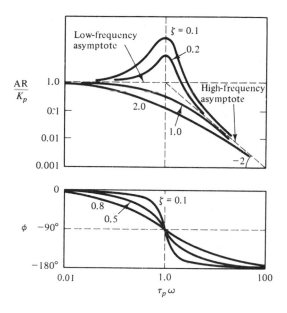

Figure 17.5 Bode plots for second-order system.

The two plots are shown in Figure 17.5 for various values of ζ when $K_p = 1$. The two asymptotes for the plot AR versus $\tau\omega$ are determined as follows:

1. As $\omega \rightarrow 0$, then log AR $\rightarrow 0$ or AR $\rightarrow 1$ (low-frequency asymptote).
2. As $\omega \rightarrow \infty$, then log AR $\rightarrow -2 \log \tau\omega$. This is the *high-frequency*

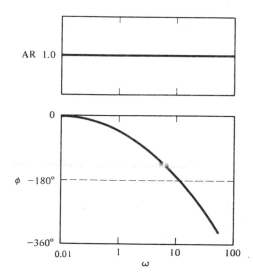

Figure 17.6 Bode plots for pure dead-time system.

asymptote. It is a straight line with a slope of -2 passing through the point

$$AR = 1, \qquad \tau\omega = 1$$

From Figure 17.5 we notice that for underdamped systems (i.e., $\zeta < 1$) the amplitude ratio can exceed significantly the value of 1. Particularly, for $\omega = \omega_r$ the AR takes its largest value *(resonance)*. ω_r is the resonant frequency and is given by $\omega_r = \sqrt{1-2\zeta^2}/\tau$ *(Note:* When $K_p \neq 1$, the low-frequency asymptote shifts vertically by the value $\log K_p$.)

Pure dead-time system

From Example 17.4 we have that

$$AR = 1 \qquad \text{and} \qquad \phi = -\tau_d\omega$$

The Bode plots for this system are easily constructed and shown in Figure 17.6.

Systems in series

Consider N systems in series with individual transfer functions

$$G_1(s), G_2(s), \ldots, G_N(s)$$

The overall transfer function is

$$G(s) = G_1(s)G_2(s) \cdots G_N(s)$$

Put $s = j\omega$ and take

$$G(j\omega) = G_1(j\omega)G_2(j\omega) \cdots G_N(j\omega)$$

or

$$G(j\omega) = |G_1(j\omega)|e^{j\phi_1}|G_2(j\omega)|e^{j\phi_2} \cdots |G_N(j\omega)|e^{j\phi_N}$$

and finally

$$|G(j\omega)|e^{j\phi} = |G_1(j\omega)||G_2(j\omega)| \cdots |G_N(j\omega)|e^{j(\phi_1+\phi_2+ \cdots +\phi_N)}$$

The last equation yields

$$|G(j\omega)| = |G_1(j\omega)||G_2(j\omega)| \cdots |G_N(j\omega)| \qquad (17.31)$$

and

$$\phi = \phi_1 + \phi_2 + \cdots + \phi_N \qquad (17.32)$$

From (17.31) we have

$$AR = (AR)_1(AR)_2 \cdots (AR)_N$$

or

$$\log(AR) = \log(AR)_1 + \log(AR)_2 + \cdots + \log(AR)_N \qquad (17.33)$$

where $(AR)_1$, $(AR)_2$, ..., $(AR)_N$ are the amplitude ratios for the individual systems in series. Equations (17.31) and (17.32) are very important and indicate certain rules for the construction of the Bode diagrams. If the transfer function of a system can be factored into the product of N transfer functions of simpler systems, use the following rules:

1. The logarithm of the overall amplitude ratio is equal to the sum of the logarithms of the amplitude ratios of the individual systems.
2. The overall phase shift is equal to the sum of the phase shifts of the individual systems.
3. The presence of a constant in the overall transfer function will move the entire AR curve vertically by a constant amount. It has no effect on the phase shift.

Example 17.6: Bode Diagrams for Two Systems in Series

Consider the following two systems in series:

$$G_1(s) = \frac{1}{2s + 1} \quad \text{and} \quad G_2(s) = \frac{6}{5s + 1}$$

The overall transfer function is

$$G(s) = \frac{1}{2s + 1} \frac{6}{5s + 1}$$

Then

$$AR = \frac{1}{\sqrt{1 + 4\omega^2}} \frac{6}{\sqrt{1 + 25\omega^2}}$$

or

$$\log AR = \log 6 + \log (AR)_1 + \log (AR)_2 \tag{17.34}$$

where $(AR)_1$ and $(AR)_2$ are the amplitude ratios of the individual systems, when their gains are 1. Figure 17.7a shows the amplitude ratios of the two systems as functions of ω. The addition of these two curves plus the factor $\log 6$ will yield the amplitude ratio of the overall system versus the frequency ω. The overall curve is also shown in Figure 17.7a without the term $\log 6$. From this curve we notice three distinct frequency regions. *The slope of the asymptote in each region is the algebraic sum of the slopes of the asymptotes for the two systems in the corresponding region.* Thus we have:

1. *Region 1:* From $\omega = 0$ to $\omega = \frac{1}{5}$. Slope of the overall asymptote $= 0 + 0$ (i.e., horizontal), going through the point AR $= 6$.
2. *Region 2:* From $\omega = \frac{1}{5}$ to $\omega = \frac{1}{2}$. Slope of the overall asymptote $= 0 + (-1) = -1$, going through the point AR $= 6$, $\omega = \frac{1}{5}$.
3. *Region 3:* For $\omega > \frac{1}{2}$. Slope of the overall asymptote $= (-1) + (-1) = -2$.

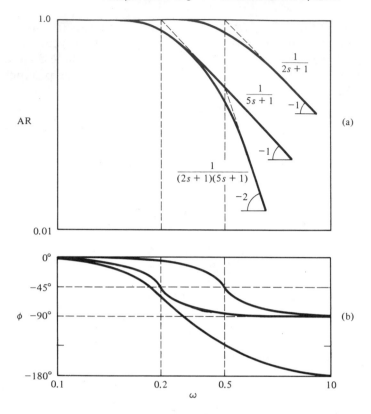

Figure 17.7 Bode plots for two capacities in series (Example 17.6).

Figure 17.7b shows the phase shift for the overall system as the algebraic sum of the phase shifts of the two individual systems:

$$\phi = \phi_1 + \phi_2 = \tan^{-1} -2\omega + \tan^{-1} -5\omega$$

It is clear that:

When $\omega \to 0$, then $\phi_1 \to 0$, $\phi_2 \to 0$, and $\phi \to 0$.
When $\omega \to \infty$, then $\phi_1 \to -90°$, $\phi_2 \to -90°$, and $\phi \to -180°$.

Feedback controllers

The Bode diagrams for various types of feedback controllers can be constructed easily using the results of Example 17.5.

1. *Proportional controller:* The Bode plots are trivial. The AR and ϕ stay constant at the values K_c and $0°$, respectively, for all frequencies.

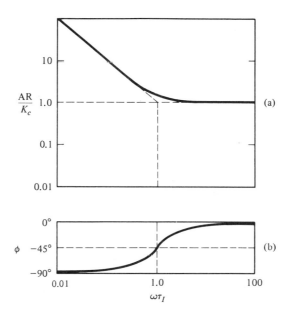

Figure 17.8 Bode plots for PI controller.

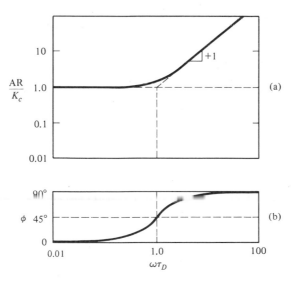

Figure 17.9 Bode plots for PD controller.

2. *Proportional-integral controller:* From eqs. (17.24) and (17.25) we take

$$\log\left(\frac{AR}{K_c}\right) = \frac{1}{2}\log\left[1 + \frac{1}{(\omega\tau_I)^2}\right] \quad \text{and} \quad \phi = \tan^{-1}\left(\frac{-1}{\omega\tau_I}\right)$$

Therefore:

Low-frequency asymptote:

$$\text{As } \omega \to 0, \quad \frac{1}{(\omega\tau_I)^2} \gg 1; \text{ then } \log\left(\frac{AR}{K_c}\right) \to -\log \omega\tau_I$$

Consequently, the low-frequency asymptote is a straight line with slope −1.

High-frequency asymptote:

$$\text{As } \omega \to \infty, \quad \frac{1}{(\omega\tau_I)^2} \to 0 \quad \text{and} \quad \log\left(\frac{AR}{K_c}\right) \to 0 \quad \left(\text{i.e., } \frac{AR}{K_c} \to 1\right)$$

The high-frequency asymptote is a horizontal line at the value $AR/K_c = 1$.

The AR/K_c versus $\omega\tau_I$ plot is shown in Figure 17.8a. For the phase shift we have the following:

$$\text{as } \omega \to 0, \quad \phi \to -90°$$

$$\text{as } \omega \to \infty, \quad \phi \to 0°$$

The ϕ versus $\omega\tau_I$ plot is shown in Figure 17.8b.

3. *Proportional-derivative controller:* The AR and ϕ are given by eqs. (17.26) and (17.27). The Bode plots can be easily constructed and are shown in Figure 17.9a and b.

4. *Proportional-integral-derivative controller:* The AR and ϕ are given by eqs. (17.28) and (17.29), respectively. The Bode plots are easily constructed and are shown in Figure 17.10a and b.

Example 17.7: Bode Plots for an Open-Loop System

Consider the feedback control system shown in Figure 17.11. The open-loop transfer function is (see Remark 2 in Section 15.2)

$$G_{OL} = G_c G_f G_p G_m$$

or

$$G_{OL} = 100K_c\left(1 + \frac{1}{\tau_I s}\right)\frac{1}{0.1s + 1}\frac{1}{(2s + 1)(s + 1)}\frac{1}{0.5s + 1} e^{-0.2s}$$

with $\tau_I = 0.25$ and $K_c = 4$. We notice that G_{OL} can be factored into a product of six transfer functions:

$$\frac{1}{2s + 1}, \quad \frac{1}{s + 1}, \quad \frac{1}{0.5s + 1}, \quad \left(1 + \frac{1}{\tau_I s}\right), \quad \frac{1}{0.1s + 1}, \quad e^{-0.2s}$$

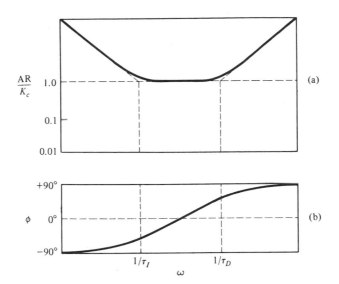

Figure 17.10 Bode plots for PID controller.

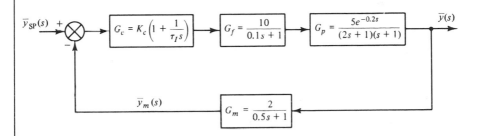

Figure 17.11 Block diagram of system in Example 17.7.

with the following corner frequencies (in the same order):

$$\omega_1 = 1/2 = 0.5, \qquad \omega_2 = 1/1 = 1, \qquad \omega_3 = 1/0.5 = 2,$$
$$\omega_4 = 1/0.25 = 4, \qquad \omega_5 = 1/0.1 = 10$$

The Bode plots of the individual transfer functions are easily constructed and are shown in Figure 17.12a and b. The Bode plots for the overall system can be constructed following the rules discussed earlier.

1. We identify the following six regions on the frequency scale:

$$0 \leq \omega < \omega_1, \qquad \omega_1 \leq \omega < \omega_2, \qquad \omega_2 \leq \omega < \omega_3,$$
$$\omega_3 \leq \omega < \omega_4, \qquad \omega_4 \leq \omega < \omega_5, \qquad \omega_5 \leq \omega < \infty.$$

2. For the AR versus ω diagram, the slope of the overall asymptote is equal to the algebraic sum of the slopes of the asymptotes of the individual transfer functions (Table 17.1). The overall asymptote is shown in Figure 17.12a.
3. The overall phase shift is equal to the algebraic sum of the phase shifts for each individual transfer function and is shown in Figure 17.12b.

335

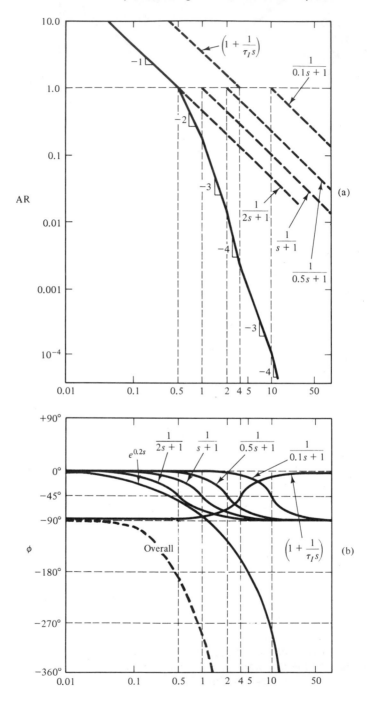

Figure 17.12 Bode plots for system in Example 17.7.

TABLE 17.1

SLOPE OF THE OVERALL ASYMPTOTE FOR THE AR VERSUS ω DIAGRAM OF EXAMPLE 17.7

Frequency region	Slopes of the asymptotes of the individual transfer functions						Slope of the overall asymptote
	$\left(\dfrac{1}{2s+1}\right)$	$\left(\dfrac{1}{s+1}\right)$	$\left(\dfrac{1}{0.5s+1}\right)$	$\left(1+\dfrac{1}{\tau_I s}\right)$	$\left(\dfrac{1}{0.1s+1}\right)$	$(e^{-0.2s})$	
$0 \leq \omega < \omega_1 = 0.5$	0	0	0	-1	0	0	-1
$\omega_1 \leq \omega < \omega_2 = 1$	-1	0	0	-1	0	0	-2
$\omega_2 \leq \omega < \omega_3 = 2$	-1	-1	0	-1	0	0	-3
$\omega_3 \leq \omega < \omega_4 = 4$	-1	-1	-1	-1	0	0	-4
$\omega_4 \leq \omega < \omega_5 = 10$	-1	-1	-1	0	0	0	-3
$\omega_5 \leq \omega < \infty$	-1	-1	-1	0	-1	0	-4

17.4   Nyquist Plots

A Nyquist plot is an alternative way to represent the frequency response characteristics of a dynamic system. It uses the Im $[G(j\omega)]$ as ordinate and Re $[G(j\omega)]$ as abscissa. Figure 17.13 shows the form of a Nyquist plot.

A specific value of the frequency ω defines a point on this plot. Thus at point 1 (Figure 17.13) the frequency has a value ω_1 and we observe the following:

1. The distance of the point 1 from the origin (0, 0) is the amplitude ratio at the frequency ω_1:

$$\text{distance} = \sqrt{[\text{Re }[G(j\omega_1)]]^2 + [\text{Im }[G(j\omega_1)]]^2} = |G(j\omega_1)| = \text{AR}$$

2. The angle ϕ with the real axis is the phase shift at the frequency ω_1:

$$\phi = \tan^{-1}\frac{\text{Im }[G(j\omega_1)]}{\text{Re }[G(j\omega_1)]} = \arg G(j\omega_1) = \text{phase shift}$$

Thus as the frequency varies from 0 to ∞, we trace the whole length of the Nyquist plot and we find the corresponding values for the amplitude ratio and phase shift. The shape and location of a Nyquist plot are characteristic for a particular system.

The *Nyquist plot contains the same information as the pair of Bode plots for the same system.* Therefore, its construction is rather easy given the corresponding Bode plots. Let us now construct the Nyquist plots of some typical systems using their Bode plots developed in the preceding section.

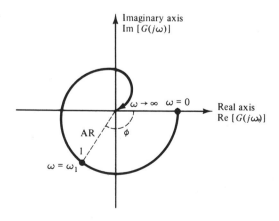

Figure 17.13 Form of a Nyquist plot.

First-order system

The corresponding Bode plots are given in Figure 17.3.

1. When $\omega = 0$, then AR $= 1$ and $\phi = 0$. Therefore, the beginning of the Nyquist plot is on the real axis where $\phi = 0$ and at a distance from the origin (0, 0) equal to 1 (see point A in Figure 17.14a).
2. When $\omega \to \infty$, then AR $\to 0$ and $\phi \to -90°$. Therefore, the end of the Nyquist plot is at the origin where the distance from it is zero (point C in Figure 17.14a).
3. Since for every intermediate frequency

$$0 < \text{AR} < 1 \quad \text{and} \quad -90° < \phi < 0$$

the Nyquist plot will be inside a unit circle and will never leave the first quadrant. Its complete shape and location are shown in Figure 17.14a.

Second-order system

The corresponding Bode plots are shown in Figure 17.5.

1. When $\omega = 0$, then AR $= 1$ and $\phi = 0$. Thus the beginning of the

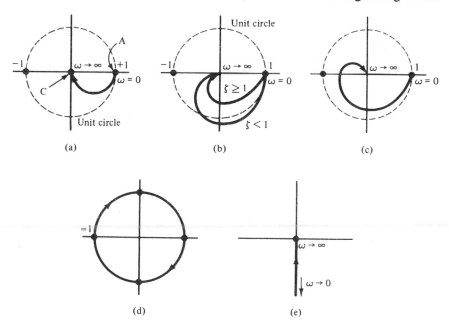

Figure 17.14 Nyquist plots for various systems: (a) first-order; (b) second-order (c) third-order; (d) pure dead-time; (e) pure capacitive.

Nyquist plot is on the real axis at a distance equal to 1 from the origin.

2. When $\omega \to \infty$, then $AR \to 0$ and $\phi \to -180°$; that is, the Nyquist plot will end at the origin and will approach it from the second quadrant.

3. When $\zeta \geq 1$, then $AR \leq 1$ and the Nyquist plot stays within a unit circle. When $\zeta < 1$, then AR becomes larger than 1 for a range of frequencies. Thus the Nyquist plot goes outside the unit circle for a certain range of frequencies. Figure 17.14b shows the Nyquist plot for a second-order system.

Third-order system

The transfer function is

$$G(s) = \frac{1}{(\tau_1 s + 1)(\tau_2 s + 1)(\tau_3 s + 1)} \qquad \text{with } \tau_1, \tau_2, \tau_3 \text{ real and positive}$$

It is easy to show that:

When $\omega = 0$, then $AR = 1$ and $\phi = 0$, while
When $\omega \to \infty$, then $AR = 0$ and $\phi \to -270°$.

Therefore, the Nyquist plot starts from the real axis at a distance 1 from the origin and ends at the origin, going through the third quadrant (Figure 17.14c).

Pure dead time

From the corresponding Bode plots (Figure 17.6) we notice that

$$AR = 1 \qquad \text{for every frequency}$$

and

$$\phi = -\tau_d \omega$$

Therefore, the Nyquist plot for this system is a circle of radius 1 and encircles the origin an infinite number of times (Figure 17.14d).

Pure capacitive process

From the corresponding Bode plots (Figure 17.4) we notice that

When $\omega \to 0$, $AR \to \infty$, while
When $\omega \to \infty$, $AR \to 0$.

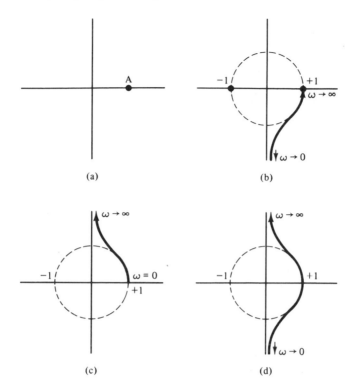

Figure 17.15 Nyquist plots for various feedback controllers: (a) P; (b) PI; (c) PD; (d) PID.

The phase lag remains constant at −90° for every frequency. Therefore, the Nyquist plot coincides with the negative part of the imaginary axis (Figure 17.14e).

Feedback controllers

In a similar manner as above we can construct the Nyquist plots for P, PI, PD, and PID controllers. They are shown in Figure 17.15a, b, c, and d, respectively. For details, consult Figures 17.8, 17.9 and 17.10.

THINGS TO THINK ABOUT

1. What are the characteristics of the ultimate response of a linear system with a transfer function $G(s)$ to a sustained sinusoidal input?

2. Define the frequency response analysis.

3. What means could you use to represent the results of the frequency response analysis for a dynamic system?

4. Define the Bode diagrams and Nyquist plots. Do you have any personal preference on one of them over the other? If yes, why?

5. The system with a transfer function

$$G(s) = \tau s + 1$$

is usually known as *first-order lead element*. Construct its Bode diagram and try to rationalize the word *lead* in its name. (*Hint*: Contrast it to the familiar first-order lag element.)

6. Construct the Bode diagram and Nyquist plot of a first-order system with dead time, having a transfer function

$$G(s) = \frac{K_p e^{-\tau_d s}}{\tau_p s + 1}$$

7. Does the Nyquist plot have a meaning for the frequencies $-\infty \le \omega \le 0$? Show that the Nyquist plot for this range of frequencies is the mirror image of the familiar Nyquist plot for the frequencies $0 \le \omega \le +\infty$.

8. Construct qualitatively the Nyquist plot of a sixth-order system with a transfer function

$$G(s) = \frac{K_p}{(\tau_1 s + 1)(\tau_2 s + 1)(\tau_3 s + 1)(\tau_4 s + 1)(\tau_5 s + 1)(\tau_6 s + 1)}$$

where τ_1, τ_2, τ_3, τ_4, τ_5, and τ_6 are all real and positive, and in order of increasing values.

9. For a system like the one in item 8, we claim that the slope of the overall asymptotes, in the log AR versus log ω plot of its Bode diagram, can be given from the algebraic sum of the slopes of the asymptotes for the individual subsystems,

$$\frac{1}{\tau_1 s + 1}, \quad \frac{1}{\tau_2 s + 1}, \quad \cdots, \quad \frac{1}{\tau_6 s + 1}$$

Explain why. Also, construct qualitatively the Bode diagram, indicating the slopes of the asymptotes for the overall system.

10. The Bode plots for a PI controller show that as $\omega \to 0$, the AR $\to \infty$. This is not physically realizable. Therefore, the transfer function

$$G(s) = K_c\left(1 + \frac{1}{\tau_I s}\right)$$

represents the behavior of an ideal PI controller. How should we modify the transfer function above so that it represents the behavior of an actual PI controller? The transfer function of the actual PI controller must be such that as $\omega \to 0$, then AR $\to$ finite value. (*Note*: Consult Ref. 13, Chapter 22.)

11. The Bode plots for a PD controller show that as $\omega \to \infty$, the AR $\to \infty$. This is, again, physically unrealizable. How should we modify the transfer function

of a PD controller so that as $\omega \rightarrow \infty$, the AR $\rightarrow$ finite value? [*Note*: Consult Ref. 13 (Chapter 22) to develop the transfer function of an actual PD controller.]

12. Based on the responses in items 10 and 11, develop the transfer function of an actual PID controller, which has the following characteristics:

$$\text{as } \omega \rightarrow 0, \quad \text{AR} \rightarrow \alpha = \text{finite}$$

$$\text{as } \omega \rightarrow \infty, \quad \text{AR} \rightarrow \beta = \text{finite}$$

Design of Feedback **18**
Control Systems Using
Frequency Response
Techniques

In Chapter 17 we studied frequency response analysis and its application to various dynamic systems. The question that may have been raised in the mind of the reader—What do we do with it?—will be answered in this chapter.

Frequency response analysis is a useful tool for designing feedback controllers. It helps the designer:

1. To study the stability characteristics of a closed-loop system, using Bode or Nyquist diagrams of the open-loop transfer function
2. To select the most appropriate values for the adjustable parameters of a controller

18.1   Bode Stability Criterion

Consider the closed-loop system shown in Figure 18.1. The open-loop transfer function is given by

$$G_{OL} = \frac{\overline{y}_m(s)}{\overline{y}_{SP}(s)} = \frac{K_c e^{-0.1s}}{0.5s + 1} \tag{18.1}$$

The Bode diagram for $G_{OL}(s)$ can be constructed easily (see Example 17.7) and is shown in Figure 18.2. We notice that when $\omega = 17.0$ rad/min, then $\phi = -180°$. The frequency where the phase lag is

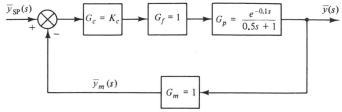

Figure 18.1 Closed-loop system.

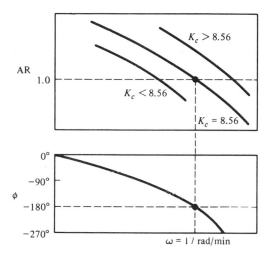

Figure 18.2 Bode plots of G_{OL} for the system of Figure 18.1.

equal to 180° is called the *crossover frequency* and is denoted by ω_{CO}. At this frequency the amplitude ratio is found from the Bode diagram to be

$$\frac{AR}{K_c} = \frac{1}{\sqrt{(0.5 \cdot 17)^2 + 1}} = 0.12 \tag{18.2}$$

Consequently, if $K_c = 1/0.12 = 8.56$, then the amplitude ratio becomes equal to 1.

Now, let us consider the "opened" loop shown in Figure 18.3a with $K_c = 8.56$. Here the measurement signal has been disconnected from the comparator of the feedback controller. If the set point changes in a sinusoidal manner with frequency $\omega = 17.0$ rad/min and an amplitude equal to 1,

$$y_{SP}(t) = 1 \sin (17.0t)$$

then the ultimate open-loop response $y_m(t)$ is given by

$$y_m(t) = \sin (17.0t - 180°) = -\sin (17.0t)$$

LIVERPOOL JOHN MOORES UNIVERSITY
LEARNING SERVICES

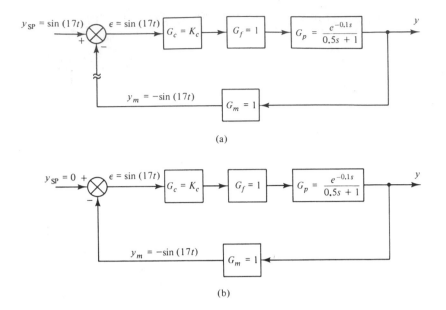

Figure 18.3 (a) Open-loop system with sinusoidal input (set point); (b) corresponding closed-loop system with sustained oscillation (zero input).

At some instant of time the set point y_{SP} is set to zero, while at the same time we "close" the loop (Figure 18.3b). Under these conditions the comparator inverts the sign of the y_m, which now plays the same role as that played by the set point in the "open" loop. Notice that the error ϵ remains the same. Theoretically, *the response of the system will continue to oscillate with constant amplitude, since AR = 1, despite the fact that both the load and the set point do not change.*
Let us examine the following cases:

1. If $K_c > 8.56$, then AR > 1 when $\phi = -180°$. Therefore, the sustained oscillation of the "closed" loop of Figure 18.3b will start exhibiting an *ever-increasing amplitude* leading to *an unstable system*.
2. On the contrary, if $K_c < 8.56$, then AR < 1 when $\phi = -180°$. Consequently, the oscillating response of the "closed" loop of Figure 18.3b will exhibit *a continuously decreasing amplitude*, leading to an eventual dying out of the oscillation.

The conclusion drawn from the observations above is the following:

A feedback control system is unstable if the AR of the corresponding open-loop transfer function is larger than 1 at the crossover frequency.

This is known as the *Bode stability criterion*.

Example 18.1: Stability Characteristics of Some Typical Dynamic Systems Using the Bode Criterion

1. *First-order open-loop response*: Consider a control system with the following dynamic components:

Process: $G_p = \dfrac{K_p}{\tau_p s + 1}$

Measuring sensor: $G_m = K_m$

Controller: $G_c = K_c$ *(i.e., proportional)*

Valve (final control element): $G_f = K_f$

The open-loop transfer function is

$$G_{OL} = G_c G_f G_p G_m = \frac{K_c K_f K_p K_m}{\tau_p s + 1} = \frac{K}{\tau_p s + 1}$$

We know (see Section 17.3) that the phase lag for a first-order system is between 0 and 90°. Therefore, according to the Bode stability criterion, the system above is always stable since there is no crossover frequency.

2. *First-order with dead time open-loop response*: Consider again the dynamic components of the loop in case 1 with the following change:

$$G_m = K_m e^{-0.5s}$$

Then the open-loop transfer function becomes

$$G_{OL} = \frac{K e^{-0.5s}}{\tau_p s + 1}$$

The phase lag for this system is

$$\phi = \tan^{-1} -\tau_p \omega + (-0.5\omega)$$

The last equation shows that phase lag is between 0° and $-\infty$. Consequently, there exists a crossover frequency ω_{CO} where $\phi = -180°$, and according to the Bode criterion the system may become unstable for a large K_c which leads to AR > 1 at this frequency. This example demonstrates a very important characteristic for the stability of chemical processes:

Dead time is a principal source of destabilizing effects in chemical process control systems.

Since most of the chemical processes exhibit an open-loop response which can be approximated by a first-order system with dead time, it is clear that the possibility for closed-loop instability will, almost always, be present. Therefore, the tuning of the feedback controller becomes a crucial task.

3. *Higher-order open-loop responses*: Consider again the control system for case 1 with the following change:

$$G_m = \frac{K_m}{\tau_m s + 1}$$

The open-loop transfer function becomes

$$G_{OL} = \frac{K}{(\tau_p s + 1)(\tau_m s + 1)}$$

and the phase lag becomes $-180°$ when $\omega = \infty$. Therefore, according to the Bode criterion, such a system is always stable since there is no finite crossover frequency. If we consider

$$G_m = \frac{K_m}{\tau_m s + 1} \quad \text{and} \quad G_f = \frac{K_f}{\tau_f s + 1}$$

then the open-loop transfer function becomes

$$G_{OL} = \frac{K}{(\tau_p s + 1)(\tau_m s + 1)(\tau_f s + 1)}$$

and the phase lag is between $0°$ and $-270°$. Therefore, there exists a finite crossover frequency ω_{CO} where $\phi = -180°$ and the system may become unstable for large enough K_c. This leads to the second important observation about the stability of chemical process control systems:

> In the absence of dead time a closed-loop system may become unstable if its open-loop transfer function is of third order or higher.

Remarks

1. All systems in Example 18.1 have an important common feature: The AR and ϕ of the corresponding open-loop transfer functions decrease continuously as ω increases. This is also true for the large majority of chemical processing systems. For such systems the Bode stability criterion leads to rigorous conclusions. Thus it constitutes a very useful tool for the stability analysis of most control systems of interest to a chemical engineer.

2. It is possible, though, that the AR or ϕ of an open-loop transfer function may not be decreasing continuously with ω. In Figure 18.4 we see the Bode plots of an open-loop transfer function where AR and ϕ increase in a certain range of frequencies. For such systems the Bode criterion may lead to erroneous conclusions and we need the more general Nyquist criterion which will be discussed in Section 18.4. Fortunately, systems with AR or ϕ like those of Figure 18.4 are very few, and consequently the Bode criterion will be applicable in most cases.

3. To use the Bode criterion, we need the Bode plots for the open-loop transfer function of the controlled system. These can be constructed in two ways: (a) numerically, if the transfer functions of the process, measuring device, controller, and final control element are known; and (b) experimentally, if all or some of the transfer functions are unknown. In the second case the system is disturbed with a sinusoidal input at various frequencies, and the amplitude and phase lag of the open-loop response are recorded. From these data we can construct the Bode plots.

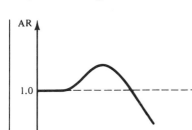

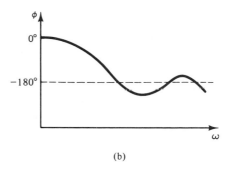

(a)   (b)

Figure 18.4 Bode plots for complex system where Bode stability criterion is not applicable.

18.2   Gain and Phase Margins

The Bode stability criterion indicates how we can establish a rational method for tuning the feedback controllers in order to avoid unstable behavior by the closed-loop response of a process.

Consider the Bode plots for the open-loop transfer function of a feedback system (Figure 18.5). The two important features of these plots are:

The crossover frequency ω_{CO}, where $\phi = -180°$
The point where AR $= 1$

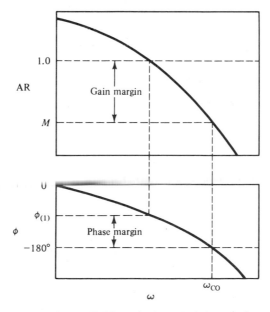

Figure 18.5 Definition of gain and phase margins.

Let M be the amplitude ratio at the crossover frequency (see Figure 18.5). According to the Bode criterion:

If $M < 1$, the closed-loop system is stable.
If $M > 1$, it is unstable.

Define

$$\text{gain margin} = \frac{1}{M} \tag{18.3}$$

Then, for a stable system $M < 1$ and

$$\text{gain margin} > 1$$

We can make the following observations on the practical significance of the gain margin:

1. It constitutes a measure of how far the system is from the brink of instability.
2. The higher the gain margin is above the value of 1, the more robust the closed-loop behavior will be and thus the safer the operation of the controlled process. In other words, the higher the gain margin, the higher the safety factor we use for controller tuning.
3. Typically, a control designer synthesizes a feedback system with gain margin larger than 1.7. This means that the AR can increase 1.7 times above the design value before the system becomes unstable.

Let us now examine the foregoing observations using an example.

Example 18.2: Gain Margin and the Tuning of a Controller

Consider the closed-loop system of Figure 18.1. The crossover frequency is $\omega_{CO} = 17$ rad/min, and the amplitude ratio at this frequency is [see eq. (18.2)]

$$\text{AR} = \frac{K_c}{\sqrt{(0.5 \cdot 17)^2 + 1}} = 0.12 K_c$$

Therefore, the gain margin is equal to

$$\text{gain margin} = \frac{1}{0.12 K_c}$$

If we require a gain margin of 1.7, we find

$$\frac{1}{0.12 K_c} = 1.7 \quad \text{or} \quad K_c = 4.9$$

Let us assume now that the dead time for the process has not been estimated accurately and that its "true" value is 0.15 instead of 0.1. Then the open-loop transfer function is given by

$$G_{OL} = \frac{K_c e^{-0.15s}}{0.5s + 1} \qquad (18.4)$$

and not by the assumed eq. (18.1). For the open-loop transfer function of eq. (18.4) we find that the crossover frequency is $\omega_{CO} = 11.6$ rad/min. At this frequency the amplitude ratio is

$$\text{AR} = \frac{K_c}{\sqrt{(\tau_p\omega)^2 + 1}} = \frac{4.9}{\sqrt{(0.5 \cdot 11.6)^2 + 1}} = 0.83$$

and *the system is still stable despite the error by 50% we made in estimating the dead time of the process*. Notice, though, that the amplitude ratio has moved closer to the value 1 (i.e., the system has moved closer to instability).

The last example demonstrates the practical significance of the gain margin in tuning feedback controllers. Two points are worth emphasizing:

1. Since process parameters such as dead times, static gains, and time constants are almost never known exactly, a gain margin larger than 1 (e.g., 1.7) is a safety factor for stable operation.
2. If the various parameters are known very well, only small safety factors are needed (i.e., gain margins in the range 1.4 to 1.7). For systems with parameters poorly known, the safety factor must increase and the recommended values for gain margins are in the range 1.7 to 3.0.

Besides the gain margin there is another safety factor which is used for the design of a feedback control system: the *phase margin*. Consider again Figure 18.5. Let $\phi_{(1)}$ be the phase lag at the frequency for which AR = 1. The phase margin is defined as follows:

$$\text{phase margin} = 180° - \phi_{(1)}$$

that is, *it is the additional phase lag needed to destabilize the system*. It is clear, therefore, that the higher the phase margin, the larger the safety factor used for designing a controller. Typical phase margins used by designers are larger than 30°.

Example 18.3: Phase Margin and the Tuning of a Controller

Consider again the closed-loop system of Figure 18.1. We know that

$$\text{AR} = \frac{K_c}{\sqrt{(0.5\omega)^2 + 1}} \qquad \text{and} \qquad \phi = \tan^{-1} -0.5\omega + (-0.1\omega)$$

Let us tune the controller using a phase margin equal to 30°. Then we have

$$K_c = \sqrt{(0.5\omega)^2 + 1} \quad \text{and} \quad 30° = 180° - |\tan^{-1} -0.5\omega + (-0.1\omega)|$$

From the second equation we find that $\omega = 12.5$ rad/min. Then the first equation gives $K_c = 6.33$.

1. Assume now that the dead time has been estimated incorrectly and that its "true" value is 0.15. Then the phase lag at the frequency $\omega = 12.5$, where AR = 1, is given by

$$\phi = \tan^{-1} -0.5\omega + (-0.15\omega) = \tan^{-1}(-0.5 \cdot 12.5) + (-0.15 \cdot 12.5)$$

$$= -188°$$

We notice that the system has become unstable; that is, *a phase margin of 30° is not enough to provide a safety factor for a 50% error in dead time.*

2. The reader can easily show that a phase margin of 45° is enough to tune the controller in case 1 and provide the necessary safety factor for absorbing a 50% error in the dead time. The value of the proportional gain K_c for a 45° phase margin is found to be $K_c = 5.05$. Assume that there is an error in the time constant which has a "true" value of 0.25 instead of the assumed 0.5. Then the crossover frequency is found from the equation

$$180° = \tan^{-1} -0.25\omega + (-0.1\omega)$$

and it is equal to $\omega_{CO} = 17.9$. At this frequency

$$AR = \frac{5.05}{\sqrt{(0.25 \cdot 17.9)^2 + 1}} = 1.1$$

and we notice again that the system is unstable. Therefore, although a phase margin of 45° was satisfactory for tuning the controller in the presence of a 50% error in dead time, it is not enough for absorbing an error of up to 50% in the time constant. A larger phase margin is needed.

18.3   Ziegler–Nichols Tuning Technique

In Section 16.5 we discussed a tuning method based on the process reaction curve. The method is primarily experimental and uses real process data from the system's response. In this section we discuss an alternative method developed by Ziegler and Nichols, which is based on frequency response analysis.

Unlike the process reaction curve method which uses data from the open-loop response of a system, the Ziegler–Nichols tuning technique is a closed-loop procedure. It goes through the following steps:

1. Bring the system to the desired operational level (design condition).
2. Using proportional control only and with the feedback loop closed, introduce a set point change and vary the proportional gain until the system oscillates continuously. The frequency of continuous oscillation is the crossover frequency, ω_{CO}. Let M be the amplitude ratio of the system's response at the crossover frequency.
3. Compute the following two quantities:

$$ultimate\ gain = K_u = \frac{1}{M}$$

$$ultimate\ period\ of\ sustained\ cycling = P_u = \frac{2\pi}{\omega_{CO}} \quad min/cycle$$

4. Using the values of K_u and P_u, Ziegler and Nichols recommended the following settings for feedback controllers:

	K_c	τ_I (min)	τ_D (min)
Proportional	$K_u/2$	—	—
Proportional-integral	$K_u/2.2$	$P_u/1.2$	—
Proportional-integral-derivative	$K_u/1.7$	$P_u/2$	$P_u/8$

The settings above reveal the rationale of the Ziegler–Nichols methodology.

1. For proportional control alone, use a gain margin equal to 2.
2. For PI control use a lower proportional gain because the presence of the integral control mode introduces additional phase lag in all frequencies (see Figure 17.8b) with destabilizing effects on the system. Therefore, lower K_c maintains approximately the same gain margin. Similar arguments were used in the process reaction curve tuning technique (see Section 16.5).
3. The presence of the derivative control mode introduces phase lead with strong stabilizing effects in the closed-loop response. Consequently, the proportional gain K_c for a PID controller can be increased without threatening the stability of the system.

Example 18.4: Controller Tuning by the Ziegler–Nichols and Cohen–Coon Methods

Consider the multicapacity process of case 2 in Example 16.4. We have

$$G_p = \frac{1}{(5s + 1)(2s + 1)} \qquad G_m = \frac{1}{10s + 1} \qquad G_f = 1.0$$

The controller settings according to the process reaction curve method were found to be:

For proportional controller: $K_c = 8.3$
For PI controller: $K_c = 7.3$ and $\tau_I = 6.6$
For PID controller: $K_c = 10.9$, $\tau_I = 5.85$, and $\tau_D = 0.89$

Let us now find the Ziegler–Nichols settings and compare them to those above.

Using proportional control only, the crossover frequency can be found from the equation

$$-180° = \tan^{-1} -5\omega_{CO} + \tan^{-1} -2\omega_{CO} + \tan^{-1} -10\omega_{CO}$$

which yields $\omega_{CO} = 0.415$ rad/min. The amplitude ratio at the crossover frequency is found from the equation

$$\log AR = \log \frac{1}{\sqrt{(5\omega_{CO})^2 + 1}} + \log \frac{1}{\sqrt{(2\omega_{CO})^2 + 1}} + \log \frac{1}{\sqrt{(10\omega_{CO})^2 + 1}}$$

and it is equal to 0.08. Therefore, the ultimate gain is

$$K_u = \frac{1}{0.08} = 12.6$$

Also, the ultimate period is found to be

$$P_u = \frac{2\pi}{\omega_{CO}} = 15.14 \text{ min/cycle}$$

Then, the Ziegler–Nichols recommended settings are:

For a proportional controller: $K_c = 12.6/2 = 6.3$
For a PI controller: $K_c = 12.6/2.2 = 5.7$ and $\tau_I = 15.14/1.2 = 12.62$
For a PID controller: $K_c = 12.6/1.7 = 7.4$, $\tau_I = 15.14/2 = 7.57$, and $\tau_D = 15.14/8 = 1.89$

Comparing the Ziegler–Nichols (Z-N) to the Cohen–Coon (C-C) settings, we observe that;

1. The proportional gains are a little larger for the C-C settings.
2. The reset and rate time constants are higher for the Z-N.

Figure 18.6a and b indicate the responses of the closed-loop system to step changes in the set point and load, respectively, using a PID controller with Z-N and C-C settings. We notice that the responses with Z-N tuning are slightly better than those with the C-C settings. It must be emphasized, though, that *no general conclusions can be drawn as to the relative superiority of one method over the other*. The only conclusion we draw is that both methods provide very good first guesses for the values of the controllers' adjustable parameters.

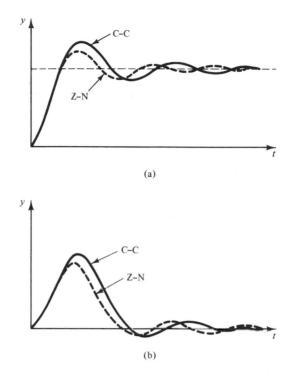

Figure 18.6 Closed-loop responses with Z-N and C-C controller settings: (a) set-point step change; (b) load step change.

18.4   Nyquist Stability Criterion

As we pointed out in Section 18.1, the Bode stability criterion is valid for systems with AR and ϕ monotonically decreasing with ω. For feedback systems with open-loop Bode plots like those of Figure 18.4 the more general Nyquist criterion is employed. In this section we present a simple outline of this criterion and its usage. For more details on the theoretical background of the methodology, the reader can consult Refs. 13 and 14.

The Nyquist stability criterion states that:

If the open-loop Nyquist plot of a feedback system encircles the point (−1, 0) as the frequency ω takes any value from −∞ to +∞, the closed-loop response is unstable.

To understand the concept of encirclement and therefore correct use of the Nyquist criterion, let us study the following examples.

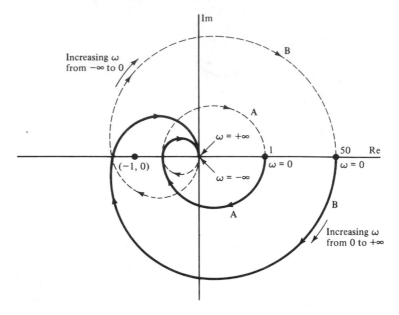

Figure 18.7 Nyquist plots for the open-loop transfer function of Example 18.5.

Example 18.5: Stability Characteristics of a Third-Order System Using the Nyquist Stability Criterion

Consider the open-loop transfer function

$$G_{OL} = \frac{K_c}{(s + 1)(2s + 1)(4s + 1)}$$

Figure 18.7 shows the Nyquist plots for G_{OL} when $K_c = 1$ (curve A) and $K_c = 50$ (curve B). For each Nyquist plot the solid line covers the frequency range $0 \leq \omega < +\infty$, and the dashed part covers the frequencies from $-\infty$ to 0. The dashed segment of the Nyquist plot is the mirror image of the solid-line segment with respect to the real axis.

Figure 18.7 shows that curve A *does not encircle* the point $(-1, 0)$, whereas curve B does. Thus, according to the Nyquist criterion, the feedback system with open-loop Nyquist plot the curve A is stable, while curve B indicates an unstable closed-loop system. This in turn implies that for $K_c = 1$ the system is stable, whereas for $K_c = 50$ it is unstable.

Example 18.6: Conditional Stability and the Nyquist Criterion

Consider the Nyquist plots shown in Figure 18.8a through c. All correspond to the same open-loop transfer function with different values for the proportional gain K_c. The plots in Figure 18.8a and c do not encircle the point $(-1, 0)$, whereas the Nyquist plot of Figure 18.8b does. Therefore,

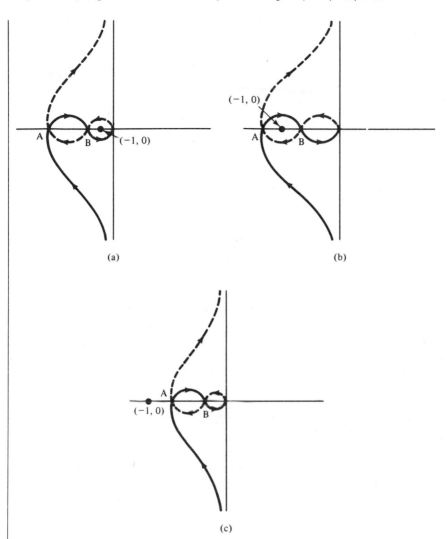

(a)

(b)

(c)

Figure 18.8 Nyquist plots for Example 18.6: (a), (c) stable; (b) unstable.

the feedback systems corresponding to the first and third Nyquist plots have stable closed-loop responses, whereas that of the second is unstable. From the plots above it is clear that the closed-loop response becomes unstable for a range of values K_c such that the point $(-1, 0)$ is between A and B of the resulting Nyquist plot. When point $(-1, 0)$ is to the left of A (Figure 18.8c) or to the right of B (Figure 18.8a), it is not encircled by the Nyquist plot and the corresponding closed-loop response is stable.

Remark.   For fast conclusions on the encirclement or not of the point

(–1, 0) by the open-loop Nyquist plot, the reader can use the following practical method:

Place a pencil at point (–1, 0). Attach one end of a thread at the pencil and with the other end trace the whole length of the Nyquist plot. If the thread has wrapped around the pencil, we can say that point (–1, 0) is encircled by the Nyquist plot.

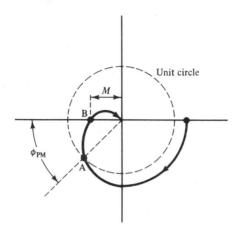

Figure 18.9 Computing gain and phase margins from Nyquist plots.

The gain margin and phase margin of an open-loop response can also be computed from a Nyquist plot. This should be expected since Bode and Nyquist plots of a system contain exactly the same information.

Consider the Nyquist plot of Figure 18.9. At the frequency of point A the Nyquist plot intersects the unit circle around the origin. Therefore, since the distance of point A from the origin is the amplitude ratio at this frequency, we conclude that the angle ϕ_{PM} represents the phase margin.

Furthermore, at the frequency of point B, the phase lag is equal to 180°. The amplitude ratio at this point is the distance between B and the origin, (i.e., AR = M). Consequently, the gain margin is easily found as $1/M$.

THINGS TO THINK ABOUT

1. Explain in your own words that by "opening" a feedback loop we place the controller in a "manual" operation, whereas by "closing" it we place the controller in the "automatic" mode.

2. What is the basis of the Bode criterion? Why is it not generally rigorous?

3. Do you think that the following modified statement of the Bode criterion is generally rigorous? Explain.

   A feedback control system is unstable, if the AR of the corresponding open-loop transfer function is larger than 1 at *any* crossover frequency.

4. Construct an open-loop transfer function whose AR or ϕ, or both, are not continuously decreasing functions of the frequency ω. Draw its Bode and Nyquist plots.

5. Identify the two major sources of instability in closed-loop responses. Elaborate on these two factors.

6. Using the Nyquist stability criterion, show that feedback systems with first- and second-order open-loop responses are always stable.

7. Define the phase and gain margins and show how you can compute them from Bode or Nyquist plots.

8. Explain in your own words what we mean when we say that phase and gain margins constitute safety margins (safety factors) in tuning a feedback controller. Why do we need a safety margin in tuning a feedback controller?

9. Describe the Ziegler–Nichols tuning methodology. This procedure is often called the "continuous cycling" tuning method. Why?

10. The Ziegler–Nichols settings result from closed-loop considerations, whereas the Cohen–Coon settings are determined from the open-loop response of the control system. Would you choose one over the other because it uses open- or closed-loop data? Explain.

11. The experimental determination of the Ziegler–Nichols settings brings the chemical process at the threshold between stable and unstable operation. Can you tolerate this in an industrial environment?

12. State the Nyquist stability criterion and give some examples of stable and unstable feedback control systems different from those presented in this chapter. Explain the concept of encirclement of the point $(-1, 0)$ by the Nyquist plot, which is so central for the Nyquist criterion.

13. Respond to the following questions and justify your answers.
    (a) A larger gain margin implies a smaller or a larger allowable controller gain?
    (b) A larger gain margin makes closed-loop response of a process faster or slower?
    (c) A larger phase margin implies faster or slower closed-loop response?
    (d) A larger phase margin implies smaller or larger allowable controller gain?

14. The discussion in Section 18.2 and Examples 18.2 and 18.3 has indicated that we could use very large phase and gain margins to guarantee closed-

loop stability in the presence of model inaccuracies. Why would you try not to use larger margins than those needed?

15. Larger uncertainty in the parameters of a model (static gain, time constant, dead time) requires larger or smaller gain and phase margins for tuning the controller's parameters?

REFERENCES FOR PART IV

Chapter 13.   There are a variety of references that the reader can consult for more information on the constructional and operational details of measuring devices, feedback controllers, transmission lines, transducers, and final control elements. The following are some typical sources:

1. *Process Instruments and Controls Handbook*, by D. M. Considine, McGraw-Hill Book Company, New York (1957).
2. *Handbook of Applied Instrumentation*, by D. M. Considine and S. D. Ross, McGraw-Hill Book Company, New York (1964).
3. *Instrument Engineers Handbook*, Vol. 1: *Process Measurement*, by B. Liptak, Chilton Book Company, Radnor, Pa. (1970).

For measuring devices Chapter 7 of the following book is very useful:

4. *Measurements and Control Applications for Practicing Engineers*, by J. O. Hougen, Cahners Books, Boston (1972).

For the dynamics of some typical sensors, the reader can consult the article:

5. "Process Dynamics: Part 2. Process Control Loops," by J. L. Guy, *Chem. Eng.*, p. 111 (Aug. 24, 1981).

For the dynamics of thermocouples, valves, pumps, piping, and so on, the following book contains useful information:

6. *Techniques of Process Control*, by P. S. Buckley, John Wiley & Sons, Inc., New York (1964).

The selection of the appropriate control valve is discussed in Ref. 6 and in the book:

7. *Process Modeling, Simulation, and Control for Chemical Engineers*, by W. L. Luyben, McGraw-Hill Book Company, New York (1973).

The distributed character of the pneumatic transmission lines dynamics is discussed in Ref. 6 and in:

8. *An Introduction to Process Dynamics and Control*, by T. W. Weber, John Wiley & Sons, Inc., New York (1973).

Chapter 15. The mathematical proof of the Routh–Hurwitz tests can be found in the classic book:

9. *Dynamics of a System of Rigid Bodies*, 3rd ed., by E. J. Routh, Macmillan, London (1877).

For an extensive discussion the reader can consult:

10. *Stability Theory of Dynamical Systems*, by J. L. Willelms, Thomas Nelson & Sons Ltd., London (1970).
11. *Mathematical Methods in Chemical Engineering*, by V. G. Jenson, and G. V. Jeffreys, Academic Press Ltd., London (1963).

The books by Willelms [Ref. 10] and Douglas [Ref. 12] can also be used for studying alternative definitions of stability and more advanced treatment on the subject.

12. *Process Dynamics and Control*, Vol. 2, by J. M. Douglas, Prentice-Hall, Inc., Englewood Cliffs, N.J. (1972).

The construction rules for the root locus of a closed-loop system can be found in the books by Douglas [Ref. 12], Luyben [Ref. 7], and in the following two classic texts:

13. *Process Systems Analysis and Control*, by D. R. Coughanowr and L. B. Koppel, McGraw-Hill Book Company, New York (1965).
14. *Modern Control Engineering*, by K. Ogata, Prentice-Hall, Inc., Englewood Cliffs, N.J. (1970).

There are a variety of references on the use of root locus for the design of closed-loop systems. The texts by Luyben [Ref. 7], Douglas [Ref. 12], Coughanowr and Koppel [Ref. 13], and Ogata [Ref. 14] offer an excellent treatment of the subject with a large number of examples.

Chapter 16. Two excellent references on the practical problems of controller design are the books by Buckley [Ref. 6] and Shinskey:

15. *Process Control Systems*, 2nd ed., by F. G. Shinskey, McGraw-Hill Book Company, New York (1979).

In these two texts the reader will find useful practical guidelines in selecting the most appropriate type of feedback controller for a particular application. In addition, one can find alternative tuning techniques employed by the industrial practice.

For an extensive discussion on the various types of performance criteria, and their advantages and shortcomings in designing feedback controllers, the reader can consult the following reference:

16. "Optimization of Closed-Loop Responses," by G. Stephanopoulos, in *Process Control*, Volume 2, T. F. Edgar (ed.), AIChE Modular Instruction, American Institute of Chemical Engineers, New York (1982).

In Ref. 16, the reader will also find various techniques for solving the controller design problems, which use time-integral performance criteria. For additional reading on this subject, the following sources are also recommended:

17. *Linear Control System Analysis and Design*, by J. J. D'Azzo and C. H. Houpis, McGraw-Hill Book Company, New York (1975).
18. *Digital Computer Process Control*, by C. L. Smith, Intext Educational Publishers, Scranton, Pa. (1972).
19. *Analytical Design of Linear Feedback Controls*, by G. C. Newton, Jr., L. A. Gould, and J. F. Kaiser, John Wiley & Sons, Inc., New York (1957).

For additional reading on the process reaction curve method and the Cohen–Coon settings, the reader can consult Refs. 8, 12, 13, and 15. The details on the development of the Cohen–Coon settings can be found in their original work:

20. "Theoretical Considerations of Retarded Control," by G. H. Cohen and G. A. Coon, *Trans. ASME*, *75*, 827 (1953).

Chapters 17 and 18. The books by Buckley [Ref. 6] and Caldwell et al. [Ref. 21] are two very good sources for in-depth study of the frequency response analysis and its ramifications in controller design.

21. *Frequency Response for Process Control*, by W. I. Caldwell, G. A. Coon, and L. M. Zoss, McGraw-Hill Book Company, New York (1959).

For systems with transfer functions that are very difficult to factor and consequently very hard to complete the frequency response analysis, Luyben [Ref. 7] discusses various numerical solution techniques. He has also included a computer program in FORTRAN which uses the "stepping" technique to develop the Bode and Nyquist plots for a distillation column. More details on the philosophy of the Ziegler–Nichols tuning method can be found in the original work:

22. "Optimum Settings for Automatic Controllers," by J. G. Ziegler and N. B. Nichols, *Trans. ASME*, *64*, 759 (1942).

In Refs. 6, 7, 13, and 15 the reader can find a large number of examples demonstrating the application of frequency response arguments in the design of feedback controllers. In particular, Refs. 6 and 15 analyze the frequency response characteristics of control systems for flow, pressure, temperature, concentration, and so on, and draw some useful general inferences according to the control system used.

PROBLEMS FOR PART IV

Chapter 13

IV.1 Consider the flow control loop shown in Figure 13.2a. The following information is also available: (1) An orifice plate is used to measure the flow; (2) a variable capacitance differential pressure transducer is employed (see Appendix 11A) to sense and transmit the pressure difference developed around the orifice plate; (3) the controller is PI and (4) the control valve is of equal percentage, with the valve flow characteristic curve given by

$$f(x) = \alpha^{x-1}$$

where $\alpha = 10$ and x = valve stem position (see Appendix 11A for the relationship between x and control signal p). Assume that the pressure drop across the valve remains constant for the range of desired flows.
(a) Derive the transfer function for each element (i.e., orifice plate, differential pressure cell, PI controller, and control valve).
(b) Derive the transfer function between the controlled flow F and the set point value F_{SP}.
(c) Derive an expression for the static gain between F and F_{SP}.

IV.2 Consider the liquid level control loop of Figure 13.2d. The differential pressure transducer cell exhibits second-order dynamics, the controller is proportional, and the control valve is linear with flow characteristic curve given by

$$f(x) = x$$

(a) Derive the transfer functions for the differential pressure cell, controller, and control valve.
(b) Derive the transfer function between the measured liquid level and the manipulated stream flow and a general expression for the static gain between these two variables.
Consult Appendix 11A for details on modeling the differential pressure cell and the control valve.

IV.3 Consider the flash drum unit shown in Figure 4.6. Develop two alternative feedback loops for
(a) The control of the liquid level in the flash drum, or
(b) The control of the pressure in the drum.
(c) Draw the corresponding block diagrams for the loops.

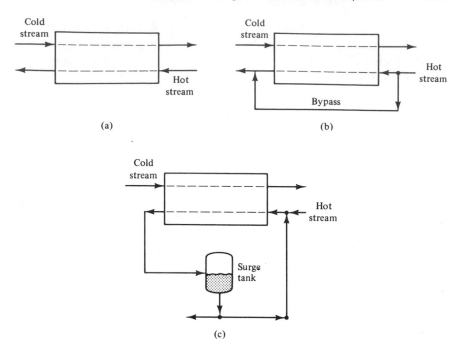

Figure PIV.1

IV.4   Consider the mixing process of Example 4.11 (see Figure 4.8). Develop two alternative feedback loops for each of the following cases:
(a) Control the liquid level in the tank.
(b) Control the concentration of A in the tank.
(c) Control the liquid temperature in the tank.

IV.5   Consider the heat exchanger shown in Figure PIV.1a with the possible piping modifications shown in Figure PIV.1b and c. The control objective is to keep the exit temperature of the cold stream at the desired set point value. Design *seven* different feedback control loops that can achieve this control objective.

Chapter 14

IV.6   We like to control the liquid level, h_2, in tank 2 of system 1 described in Problem II.1 (see Figure PII.1). There exist three alternative manipulated variables, F_1, F_2, and F_3.
(a) Draw the block diagram of the closed-loop system using a PI controller and F_1, F_2, or F_3 as manipulated variable.
(b) Derive the corresponding closed-loop responses to load or set point changes.
(c) Derive a general expression for the closed-loop static gains for each of the corresponding three cases.

(d) Identify the corresponding closed-loop transfer functions for changes in the load (G_{load}) or set point (G_{SP}).

Assume that the transfer functions of the measuring devices and control valves are equal to unity. Also, assume that the flow rates of the free streams are linear functions of the liquid level.

IV.7 Repeat Problem IV.6 assuming that (1) a proportional controller is used instead of PI, and (2) the transfer functions of the measuring device and final control element are given, respectively, by

$$G_m = K_m \qquad G_f = \frac{K_f}{\tau_f s + 1}$$

IV.8 Consider system 2 of Problem II.1 (Figure PII.1). We can control the liquid level h_2 of tank 2 by manipulating flow rate F_1 or F_3. For each of these two cases, do the following:

(a) Draw the corresponding block diagram.
(b) Derive the corresponding closed-loop responses and identify the closed-loop transfer functions to load or set point changes.
(c) Derive general expressions for the corresponding closed-loop static gains.

Assume that a proportional controller is used and that the transfer functions of the measuring sensor and control valve are equal to unity.

IV.9 Repeat Problem IV.8 assuming a PI controller and the following transfer functions for the measuring sensor and control valve:

$$G_m = \frac{K_m}{\tau_m^2 s^2 + 2\zeta_m \tau_m s + 1} \qquad G_f = K_f$$

IV.10 Consider the two stirred tank heaters of Problem II.3 (see Figure PII.3). We would like to control temperature T_3 by manipulating the steam flow rate in either the first or the second heater (i.e., Q_1 or Q_2). The inlet flow rate F_1 remains constant, while the inlet temperature T_1 changes, thus causing the control problem. For each of the two manipulated variables above, do the following:

(a) Draw the corresponding closed-loop block diagram.
(b) Derive the closed-loop response and identify the closed-loop transfer functions to load or set point changes.
(c) Derive general expressions for the corresponding closed-loop static gains.

Assume that we use a PI controller and that the transfer functions for the measuring sensor (thermocouple) and control valve are equal to unity.

IV.11 Consider the closed-loop block diagram of the feedback system shown in Figure PIV.2a. For a set point step change of magnitude 2, do the following:

(a) Derive an expression for the closed-loop response in the Laplace domain.

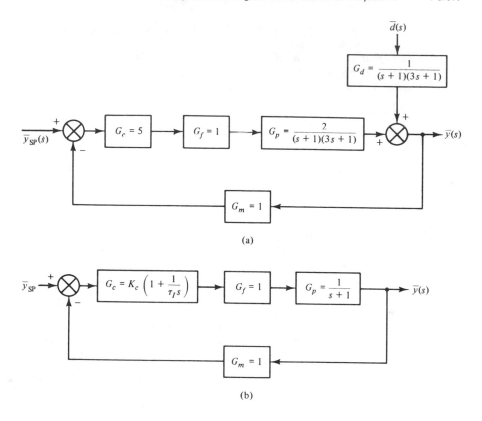

Figure PIV.2

(b) Find how the closed-loop output responds with time to the set point step change above.

(c) Compute the maximum value of $y(t)$ and state when it occurs.

(d) Compute the offset of the final steady state.

(e) Compute the period of oscillation of the closed-loop response.

(f) Give a qualitative sketch of the closed-loop response.

IV.12 Repeat Problem IV.11 but now consider a load step change of magnitude 1.5.

IV.13 Consider the block diagram of the closed-loop system shown in Figure PIV.2b. For a unit step change in the set point, do the following:

(a) Compute the overshoot, decay ratio, and period of oscillation when (1) $K_c = 1$, $\tau_I = 0.5$; (2) $K_c = 20$, $\tau_I = 0.5$; and (3) $K_c = 1$, $\tau_I = 0.1$.

(b) Compare the responses found in part (a) and discuss the effect of K_c and τ_I on the closed-loop response of a system.

(c) Sketch qualitatively the responses for the three cases of K_c and τ_I values given in part (a).

IV.14 Repeat Problem IV.13 but assume a first-order process with a transfer function

$$G_p(s) = \frac{10}{5s + 1}$$

IV.15 Consider the feedback control of a first-order process with a transfer function

$$G_p(s) = \frac{5}{2s + 1}$$

Let the controller be proportional with gain $K_c = 1$, while the transfer function of the measuring device is given by

$$G_m(s) = \frac{K_m}{\tau_m s + 1}$$

Assuming that the transfer function of the final control element is equal to unity, do the following:

(a) Examine the effect of K_m on the quality of the closed-loop response (i.c., for $\tau_m = 1$ compute τ and ζ of the closed-loop response for various values of K_m).

(b) Examine the effect of τ_m on the quality of the closed-loop response (i.e. for $K_m = 1$ compute τ and ζ of the closed-loop response for various values of τ_m).

(c) Sketch qualitatively the closed-loop response for various values of K_m and τ_m. Based on the characteristics of these plots, discuss the effect that a measuring device may have on the closed-loop response of a system (i.e. discuss the effect that K_m and τ_m have on the overshoot, decay ratio, and period of oscillation).

IV.16 Examine the effect that various values of the gain K_m of a measuring device will have on the closed-loop response of a process with the following transfer function:

$$G_p(s) = \frac{1}{(s + 1)(2s + 1)}$$

Assume that, $G_m = K_m$, $G_f = 1$ and the controller is proportional with $K_c = 1$.

IV.17 Consider the two noninteracting tanks of system 1 in Figure PII.1. We want to control the liquid level h_2 of tank 2 by manipulating flow rate F_1 through a proportional controller. Assume that the cross-sectional areas of the two tanks are equal to 5 ft^2. Initially, the system is at steady state with $F_1 = 1$ ft^3/min and $h_1 = h_2 = 3$ ft. Find the values of the controller gain which

(a) Produce a critically damped response, or

(b) Produce an underdamped response with decay ratio 1/4, for h_2.

(c) For each of the two cases above, describe the dynamic response of liquid level h_1 in tank 1, for a unit step change in the set point of h_2. Sketch qualitatively these two responses.

IV.18 Consider the two interacting tanks of system 2 in Figure PII.1. We want to control the liquid level h_2 of tank 2 by manipulating flow rate F_1 through a proportional controller. Assume that tank 1 has a cross-sectional area of 5 ft^2, while for tank 2 the cross-sectional area is 2 ft^2. Initially, the system is at steady state with $F_1 = 1$ ft^3/min, $h_1 = 4$ ft, and $h_2 = 3$ ft. Find the values of the controller gain that produce
(a) A critically damped response, or
(b) An underdamped response with decay ratio equal to 1/4 for h_2.
(c) For each of the two cases above, describe the dynamic response of liquid level h_1 in tank 1 for a unit step change in the set point of h_2. Sketch qualitatively these two responses.

IV.19 Consider the block diagram of Figure Q14.1, which includes two control loops. Assume that $G_{m1} = G_{m2} = 1$ and

$$G_p = \frac{10}{(s + 1)(2s + 1)}$$

(a) Derive an expression for the closed-loop response to a unit step change in the set point, assuming that both controllers are proportional with gains K_{c1} and K_{c2}.
(b) Examine if the closed-loop response exhibits an offset to a unit step change in the set point. If it does, compute the value of the offset. If it does not, explain why.
(c) Suppose that $K_{c2} = 1$. Find the value of K_{c1} which produces (1) a critically damped response, and (2) an underdamped response with a decay ratio 1/4.
(d) Sketch the closed-loop response for each of the two cases in part (c).
(e) Compute the closed-loop poles for the two cases in part (c). What do you observe?

IV.20 Repeat Problem IV.19 assuming that the process is first order with transfer function

$$G_p(s) = \frac{5}{s + 1}$$

The controller of the inner loop is PI with K_{c2} and $\tau_I = 0.5$, while the controller of the outer loop is proportional with gain K_{c1}.

IV.21 Consider the liquid-level control in a simple storage tank. As the manipulated variable we can use either the effluent flow rate, F_o, or the inlet flow rate, F_i. Initially, the system is at steady state with $F_i = F_o = 10$ ft^3/min and liquid level at 2 ft. The cross-sectional area of the tank is 6 ft^2.
(a) Compute the closed-loop response to a unit step increase in the desired set point when F_i is used as the manipulated variable.

(b) Do the same as in part (a), but consider F_o as the manipulated variable.

(c) Sketch the two responses above and qualitatively explain whether you would use F_i or F_o, as the manipulated variable, or it makes no difference which one you use.

Assume a proportional controller with $K_c = 10$ and that the transfer functions for the measuring device and control valve are equal to unity. (Note that F_i is incoming whereas F_o is outflowing).

IV.22 Repeat Problem IV.21 with the following modifications:

(1) The control valve for stream F_i has the transfer function

$$G_{f,i}(s) = 1$$

while the transfer function of the control valve for F_o is given by

$$G_{f,o}(s) = \frac{10}{3s + 1}$$

(2) The proportional gain is $K_c = 1.0$.

Chapter 15

IV.23 Each of the following systems is feedback controlled with a proportional controller. Find the range of values of the proportional gain K_c that produce stable (if it is possible) closed-loop responses. Also, identify the characteristic equations. Assume that $G_m = G_f = 1$.

(a) $G_p(s) = \dfrac{10}{2s - 1}$

(b) $G_p(s) = \dfrac{2}{0.1s + 1}$

(c) $G_p(s) = \dfrac{10}{2s^2 + 3s - 4}$

(d) $G_p(s) = \dfrac{1}{3s^3 + 2s^2 + s - 5}$

(e) $G_p(s) = \dfrac{1}{10s^3 + 2s^2 + s - 5}$

IV.24 Repeat Problem IV.23, but assume that a PI controller is used with gain K_c and $\tau_I = 1$.

IV.25 A first-order system with a transfer function

$$G_p(s) = \frac{5}{0.1s + 1}$$

is controlled with a feedback PI controller

$$G_c(s) = K_c\left(1 + \frac{1}{\tau_I s}\right)$$

Assuming that the final control element has a transfer function $G_f = 1$ and that the transfer function of the measuring device is

$$G_m(s) = \frac{K_m}{\tau_m s + 1}$$

do the following:
(a) Set $K_m = 1$, $\tau_m = 1$, and using the Routh–Hurwitz criterion, find a pair of values K_c and τ_I which yield stable closed-loop response.
(b) Using the values K_c and τ_I found in part (a), examine the effect of changing K_m on the stability of the closed-loop response.
(c) Do the same with τ_m.
(d) Based on the results above, discuss the effect that measurement dynamics have on the stability of the closed-loop response.

IV.26 Examine the effect that a measuring device has on the closed-loop stability of a second-order process,

$$G_p(s) = \frac{1}{s^2 + 4s + 1}$$

Assume that the controller is proportional with gain $K_c = 2$, the final control element has $G_f = 1$, and that the transfer function of the measuring device is

$$G_m = \frac{K_m}{\tau_m s + 1}$$

(*Note*: Examine the effect of K_m and τ_m on the stability of the closed-loop response.)

IV.27 Prove the first test of the Routh–Hurwitz criterion, that is, if any of the coefficients $a_1, a_2, \ldots, a_{n-1}, a_n$ is negative, there is at least one root of the characteristic equation which has a positive real part and the system is unstable. [*Note*: It will help you in the proof to show first that a_1/a_0 is equal to *minus* the sum of all roots of the polynomial, a_2/a_0 is *plus* the sum of all possible products of two roots, ..., a_k/a_0 is equal to $(-1)^k$ times the sum of all possible products of k roots, and finally that a_n/a_0 is equal to $(-1)^n$ times the product of all n roots.]

IV.28 Consider a first-order system with dead time having the following transfer function:

$$G_p(s) = \frac{10 \, e^{-t_d s}}{s + 1}$$

The system is controlled with a proportional controller of gain K_c.
(a) Approximate $e^{-t_d s}$ by a first-order Padé approximation (see Section 12.2) and find the relationship between K_c and t_d that leads to a stable closed-loop response.
(b) Approximate $e^{-t_d s}$ by a second-order Padé approximation (see Section 12.2) and find the relationship between K_c and t_d which leads to a stable closed-loop response. Assume that $G_m = G_f = 1$.

IV.29 Repeat Problem IV.28 but for the following second-order process:

$$G_p(s) = \frac{e^{-t_d s}}{(s + 1)(2s + 1)}$$

IV.30 An unstable first-order process with transfer function

$$G_p(s) = \frac{2}{s - 4}$$

is controlled by a proportional controller with gain K_c. Assuming that $G_m = G_f = 1$, do the following:

(a) Sketch the location of the closed-loop pole as K_c goes from 0 to 10. This is the root locus of the closed-loop system.

(b) Determine the range of K_c values which yield stable closed-loop responses.

(c) Find the value of K_c for which the closed-loop pole is zero. Using this value of K_c, compute the closed-loop response to a unit step change in the set point. Is the response stable or unstable?

IV.31 Draw the root locus of a closed-loop system with the following characteristics:

Process: $G_p(s) = \dfrac{1}{(s + 1)(2s + 1)}$

Controller: $G_c(s) = K_c$

Measuring device: $G_m(s) = 1$

Final control element: $G_f(s) = 1$

Indicate what segments of the root locus (i.e., values of K_c) yield (a) overdamped, (b) critically damped, and (c) underdamped closed-loop responses.

IV.32 Draw the root locus of a closed-loop system with

$$G_p(s) = \frac{2}{s - 4} \qquad G_c(s) = K_c\left(1 + \frac{1}{s}\right) \qquad G_m(s) = G_f(s) = 1$$

IV.33 Draw the root locus of the systems in Problem IV.23 using a proportional controller. The transfer functions of the measuring device and final control element are $G_m(s) = G_f(s) = 1$.

IV.34 Draw an approximate sketch of the root locus for the closed-loop system with the following characteristics:

$$G_p(s) = \frac{2(s + 1)}{s(s + 2)(s^2 + 1)} \qquad G_c(s) = K_c \qquad G_m(s) = G_f(s) = 1$$

Using the information from the root locus, sketch qualitatively the closed-loop response of the process to a unit step in the set point when $K_c = 0$, $K_c = 1$, $K_c = 10$, and $K_c = 100$.

Chapter 16

IV.35 A first-order process is controlled with a PI controller. Find the values of the controller gain K_c and reset time τ_I so that (a) the closed-loop gain to load changes is 10 and (b) the decay ratio of the closed-loop response is 1/4. The following information is given:

$$G_p(s) = G_d(s) = \frac{1}{s + 3} \qquad G_m(s) = G_f(s) = 1$$

IV.36 Find the gain of a proportional controller that produces a closed-loop response for a second-order system with decay ratio equal to 1/4. The process is described by

$$G_p(s) = \frac{1}{s^2 + 3s + 1}$$

and

$$G_m(s) = G_f(s) = 1$$

IV.37 Select the gain and reset time settings of a PI controller, employing the minimum ISE criterion for a unit step change in the set point. The process is first-order with $K_p = 10$, and $\tau_p = 1.0$. Assume that $G_m(s) = G_f(s) = 1$. The selected settings must satisfy the restrictions

$$100 \geq K_c \geq 1 \qquad \text{and} \qquad 10 > \tau_I \geq 0.1$$

IV.38 Repeat Problem IV.37 but consider a step change in the set point of magnitude 5. Are the new controller settings different from those found in Problem IV.37? Explain why they are the same or not the same.

IV.39 Select the gain of a proportional controller using the minimum ISE criterion. Consider a unit step change in the set point. The process is second-order with $K_p = 5$, $\tau = 2$, and $\zeta = 3$. Assume that $G_m(s) = G_f(s) = 1$. The selected setting must satisfy the restriction

$$100 \geq K_c \geq 1$$

[You should realize that ISE is a very "difficult" criterion to use for tuning proportional controllers and that another criterion, far simpler than ISE, could be used instead, with equivalent results.]

IV.40 Repeat Problem IV.39 but consider a step change in the set point of magnitude 5. Is the new gain setting different from that found in Problem IV.39? Explain why it is the same or not the same.

IV.41 Select the gain of a proportional controller using the one-quarter decay ratio criterion. The process is described by

$$G_p(s) = \frac{10}{(s + 2)(2s + 1)}$$

Assume that $G_m(s) = G_f(s) = 1$. Also, select the gain using the minimum

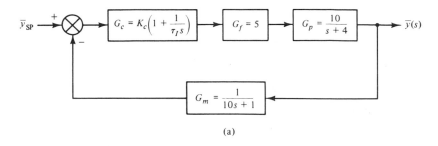

(a)

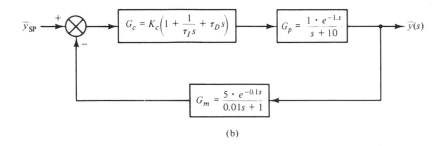

(b)

Figure PIV.3

ISE criterion and a unit step change in the set point. The condition $100 \geq K_c \geq 0.1$ must be satisfied by the gain values selected. Compare the settings computed by the two methods and explain the difference between them. [The reader should make a note of the remark made at the end of Problem II.39].

IV.42 Repeat Problem IV.41 but consider the process

$$G_p(s) = \frac{1}{s^2 + 0.5s + 1}$$

IV.43 The process reaction curve of a temperature control system gave the values

$$K = 10 \qquad \tau = 2 \text{ min} \qquad t_d = 0.1 \text{ min}$$

Compute the settings of a PID controller using the Cohen–Coon tuning methodology.

IV.44 Consider the feedback loop shown in Figure PIV.3a. Select the settings of the PI controller using the Cohen–Coon tuning technique. In a graph paper display the actual process reaction curve and its first-order plus dead time approximation.

IV.45 Select the controller settings for the PID controller of the feedback loop shown in Figure PIV.3b. Use the Cohen–Coon tuning technique. In a graph paper display the actual process reaction curve and its first-order plus dead time approximation.

IV.46 Table PIV.1 shows the experimental process reaction curve of an open-loop system with a PI controller. Using these values:
(a) Approximate the open-loop response with that of a first-order system plus dead time (i.e., find the values of K, τ, and t_d).
(b) Select the controller settings using the Cohen–Coon technique.

IV.47 Repeat Problem IV.46 using (1) a PID controller instead of the PI, and (2) the process reaction curve data of Table PIV.2.

IV.48 In this problem we see how to select the better manipulated variable among two alternatives. Consider the two noninteracting liquid storage tanks of system 1 shown in Figure PII.1. We can control liquid level h_2 by manipulating F_1 or F_3. Assume that the cross-sectional areas of the two tanks are $A_1 = 2$ ft^2 and $A_2 = 5$ ft^2. Initially, the system is at steady state with $F_1 = 3$ ft^3/min and liquid levels $h_1 = 3$ ft and $h_2 = 6$ ft.
(a) Using the minimum ISE tuning criterion, find the gain K_c of a proportional controller when the manipulated variable is either F_1 or F_3. Assume a 20% step increase in the set point of liquid level h_2. Note that gain K_c should not exceed the value 100.
(b) For each of the two cases above, compute the value of ISE and select that manipulated variable that yields the smaller ISE.
In the analysis above, assume linear flow resistance and $G_m = G_f = 1$.
[The reader should make a note of the remark made at the end of

<table>
<tr><th colspan="3">TABLE PIV.1</th></tr>
<tr><th>Time
(min)</th><th>Manipulated
input</th><th>Measurement
of output</th></tr>
<tr><td>-2</td><td>100</td><td>200</td></tr>
<tr><td>-1</td><td>100</td><td>200</td></tr>
<tr><td>0</td><td>150</td><td>200.1</td></tr>
<tr><td>0.2</td><td>150</td><td>201.1</td></tr>
<tr><td>0.4</td><td>150</td><td>204.0</td></tr>
<tr><td>0.6</td><td>150</td><td>227.0</td></tr>
<tr><td>0.8</td><td>150</td><td>251.0</td></tr>
<tr><td>1.0</td><td>150</td><td>280.0</td></tr>
<tr><td>1.2</td><td>150</td><td>302.5</td></tr>
<tr><td>1.4</td><td>150</td><td>318.0</td></tr>
<tr><td>1.6</td><td>150</td><td>329.5</td></tr>
<tr><td>1.8</td><td>150</td><td>336.0</td></tr>
<tr><td>2.0</td><td>150</td><td>339.0</td></tr>
<tr><td>2.2</td><td>150</td><td>340.5</td></tr>
<tr><td>2.4</td><td>150</td><td>341.0</td></tr>
</table>

<table>
<tr><th colspan="3">TABLE PIV.2</th></tr>
<tr><th>Time
(min)</th><th>Manipulated
input</th><th>Measurement
of output</th></tr>
<tr><td>-10</td><td>10</td><td>0.650</td></tr>
<tr><td>-5</td><td>10</td><td>0.650</td></tr>
<tr><td>0</td><td>15</td><td>0.650</td></tr>
<tr><td>5</td><td>15</td><td>0.651</td></tr>
<tr><td>10</td><td>15</td><td>0.652</td></tr>
<tr><td>15</td><td>15</td><td>0.668</td></tr>
<tr><td>20</td><td>15</td><td>0.735</td></tr>
<tr><td>25</td><td>15</td><td>0.817</td></tr>
<tr><td>30</td><td>15</td><td>0.881</td></tr>
<tr><td>35</td><td>15</td><td>0.979</td></tr>
<tr><td>40</td><td>15</td><td>1.075</td></tr>
<tr><td>45</td><td>15</td><td>1.151</td></tr>
<tr><td>50</td><td>15</td><td>1.213</td></tr>
<tr><td>55</td><td>15</td><td>1.239</td></tr>
<tr><td>60</td><td>15</td><td>1.262</td></tr>
<tr><td>65</td><td>15</td><td>1.311</td></tr>
<tr><td>70</td><td>15</td><td>1.329</td></tr>
<tr><td>75</td><td>15</td><td>1.338</td></tr>
<tr><td>80</td><td>15</td><td>1.350</td></tr>
<tr><td>85</td><td>15</td><td>1.351</td></tr>
<tr><td>90</td><td>15</td><td>1.350</td></tr>
</table>

Problem IV.39. It simplifies the solution of the present problem significantly.]

IV.49 In this problem we try to select the better manipulated variable among two alternatives, F_1 or F_3, in order to control the liquid level h_2 of two interacting tanks (see system 2 of Figure PII.1). The following information is given: Cross-sectional areas of the two tanks are $A_1 = 2$ ft^2 and $A_2 = 5$ ft^2; initial steady state, $h_1 = 6$ ft, $h_2 = 4$ ft, and $F_1 = 3$ ft^3/min.
   (a) Using the minimum offset as tuning criterion, find the gain K_c of a proportional controller when the manipulated variable is either F_1 or F_3. Assume a 20% step increase in the set point of liquid level h_2. The gains selected for each case should not be larger than 100.
   (b) For each of the two cases above, compute the value of the offset and select that manipulation which yields the smaller one. In the analysis above, assume linear flow resistance and $G_m = G_f = 1$.

IV.50 Consider the system of the two stirred tank heaters shown in Figure PII.3. We want to maintain temperature T_3 at the desired set point value using one of the steam flow rates Q_1 or Q_2 as the manipulated variable. Make the necessary computations and decide which manipulated variable is better using the minimum offset as selection criterion. The following information is given:
   (a) Flow rates F_1, F_2, and F_3 remain constant at their steady state values.
   (b) Temperature T_1 is the main disturbance.
   (c) The dynamics of the two heaters are given by the following equations:

$$\overline{T}_2(s) = \frac{1}{4s + 1} \overline{T}_1(s) + \frac{0.2}{4s + 1} \overline{Q}_1(s)$$

$$\overline{T}_3(s) = \frac{1}{10s + 1} \overline{T}_2(s) + \frac{0.1}{10s + 1} \overline{Q}_2(s)$$

   (d) Consider a 10% step increase in the value of the disturbance T_1 in order to compute the offset of the closed-loop responses.
   (e) The controller is proportional, while the thermocouples and control valves have transfer functions equal to unity.
   (f) The following restrictions must be satisfied by the controller gain

$$100 \geqslant K_c \geqslant 0.1$$

IV.51 Repeat Problem IV.50 but use the following dynamics for the two heaters:

$$\overline{T}_2(s) = \frac{1}{5s + 1} \overline{T}_1(s) + \frac{0.1}{5s + 1} \overline{Q}_1(s)$$

$$\overline{T}_3(s) = \frac{1}{s + 1} \overline{T}_2(s) + \frac{0.2}{s + 1} \overline{Q}_2(s)$$

Chapter 17

IV.52 Prove analytically the expressions for the amplitude ratio and phase shift of the ultimate response for the following systems:
(a) Two first-order systems in series
(b) Second-order system
(c) PID controller
(d) Pure capacitive process

IV.53 Sketch the Bode plots and Nyquist diagrams for the systems with the following transfer functions:

(a) $G(s) = \dfrac{10}{s(s + 5)}$ 　　　　　(b) $G(s) = \dfrac{s + 1}{s(s + 5)}$

(c) $G(s) = \dfrac{50}{(s + 2)^2}$ 　　　　　(d) $G(s) = \dfrac{s - 2}{(s + 2)^2}$

(e) $G(s) = \dfrac{10e^{-s}}{(s + 1)(s + 3)}$ 　　　(f) $G(s) = \dfrac{10e^{-0.01s}}{(s + 1)(s + 3)}$

(g) $G(s) = \dfrac{1}{s^2 + 3s + 1}$ 　　　　(h) $G(s) = \dfrac{1}{s^2 + 4s + 6}$

IV.54 Sketch the Bode plots and Nyquist diagrams of open-loop transfer functions with the following dynamic components:

(a) $G_p(s) = \dfrac{1}{s + 1}$, $G_m(s) = \dfrac{0.95}{0.01s + 1}$, $G_c(s) = 10$, $G_f(s) = 1$

(b) $G_p(s) = \dfrac{10}{s + 2}$, $G_m(s) = 0.95$, $G_c(s) = 10\left(1 + \dfrac{1}{0.1s}\right)$, $G_f(s) = 1$

(c) $G_p(s) = \dfrac{5}{(s + 1)(s + 2)}$, $G_m(s) = \dfrac{1}{0.1s + 1}$, $G_c(s) = 1$, $G_f(s) = 10$

(d) $G_p(s) = \dfrac{2}{s^2 + 4s + 1}$, $G_m(s) = 1$, $G_c(s) = 10\left(1 + \dfrac{1}{0.5s}\right)$, $G_f(s) = 5$

(e) $G_p(s) = \dfrac{1}{5s^2 + 2s + 6}$, $G_m(s) = \dfrac{1}{s + 1}$, $G_c(s) = 10\left(1 + \dfrac{1}{0.5s}\right)$,
$G_f(s) = \dfrac{10}{s + 10}$

IV.55 Table PIV.3 shows the amplitude ratio and phase lag values of an unknown system at various frequencies. (a) Determine the order of the unknown system. (b) Examine it for dead time. (c) Compute the values of the system parameters (e.g., static gain, time constant; or K_p, τ, ζ, etc.) and the value of dead time if the system possesses dead time.

IV.56 Repeat Problem IV.55 using the amplitude ratio and phase lag values given in Table PIV.4.

TABLE PIV.3			TABLE PIV.4			TABLE PIV.5		
Frequency (cycles/min)	AR	ϕ (deg)	Frequency (cycles/min)	AR	ϕ (deg)	Frequency (cycles/min)	AR	ϕ (deg)
0.01	10	−0.63	0.01	5.00	−0.23	0.01	17	−1.49
0.05	9.99	−3.15	0.05	5.05	−1.13	0.02	16.99	−2.98
0.10	9.99	−6.30	0.10	5.20	−2.39	0.10	16.67	−14.75
1.0	9.95	−63.01	0.20	5.93	−5.44	0.30	14.42	−41.21
3.0	9.58	−188.60	0.30	7.68	−11.62	0.50	11.66	−61.90
5.0	8.94	−313.04	0.40	12.69	−23.96	0.70	9.33	−77.76
7.0	8.19	−436.06	0.50	25.00	−90.00	1.00	6.80	−95.73
9.0	7.43	−557.65	0.60	9.98	−151.39	1.50	4.30	−117.03
10.0	7.04	−617.96	0.70	5.00	−163.74	2.00	2.92	−132.42
12.0	6.40	−737.74	0.80	3.25	−168.10	2.50	2.07	−144.53
15.0	5.55	−915.75	0.90	2.20	−170.87	3.00	1.55	−154.04
20.0	4.47	−1209.35	1.10	1.29	−173.46	4.00	0.94	−169.23
30.0	3.16	etc.	1.50	0.62	−175.71	8.00	0.26	−208.22
40.0	2.43		2.00	0.33	−176.95	10.00	0.17	−223.12
50.0	1.96		5.00	0.05	−178.84	20.00	0.04	−287.45

IV.57 Repeat Problem IV.55 using the amplitude ratio and phase lag values given in Table PIV.5.

Chapter 18

IV.58 Using the Bode stability criterion, find which of the control systems with the following open-loop transfer functions are stable and which are unstable:

(a) $G_{OL} = \dfrac{1}{s - 1}$

(b) $G_{OL} = \dfrac{10e^{-3s}}{4s + 1}$

(c) $G_{OL} = \dfrac{5e^{-5s}}{(2s + 1)(s + 1)}$

(d) $G_{OL} = \dfrac{1}{0.2s^2 + 0.8s - 1}$

IV.59 Consider the processes with the transfer functions given in Problem IV.23. Each of these processes is feedback controlled with a proportional controller. Assume that $G_m = G_f = 1$. Using the Bode stability criterion, find the range of values of the proportional gain K_c which produce stable (if it is possible) closed-loop responses.

IV.60 A first-order process with dead time has the transfer function

$$G_p(s) = \frac{10e^{-t_d s}}{2s + 1}$$

This process is to be controlled with a PI controller. Use the Bode stability criterion to find the range of values for the gain K_c as a function of t_d so that the closed-loop response is stable. Assume that $G_m = G_f = 1$ and that the reset time for the PI controller is $\tau_I = 0.5$ min.

IV.61 Repeat Problem IV.25 using the Bode stability criterion.

IV.62 Repeat Problem IV.26 using the Bode stability criterion.

IV.63 Using the Bode stability criterion, find the range of K_c values that stabilize the unstable process

$$G_p(s) = \frac{1}{s - 5}$$

The controller is proportional and $G_m = G_f = 1$.

IV.64 Compute the phase and gain margins for the feedback systems with the open-loop transfer functions given in Problem IV.58. On the basis of these values, find which systems are stable and which are not.

IV.65 Suppose that the gain, time constant, and dead time of a process with

$$G_p(s) = \frac{10e^{-0.1s}}{0.5s + 1}$$

are known with a possible error of ±20% of their values. Compute the largest permissible gain K_c of a proportional controller so that the closed-loop response is always stable. Assume that $G_m = G_f = 1$.

IV.66 Repeat Problem IV.39 but use the Ziegler–Nichols tuning technique instead of the ISE criterion. Compare the Ziegler–Nichols settings found here with those found in Problem IV.39.

IV.67 Using the Ziegler–Nichols settings, compute the gain of the proportional controller of Problem IV.41.

IV.68 Compute the Ziegler–Nichols settings for the PID controller of Problem IV.43. Compare them to the Cohen–Coon settings found in Problem IV.43. Assume that the approximate values of K, τ, and t_d found from the process reaction curve are not very reliable. Compute the percent error in the values of K, τ, and t_d which can be tolerated by the Ziegler–Nichols settings before the closed-loop response becomes unstable.

IV.69 (a) Compute the Ziegler–Nichols settings for the controller of the feedback loop shown in Figure PIV.3a.
    (b) Compare them to the Cohen–Coon settings found in Problem IV.44
    (c) Which settings are better? To answer this question, compute the ISE of the closed-loop response to a unit step change in set point for the two different sets of controller settings.
    (d) Compute the tolerance of the Ziegler–Nichols settings to an error in the process static gain.
    [Question (c) involves lengthy analytic computations and can be omitted.]

IV.70 Repeat Problem IV.69 but for the controller of the feedback loop shown in Figure PIV.3b. Compute the tolerance of the Ziegler–Nichols settings to errors in the process or measurement dead times.

IV.71 Table PIV.1 shows the experimental process reaction curve of an open-loop system with a PI controller.

  (a) Approximate the process by a first-order system with dead time (i.e., find the values of the static gain K, time constant τ, and dead time t_d).

  (b) On the basis of the approximation above, compute the Ziegler–Nichols settings of the PI controller.

  (c) Compare them to the Cohen–Coon settings which can be computed from the process reaction curve and find which settings yield the smaller ISE of the closed-loop response to a unit step change in set point.

  (d) Compute the tolerance of the Ziegler–Nichols and Cohen–Coon settings to errors in static gain K, time constant τ, or dead time t_d. Which settings possess larger tolerance?

IV.72 Repeat Problem IV.71 but use the process reaction curve data given in Table PIV.2.

IV.73 Using the Nyquist stability criterion, find which of the closed-loop systems of Problem IV.54 are stable and which are not.

IV.74 Examine the stability of the closed-loop systems whose open-loop transfer functions are given in Problem IV.58. Employ the Nyquist criterion for your analysis.

IV.75 Use the Nyquist criterion and find the range of K_c values that yield stable closed-loop response for

  (a) The feedback loop described in Problem IV.60.

  (b) The proportionally controlled unstable process of Problem IV.63.

IV.76 Consider the open-loop Bode plots described by the data of Tables PIV.3 through PIV.5.

  (a) Draw the corresponding Nyquist plots.

  (b) Using the Nyquist criterion, examine the stability of the corresponding closed-loop systems.

  (c) Compute the corresponding gain and phase margins.

Analysis and Design of Advanced Control Systems

V

A good theory must be useful to process control engineers and should be developed to accommodate the needs and skills of the potential users....It was once said that only Frank Lloyd Wright can design a house for a family without asking about the number of children, or the family budget.
                                                    W. Lee and V. W. Weekman, Jr.*

Although feedback control is the type encountered most commonly in chemical processes, it is not the only one. There exist situations where feedback control action is insufficient to produce the desired response of a given process. In such cases other control configurations are used, such as feedforward, ratio, multivariable, cascade, override, split range, and adaptive control.

In the following four chapters of Part V we will study the static and dynamic characteristics, as well as methods for the design, of the following advanced control systems:

1. Compensatory control for processes with large dead time or inverse response
2. Multiple loop control (cascade, selective, split range)
3. Feedforward and ratio control
4. Adaptive and inferential control

---

*"Advanced Control Practice in the Chemical Industry: A View from Industry," *AIChE J.*, *22*(1), 27 (1976).

The material of the subsequent four chapters (Chapter 19, 20, 21, and 22) should be viewed as an introduction to the analysis and design of the control systems above. The subject is quite involved, and the interested reader should consult the references at the the end of Part V. In particular, the discussion on the adaptive and inferential control is limited to a simple qualitative presentation of these control systems, since a more rigorous presentation goes beyond the scope of this text. Nevertheless, in Chapter 31, the interested reader will find a mathematical treatment of the adaptive control system design.

Feedback Control **19**
of Systems
with Large Dead Time
or Inverse Response

All the chapters of Part IV were devoted to the analysis and design of feedback control systems for rather simple processes. In this chapter we are concerned with the feedback control of two special types of systems: those with large dead times or inverse responses. We will see that for such systems, conventional P, PI, or PID controllers may not be sufficient to yield the desired response.

19.1 Processes with Large Dead Time

Consider the general feedback control system of Figure 14.1. All the dynamic components of the loop may exhibit significant time delays in their response. Thus:

1. The main process may involve transportation of fluids over long distances or include phenomena with long incubation periods.
2. The measuring device may require long periods of time for completing the sampling and the analysis of the measured output (a gas chromatograph is such a device).
3. The final control element may need some time to develop the actuating signal.
4. A human controller (decision maker) may need significant time to think and take the proper control action.

In all of the situations noted above, a conventional feedback controller would provide quite unsatisfactory closed-loop response, for the following reasons:

1. A disturbance entering the process will not be detected until after a significant period of time.
2. The control action that will be taken on the basis of the last measurement will be inadequate because it attempts to regulate a situation (eliminate an error) that originated awhile back in time.
3. The control action will also take some time to make its effect felt by the process.
4. As a result of all the factors noted above, significant dead time is a significant source of instability for closed-loop responses.

Example 19.1: Dead Time as a Main Source of Closed-Loop Instability

Consider the open-loop transfer function

$$G_{\mathrm{OL}} = \frac{K_c e^{-t_d s}}{0.5s + 1}$$

1. If t_d = 0.01 min (i.e., very small), crossover frequency = 160 rad/min and the ultimate gain = 80.01.
2. Suppose that the dead time is increased to t_d = 0.1. Then the crossover frequency = 17 rad/min and the ultimate gain = 8.56. We notice that the *increase in dead time has introduced significant additional phase lag, which reduces the crossover frequency and the maximum allowable gain*. In other words, the increase in dead time has made the closed-loop response more sensitive to periodic disturbances and has brought the system closer to the brink of instability.
3. A further increase in dead time, (i.e., t_d = 1.0) yields a crossover frequency = 2.3 rad/min and an ultimate gain = 1.52. We see the same trends as above.

We see from Example 19.1 that as the dead time of an open-loop transfer function increases, the following two undesirable effects take place:

1. *The crossover frequency decreases.* This implies that the closed-loop response will be sensitive even to lower-frequency periodic disturbances entering the system.
2. *The ultimate gain decreases.* Therefore, to avoid the instabilities of the closed-loop response, we must reduce the value of the proportional gain K_c, which leads to sluggish response.

Figure 19.1 graphically depicts these results.

The discussion above indicates that a control system different from

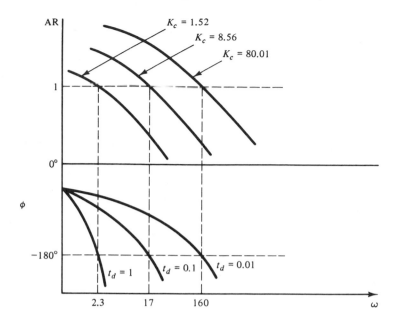

Figure 19.1 Effect of dead time on crossover frequency (Example 19.1).

the conventional feedback loop is needed to compensate for dead-time effects.

19.2 Dead-Time Compensation

In the preceding section we identified the critical need for more effective control of processes with significant dead time. In this section we discuss a modification of the classical feedback control system which was proposed by O. J. M. Smith for the compensation of dead-time effects. It is known as the *Smith predictor* or the *dead-time compensator*.

To understand the nature of the dead-time compensation proposed by Smith, consider the simple feedback loop with set point changes only shown in Figure 19.2a. We have assumed that all the dead time is caused by the process:

$$G_p(s) = G(s)e^{-t_d s}$$

and that for simplicity, $G_m(s) = G_f(s) = 1$. The open-loop response to a change in the set-point is equal to

$$\bar{y}(s) = G_c(s)[G(s)\, e^{-t_d s}]\, \bar{y}_{SP}(s) \tag{19.1}$$

(i.e., it is delayed by t_d minutes).

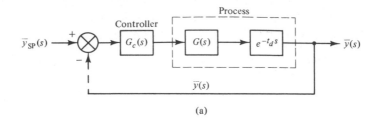

(a)

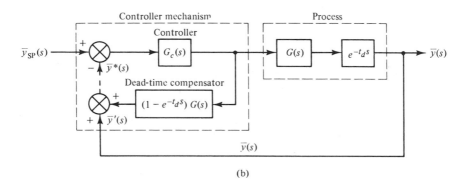

(b)

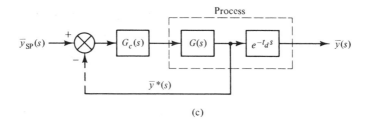

(c)

Figure 19.2 (a) Feedback system with process dead time; (b) feedback with complete dead-time compensation; (c) net result of dead-time compensation.

In order to eliminate the undesired effects, we would like to have an open-loop feedback signal that carries *current and not delayed information*, such as

$$\bar{y}^*(s) = G_c(s)G(s)\bar{y}_{SP}(s) \qquad (19.2)$$

This is possible if in the open-loop response $\bar{y}(s)$ we add the quantity

$$\bar{y}'(s) = (1 - e^{-t_d s})G_c(s)G(s)\bar{y}_{SP}(s) \qquad (19.3)$$

It is easy to verify that

$$\bar{y}'(s) + \bar{y}(s) = \bar{y}^*(s)$$

The implication of adding $\bar{y}'(s)$ to the signal $\bar{y}(s)$ is shown in Figure 19.2b. There we notice that the signal $\bar{y}'(s)$ can be taken by a simple local loop around the controller, which is called the *dead-time compensator* or *Smith predictor*. The simplified loop of Figure 19.2c is completely equivalent to that of Figure 19.2b and indicates the real effect of the dead-time compensator:

It moves the effect of dead time outside the loop.

Remarks

1. In the block diagram of Figure 19.2c it is not correct to think that we take a measurement signal after the block $G(s)$ because such a signal is not measurable in a real process with dead time. The only measurable signals are the process output, $\bar{y}(s)$, and the manipulated variable. Therefore, the block diagram of Figure 19.2c is meant to give only a schematic representation of what is the effect of the dead-time compensator, not to depict physical reality.

2. The dead-time compensator *predicts* the delayed effect that the manipulated variable will have on the process output. This prediction led to the term *Smith predictor* and it is possible only if we have a model for the dynamics of the process (transfer function, dead time).

3. In most process control problems the model of the process is not perfectly known; that is, $G(s)$ and t_d are known only approximately. Consider that $G(s)$ and t_d represent the "true" characteristics of the process, while $G'(s)$ and t'_d represent their approximations, as these are given by some mathematical model for the process. Then, using $G'(s)$ and t'_d to construct the Smith predictor, we take the system shown in Figure 19.3. In this case the composite open-loop feedback signal is

$$\bar{y}^*(s) = \bar{y}(s) + \bar{y}'(s)$$

$$= [G_c G e^{-t_d s} + (1 - e^{-t'_d s}) G_c G'] \bar{y}_{SP}(s)$$

or

$$\bar{y}^*(s) = G_c [G' + (G e^{-t_d s} - G' e^{-t'_d s})] \bar{y}_{SP}(s) \qquad (19.4)$$

The equation above indicates some important features of dead-time compensators:

(a) *Only for perfectly known processes will we have perfect compensation (i.e., for $G = G'$ and $t_d = t'_d$).*

(b) *The larger the modeling error [i.e., the larger the differences $(G - G')$ and $(t_d - t'_d)$], the less effective is the compensation.*

(c) *The error in estimating the dead time is more detrimental for effective dead-time compensation [i.e., $(t_d - t'_d)$ is more crucial than $(G - G')$], because of the exponential function.*

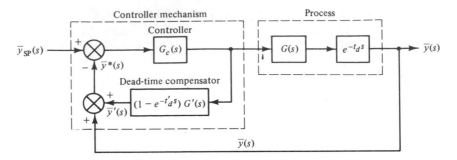

Figure 19.3 Dead-time compensation with inaccurate knowledge of process transfer function and dead time.

4. The dead time in a chemical process is usually caused by material flows. Since the flow rate is not normally constant but shows variations during the operation of a plant, *the value of the dead time changes*. Therefore, if the dead-time compensator is designed for a certain value of the dead time, then when t_d takes on a new value the compensation will not be as effective.

Example 19.2: Dead-Time Compensation and the Effect of Modeling Error

Consider the feedback loop shown in Figure 19.2a. Let the controller be simple proportional and the "true" transfer function of the process be

$$G_p(s) = \frac{1}{0.5s + 1} e^{-1s}$$

It is easy to recognize that

$$G(s) = \frac{1}{0.5s + 1} \quad \text{and} \quad t_d = 1$$

1. Suppose that we use simple feedback control. For this system it was found in Example 19.1 that the open-loop transfer function has crossover frequency $\omega_{CO} = 2.3$ rad/min and ultimate gain $K_c = 1.52$. The fact that the ultimate gain is 1.52 forced us to use $K_c < 1.52$. Nevertheless, the system is very close to the brink of instability and has a rather unacceptable offset (see Section 14.2):

$$\text{offset} = \frac{1}{1 + K_p K_c} = \frac{1}{1 + 1 \cdot 1.52} = 0.4$$

Curve A (Figure 19.4) shows the response of the system to a unit step change in the set point.
2. Let us introduce "perfect" dead-time compensation. This is possible if the "true" transfer function of the process is known. Then the

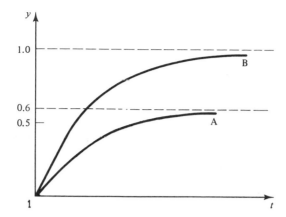

Figure 19.4 Closed-loop responses for the system of Example 19.2.

control system is given by the block diagram of Figure 19.2b. The open-loop transfer function is [see eq. (19.2)]

$$\frac{\overline{y}^*(s)}{\overline{y}_{SP}(s)} = G_c G = \frac{K_c}{0.5s + 1}$$

which has no crossover frequency. Consequently, we can use arbitrarily large proportional gain to reduce the offset without endangering the stability of the system. Curve B (Figure 19.4) shows the response of the closed-loop system with $K_c = 50$. The offset has been greatly reduced:

$$\text{offset} = \frac{1}{1 + K_p K_c} = \frac{1}{1 + 1 \cdot 50} = 0.0196$$

3. Suppose that the process gain and time constant are perfectly known but not the dead time. In such a case, $G' = G = 1/(0.5s + 1)$. The dead time of the process can only be approximated. Let $t'_d = 0.8$.

Let us examine a common error in the design of a process control system which we must avoid. Were we to consider the value 0.8 as the "true" value of the process dead time, we could design a dead-time compensator as in case 2 above. Since there would be no crossover frequency, we could use an arbitrarily large K_c in order to reduce the offset. Let $K_c = 100$.

Since the "true" value of the process dead time *is not 0.8* but 1.0, the compensation is not perfect. There is a residual dead time equal to $1.0 - 0.8 = 0.2$, which has not been compensated by the dead-time compensator. Thus uncompensated dead time gives rise to additional phase lag and leads eventually to a crossover frequency. If the ultimate gain is smaller than 100, the system with $K_c = 100$ is unstable. Indeed, for the present example and an uncompensated dead time, the crossover frequency is $\omega_{CO} = 9$ rad/min and the ultimate gain is 4.6. Therefore, *if we are not certain of the value of dead time, we must be conservative in selecting the value of K_c, even with partial dead-time compensation.*

19.3 Control of Systems with Inverse Response

In Section 12.3 we analyzed the behavior of a special class of systems with inverse response. There we saw that the net result of two opposing effects may produce an initial response which is in the opposite direction to where it will eventually end up (see Figures 12.4b and 12.5b).

The most common case of a process with inverse response is that resulting from the conflict of two first-order systems with opposing effects (Figure 12.5). In this section we limit our attention to the regulation of such processes. Extensions to more complex systems such as those of Table 12.1 are easy and straightforward.

There are two very popular ways to control systems with inverse response; the first uses a PID feedback controller with Ziegler–Nichols tuning and the second uses an inverse response compensator.

Simple PID control

From all types of feedback controllers only PID can be used effectively, for the following simple reason:

> The derivative control mode by its nature will anticipate the "wrong" direction of the system's response and will provide the proper corrective action to limit (never eliminate) the inverse shoot.

Waller and Nygårdas [Ref. 6] have demonstrated numerically that the Ziegler–Nichols classical tuning of a PID controller yields very good control of systems with inverse response.

Inverse response compensator

In Section 19.2 we discussed how we can develop a Smith predictor (dead-time compensator) which cancels the effect of dead time. The same general concept of the predictor (compensator) can be used to cope with the inverse response of a process and was proposed by Iinoya and Altpeter [Ref. 5].

Consider the feedback system of Figure 19.5a. The controlled process exhibits inverse response when (see Example 12.4)

$$\frac{\tau_1}{\tau_2} > \frac{K_1}{K_2} > 1$$

The open-loop response of the system is

$$\bar{y}(s) = G_c(s) \frac{(K_1\tau_2 - K_2\tau_1)s + (K_1 - K_2)}{(\tau_1 s + 1)(\tau_2 s + 1)} \bar{y}_{SP}(s) \qquad (19.5)$$

and has a positive zero at the point (see also Example 12.4)

$$z = -\frac{K_1 - K_2}{K_1\tau_2 - K_2\tau_1} > 0$$

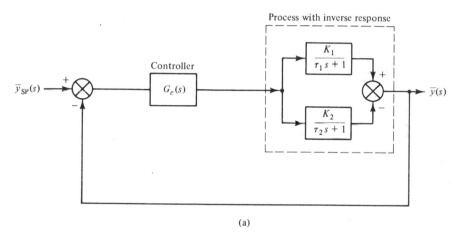

(a)

(b)

Figure 19.5 Feedback control of process with inverse response: (a) conventional; (b) with compensator.

To eliminate the inverse response it is enough to eliminate the positive zero of the above open-loop transfer function. This is possible if in the open-loop response $\overline{y}(s)$ we add the quantity $\overline{y}'(s)$ given by

$$\overline{y}'(s) = G_c(s)k\left(\frac{1}{\tau_2 s + 1} - \frac{1}{\tau_1 s + 1}\right)\overline{y}_{SP}(s) \tag{19.6}$$

Then, from eqs. (19.5) and (19.6) we can easily find that

$$\overline{y}^*(s) = \overline{y}(s) + \overline{y}'(s) = G_c(s)\frac{[(K_1\tau_2 - K_2\tau_1) + k(\tau_1 - \tau_2)]s + (K_1 - K_2)}{(\tau_1 s + 1)(\tau_2 s + 1)}\overline{y}_{SP}(s)$$

and for

$$k \geq \frac{K_2\tau_1 - K_1\tau_2}{\tau_1 - \tau_2} \tag{19.7}$$

we find that the zero of the resulting open-loop transfer function is nonpositive:

$$z = -\frac{K_1 - K_2}{(K_1\tau_2 - K_2\tau_1) + k(\tau_1 - \tau_2)} \leq 0$$

Adding the signal $\overline{y}'(s)$ to the main feedback signal $\overline{y}(s)$ means the creation of the local loop around the controller as shown in Figure 19.5b. The system in this local loop is the *modified Smith predictor* and *the actual compensator of the inverse response*. As can be seen from eq. (19.6), its transfer function is

$$G_{\text{compensator}} = k\left(\frac{1}{\tau_2 s + 1} - \frac{1}{\tau_1 s + 1}\right) \tag{19.8}$$

where k must satisfy condition (19.7).

Remarks

1. The inverse response compensator *predicts* the inverse behavior of the process and provides a corrective signal to eliminate it. The prediction is based on a model for the process. The ideal prediction comes if the transfer function of the process is completely known. In such case the compensator is given by

$$G_{\text{compensator}} = \frac{K_2}{\tau_2 s + 1} - \frac{K_1}{\tau_1 s + 1}$$

   Therefore, the compensator given by eq. (19.8) is only an approximation of the process's transfer function.
2. Modeling inaccuracies in terms of τ_1 and τ_2 will deteriorate the performance of an inverse response compensator (i.e., they will cause increased inverse shoots and sluggish responses).
3. For the controller, PI is the most common choice.

THINGS TO THINK ABOUT

1. What is the effect of dead time in the response of simple feedback control loops? Explain in physical terms.

2. Why is the controller design of processes with dead time a particularly sensitive and difficult problem? Demonstrate using a practical example.

3. Describe in physical terms the concept of dead-time compensation. Why is such a system also called a predictor?

4. Show that the dead-time and inverse response compensators are based on the same logic. What are their implementational difficulties?

5. Consider the feedback loop of Figure Q19.1 with load changes only. Construct a dead-time compensator assuming $G(s)$ and t_d perfectly known. Is it the same as the dead-time compensator constructed for set point changes?

6. What is the impact of model inaccuracies on the effectiveness of dead-time compensators?

7. Is the dead time of a process constant or does it vary with time? If it varies, give three relevant physical examples. What is the effect of changing dead time on the design of a dead-time compensator?

8. What is our goal when designing a controller for a system with inverse response? Describe what an inverse response compensator does.

9. Consider the system with inverse response described in Section 19.3. Identify the transfer function of the compensator. Notice that it is a function of the parameter k which must satisfy condition (19.7). Do you have any ideas on how k would affect the quality of the controlled response? (See also the numerical example in Ref. 5.)

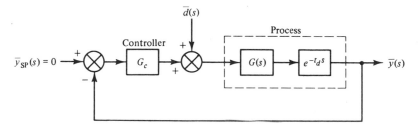

Figure Q19.1

Control Systems **20** with Multiple Loops

The feedback control configuration involves *one* measurement (output) and *one* manipulated variable in a single loop. There are, however, other simple control configurations which may use:

More than one measurement and one manipulation, or
One measurement and more than one manipulated variables

In such cases control systems with multiple loops may arise. Typical examples of such configurations, that we will study in the present chapter, are the following:

1. Cascade control
2. Various types of selective control
3. Split-range control

Before proceeding we should emphasize that these control systems involve loops that are not separate but share either the single manipulated variable or the only measurement. In this respect the multiple-loop control systems of this chapter are generically different from those we will study in Chapters 23 and 24.

20.1   Cascade Control

In a cascade control configuration we have one manipulated variable and more than one measurement. It is clear that *with a single manipulation we can control only one output*. Let us now examine the motivation behind cascade control and its typical characteristics using an example from the chemical processes.

Example 20.1: Cascade Control for a Jacketed CSTR

Consider the CSTR shown in Figure 1.7. The reaction is exothermic and the heat generated is removed by the coolant, which flows in the jacket around the tank. The control objective is to keep the temperature of the reacting mixture, T, constant at a desired value. Possible disturbances to the reactor include the feed temperature T_i and the coolant temperature T_c. The only manipulated variable is the coolant flow rate F_c.

Simple feedback control. If we use simple feedback, we will take the control configuration shown in Figure 20.1a (i.e., measure temperature T and manipulate coolant flow rate F_c). It is clear that T will respond much faster to changes in T_i than to changes in T_c. Therefore, the simple feedback control of Figure 20.1a will be very effective in compensating for changes in T_i and less effective in compensating for changes in T_c.

Cascade control. We can improve the response of the simple feedback control to changes in the coolant temperature by measuring T_c and taking control action *before its effect has been felt by the reacting mixture*. Thus, if T_c goes up, increase the flow rate of the coolant to remove the same amount of heat. Decrease the coolant flow rate when T_c decreases.

We notice, therefore, that we can have two control loops *using two different measurements*, T and T_c, but *sharing a common manipulated variable*, F_c. How these loops are related is shown in Figure 20.1b. There we notice that:

(a) *The loop that measures T (controlled variable) is the dominant or primary, or master control loop and uses a set point supplied by the operator, while*
(b) *The loop that measures T_c uses the output of the primary controller as its set point and is called the secondary or slave loop.*

The control configuration with these two loops is known as *cascade control* and is very common in chemical processes.

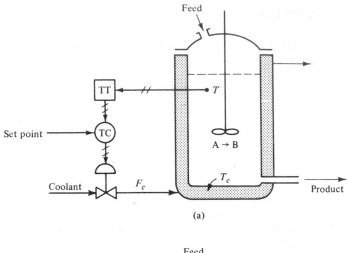

(a)

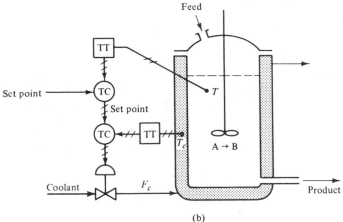

(b)

TT, temperature sensor and transmitter
TC, temperature controller

Figure 20.1 Temperature control of jacketed CSTR: (a) conventional feedback; (b) cascade.

Let us generalize the discussion above. Consider a process consisting of two parts, as shown in Figure 20.2a: process I and process II. Process I (primary) has as its output the variable we want to control. Process II (secondary) has an output that we are not interested in controlling but which affects the output we want to control. For the CSTR system of Example 20.1, process I is the reaction in the tank and the controlled output is the temperature T. Process II is the jacket and its output T_c affects process I (reactor) and consequently T.

Figure 20.2b shows the typical simple feedback control system, and

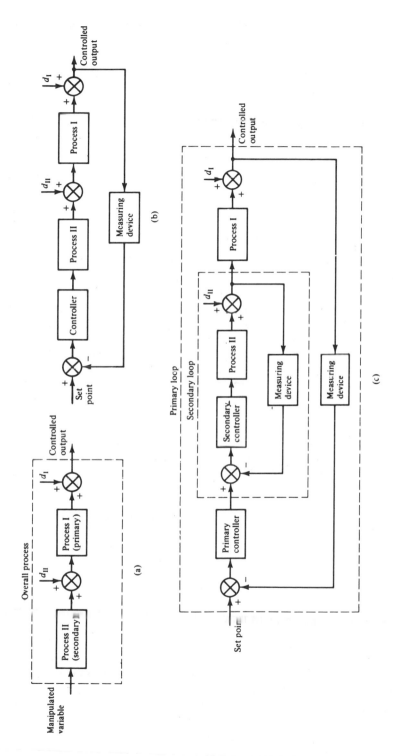

Figure 20.2 Schematic representation of: (a) open-loop process; (b) conventional feedback; (c) cascade control.

397

Figure 20.2c indicates the general form of cascade control. The last figure demonstrates very clearly the major benefit to be gained by cascade control:

> Disturbances arising within the secondary loop are corrected by the secondary controller before they can affect the value of the primary controlled output.

This important benefit has led to the extensive use of cascade control in chemical processes.

Example 20.2: Cascade Control for Various Processes

Let us describe the use of cascade control in various typical processing systems.

1. *Heat exchangers*: The typical configuration is shown in Figure 20.3a. The control objective is to keep the exit temperature of stream 2 at a desired value. The secondary loop is used to compensate for changes in the flow rate of stream 1.
2. *Distillation columns*: Cascade control is usually employed to regulate the temperature (and consequently the concentration) at the top or bottom of a distillation column. Figure 20.3b and c show two such typical cascade control systems. In both cases the secondary loop is used to compensate for flow rate changes.
3. *Furnaces*: Cascade control can be used to regulate the temperature of a process stream (e.g., feed to a reactor) exiting from a furnace. Figure 20.3d shows the resulting cascade configuration. Again, the secondary loop is used to compensate for flow rate changes (fuel flow rate).

The reader should notice that in all the cascade configurations of Example 20.2, the secondary loop is used to compensate for flow rate changes. This observation is quite common in chemical processes and we could state:

> In chemical processes, flow rate control loops are almost always cascaded with other control loops.

Let us now turn our attention to the closed-loop behavior of cascade control systems. Consider the block diagram of a general cascade system shown in Figure 20.4a. To simplify the presentation we have assumed that the transfer functions of the measuring devices are both equal to 1.

The *closed-loop response of the primary loop is influenced by the*

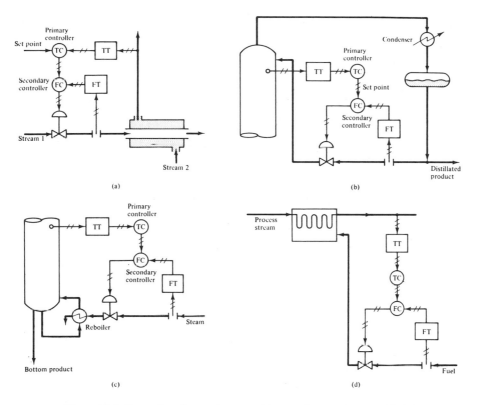

Figure 20.3 Examples of cascade control for: (a) heat exchanger; (b), (c) distillation column; (d) process furnace.

dynamics of the secondary loop, whose open-loop transfer function is equal to

$$G_{\text{secondary}} = G_{c,\text{II}} G_{p,\text{II}} \tag{20.1}$$

The stability of the secondary loop is determined by the roots of its characteristic equation

$$1 + G_{c,\text{II}} G_{p,\text{II}} = 0 \tag{20.2}$$

Figure 20.4b shows a simplified form of the general block diagram (Figure 20.4a), where the secondary loop has been considered as a dynamic element.

For the primary loop the overall open-loop transfer function is

$$G_{\text{primary}} = G_{c,\text{I}} \left(\frac{G_{c,\text{II}} G_{p,\text{II}}}{1 + G_{c,\text{II}} G_{p,\text{II}}} \right) G_{p,\text{I}} \tag{20.3}$$

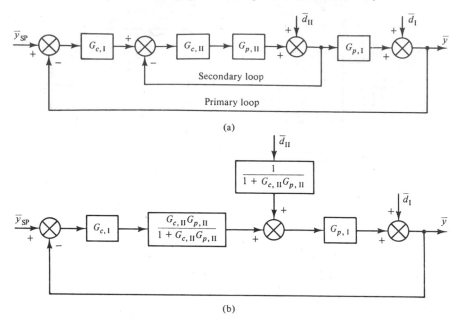

Figure 20.4 (a) Block diagram of a cascade control system; (b) simplified, but equivalent form.

and consequently the characteristic equation whose roots determine the stability of the primary loop is the following:

$$1 + G_{c,I}\left(\frac{G_{c,II}G_{p,II}}{1 + G_{c,II}G_{p,II}}\right)G_{p,I} = 0 \qquad (20.4)$$

Remarks

1. The two controllers of a cascade control system are standard feedback controllers (i.e., P, PI, PID). Generally, a proportional controller is used for the secondary loop, although a PI controller with small integral action is not unusual. *Any offset caused by P control in the secondary loop is not important since we are not interested in controlling the output of the secondary process.*

2. The dynamics of the secondary loop are much faster than those of the primary loop. Consequently, the *phase lag of the closed secondary loop will be less than that of the primary loop.* This feature leads to the following important result, which constitutes the rationale behind the use of cascade control: *The crossover frequency for the secondary loop is higher than that for the primary loop.* This allows us to use higher gains in the secondary controller in order to regulate more effectively the effect of a disturbance

occurring in the secondary loop without endangering the stability of the system.

3. The tuning of the two controllers of a cascade control system proceeds in two steps:

(a) First, we determine the settings for the secondary controller using one of the methods that we studied in Chapters 16 and 18: Cohen–Coon or Ziegler–Nichols or others employing time-integral criteria or phase and gain margin considerations. The open-loop transfer function that we can use for tuning the secondary controller is given by eq. (20.1).

(b) Second, from the Bode plots of the overall system we determine the crossover frequency using the settings for the secondary loop we found above. Then, using the frequency response techniques described in Chapter 18, we choose the settings for the primary controller. The open-loop transfer function of the primary loop, needed for the construction of the Bode plots, is given by eq. (20.3).

Example 20.3: Dynamic Characteristics of a Cascade Control System

Consider a process with the following transfer functions for its primary and secondary elements:

$$G_{p,I} = \frac{1}{(0.5s + 1)(s + 1)} \quad \text{and} \quad G_{p,II} = \frac{1}{0.1s + 1}$$

The secondary process is faster than the primary, as can be seen from the corresponding time constants.

Simple feedback control. Were we to use simple feedback control, the open-loop transfer function with PI control ($\tau_I = 1$) would be

$$G_{c,I}G_{p,II}G_{p,I} = K_{c,I}\left(1 + \frac{1}{s}\right)\frac{1}{(0.1s + 1)}\frac{1}{(0.5s + 1)(s + 1)}$$

The crossover frequency can be found from the equation that sets the total phase lag equal to $-180°$:

$$\tan^{-1}\left(\frac{-1}{\omega_{co}}\right) + \tan^{-1} -0.1\omega_{co} + \tan^{-1} -0.5\omega_{co} + \tan^{-1} -\omega_{co} = -180°$$

and it is equal to

$$\omega_{co} = 4.45 \text{ rad/min}$$

Also, the overall amplitude ratio is given by

$$AR = K_{c,I}\sqrt{1 + \frac{1}{\omega^2}} \frac{1}{\sqrt{(0.1\omega)^2 + 1}} \frac{1}{\sqrt{(0.5\omega)^2 + 1}\sqrt{\omega^2 + 1}}$$

The ultimate value of the gain $K_{c,\mathrm{I}}$ can be found from the condition

$$AR = 1 \qquad \text{at} \qquad \omega = \omega_{CO}$$

Thus

$$1 = K_{c,\mathrm{I}} \sqrt{1 + \frac{1}{(4.45)^2}} \; \frac{1}{\sqrt{(0.1 \cdot 4.45)^2 + 1}} \; \frac{1}{\sqrt{(0.5 \cdot 4.45)^2 + 1} \; \sqrt{(1 \cdot 4.45)^2 + 1}}$$

and we find that

$$K_{c,\mathrm{I}} = 11.88$$

Therefore, when the disturbance d_{II} (of the secondary process) changes, the simple feedback controller can use a gain up to 11.88 before the system becomes unstable. Also, given the fact that the overall process is of third order, we expect that the closed-loop response of $y(t)$ to changes in d_{II} will be rather sluggish.

Cascade control. Consider a cascade control system similar to that of Figure 20.4a. The open-loop transfer function for the secondary loop is given by eq. (20.1) and assuming a simple proportional controller, we find that

$$G_{c,\mathrm{II}} G_{p,\mathrm{II}} = K_{c,\mathrm{II}} \frac{1}{0.1s + 1}$$

There is no crossover frequency for the secondary control loop. Therefore, we can use large values for the gain $K_{c,\mathrm{II}}$, which produce a very fast closed-loop response, to compensate for any changes in the disturbance d_{II} arising within the secondary process.

Once we have selected the value of $K_{c,\mathrm{II}}$ for the secondary loop, we can find the crossover frequency for the overall open-loop transfer function given by eq. (20.3). Then we can select the value of $K_{c,\mathrm{I}}$ for the primary controller, using the Ziegler–Nichols methodology. Quite often we will not arbitrarily select a very large $K_{c,\mathrm{II}}$, but rather select it in coordination with the resulting values of $K_{c,\mathrm{I}}$.

20.2 Selective Control Systems

These are control systems that involve one manipulated variable and several controlled outputs. Since with one manipulated variable we can control only one output, the selective control systems transfer control action from one controlled output to another according to need. There are several types of selective control systems, but in this section we discuss only two:

1. Override control for the protection of process equipment
2. Auctioneering control

Override control

During the normal operation of a plant or during its startup or shutdown it is possible that dangerous situations may arise which may lead to destruction of equipment and operating personnel. In such cases it is necessary to change from the normal control action and attempt to prevent a process variable from exceeding an allowable upper or lower limit. This can be achieved through the use of special types of switches. The *high selector switch* (HSS) is used whenever a variable should not exceed an upper limit, and the *low selector switch* (LSS) is employed to prevent a process variable from exceeding a lower limit.

Example 20.4: Examples of Override Control

1. *Protection of a boiler system*: Usually, the steam pressure in a boiler is controlled through the use of a pressure control loop on the discharge line (loop 1 in Figure 20.5). At the same time the water level in the boiler should not fall below a lower limit which is necessary to keep the heating coil immersed in water and thus prevent its burning out. Figure 20.5 shows the override control system using a low switch selector (LSS). According to this system, whenever the liquid level falls below the allowable limit, the LSS switches the control action from pressure control to level control (loop 2 in Figure 20.5) and closes the valve on the discharge line.

2. *Protection of a compressor system*: The discharge of a compressor is controlled with a flow control system (loop 1 in Figure 20.6). To prevent the discharge pressure from exceeding an upper limit, an override control with a high switch selector (HSS) is introduced. It transfers control action from the flow control to the pressure control loop (loop 2 in Figure 20.6) whenever the discharge pressure

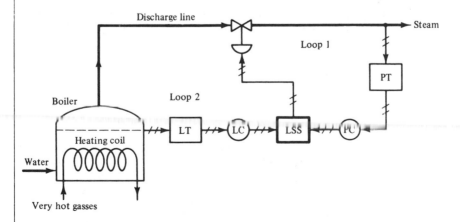

Figure 20.5 Override control to protect a boiler system.

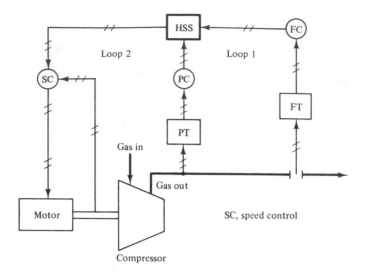

Figure 20.6 Override control to protect a compressor.

exceeds the upper limit. Notice that flow control or pressure control is actually cascaded to the speed control of the compressor's motor.

3. *Protection of a steam distribution system*: In any chemical process there is a network distributing steam, at various pressure levels, to the processing units. High-pressure steam is "let down" to lower

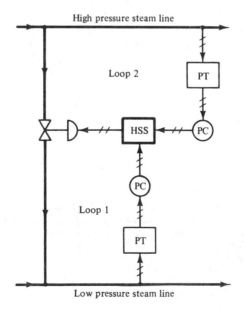

Figure 20.7 Override control for steam distribution system.

pressure levels at the let-down stations. The amount of steam let down at such stations is controlled by the demand on the low-pressure steam line (loop 1 in Figure 20.7). To protect the high-pressure line from excessive pressures, we can install an override control system with a HSS, which transfers control action from loop 1 to loop 2 when the pressure in the high-pressure line exceeds an upper limit.

Auctioneering control systems

Such control configurations select among several similar measurements the one with the highest value and feed it to the controller. Thus it is a selective controller, which possesses several measured outputs and one manipulated input.

Example 20.5: Examples of Auctioneering Control

1. *Catalytic tubular reactors with highly exothermic reactions*: Several highly exothermic reactions take place in tubular reactors filled with a catalyst bed. Typical examples are hydrocarbon oxidation reactions such as the oxidation of o-xylene or naphthalene to produce phthalic anhydride. Figure 20.8 shows the temperature profile along the length of the tubular reactor. The highest temperature is called the *hot spot*. The location of the hot spot moves along the length of the reactor depending on the feed conditions (temperature, concentration, flow rate) and the catalyst activity (Figure 20.8). The value of the hot-spot temperature also depends on the factors listed above and the temperature and flow rate of the cool-

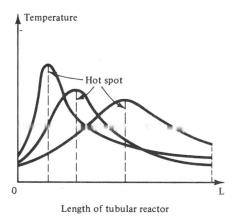

Length of tubular reactor

Figure 20.8 Temperature profiles in a tubular catalytic reactor.

ant. The control of such systems is a real challenge for the chemical engineer.

The primary control objective is to keep the hot-spot temperature below an upper limit. Therefore, we need a control system that can identify the location of the hot spot and provide the proper control action. This can be achieved through:

Placement of several thermocouples along the length of the reactor

Use of an auctioneering system to select the highest temperature which will be used to control the flow rate of the coolant (Figure 20.9)

2. *Regeneration of catalytic reactors*: The catalyst in catalytic reactors undergoes deactivation as the reaction proceeds, due to carbonaceous deposits on it. It can be regenerated by burning off these deposits with air or oxygen. To avoid destruction of the catalyst, due to excessive temperatures during the combustion of the deposits, we can use an auctioneering system which:

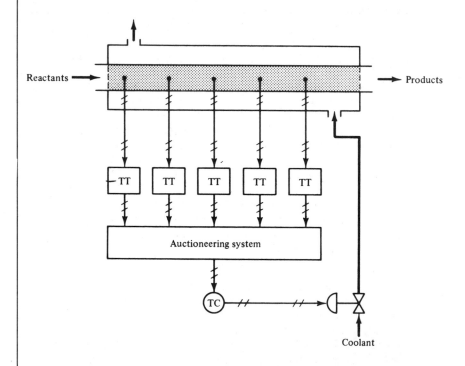

Figure 20.9 Auctioneering control system for a tubular catalytic reactor.

Takes the temperature measurements from various thermocouples along the length of the reactor

Selects the highest temperature that corresponds to the combustion front as it moves through the bed

Controls appropriately the amount of air

20.3 Split-Range Control

Unlike the cascade and selective control schemes examined in Sections 20.1 and 20.2, the split-range control configuration has *one measurement only* (controlled output) and more than one manipulated variable.

Since there is only one controlled output, we have only one control signal, which is thus split into several parts, each affecting one of the available manipulations. In other words, *we can control a single process output by coordinating the actions of several manipulated variables, all of which have the same effect on the controlled output*. Such systems are not very common in chemical processes but provide added safety and operational optimality whenever necessary, as the following examples demonstrate.

Example 20.6: Split-Range Control of a Chemical Reactor

Consider the reactor shown in Figure 20.10a, where a gas phase reaction takes place. Two control valves manipulate the flows of the feed and the reaction product. It is clear that in order to control the pressure in the reactor, the two valves cannot act independently but should be coordinated. Figure 20.10b indicates the coordination of the two valves' actions as a function of the controller's output signal (see also Table 20.1).

Let the controller's output signal corresponding to the desired operation of the reactor be 6 psig. From Figure 20.10b we see that valve V_2 is partly open while valve V_1 is completely open. When for various reasons the pressure in the reactor increases, the controller's output signal also increases. Then *it is split* into two parts, affects the two valves simultaneously, and the following actions take place:

1. As the controller output increases from 6 psig to 9 psig, valve V_2 opens continuously while V_1 remains completely open. Both actions lead to a reduction in the pressure.
2. For large increases in the reactor's pressure, the control output may exceed 9 psig. In such a case, as we can see from Figure 20.10b, valve V_2 is completely open while V_1 starts closing. Both actions again lead to a reduction in pressure until the reactor has returned to the desired operation.

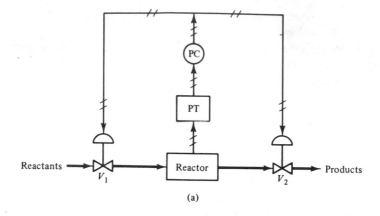

(a)

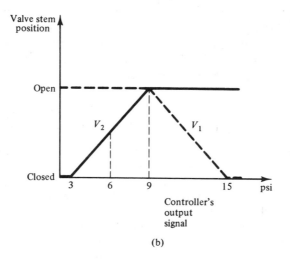

(b)

Figure 20.10 (a) Reactor system with split-range control; (b) action of two valves.

TABLE 20.1
OUTPUT SIGNAL AND VALVE COORDINATION

Controller's output signal	Valve V₁ stem position	Valve V₂ stem position
3 psig	Open	Closed
9 psig	Open	Open
15 psig	Closed	Open

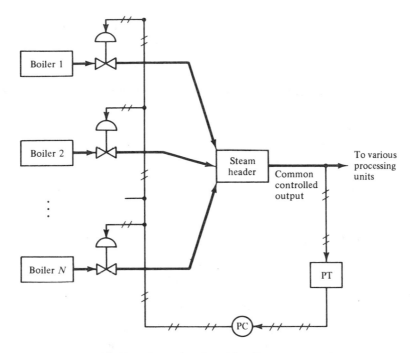

Figure 20.11 Steam header with split-range control.

Example 20.7: Split-Range Control of the Pressure in a Steam Header

Let us consider another example of split-range control, which is encountered very often in chemical plants. Several parallel boilers discharge steam in a common steam header and from there to the process needs (Figure 20.11). The control objective is to maintain constant pressure in the steam header when the steam demand at the various processing units changes. There are several manipulated variables (steam flow from every boiler) which can be used simultaneously. Figure 20.11 also shows the structure of the resulting control system. It should be noted that instead of controlling the steam flow from each boiler, we could control the firing rate and thus the steam production rate at each boiler.

Similar structures can be developed for the pressure control of a common discharge or suction header for N parallel compressors.

THINGS TO THINK ABOUT

1. Consider a process with one manipulated input and two measured outputs. Can you keep both outputs at the desired values using only the single manipulated variable? If not, explain why.

2. Starting from the premise that the answer to item 1 is negative, explain how it is possible to have (a) a cascade control system or (b) a selective control system, both of which have a single manipulated variable and two measured outputs.

3. Discuss the rationale of a cascade control system and demonstrate why it provides better response than simple feedback.

4. In Section 20.1 we assumed that the secondary process (process II, Figure 20.2a) in a cascade control system is faster than the primary process (process I, Figure 20.2a). Is this necessary to justify the use of a cascade control configuration? In other words, would you still recommend cascade control for a process (like that of Figure 20.2a) with a secondary process much slower than the primary?

5. What are the main advantages and disadvantages of cascade control? For what kind of processes can you employ cascade control?

6. In chemical processes, flow rate control loops are almost always cascaded with other control loops. Why does this happen? [*Note*: Take into account the following two facts: (a) The flow rate itself is subject to changes and is regulated by the flow control loop, and (b) flow rates are the most common manipulated variables in chemical processes.]

7. What types of controllers would you use for the two controllers of a cascade system? How would you tune them? Discuss a methodology to select the adjusted parameters of the two controllers.

8. Are the stability characteristics of the closed-loop response of a cascade system better than those of a simple feedback? Elaborate on your answer.

9. What is meant by selective control systems? How many different types of selective control systems are available? Discuss their characteristics.

10. Discuss the rationale behind an override control system. Why is it very useful, and what situations is it called upon to control?

11. Describe two or three situations (different from those discussed in Example 20.4) where you should use override control systems.

12. What is an auctioneering control system, and where would you use it? Describe a situation (different from those of Example 20.5) where you could use auctioneering control.

13. Consider a process with one controlled output and two active manipulated variables. Under what conditions could you use both manipulated variables to control the single output?

14. What is split-range control? In Example 20.6 we have a situation with split-range control. To control the pressure in the reactor we could use valve V_1 or valve V_2 with simple control configurations or both valves in a split-range control configuration. Which of the three is better? Why?

Feedforward and Ratio **21** *Control*

Feedback control loops can never achieve perfect control of a chemical process, that is, keep the output of the process *continuously* at the desired set point value in the presence of load or set point changes. The reason is simple: A feedback controller reacts only after it has detected a deviation in the value of the output from the desired set point.

Unlike the feedback systems, a feedforward control configuration measures the disturbance (load) directly and takes control action to eliminate its impact on the process output. Therefore, feedforward controllers have the theoretical potential for perfect control.

In this section we study the characteristics of feedforward control systems and describe the techniques that are used for their design. In the final section we examine a special case of feedforward control, ratio control.

21.1 Logic of Feedforward Control

Consider the stirred tank heater shown in Figure 1.1. The control objective is to keep the temperature of the liquid in the tank at a desired value (set point) despite any changes in the temperature of the inlet stream. Figure 1.2 shows the conventional feedback loop, which measures the temperature in the tank and after comparing it with the desired value, increases or decreases the steam pressure, thus providing more or less heat into the liquid. A feedforward control system uses a differ-

411

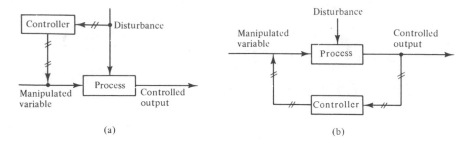

Figure 21.1 Structure of (a) feedforward, and (b) feedback control schemes.

ent approach. It measures the temperature of the inlet stream (disturbance) and adjusts appropriately the steam pressure (manipulated variable). Thus it increases the steam pressure if the inlet temperature decreases and decreases the steam pressure when the inlet temperature goes up. Figure 1.4 shows the feedforward control configuration.

In Figure 21.1a we can see the general form of a feedforward control system. It measures the disturbance directly and then it anticipates the effect that it will have on the process output. Subsequently, it changes the manipulated variable by such an amount as to eliminate completely the impact of the disturbance on the process output (controlled variable). Control action starts immediately after a change in the disturbance(s) has been detected. In Figure 21.1b we have repeated the schematic of a typical feedback loop so that you can contrast the two control systems directly. It is clear that feedback acts *after the fact* in a *compensatory manner*, whereas feedforward acts *beforehand* in an *anticipatory manner*.

Let us now look at some common feedforward control systems used in chemical processes.

Example 21.1: Feedforward Control of Various Processing Units

1. *Feedforward control of a heat exchanger* (shown in Figure 21.2a): The objective is to keep the exit temperature of the liquid constant by manipulating the steam pressure. There are two principal disturbances (loads) that are measured for feedforward control: liquid flow rate and liquid inlet temperature.
2. *Feedforward control of a drum boiler* (shown in Figure 21.2b): Here the objective is to keep the liquid level in the drum constant. The two disturbances are the steam flow from the boiler, which is dictated by varying demand elsewhere in the plant, and the flow of the feedwater. The last is also the principal manipulated variable.
3. *Feedforward control of a distillation column* (shown in Fig-

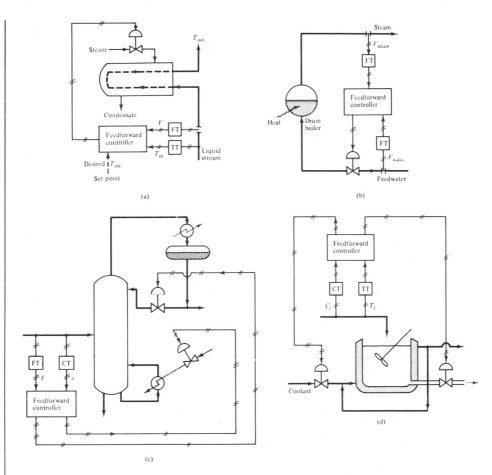

Figure 21.2 Examples of feedforward control: (a) heat exchanger; (b) drum boiler; (c) distillation column; (d) CSTR.

ure 21.2c): The two disturbances here are the feed flow rate F and the composition c. The available manipulated variables are the steam pressure in the reboiler and the reflux ratio. The composition of overhead or bottom product is the control objective. Feedforward control is particularly useful for a distillation column, because its response time can be measured in hours leading to large amounts of off-specification products.

4. *Feedforward control of a CSTR* (shown in Figure 21.2d): Inlet concentration and temperature are the two disturbances, and the product withdrawal flow rate and the coolant flow rate are the two manipulations. There are two objectives: to maintain constant temperature and composition within the CSTR.

Remarks

1. As the examples above have indicated, feedforward control systems can be developed for more than one disturbance. The controller acts according to which disturbance changed value. Therefore, the schematic of Figure 21.1a with several disturbances represents the general case of feedforward control with a single controlled variable.
2. The feedforward control of a CSTR, in Example 21.1, indicates that the extension to systems with multiple controlled variables should be rather straightforward.
3. With the exception of the controller, all the other hardware elements in a feedforward loop are the same as for a feedback loop (sensors, transducers, transmitters, final control elements).

21.2　Problem of Designing Feedforward Controllers

The question that arises is: How do we design feedforward controllers? The reader may have suspected already that conventional P, PI, or PID controllers will not be appropriate. Let us start with an example, the design of feedforward controllers for a stirred tank heater.

Example 21.2: Designing Feedforward Controllers for a Stirred Tank Heater

In Example 4.4 we developed the dynamic mass and energy balances for the stirred tank heater of Figure 1.1. They are given by eqs. (4.4a) and (4.5b).

$$A \frac{dh}{dt} = F_i - F \qquad (4.4a)$$

$$Ah \frac{dT}{dt} = F_i(T_i - T) + \frac{Q}{\rho c_p} \qquad (4.5b)$$

Assume that F_i does not change and that $F_i = F$. Then $dh/dt = 0$ and we have only the heat balance, eq. (4.5b). The inlet temperature T_i is the disturbance and the amount of heat Q supplied by steam is the manipulated variable. The control objective is to keep the liquid temperature, T, at the desired set point value, T_{SP}.

1. *Steady-state feedforward controller:* The simplest form of feedforward controller can be developed if we consider the steady-state heat balance,

$$0 = F_i(T_i - T) + \frac{Q}{\rho c_p}$$

or

$$T = T_i + \frac{Q}{F_i \rho c_p} \tag{21.1}$$

From eq. (21.1) we find that in order to keep $T = T_{SP}$, the manipulated variable Q should change according to the equation

$$Q = F_i \rho c_p (T_{SP} - T_i) \tag{21.2}$$

Equation (21.2) is the *design equation for the steady-state feedforward controller.* It shows how Q should change in the presence of disturbance or set point changes. Figure 21.3a depicts the resulting control system.

The steady-state feedforward controller will always achieve the desired steady-state performance of the heater (i.e., $T = T_{SP}$ at steady state). This will not be true, in general, during transient response.

2. *Dynamic feedforward controller:* To improve the quality of control during the transient response we will design a feedforward controller using the dynamic heat balance and not its equivalent steady state, as above. Equation (4.5b) can be written as follows:

$$\frac{V}{F_i} \frac{dT}{dt} + T = T_i + \frac{Q}{F_i \rho c_p} \tag{21.3}$$

where $V = Ah$ = liquid volume in the tank. Put eq. (21.3) into a form with deviation variables and take

$$\frac{V}{F_i} \frac{dT'}{dt} + T' = T_i' + \frac{Q'}{F_i \rho c_p} \tag{21.3a}$$

Take the Laplace transforms of eq. (21.3a):

$$\overline{T}'(s) = \frac{\overline{T_i'}(s)}{\tau s + 1} + \frac{1}{F_i \rho c_p} \frac{1}{\tau s + 1} \overline{Q}'(s) \tag{21.4}$$

where $\tau = V/F_i$ = retention time of liquid in the tank. The feedforward controller should make sure that $\overline{T}'(s) = \overline{T_{SP}'}(s)$ = set point, despite any changes in the disturbance T_i' or set point T_{SP}'. Therefore, from eq. (21.4) we find that Q should be given by

$$\overline{Q}'(s) = F_i \rho c_p [(\tau s + 1)\overline{T_{SP}'}(s) - \overline{T_i'}(s)] \tag{21.5}$$

Equation (21.5) is the *design equation for the dynamic feedforward controller* and Figure 21.3b shows the resulting control mechanism. As can be seen from Figure 21.3a and b, the only difference between the steady-state and dynamic feedforward controllers for the tank heater is the transfer function $(\tau s + 1)$ multiplying the set point.

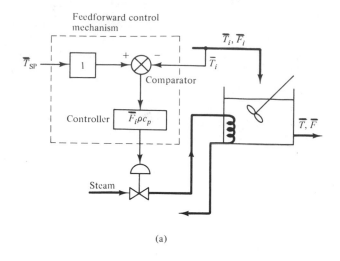

(a)

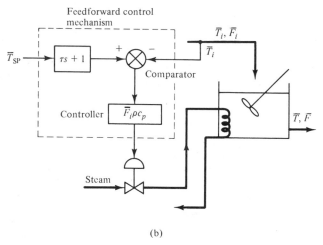

(b)

Figure 21.3 Block diagram for feedforward temperature control of a stirred tank heater: (a) steady state; (b) dynamic.

Therefore, we expect that for load (disturbance) changes the two controllers will be equivalent. On the contrary, dynamic feedforward control will be better for set point changes. Figure 21.4a and b demonstrate this point.

Example 21.2 has pointed out a very essential characteristic in feedforward control:

The design of a feedforward controller arises directly from the model of a process.

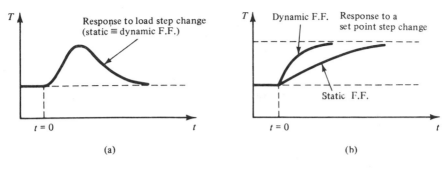

Figure 21.4 Comparison of static and dynamic feedforward control for Example 21.2.

Thus the steady-state design came out from the steady-state heat balance, and the dynamic controller from the dynamic heat balance. It is obvious that:

The better a model represents the behavior of a process, the better the resulting feedforward controller will be.

Let us now generalize the design procedure outlined in Example 21.2.

Consider the block diagram of an uncontrolled process (Figure 21.5a). The process output is given by

$$\overline{y}(s) = G_p(s)\overline{m}(s) + G_d(s)\overline{d}(s) \tag{21.6}$$

Let $\overline{y}_{SP}(s)$ be the desired set point for the process output. Then eq. (21.6) for $\overline{y}(s) = \overline{y}_{SP}(s)$ yields

$$\overline{y}_{SP}(s) = G_p(s)\overline{m}(s) + G_d(s)\overline{d}(s) \tag{21.7}$$

We can solve eq. (21.7) with respect to $\overline{m}(s)$ and find the value that the manipulated variable should have in order to keep $\overline{y}(s) = \overline{y}_{SP}(s)$ in the presence of disturbance or set point changes. Thus we take

$$\overline{m}(s) = \left[\frac{1}{G_d(s)}\overline{y}_{SP}(s) - \overline{d}(s)\right]\frac{G_d(s)}{G_p(s)} \tag{21.8}$$

Equation (21.8) determines the form that the feedforward control system should have, which is shown in Figure 21.5b. It also determines the two transfer functions, G_c and G_{SP}, which complete the design of the control mechanism:

$$G_c(s) = \frac{G_d(s)}{G_p(s)} \equiv \text{transfer function of the main} \tag{21.9}$$
$$\qquad\qquad\qquad\quad \text{feedforward controller}$$

$$G_{SP}(s) = \frac{1}{G_d(s)} \equiv \text{transfer function of the set point element} \tag{21.10}$$

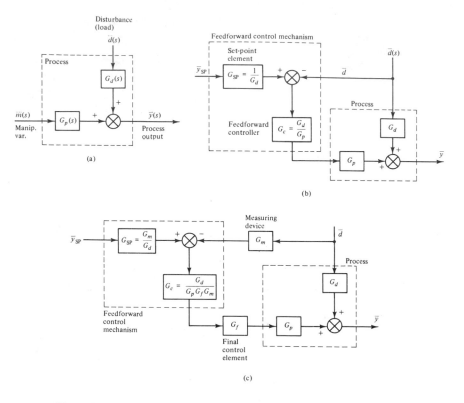

Figure 21.5 Block diagrams for: (a) process; (b) feedforward loop; (c) feedforward loop with measuring device and final control element.

Remarks

1. From Figure 21.5b we notice that the feedforward loop retains all the external characteristics of a feedback loop. Thus it has a primary measurement which is compared to a set point signal and the result of the comparison is the actuating signal for the main controller. In substance, though, the two control systems differ significantly, as pointed out in Section 21.1.

2. From the design equations (21.9) and (21.10) it is clear that a *feedforward controller cannot be a conventional feedback controller* (P, PI, PID). Instead, *it should be viewed as a special-purpose computing machine.* This is the reason it is sometimes referred to as a *feedforward computer.*

3. The design equations (21.9) and (21.10) demonstrate again that *feedforward control depends heavily on a good knowledge of the process model* (G_p, G_d). Perfect control necessitates perfect

knowledge of G_p and G_d, which is not practically possible. This is the main drawback of feedforward control.

4. In the control system of Figure 21.5b we left out the sensor that measures the disturbance and the final control element. The inclusion of these two elements alters the design of the transfer functions $G_c(s)$ and $G_{SP}(s)$. Consider the more general feedforward control system of Figure 21.5c, including the measuring sensor and the final control element. We can easily show that

$$\bar{y} = G_p G_f G_c G_{SP} \bar{y}_{SP} + [G_d - G_p G_f G_c G_m]\bar{d} \qquad (21.11)$$

The design transfer functions G_c and G_{SP} can now be identified by the following two requirements:

Disturbance rejection. The controller should be capable of completely eliminating the impact of a disturbance change on the process output. This implies that the coefficient of $\bar{d}$ in eq. (21.11) should be zero:

$$G_d - G_p G_f G_c G_m = 0$$

or

$$G_c = \frac{G_d}{G_p G_f G_m} \qquad (21.12)$$

Set point tracking. The control mechanism should be capable of making the process output track exactly any changes in the set point (i.e., keep $\bar{y} = \bar{y}_{SP}$). This implies that the coefficient of $\bar{y}_{SP}$ in eq. (21.11) should be equal to 1:

$$G_p G_f G_c G_{SP} = 1$$

or

$$G_p G_f \left(\frac{G_d}{G_p G_f G_m} \right) G_{SP} = 1$$

and finally,

$$G_{SP} = \frac{G_m}{G_d} \qquad (21.13)$$

Equations (21.12) and (21.13) are more general than (21.9) and (21.10), with the latter resulting from the former for $G_m = G_f = 1$.

21.3   Practical Aspects on the Design of Feedforward Controllers

The design equations (21.9) and (21.10), or their more general counter-parts (21.12) and (21.13), indicate that the feedforward controller will be a special-purpose computational machine. Its practical implementation is rather easy if we use a digital computer as the controller (see Part VII), but for analog controllers it is rather difficult and expensive to build these special-purpose machines. In this section we examine some simplifications that lead to practical implementations of the feedforward control concept.

To simplify the presentation, let us first assume that $G_m = G_f = 1$. Then eqs. (21.9) and (21.10) will be the basis of the controller design. Each of the two process transfer functions, $G_p(s)$ and $G_d(s)$, has two elements: (1) *the static element*, which corresponds to the static gain, and (2) *the purely dynamic element*, which is a function of s. Thus

$$G_p(s) = K_p G_p'(s) \qquad \text{and} \qquad G_d(s) = K_d G_d'(s)$$

For instance, in Example 21.2, for the stirred tank heater we can easily identify the static and dynamic parts of the process transfer functions [see eq. (21.4)]:

$$G_p(s) = \frac{1}{F_i \rho c_p} \frac{1}{\tau s + 1} \quad \text{indicates that} \quad K_p = \frac{1}{F_i \rho c_p} \quad \text{and} \quad G_p'(s) = \frac{1}{\tau s + 1}$$

$$G_d(s) = \frac{1}{\tau s + 1} \quad \text{indicates that} \quad K_d = 1 \quad \text{and} \quad G_d'(s) = \frac{1}{\tau s + 1}$$

Design of steady-state feedforward controllers

The simplest feedforward controller and the easiest to implement is the steady-state one. As demonstrated in Example 21.2, we use simple steady-state balances for design. How does this modify the design equations (21.9) and (21.10)?

At steady state, we retain only the static elements of the process transfer functions. Thus

$$G_p = K_p \qquad \text{and} \qquad G_d = K_d$$

Then the design transfer functions, G_c and G_{SP}, are given by

$$G_c = \frac{K_d}{K_p} \qquad\qquad\qquad (21.14a)$$

and

$$G_{SP} = \frac{1}{K_d}$$ (21.14b)

that is, they are simple constants. Therefore, the elements G_c and G_{SP} can be constructed easily in the same way as a proportional controller, which has only proportional gain. This is the reason the design elements given by eqs. (21.14a) and (21.14b) are called *gain-only* elements.

Design of simple dynamic feedforward controllers

Instead of using the exact transfer functions $G_p(s)$ and $G_d(s)$, it is possible to use approximations to them and still obtain very good results. Although they are approximations, they are expected to give improved results over the steady-state feedforward controller.

Consider that $G_p(s)$ and $G_d(s)$ are approximated by first-order lags. Then

$$G_c(s) = \frac{G_d(s)}{G_p(s)} = \frac{1/(\alpha s + 1)}{1/(\beta s + 1)} = \frac{\beta s + 1}{\alpha s + 1}$$ (21.15a)

and

$$G_{SP}(s) = \frac{1}{G_d(s)} = \alpha s + 1$$ (21.15b)

The controller given by eq. (21.15a) is called a *lag–lead element* because $\beta s + 1$ introduces phase lead and the $1/(\alpha s + 1)$ adds phase lag. α and β are adjustable parameters for the controller. For the set point element $G_{SP}(s)$, eq. (21.15b) indicates that we should use a *lead element*.

The lag–lead element is the most commonly used in dynamic feedforward control. It is quite versatile because the two adjustable parameters α and β allow its use as a *lead element* (when β is much larger than α) or as a *lag element* (when α is much larger than β). Finally, lag–lead elements can be bought easily and they are not expensive as are special-purpose analog computational devices.

Physical realizability of feedforward controllers

Let us consider a process with the following transfer functions

$$G_p = \frac{10 e^{-0.5s}}{2s + 1} \quad \text{and} \quad G_d = \frac{2 e^{-0.1s}}{2s + 1}$$

Then, the transfer function of the main feedforward controller is given by [see eq. (21.9)]

$$G_c \equiv \frac{G_d}{G_p} = \frac{1}{5} e^{0.4s}$$

The exponential term implies that *we need future values of the disturbance in order to compute the current value of the manipulated variable*. But, such future values of the disturbance cannot be available. In such case the feedforward controller, described by the last equation above, is characterized as *physically unrealizable* and cannot be applied in real situations.

Example 21.3: Designing Feedforward Controllers for a CSTR

Consider the CSTR system described in Example 4.10. In Example 9.2 we developed the transfer functions for the linearized model of the system:

$$\overline{c}_A'(s) = \frac{b_1(s + a_{22})}{P(s)} \overline{c}_{A_i}'(s) - \frac{a_{12}b_1}{P(s)} \overline{T}_i'(s) - \frac{a_{12}b_2}{P(s)} \overline{T}_c'(s) \qquad (9.15a)$$

$$\overline{T}'(s) = -\frac{a_{21}b_1}{P(s)} \overline{c}_{A_i}'(s) + \frac{b_1(s + a_{11})}{P(s)} \overline{T}_i'(s) + \frac{b_2(s + a_{11})}{P(s)} \overline{T}_c'(s) \quad (9.15b)$$

where

$$P(s) \equiv s^2 + (a_{11} + a_{22})s + (a_{11}a_{22} - a_{12}a_{21})$$

For the definition of the constant parameters a_{11}, a_{12}, a_{21}, a_{22} and b_1, b_2, see Example 9.2. All variables are in deviation form.

Let us examine two different control problems and develop the necessary feedforward control systems.

Problem 1. Control the concentration c_A in the presence of changes in the inlet concentration and temperature. The temperature of the coolant, T_c, is the manipulated variable. Since we have two disturbances, we need two distinct feedforward controllers. To develop the design equations for the two controllers, put $\overline{c}_A'(s) = 0$ in eq. (9.15a). Then we take

$$\overline{T}_c'(s) = \frac{b_1}{a_{12}b_2}(s + a_{22})\overline{c}_{A_i}'(s) - \frac{b_1}{b_2} \overline{T}_i'(s) \qquad (21.16)$$

Equation (21.16) indicates that the first controller (G_{c1}) is a *lead element* while the second (G_{c2}) is a *gain-only element*. The resulting feedforward system is shown in Figure 21.6a.

Problem 2. Control the temperature T, considering c_{A_i} and T_i as the two disturbances and T_c as the manipulated variable. Setting $\overline{T}'(s) = 0$, eq. (9.15b) yields

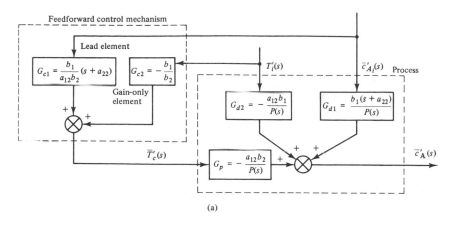

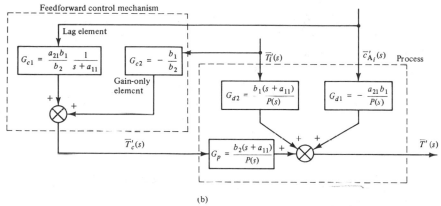

Figure 21.6 Feedforward schemes for CSTR of Example 21.3: (a) concentration control; (b) temperature control.

$$\overline{T}'_c(s) = \frac{a_{21}b_1}{b_2} \frac{1}{s + a_{11}} \overline{c}'_{A_i}(s) - \frac{b_1}{b_2} \overline{T}'_i(s) \qquad (21.17)$$

Equation (21.17) shows that the first controller (G_{c1}) is a *lag element* while the second (G_{c2}) is a *gain-only element*. The resulting feedforward system is shown in Figure 21.6b.

21.4   Feedforward–Feedback Control

Feedforward control has the potential for perfect control but it also suffers from several inherent weaknesses. In particular:

1. It requires the identification of all possible disturbances and their

direct measurement, something that may not be possible for many processes.

2. Any changes in the parameters of a process (e.g., deactivation of a catalyst with time, reduction of a heat transfer coefficient due to fouling, etc.) cannot be compensated by a feedforward controller because their impact cannot be detected.

3. Feedforward control requires a very good model for the process, which is not possible for many systems in chemical industry.

On the other hand, feedback control is rather insensitive to all three of these drawbacks but has poor performance for a number of systems (multicapacity, dead time, etc.) and raises questions of closed-loop stability. Table 21.1 summarizes the relative advantages and disadvantages of the two control systems.

We would expect that a combined feedforward–feedback control system will retain the superior performance of the first and the insensitivity of the second to uncertainties and inaccuracies. Indeed, any deviations caused by the various weaknesses of the feedforward control will be corrected by the feedback controller. This is possible because a feedback control loop directly monitors the behavior of the controlled process (measures process output). Figure 21.7 shows the configuration of a combined feedforward–feedback control system.

Let us now develop an equation for the closed-loop response of the

TABLE 21.1
RELATIVE ADVANTAGES AND DISADVANTAGES OF FEEDFORWARD AND FEEDBACK CONTROL

Advantages	Disadvantages
Feedforward	
1. Acts before the effect of a disturbance has been felt by the system.	1. Requires identification of all possible disturbances and their direct measurement.
2. Is good for slow systems (multicapacity) or with significant dead time.	2. Cannot cope with unmeasured disturbances.
3. It does not introduce instability in the closed-loop response.	3. Sensitive to process parameter variations.
	4. Requires good knowledge of the process model.
Feedback	
1. It does not require identification and measurement of any disturbance.	1. It waits until the effect of the disturbances has been felt by the system, before control action is taken.
2. It is insensitive to modeling errors.	2. It is unsatisfactory for slow processes or with significant dead time.
3. It is insensitive to parameter changes.	3. It may create instability in the closed-loop response.

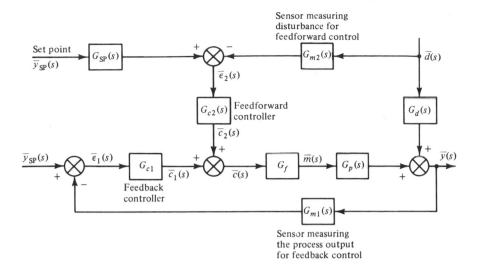

Figure 21.7 Generalized block diagram for feedforward–feedback control.

feedforward–feedback system of Figure 21.7. First, recall that (we have dropped the argument s to simplify the presentation)

$$\overline{y} = G_p \overline{m} + G_d \overline{d} \qquad (21.6)$$

The value of the manipulated variable is given by

$$\overline{m} = G_f \overline{c} = G_f(\overline{c}_1 + \overline{c}_2) = G_f G_{c1} \overline{\epsilon}_1 + G_f G_{c2} \overline{\epsilon}_2$$

or

$$\overline{m} = G_f G_{c1}(\overline{y}_{SP} - G_{m1}\overline{y}) + G_f G_{c2}(G_{SP}\overline{y}_{SP} - G_{m2}\overline{d}) \quad (21.18)$$

Replace $\overline{m}$ in eq. (21.6) by its equal from eq. (21.18) and after algebraic rearrangements take

$$\overline{y} = \frac{G_p G_f(G_{c1} + G_{c2}G_{SP})}{1 + G_p G_f G_{c1} G_{m1}} \overline{y}_{SP} + \frac{G_d - G_p G_f G_{c2} G_{m2}}{1 + G_p G_f G_{c1} G_{m1}} \overline{d} \qquad (21.19)$$

A close examination of eq. (21.19), which yields the closed-loop process output under feedforward-feedback control, reveals the following characteristics:

1. The stability of the closed-loop response is determined by the roots of the characteristic equation

$$1 + G_p G_f G_{c1} G_{m1} = 0$$

which depends on the transfer functions of the feedback loop only. Therefore:

The stability characteristics of a feedback system will not change with the addition of a feedforward loop.

2. The transfer functions of the feedforward loop, G_{c2} and G_{SP}, will be given by the design equations (21.12) and (21.13):

$$G_{c2} = \frac{G_d}{G_p G_f G_{m2}} \quad \text{and} \quad G_{SP} = \frac{G_{m2}}{G_d}$$

If G_p, G_d, G_f, and G_{m2} are known exactly, the feedforward loop compensates completely for disturbance or set point changes and the feedback controller remains idle since ϵ_1 stays continuously zero.

3. If any of G_p, G_d, G_f, and G_{m2} are known only approximately, then

$$G_d - G_p G_f G_{c2} G_{m2} \neq 0 \quad \text{and/or} \quad G_p G_f G_{c2} G_{SP} \neq 1$$

In such case the feedforward loop does not provide perfect control (i.e., $\overline{y} \neq \overline{y}_{SP}$). Then $\overline{\epsilon}_1 \neq 0$ and the feedback loop is activated and offers the necessary compensation.

Example 21.4: Feedforward–Feedback Control of the Tank Heater

Consider again the tank heater of Example 21.2. Under feedforward control only, we have the configuration shown in Figure 21.3b. The design transfer functions are

$$G_c = F_i \rho c_p \quad \text{and} \quad G_{SP} = \tau s + 1$$

Assume that the density ρ or the heat capacity c_p are not known exactly. Then the feedforward loop does not provide perfect control. Figure 21.8a

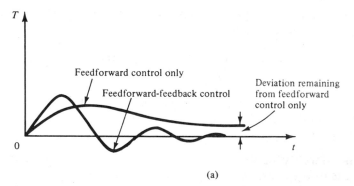

Feedforward control only

Feedforward-feedback control

Deviation remaining from feedforward control only

0

t

(a)

Figure 21.8 (a) Temperature response of a stirred tank heater under feedforward control alone, and with feedback trimming; (b) corresponding block diagram.

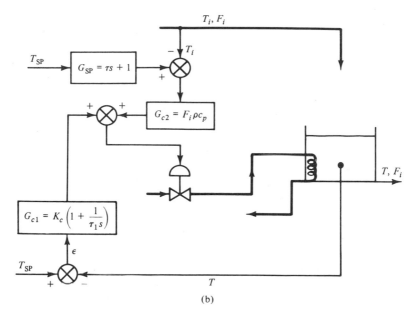

Figure 21.8 (Continued)

shows the temperature in the tank after a step change in the inlet temperature. Notice the remaining deviation (offset).

Introduce now in the system a feedback loop with PI controller (Figure 21.8b). In Figure 21.8a we have plotted again the temperature of the
liquid in the tank, for the same step change in the inlet temperature.
Notice that the deviation has disappeared.

21.5   Ratio Control

Ratio control is a special type of feedforward control where two disturbances (loads) are measured and held in a constant ratio to each other.
It is mostly used to control the ratio of flow rates of two streams. Both
flow rates are measured but only one can be controlled. The stream
whose flow rate is not under control is usually referred to as *wild
stream.*

Figure 21.9a and b show two different ratio control configurations
for two streams. Stream A is the wild stream.

1. In configuration 1 (Figure 21.9a) we measure both flow rates and
   take their ratio. This ratio is compared to the desired ratio (set
   point) and the deviation (error) between the measured and
   desired ratios constitutes the actuating signal for the ratio controller.

2. In configuration 2 (Figure 21.9b) we measure the flow rate of the

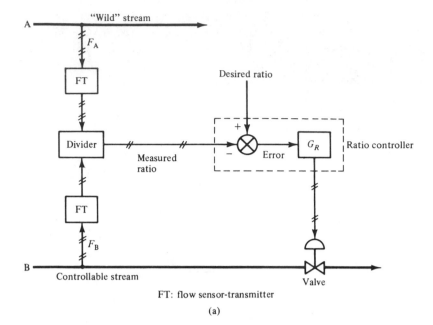

(a)

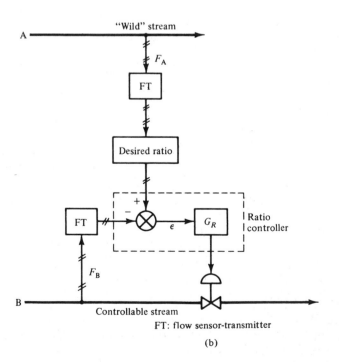

(b)

Figure 21.9 Alternative configurations of ratio control systems.

wild stream A and multiply it by the desired ratio. The result is the flow rate that the stream B should have and constitutes the set point value which is compared to the measured flow rate of stream B. The deviation constitutes the actuating signal for the controller, which adjusts appropriately the flow of B.

Ratio control is used extensively in chemical processes with the following as the most commonly encountered examples:

1. Keep a constant ratio between the feed flow rate and the steam in the reboiler of a distillation column.
2. Hold constant the reflux ratio in a distillation column.
3. Control the ratio of two reactants entering a reactor at a desired value.
4. Hold the ratio of two blended streams constant, in order to maintain the composition of the blend at the desired value.
5. Hold the ratio of a purge stream to the recycle stream constant.
6. Keep the ratio of fuel/air in a burner at its optimum value (most efficient combustion).
7. Maintain the ratio of the liquid flow rate to vapor flow rate in an absorber constant, in order to achieve the desired composition in the exit vapor stream.

THINGS TO THINK ABOUT

1. Define the concept of feedforward control on physical grounds.

2. Is driving a car mostly feedforward or feedback control? What about riding a bicycle?

3. Discuss the relative advantages and disadvantages of feedforward and feedback control systems. Why does the addition of feedback control improve the performance of a feedforward system?

4. What kinds of processes stand to benefit the most from feedforward control? Why?

5. Do the stability characteristics of a process change with feedforward control as they do with feedback? Elaborate on your answer.

6. In Section 21.4 it was claimed that the stability characteristics of a feedforward–feedback control system are affected only by the feedback loop. Explain why.

7. Draw three different feedforward control configurations for the mixing process of Example 4.11.

8. Under what conditions will a steady-state feedforward control system yield

the same performance as a dynamic feedforward controller in rejecting the effect of a disturbance?

9. What is a lag–lead element, and why is it considered to be a versatile component for feedforward control?

10. Consider the feedforward control of a distillation column. What kind of dynamic feedforward element will be needed: lag–lead, lag only, lead only, gain only? Give a rather qualitative explanation.

11. What is the ratio control, and why is it useful in process control? Give three specific examples.

12. How do you select the desired value of the ratio in a ratio control system?

13. Draw the feedforward and feedback control systems that regulate the flow through a pipe. Do you expect one of them to be significantly better than the other in maintaining the desired flow?

14. In Figure 21.9a and b we see two different ratio control configurations. Which one would you prefer and why? (*Hint*: Examine the static gain of the control loop in Figure 21.9a and consult Ref. 3 for details.)

Adaptive and Inferential **22**
Control Systems

In this chapter we examine two additional types of control systems: (1) adaptive, and (2) inferential. Although their basic objectives and functions can be easily described in a qualitative manner, their practical implementation is rather complicated, involves extensive computations, and is accomplished through the use of digital computers. For this reason we postpone a detailed quantitative discussion on these two control systems until Part VII, where we will study the use of digital computers for process control. For the time being we will only make a qualitative presentation of each control system, discussing their basic logic and giving examples of their practical applications.

22.1 Adaptive Control

Adaptive is called a control system, which can adjust its parameters automatically in such a way as to compensate for variations in the characteristics of the process it controls. The various types of adaptive control systems differ only in the way the parameters of the controller are adjusted.

But why are adaptive controllers needed in chemical processes? There are two main reasons. First, most chemical processes are nonlinear. Therefore, the linearized models that are used to design linear controllers depend on the particular steady state (around which the process is linearized). It is clear, then, that as the desired steady-state operation of a process changes, the "best" values of the controller's parameters change. This implies the need for controller adaptation. Example 10.5 demonstrates how the time constant and process gain of a simple liquid storage tank depend on the value of the steady-state liquid level.

Second, most of the chemical processes are *nonstationary* (i.e., their characteristics change with time). Typical examples are the decay of the catalyst activity in a reactor and the decrease of the overall heat transfer coefficient in a heat exchanger due to fouling (Example 10.6). This change leads again to a deterioration in the performance of the linear controller, which was designed using some nominal values for the process parameters, thus requiring adaptation of the controller parameters.

What is the objective of the adaptation procedure? Clearly, it is not to keep the controlled variable at the specified set point. This will be accomplished by the control loop, however badly. We need an additional *criterion*, an *objective function* that will guide the adaptation mechanism to the "best" adjustment of the controller parameters. To phrase it differently, we need a criterion to guide the *adaptive tuning* of the controller. Any of the performance criteria we discussed in Chapters 16 and 18 could be used:

One-quarter decay ratio
Integral of the square error
Gain or phase margins, etc.

For example, variations in process parameters may lead to decay ratios higher than the one-quarter. In such case, the adaptive mechanism will adjust the controller parameters in such a way that the process response exhibits again decay ratio equal to one-quarter.

There are two different mechanisms for the adaptation of the controller parameters.

Programmed or scheduled adaptive control

Suppose that the process is well known and that an adequate mathematical model for it is available. If there is an auxiliary process variable which correlates well with the changes in process dynamics, we can relate ahead of time the "best" values of the controller parameters to the value of the auxiliary process variable. Consequently, by measuring the value of the auxiliary variable we can *schedule* or *program* the adaptation of the controller parameters. Figure 22.1 shows the block diagram of a programmed adaptive control system. We notice that it is composed of two loops. The inner loop is an ordinary feedback control loop. The outer loop includes the parameter adjustment (adaptation)

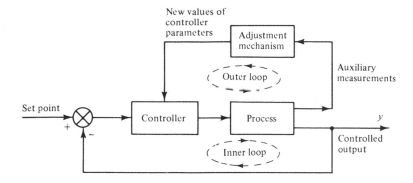

Figure 22.1 Programmed adaptive control system.

mechanism and *it is comparable to feedforward compensation*. A typical example is *gain scheduling adaptive control*.

Example 22.1: Gain Scheduling Adaptive Control

In a normal feedback control loop (Figure 22.2a) the control valve or another of its components may exhibit a nonlinear character. In such a case the gain of the nonlinear component will depend on the current steady state. Suppose that we want to keep the total gain of the overall system constant. From Figure 22.2a we find easily that the open-loop gain is given by

$$K_{overall} = K_p K_m K_c K_f = \text{constant}$$

It is clear then that as the gain K_f of the nonlinear valve changes, the gain of the controller, K_c, should change as follows:

$$K_c = \frac{\text{constant}}{K_p K_m K_f} \tag{22.1}$$

We assume that the gains K_p and K_m are known exactly. Furthermore, if the characteristics of the control valve are known well, then its gain, K_f,

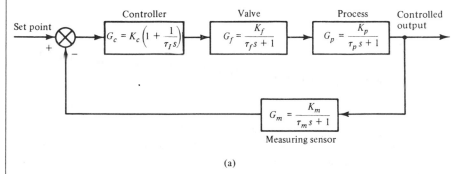

(a)

Figure 22.2 (a) Feedback control loop for Example 22.1; (b) corresponding gain scheduling adaptive control.

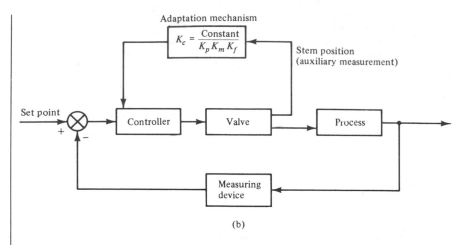

(b)

Figure 22.2 (Continued)

can be calculated from the stem position. Therefore, by measuring the stem position (auxiliary measurement) we can compute K_f. Then eq. (22.1) yields the adaptation mechanism of this simple gain scheduling adaptive controller. Figure 22.2b shows the resulting control structure.

Remark.   Notice that the gain scheduling is comparable to feedforward compensation. There is no feedback to compensate for incorrect adaptation.

Example 22.2: Programmed Adaptive Control of a Combustion System

Consider a burner where the fuel/air ratio is kept at its optimal value to achieve the highest efficiency of combustion. Excess fuel or air will reduce the efficiency. The optimal fuel/air ratio is maintained through a ratio control mechanism (Section 21.5). The control system is shown in Figure 22.3a.

The optimal value of the fuel/air ratio, which maximizes the combustion efficiency, depends on the conditions prevailing within the process (e.g., the temperature of the air). Consequently, as the temperature of the air changes, so does the optimal value of the fuel/air ratio.

From previous experimental data we know how the optimal fuel/air ratio changes with air temperature for maximum efficiency. Therefore, to maintain the ratio continuously at its optimal value despite any changes in the air temperature, we can use a programmed adaptive control system. Such a system is shown in Figure 22.3b. It measures the temperature of the air (auxiliary measurement) and adjusts the value of the fuel/air ratio. Notice again that the *ratio adjustment mechanism is like feedforward compensation*.

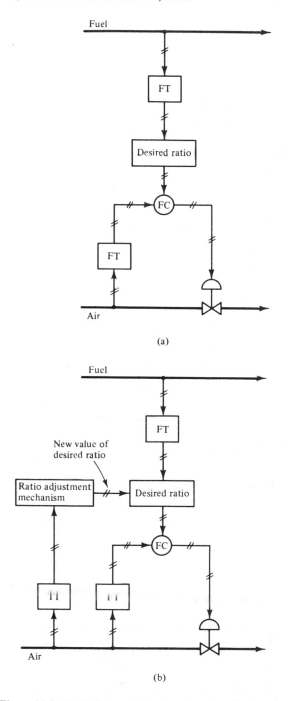

(a)

(b)

Figure 22.3 (a) Ratio controller for a combustion system; (b) corresponding ratio adjustment mechanism.

Self-adaptive control

If the process is not known well, we need to evaluate the objective function on-line (while the process is operating) using the values of the controlled output. Then the adaptation mechanism will change the controller parameters in such a way as to optimize (maximize or minimize) the value of the objective function (criterion). In the following two examples we examine the logic of two special self-adaptive control systems: *model reference adaptive control* (MRAC) and *self-tuning regulators* (STRs).

Example 22.3: Model-Reference Adaptive Control (MRAC)

Figure 22.4 illustrates a different way to adjust the parameters of the controller. We postulate a *reference model* which tells us how the controlled process output ideally should respond to the command signal (set point). The model output is compared to the actual process output. The difference (error ϵ_m) between the two outputs is used through a computer to adjust the parameters of the controller in such a way as to minimize the integral square error:

$$\text{minimize ISE} = \int_0^t [\epsilon_m(t)]^2 \, dt$$

The model chosen by the control designer for reference purposes is to a certain extent arbitrary. Most often a rather simple linear model is used.

We notice that the model-reference adaptive control is composed of two loops. The inner loop is an ordinary feedback control loop. The outer loop includes the adaptation mechanism and also looks like a feedback loop. The model output plays the role of the set point while the process

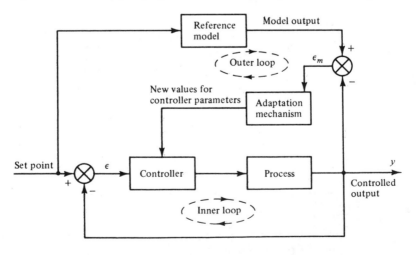

Figure 22.4 Model-reference adaptive control.

output is the actual measurement. There is a comparator whose output (error ϵ_m) is the input to the adjustment mechanism. *The key problem is to design the adaptation mechanism in such a way as to provide a stable system (i.e., bring the error ϵ_m to zero).* This is not a trivial problem, but an extensive discussion on how it can be solved requires mathematical analysis that goes beyond the scope of the present text.

Example 22.4: Self-Tuning Regulator (STR)

Consider the block diagram of Figure 22.5. It represents the structure of a self-tuning regulator, which constitutes another way of adjusting the parameters of a controller.

The STR is composed, again, of two loops. The inner loop consists of the process and an ordinary linear feedback controller. The outer loop is used to adjust the parameters of the feedback controller and is composed of (1) a recursive parameter estimator and (2) an adjustment mechanism for the controller parameters.

The parameter estimator assumes a simple linear model for the process:

$$\frac{\overline{y}(s)}{\overline{m}(s)} = \frac{K_p e^{-t_d s}}{\tau s + 1}$$

Then, using measured values for the manipulated variable m and the controlled output y, it estimates the values of the parameters K_p, τ, and t_d, employing a least-squares estimation technique (see Chapter 31). Once the values of the process parameters K_p, τ, and t_d are known, the adjustment mechanism can find the "best" values for the controller parameters using various design criteria, such as:

Phase or gain margins
Integral of the squared error, etc.

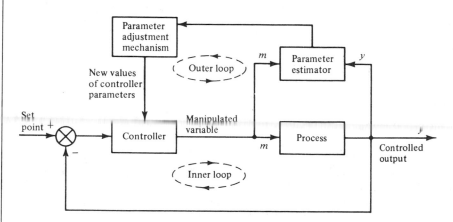

Figure 22.5 Self-tuning regulator.

Both the parameter estimator and the adjustment mechanism require involved computations. For this reason the STR can be implemented only through the use of digital computers.

Adaptive control systems have been applied in chemical processes. The range of their applicability has expanded with the introduction of digital computers for process control. Several theoretical and experimental studies have appeared in the chemical engineering literature, while the number of industrial adaptive control mechanisms increases continuously. Most of the adaptive control systems require extensive computations for parameter estimation and optimal adjustment of controller parameters, which can be performed on-line only by digital computers. Therefore, we will delay any discussion on the quantitative design of such systems until Chapter 31, after we have studied the use of digital computers for control.

22.2   Inferential Control

Quite often, the controlled output of a processing unit cannot be measured directly. Consequently, we cannot use feedback control or any other configuration which necessitates direct measurement of the controlled variable. If the disturbances that create the control problems can be measured and an adequate process model is available, we could use feedforward control to keep the unmeasured output at its desired value (see Chapter 21).

What happens, though, if the disturbances cannot be measured? *None of the control configurations studied so far can be used to control an unmeasured process output in the presence of unmeasured disturbances.* This is the type of control problems where inferential control is the only solution. Let us now examine the structure of an inferential control system.

Consider the block diagram of the process shown in Figure 22.6a, with one unmeasured controlled output, y, and one secondary measured output, z. The manipulated variable m and the disturbance d affect both outputs. The disturbance is considered to be unmeasured. The transfer functions in the block diagram indicate the relationships between the various inputs and outputs, and they are assumed to be perfectly known.

From Figure 22.6a we can easily derive the following input–output relationships;

$$\overline{y} = G_{p1}\overline{m} + G_{d1}\overline{d} \tag{22.2}$$

$$\overline{z} = G_{p2}\overline{m} + G_{d2}\overline{d} \tag{22.3}$$

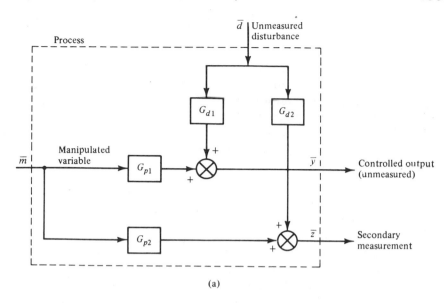

(a)

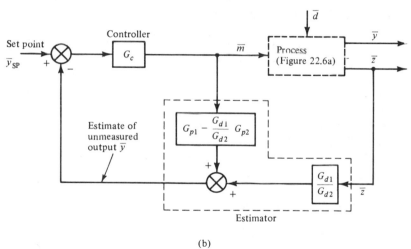

(b)

Figure 22.6 (a) Process with need for inferential control; (b) corresponding inferential control system.

From eq. (22.3) we can solve with respect to $\overline{d}$ and find the following estimate of the unmeasured disturbance:

$$\overline{d} = \frac{1}{G_{d2}}\,\overline{z} - \frac{G_{p2}}{G_{d2}}\,\overline{m} \tag{22.4}$$

Substitute the estimate above into eq. (22.2) and find the following relationship:

$$\overline{y} = \left[G_{p1} - \frac{G_{d1}}{G_{d2}} G_{p2} \right] \overline{m} + \frac{G_{d1}}{G_{d2}} \overline{z} \tag{22.5}$$

Equation (22.5) provides the needed estimator which relates the unmeasured controlled output to measured quantities like m and z. Figure 22.6b shows the structure of the resulting inferential control system. Notice that the estimated value of the unmeasured output plays the same role as a regular measured output; that is, it is compared to the desired set point and the difference is the actuating signal for the controller.

Remarks

1. It is important to emphasize that the success of an inferential control scheme depends heavily on the availability of a good estimator, which in turn depends on how well we know the process. Thus if the process transfer functions G_{p1}, G_{p2}, G_{d1}, and G_{d2} are perfectly known, a perfect estimator can be constructed and consequently we will have perfect control. When the process transfer functions are only approximately known (which is usually the case), the inferential scheme provides control of varying quality, depending on how well the process is known.
2. In chemical process control the variable that is most commonly inferred from secondary measurements is composition. This is due to the lack of reliable, rapid, and economical measuring devices for a wide spectrum of chemical systems. Thus inferential control may be used for the control of chemical reactors, distillation columns, and other mass transfer operations such as driers and absorbers. Temperature is the most common secondary measurement used to infer the unmeasured composition.

Example 22.5: Inferential Control of a Distillation Column

Consider a distillation column with 16 trays, which separates a mixture of propane–butane into two products. The feed composition is the unmeasured disturbance and the control objective is to maintain the overhead product molar composition 95% in propane. The reflux ratio is the manipulated variable.

Since the feed and overhead compositions are considered unmeasured, we can only use inferential control. The secondary measurement employed to infer the overhead composition is the temperature at the top tray. Let us now examine how we can develop and design the inferential control mechanism.

The process as defined above has two inputs and two outputs:

Inputs: feed composition (disturbance), reflux ratio (manipulation)

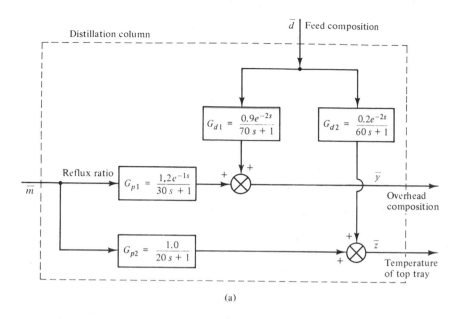

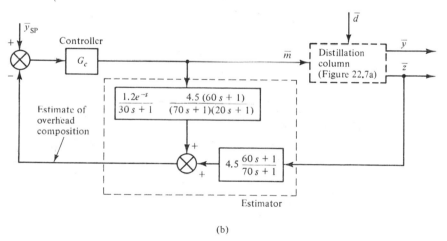

Figure 22.7 (a) Block diagram of a distillation column; (b) corresponding inferential control system.

Outputs: overhead propane composition (unmeasured controlled output) and temperature of top tray (secondary measurement)

How can we identify the four process transfer functions? In Example 4.13 we saw that a rigorous approach leads to an overwhelming mathematical model. The process reaction curve method, which was discussed in Section 16.5, is a simpler approach and yields the transfer functions between the various inputs and outputs. Following this methodology, we developed the input–output relationships (see also Figure 22.7a):

$$\overline{y}(s) \simeq \frac{0.90e^{-2s}}{70s + 1}\, \overline{d}(s) + \frac{1.20e^{-1s}}{30s + 1}\, \overline{m}(s)$$

$$\overline{z}(s) \simeq \frac{0.20e^{-2s}}{60s + 1}\, \overline{d}(s) + \frac{1.0}{20s + 1}\, \overline{m}(s)$$

Having developed the four process transfer functions it is easy to design the inferential control system (Figure 22.7b).

Remarks

1. The temperature of the top tray was selected arbitrarily to be the secondary measurement. But why did we not select the temperature of the second, third,...tray from the top? The answer is rather complex and beyond the scope of this text. Nevertheless, we expect that the temperature at the top tray will reflect better the condition of the overhead product.
2. Were we to control the purity of the bottoms product, a different temperature would be needed. Most likely the temperature would be close to the bottom of the column.
3. Recall that the effectiveness of an inferential control scheme depends heavily on the goodness of the estimator, which in turn depends on the model that is available for the process. Assume that the overhead composition can be measured intermittently, either by taking samples manually and analyzing them or even better using a gas chromatograph on-line. From the composition measurements we can take the useful information needed to judge how effective the inferential control has been. Thus if the measured

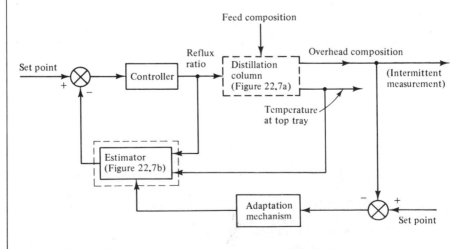

Figure 22.8 Adaptive inferential control for a distillation column.

steady-state value of the overhead composition deviates significantly from the desired set point value, we can use this deviation
(error) through an adaptive mechanism to correct the estimator.
The resulting system is shown in Figure 22.8.

THINGS TO THINK ABOUT

1. What is adaptive control, and why is it needed in chemical process control?

2. Give two examples of adaptive control for processing units, different from
   those described in Section 22.1. Describe qualitatively the functions of the
   adaptive control schemes you proposed.

3. What is programmed adaptive control, and what is self-adaptive control?
   Give one example of each, different from those in Section 22.1. When
   would you recommend the programmed and when the self-adaptive
   scheme?

4. What is gain scheduling control, and why can you use it in chemical process
   control? It was claimed in Example 22.1 that it resembles feedforward
   compensation. Explain why. What are its advantages and disadvantages?

5. Discuss the logic of model-reference adaptive control and self-tuning regulators. Find the similarities and differences between the two configurations.

6. Show qualitatively that the structure of a self-tuning regulator can be
   derived from that of a model-reference adaptive control if the parameter
   estimation is done by updating the reference model.

7. Are the outer loops in the MRAC and STR configurations of feedforward
   or feedback nature?

8. Consider the neutralization of an acidic effluent industrial waste with a
   caustic solution. The titration curve of the waste being neutralized is nonlinear and changes with time due to unmeasured disturbances. Develop a
   qualitative self-adaptive control scheme and describe the functions of its
   components. (You can consult Ref. 3.)

9. As discussed in Section 22.1, the purpose of an adaptive controller is not to
   keep the controlled output at its desired set point; this is accomplished by
   the regular feedback loop. What, then, is the criterion or objective function
   that guides parameter adjustment of an adaptation mechanism? How is
   this objective evaluated by the programmed and self-adaptive schemes?

10. As discussed in Chapter 21, the effectiveness of a feedforward control loop
    depends heavily on the quality of the model available for the process.
    Develop an adaptive control mechanism for a feedforward controller which
    will compensate for any variations in the model of the process.

11. Discuss the logic of an inferential control scheme. Why is this control

scheme needed? Describe two examples of inferential control different from those of Section 22.2.

12. What do we mean when we say that a process variable is "unmeasured"?

13. Consider two processes: one (process A) slow with time constant 5 hours and another (process B) faster with time constant 1 hour. The composition of the output streams from the two processes is measured every 2 to 3 hours. Which of the two process outputs can be controlled by conventional feedback and which one will require inferential control?

14. Show that the inferential control employed for process A or B in item 13 (above) can be improved through an adaptive mechanism that uses the direct composition measurement every 2 to 3 hours. (Consult Example 22.5.)

15. Develop an inferential control scheme that can be used to control the drying of solids with warm air. Discuss how you would develop the estimator of the inferential structure. (Consult a reference on solids drying with air and Ref. 3.)

16. If, in addition to the unmeasured disturbance there are measured disturbances in a system, we can develop a combined inferential–feedforward configuration. Develop such a configuration for a system of your choice.

REFERENCES FOR PART V

Chapter 19.   The dead-time compensator can be found in the original paper by O. J. M. Smith [Ref. 1], while a tutorial treatment is also presented in the text by C. L. Smith [Ref. 2].

1. "Close Control of Loops with Dead Time," by O. J. M. Smith, *Chem. Eng. Prog., 53*(5), 217 (1957).
2. *Digital Computer Process Control*, by C. L. Smith, Intext Educational Publishers, Scranton, Pa. (1972).

Shinskey [Ref. 3] uses the term *complementary feedback* for the dead-time compensator. He also discusses several practical considerations which should guide the design of dead-time compensators for various processing units. The terms "complementary feedback" was introduced in an interesting paper by Giloi [Ref. 4].

3. *Process Control Systems*, 2nd ed., by F. G. Shinskey, McGraw-Hill Book Company, New York (1979).
4. "Optimized Feedback Control of Dead-Time Plants by Complementary Feedback," by W. Giloi, *IEEE, Trans. Appl. and Ind., 83*, 183 (1964).

The inverse response compensator was first proposed by Iinoya and Altpeter in the following paper:

5. "Inverse Response in Process Control," by K. Iinoya and R. J. Altpeter, *Ind. Eng. Chem.*, *54*(7), 39 (1962).

There they also gave some loose recommendations for selecting the value of k in the transfer function of the compensator. Waller and Nygårdas [Ref. 6] experimented numerically with Iinoya and Altpeter's compensator as well as with simple PID loops using Ziegler–Nichols settings. They found that the second performed relatively better.

6. "On Inverse Response in Process Control," by K. V. T. Waller and C. G. Nygårdas, *Ind. Eng. Chem. Fundam.*, *14*(3), 221 (1975).

Chapter 20. Chapter 6 of Shinskey's book [Ref. 3] is an excellent reference for multiple-loop control systems. It treats cascade, selective control loops, and adaptive systems. Besides the general treatment of each control configuration, it discusses the practical considerations guiding the selection and design of such control systems. In particular, it covers the following items which could attract the interest of the reader:

Cascade control: discusses the properties of the commonly used cascade flow loop, as well as the use of temperature measurement in the secondary loop and how to protect the primary controller against windup.

Selective control loops: covers a large number of diversified selective control systems for protection of equipment, auctioneering, redundant instrumentation, variable structuring, and so on.

Split-range control: discusses several interesting practical systems used in chemical processes, such as several boilers discharging into a common steam header, and parallel compressors discharging into a common header.

The polyglot reader will find the following book by Van der Grinten a very valuable reference source for the control systems of this chapter and others:

7. *Proces regelingen*, by P. M. E. M. Van der Grinten, Het Spectrum, Utrecht (1970).

Chapter 21. Chapter 7 in Shinskey [Ref. 3] is again an excellent reference for the practical considerations guiding the design of feedforward and ratio control systems. It also discusses the use of feedforward schemes for optimizing control of processing systems. Good tutorial references are the books by Smith [Ref. 2], Murrill [Ref. 8], and Luyben [Ref. 9]. The last one has a simple but instructive example on the nonlinear feedforward control of a CSTR.

8. *Automatic Control of Processes*, by P. W. Murrill, Intext Educational Publishers, Scranton, Pa. (1967).

9. *Process Modeling, Simulation, and Control for Chemical Engineers*, by W. L. Luyben, McGraw-Hill Book Company, New York (1973).

Chapter 22. The book by Shinskey is once more a valuable guide for the design of useful adaptive and inferential control systems. The mathematical treatment of the subject is simple and to the point. The general reader will find very instructive the following papers on adaptive control:

10. "Making Sense Out of the Adaptive Principle," by J. E. Gibson, *Control Eng.*, p. 113 (Aug. 1960).
11. "Mechanizing the Adaptive Principle," by J. E. Gibson, *Control Eng.*, p. 109 (Oct. 1960).
12. "Generalizing the Adaptive Principle," by J. E. Gibson, *Control Eng.*, 7(12), 93 (Dec. 1960).
13. "Understanding Adaptive Control," by J. Adams, *Automation, 17*, 108 (Mar. 1970).
14. "Adaptive Process Control: Versatile On-Line Tool," by E. H. Bristol, *Control Eng., 20*(4), 41 (Apr. 1973).
15. "A New Look at Adaptive Control," by M. P. Groover, *Automation, 20*, 60 (Apr. 1973).

The following references on adaptive control involve mathematical analyses which employ more advanced tools than those covered so far in this text. The interested reader can return to these sources after he or she has had some exposure to multivariable systems theory (control and estimation).

16. *Adaptive Processes—A Guided Tour*, by R. Bellman, Princeton University Press, Princeton, N. J. (1961).
17. *Adaptive Control Systems*, by E. Mishkin and L. Braun, McGraw-Hill Book Company, New York (1961).
18. *Methods and Applications in Adaptive Control*, H. Unbehauen (ed.), Springer-Verlag, Berlin (1980).
19. *Applications of Adaptive Control*, K. S. Narendra and R. V. Monopoli (eds.), Academic Press, Inc., New York (1980).

The model-reference adaptive control was originally proposed by Whitaker et al. in 1958 and was developed for servo problems:

20. "Design of Model-Reference Adaptive Control Systems for Aircraft," by H. P. Whitaker, J. Yamron, and A. Kezer, Report R-164, Instrumentation Laboratory, MIT, Cambridge, Mass. (1958).

For a comprehensive treatment of the work up to 1977, the reader can consult:

21. *Adaptive Control—The Model Reference Approach*, by I. D. Landau, Marcel Dekker, Inc., New York (1979).

The self-tuning regulator was originally proposed by Kalman, who built a special-purpose computer to implement it. Subsequently, the STR received a lot of attention because it is flexible and easy to understand and implement it. In Section 31.2 the reader can find a very simplistic quantification of STR. The work of Åström and his coworkers represents the evolution of STR development, and the following two references would be very instructive for the motivated reader:

22. "On Self-Tuning Regulators," by K. J. Åström and B. Wittenmark, *Automatica*, *9*, 185 (1973).

23. "Self-Tuning Regulators—Design Principles and Applications," by K. J. Åström. This paper can be found in Ref. 19.

In "References for Part VII" the reader can find additional entries on the applications of self-tuning regulators. A very good source for the problems and design of inferential control systems is the work of C. Brosilow and his coworkers:

24. "The Use of Secondary Measurements to Improve Control," by R. Weber and C. Brosilow, *AIChE J., 18*, 614 (1972).

25. "The Structure and Dynamics of Inferential Control Systems," by C. Brosilow and M. Tong, *AIChE J., 24*, 492 (1978).

In the following two entries the reader can find the application of inferential systems to control product dryness without measuring it:

26. "How to Control Product Dryness without Measuring It," by F. G. Shinskey, Preprint volume of Joint Automatic Control Conference, University of Michigan (1978).

27. *Energy Conservation through Control*, by F. G. Shinskey, Academic Press, Inc., New York (1978).

PROBLEMS FOR PART V

Chapter 19

V.1 Consider a first-order process with dead time

$$G_p(s) = \frac{10\,e^{-t_d\,s}}{s + 1}$$

which is controlled by a proportional controller with gain K_c. For the measuring device and control valve, assume that $G_m = G_f = 1$.

(a) Show that as dead time t_d increases, the ultimate gain of the proportional controller decreases.

(b) Assuming that $t_d = 0.5$ min, compute the gain of the proportional controller so that the gain margin is 1.7 or the phase margin is 30°, whichever is safer.

(c) Sketch the closed-loop response of the system to a unit step change in the set point, using the gain found in part (b). Compute the steady-state offset.

(d) Design a dead-time compensator for the system when $t_d = 0.5$. Compute the closed-loop response of the system to a unit step in set point using a gain 100 times larger than that found in part (b). Compare the closed-loop response with dead-time compensator to that without a compensator.

(e) Show that with such a large gain the tolerance to an error in the value of dead time is very small.

V.2   Repeat Problem V.1 but use the following second-order process:

$$G_p(s) = \frac{5e^{-t_d\,s}}{(0.1s + 1)(20s + 1)}$$

(*Note:* Notice the large difference between the two time constants: 0.1 versus 20.)

V.3   (a) Design a dead-time compensator for the system with process reaction curve given by the data of Table PIV.1. Consider a proportional controller and $G_m = G_f = 1$.

(b) Tune the proportional controller in such a way that it can tolerate a 50% error in dead time.

(c) Compute the steady-state offset when a unit step change is introduced in the system.

V.4   Repeat Problem V.3 but use the process reaction curve data given in Table PIV.2.

V.5   Repeat Problem V.3 but instead of using process reaction data, use the Bode plots data given in Table PIV.3.

V.6   Design a perfect compensator for the system with inverse response described in Problem III.59.

V.7   Design a perfect compensator for the system with inverse response described in Problem III.60.

V.8   Consider a process with the transfer function

$$G_p(s) = \frac{10}{3s + 1} - \frac{5}{s + 1}$$

(a) Show that this process exhibits inverse response.

(b) Design a perfect inverse response compensator. Compute and sketch the closed-loop response to a unit step change in the set point, using a PI controller with $K_c = 10$ and $\tau_I = 0.5$.

(c) Assuming that the process gains 10 and 5 are poorly known, design an approximate compensator of the type given by eq. (19.8). Evaluate the effect of the tuning parameter k on the closed-loop response, assuming unit step changes in the set point.

(d) Compute the closed-loop response of the process to a unit step change in set point, using a PID controller with Ziegler–Nichols settings. Compare the response with those found in parts (b) and (c).

V.9 Repeat Problem V.8 but consider the process with the transfer function

$$G_p(s) = \frac{10}{s^2 + 3s + 1} - \frac{5}{s + 1}$$

Chapter 20

V.10 Consider the process with two different disturbances d_1 and d_2 shown in Figure PV.1.

(a) Draw the block diagram of a simple proportional feedback loop for controlling output y, using m as the manipulated variable. Compute the offset of the closed-loop response to a unit step change in d_1 and sketch its behavior with time. Use a gain for the proportional controller so that the steady state offset is the minimum possible.

(b) Construct a cascade control system for the process and draw its block diagram. Compute the offset of the closed-loop response to a unit step change in d_1, when both controllers of the cascade system are proportional, and the gains have taken their maximum possible values, without destabilizing the closed-loop response.

(c) Compare the closed-loop responses produced by simple feedback and cascade control systems and show that the steady-state offset in the cascade system can be greatly reduced without endangering the stability of the closed-loop response. Also the speed of closed-loop response can be improved with a cascade control system.

V.11 Figure PV.2 shows the block diagrams of two dynamic processes.

(a) Which of the two processes should be controlled by a cascade system in order to improve its closed-loop behavior to changes in disturbance d_1? Why?

(b) Construct the cascade system for the process you selected in part (a)

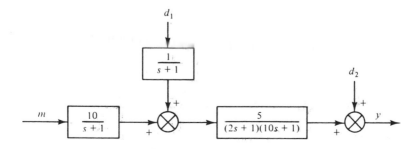

Figure PV.1

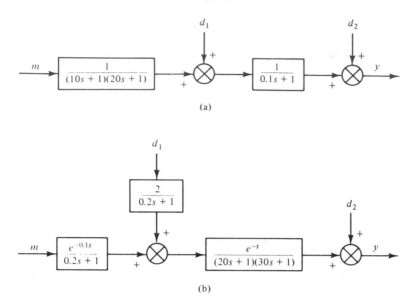

Figure PV.2

for cascade control. Draw the corresponding block diagram using two proportional controllers.

(c) Select the gains of the two proportional controllers so that you can have the "best" closed-loop response to changes in load d_1 without endangering the stability of the system.

V.12 Consider the series of four stirred tanks shown in Figure PV.3. We like to control the water exit temperature T_4 at the desired set point against any changes in the loads T_0 and T_ℓ by manipulating the heat input Q in the first tank.

(a) Show that a cascade control system will produce a better closed-loop response to any changes in T_o and T_ℓ than will a simple feedback loop.

(b) Where is the best place to take the secondary measurement for the secondary loop of the cascade system? Why?

(c) Assume that all tanks are identical with cross-sectional area 5 ft² and steady-state liquid levels of 3 ft for the last three tanks and 2 ft for the first. The water flows are constant with $F_o = 4$ ft³/min and $F_\ell = 2$ ft³/min. Compute the ultimate gains of the primary and secondary PI controllers of the cascade system so that they yield gain margins equal to 1.7. Assume reset times equal to 0.5 min.

(d) Using the gains found in part (c), compute the closed-loop responses resulting from a 10°F step change in T_o or T_ℓ.

(e) Suppose that the secondary measurement is moved upstream or downstream one tank from the tank identified in part (b). Examine how the quality of the closed-loop responses computed in part (d) is affected by this change in measurement location.

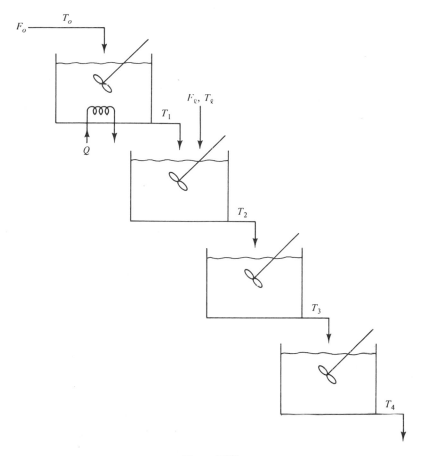

Figure PV.3

V.13 Consider the jacketed mixing tank shown in Figure PV.4. The content of the tank is heated with hot water which enters the jacket with a temperature $T_{h,i}$. The objective is to control the temperature of the mixture in the tank by manipulating the flow rate of hot water F_h.
  (a) Show that a cascade control system that uses the temperature of the hot water in the jacket as a secondary measurement is always better than a simple feedback control system which uses the temperature of the tank's content as the only measurement.
  (b) Draw the block diagram of the cascade control system and place the appropriate transfer functions for each block.
  The following information is given: V is the constant volume of the mixture in the tank; F_1, F_2, and F are constant flow rates; and ρ and c_p are the density and heat capacity of the liquid mixture in the tank. It is assumed that these properties are the same for both streams entering the tank. U is the overall heat transfer coefficient between the heating medium and the mixture in the tank. A_t is the heat transfer area of the

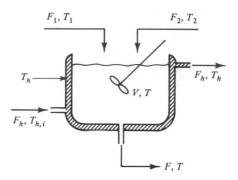

Figure PV.4

jacket. Also, ρ_h, c_{p_h}, and V_h are the density, heat capacity, and volume of the heating medium in the jacket, and are assumed to remain constant during operation.

V.14 The flow F_2 of a gas from a storage tank (Figure PIII.2a) is to be maintained at a desired fixed rate as long as the pressure in the tank is within certain lower and upper limits. Design a control system that can regulate the gas flow while at the same time maintaining gas pressure p of the tank within the limits.

V.15 Repeat Problem V.14 but for the liquid flow from a storage tank with specified upper and lower limits on the liquid level of the tank.

Chapter 21

V.16 Consider the four noninteracting tanks shown in Figure PV.3 of Problem V.12. Move the heating from the first to the last tank.
   (a) Develop a steady-state feedforward control system which maintains T_4 at the desired set point value in the presence of changes in the loads T_o and T_t. Draw the resulting block diagram and place the corresponding transfer functions in each block.
   (b) Design simple dynamic feedforward controllers for the system above. Draw the resulting block diagram and identify all necessary transfer functions.
   (c) Sketch qualitatively the expected responses to unit step changes in T_o and T_t for the steady-state and dynamic feedforward control systems.

V.17 Design the steady-state and dynamic feedforward controllers (for disturbance rejection and set point tracking) for the systems with the following transfer functions (G_p is the transfer function between manipulation and controlled output; G_d is the transfer function between disturbance and controlled output). Assume that $G_m = G_f = 1$.

(a) $G_p(s) = \dfrac{10}{s + 1}$ $G_d(s) = \dfrac{2}{(s + 1)(5s + 1)}$

(b) $G_p(s) = \dfrac{5}{s^2 + 3s + 2}$ $G_d(s) = \dfrac{1}{5s + 1}$

(c) $G_p(s) = \dfrac{10e^{-0.1s}}{2s + 1}$ $G_d(s) = \dfrac{2}{2s + 1}$

(d) $G_p(s) = \dfrac{e^{-0.5s}}{(s + 1)(3s + 1)}$ $G_d(s) = \dfrac{e^{-s}}{s + 1}$

Which of these dynamic feedforward controllers are *physically unrealizable*, that is, require future values of the disturbance in order to compute the current value of the manipulated variable?

V.18 Consider the jacketed mixing tank of Problem V.13 with temperatures T_1 and T_2 as the two external disturbances. Design a feedforward control system which can reject any changes in disturbances T_1 or T_2 and can track any set point changes in the temperature of the mixture T. The inlet temperature of the heating medium, $T_{h,i}$, is the manipulated variable. Assume that $G_m = G_f = 1$. Draw the corresponding block diagram with all relevant transfer functions.

V.19 Repeat Problem V.18 but expand the feedforward control system so that it can reject changes in the feed flow rates F_1 and F_2. Draw the block diagram of the expanded system with all relevant transfer functions.

V.20 In Example 6.4 we developed the linearized model of a nonisothermal CSTR. Develop a nonlinear steady-state feedforward controller which maintains the value of c_A at the desired set point in the presence of changes in c_{A_i}, T_i. The coolant temperature T_c is the manipulated variable.

V.21 Derive the nonlinear, steady-state feedforward control system that will keep the exit temperature of a stirred tank heater at the desired set point despite any changes in the inlet temperature or flow rate, T_i and F_i. The feedforward control system should be capable of (1) rejecting the effect of disturbance changes, and (2) tracking any set point changes. Identify all relevant transfer functions.

V.22 The following transfer functions have been identified for a distillation column:

$$\frac{\overline{x}_D}{F} = \frac{10e^{-10s}}{10s + 1} \qquad \frac{\overline{x}_D}{x_F} = \frac{25e^{-5s}}{5s + 1} \qquad \frac{x_D}{R} = \frac{5e^{-t_d s}}{\tau s + 1}$$

where F and x_F are the feed flow rate and feed composition (the two main disturbances); x_D is the composition of the distillate product (controlled variable) and R is the reflux flow (manipulated variable).
(a) Design a feedforward control system that can reject any changes in the feed flow rate and composition.
(b) What values of t_d and τ yield physically realizable feedforward controllers (i.e., controllers that do not require future values of the

disturbances to compute the current value of the manipulated variable)?

(c) Assuming that $t_d = 0.5$ min and $\tau = 1$ min, compute the response of x_D to a unit step in F using the feedforward controller designed in part (a). Sketch the time behavior of x_D.

(d) Do the same with a 0.1 step change in x_F.

V.23 Suppose that the transfer functions $\overline{x}_D/\overline{F}$ and $\overline{x}_D/\overline{x}_F$ for the distillation column of Problem V.22 are not precisely known. Consequently, the feedforward controllers designed in part (a) of Problem V.22 give imperfect control with a certain amount of residual difference between x_D and the desired set point. Assume that $t_d = 0.5$ min, $\tau = 1$ min, and the following new expressions for the actual transfer functions:

$$\frac{\overline{x}_D}{\overline{F}} = \frac{7e^{-12s}}{8s + 1} \quad \text{and} \quad \frac{\overline{x}_D}{\overline{x}_F} = \frac{20e^{-4s}}{3s + 1}$$

(a) Compute the response of x_D to a unit step change in F and a 0.1 step change in x_F, using the feedforward controllers designed in part (a) of Problem V.22. Sketch the time behavior of these two responses and compare them with the corresponding responses sketched in parts (c) and (d) of Problem V.22. Discuss the effect of imperfect feedforward control.

(b) Introduce a compensatory feedback loop to the feedforward control system. Compute and sketch the resulting closed-loop response of x_D to a unit step change in F using the combined feedforward–feedback control system. Discuss the improvement in the x_D response due to the feedback control action.

V.24 Consider a process with the transfer functions

$$\frac{\overline{y}(s)}{\overline{m}(s)} = G_p(s) = \frac{K_p}{\tau s + 1} \quad \text{and} \quad \frac{\overline{y}(s)}{\overline{d}(s)} = G_d(s) = \frac{K_d}{\tau s + 1}$$

(a) Design a feedforward control system with (1) disturbance rejection and (2) set point tracking capabilities, assuming the following values:

$$K_p = 10 \qquad K_d = 5 \qquad \tau = 2$$

(b) Compute and sketch the response of the controlled output y to a unit step change in the disturbance d using the feedforward controller designed in part (a).

(c) Suppose that the actual value of K_p is 15 instead of 10 which was assumed in part (a) for the design of feedforward controllers. Compute and sketch the response of y to a unit step in disturbance d. Use the value $K_p = 15$ for the process and employ the feedforward system designed in part (a) with $K_p = 10$. Discuss the effect of modeling error on the quality of the controlled response, by comparing the response computed with $K_p = 15$ to that computed in part (b), where $K_p = 10$.

(d) Introduce a feedback loop with a PI controller to compensate for any modeling errors. Let $K_c = 1$ and $\tau_I = 1$. Compute the response of y to a unit step in disturbance d, using the combined feedforward–feedback control system. Again $K_p = 15$ for the process. Discuss the effect of feedback trimming on the closed-loop response.

(e) Repeat parts (c) and (d) but assume that τ is approximately known. Its actual value is 1 instead of the assumed 2.

V.25 (a) Develop a feedforward–feedback control system for a process with the following transfer functions;

$$G_p(s) = \frac{\overline{y}(s)}{\overline{m}(s)} = \frac{s+1}{(s+2)(2s+3)} \qquad G_d(s) = \frac{\overline{y}(s)}{\overline{d}(s)} = \frac{5}{s+2}$$

The following specifications are also given: (1) use a PI controller for the feedback loop, and (2) the feedforward system should have both disturbance rejection and set point tracking capabilities

(b) Show how you would tune the feedback PI controller.

(c) Derive the conditions that must be satisfied in order to have stable closed-loop response. Do these conditions depend on both feedforward and feedback characteristics?

V.26 (a) Develop a ratio control system to regulate the temperature T of the resulting mixture of Figure PV.5. The following specifications are provided: (1) Stream 1 is the "wild" stream. Its temperature remains constant but its flow rate changes in an unpredictable manner. (2) The temperature of stream 2 remains constant.

(b) Draw the corresponding block diagram and place all relevant transfer functions. Assume that the thermocouples have first-order dynamics with $K_m = 0.9$ and $\tau_m = 0.10$, while the control valves are constant gain elements with $K_f = 0.5$.

V.27 Suppose that our control objective is to regulate the flow rate of the resulting mixture in the mixing process of Figure PV.5.

(a) Develop a ratio control system if stream 1 is the "wild" stream.

(b) Draw the corresponding block diagram and place all relevant transfer functions if all the measuring devices and control valves are constant-gain elements with gains K_m and K_f, respectively.

(c) Modify the ratio control system so that it is cascaded with a local flow loop which controls the flow of stream 2 against any disturbances in the flow of this stream.

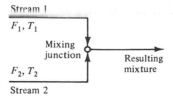

Figure PV.5

Chapter 22

V.28 (a) Show how would you construct an inferential control system for a nonisothermal CSTR using temperature measurements to regulate the exit concentration at a desired value. The reaction is exothermic and the reacting mixture is cooled with water which flows through a jacket around the reactor.

(b) Suppose that the overall heat transfer coefficient between the cooling water and the reacting mixture drops with time due to fouling. Construct an adaptive scheme which uses intermittently exit composition measurements to adjust the parameters of the inferential controller.

(c) Identify the transfer functions that must be evaluated to design the foregoing adaptive and inferential controllers.

V.29 (a) Construct an inferential control system that uses temperature measurements along the length of a tubular catalytic reactor in order to control the exit concentration. The reaction is exothermic and is cooled by a coolant (see Figure PII.9).

(b) Since the catalyst activity decays, add an adaptation mechanism to the inferential control system. The adaptive system will adjust the parameters of the inferential controller using as information exit composition measurements which are taken periodically with a gas chromatograph.

(c) What information on the dynamics of the catalytic reactor do you need in order to design the foregoing inferential and adaptive control schemes?

V.30 Consider the drying of solids with warm air. Since it is very difficult to measure the moisture of solids directly, we should use an inferential control scheme. Construct such an inferential controller which uses temperature measurements of the inlet and outlet airstreams. Draw the resulting block diagram. The following information is given:

(a) The moisture of the solids can be computed from the following relationship derived by Shinskey [Ref. 3]:

$$x = \frac{x_c G c_p}{H_v \gamma A} \ln \frac{T_i - T_w}{T_o - T_w}$$

where T_i and T_o are the inlet and outlet dry-bulb temperatures of the warm air, T_w the wet-bulb temperature, G and c_p the airflow and heat capacity, H_v the latent heat of water, γ the mass transfer coefficient, and A the surface of solids. x_c is the critical moisture content which determines the lower end of the constant-rate drying. We can assume that $x_c G c_p / H_v \gamma A$ remains constant.

(b) There exist diagrams that relate T_i to T_o for various values of the ratio $(T_i - T_w)/(T_o - T_w)$.

V.31 Construct an adaptive control system that can adjust the parameters of a PI controller used to regulate the exit temperature of a stirred tank.

The adaptive controller is needed to compensate for the decrease of the overall heat transfer coefficient due to fouling.

V.32 (a) Construct an inferential control system to regulate the distillate composition of a distillation column when the following transfer functions are provided:

$$\frac{\bar{x}_D}{\bar{F}} = \frac{10e^{-10s}}{10s+1} \qquad \frac{\bar{x}_D}{\bar{x}_F} = \frac{25e^{5s}}{5s+1}$$

$$\frac{\bar{x}_D}{\bar{R}} = \frac{5e^{-0.5s}}{s+1} \qquad \frac{\bar{x}_D}{\bar{T}_1} = \frac{0.05}{0.5s+1}$$

x_F and F are the composition and flow rate of the feed, x_D the distillate composition, R the reflux flow, and T_1 the temperature at the top tray. Assume that concentrations cannot be measured and that R is the manipulated variable.

(b) Draw the resulting block diagram and place all relevant transfer functions.

(c) Since the parameters of the transfer functions above may change, augment the inferential control system with an adaptation mechanism. This mechanism can use intermittent measurements of x_D and adjust accordingly the parameters of the inferential controller.

Design of Control Systems for Multivariable Processes: Introduction to Plant Control

VI

Operational research people have taught us to see the control objective in relation to the overall economy of the company. This means that the control of an individual unit is influenced by the transport of material to and from other units or plants and by the distribution and marketing of the product.

*J. E. Rijnsdorp**

The control configurations we have examined so far were confined to processes with a single controlled output, requiring a single manipulated input. Such single-input, single-output (SISO) systems are very simple and not the typical processing units encountered by a chemical engineer. Chemical processes usually have two or more controlled outputs, requiring two or more manipulated variables. The design of control systems for such multiple-input, multiple-output (MIMO)

*"Chemical Process Systems and Automatic Control," *Chem. Eng. Prog., 63*(7), 97 (1967).

processes will be the subject of the three chapters in Part VI. In particular:

1. Chapter 23 will discuss the new questions that must be answered for the controller design of MIMO systems. It will also present a methodology for the development of alternative control configurations for such systems, based on their degrees of freedom.
2. In Chapter 24 we will examine the selection of the appropriate measurements and manipulations in order to "close the loops." Furthermore, we will study the design of "decoupled loops" for MIMO systems.
3. Finally, in Chapter 25 we will present an introduction to the design of control systems for complete plants, which constitute the most complex MIMO systems to be encountered by a chemical engineer.

Synthesis of Alternative 23
Control Configurations
for Multiple-Input,
Multiple-Output
Processes

The presence of multiple controlled outputs and multiple manipulated inputs creates a situation that we have not confronted so far: *There are more than one possible control configurations for a MIMO process.* In this chapter we develop a concise methodology for the development of all feasible control systems for (1) single processing units and (2) processes composed of more than one interacting units.

23.1 Design Questions for MIMO Control Systems

Consider a general process with several inputs and outputs (Figure 2.1). Several questions must be answered before we attempt to design a control system for such a process.

1. *What are the control objectives?* In other words, how many and which of all possible variables should be controlled at desired values? This seemingly simple question is critical for the design of efficient control systems.
2. *What outputs should be measured?* Once the control objectives have been identified we need to select the measurements necessary to monitor the operation of the process. We can classify the measured outputs into two categories:
   (a) *Primary measurements:* These are the controlled outputs

through which we can determine directly if the control objectives are satisfied.

(b) *Secondary measurements:* These are not used to monitor directly the control objectives but are auxiliary measurements employed for cascade, adaptive, or inferential control (see Figures 20.2, 22.1, and 22.6).

3. *What inputs can be measured?* We assume that all the manipulated variables are measurable and therefore can be employed for adaptive (model reference or self-tuning regulator) and inferential control (see Figures 22.1 and 22.6). With respect to the disturbances only a few can be measured easily, rapidly, and reliably. These measurable disturbances can be used to construct feedforward (Figure 21.2), feedforward–feedback (Figure 21.7), and ratio control configurations (Figure 21.9).

4. *What manipulated variables should be used?* A multiple-input, multiple-output system possesses several manipulated variables which can be used for the design of a control system. The selection of the most appropriate manipulations is a very critical problem and should be approached with care. Some manipulations have a direct, fast, and strong effect on the controlled outputs; others do not. Furthermore, some variables are easy to manipulate in real life (e.g., liquid flows); others are not (e.g., flow of solids, slurries, etc.).

5. *What is the configuration of the control loops?* Once all the possible measurements and manipulations have been identified, we need to decide how they are going to be interconnected through the control loops. In other words, what measurement will actuate a given manipulated variable or what manipulation will be used to regulate a given controlled output at its desired value?

For MIMO systems there is a large number of alternative control configurations. The selection of the most appropriate is the central and critical question to be resolved.

Let us now examine the foregoing design questions in more detail and develop systematic approaches to answer them.

23.2   Degrees of Freedom and the Number of Controlled and Manipulated Variables

We have defined the degrees of freedom for a given process (see Section 5.2) as *the independent variables that must be specified in order to define the process completely.* The number of degrees of freedom was also found to be given by the equation

$$f = V - E \qquad (23.1)$$

where V = number of independent variables describing a process
$\quad\quad E$ = number of independent equations physically relating the V variables

It is clear that in order to have a completely determined process the number of its degrees of freedom should be zero. There are two sources that provide the additional equations needed to reduce the number of degrees of freedom to zero.

1. *The external world*, which specifies the values of certain input variables. By external world we mean everything outside the process, such as:

   The general surroundings influencing the operating conditions
   An up-stream unit that feeds the process
   A down-stream unit when the outflow of the process is a manipulated inflow for the down-stream unit

2. *The control system*, which imposes certain relationships between the controlled outputs and the manipulated inputs (feedback) or between the measured disturbances and the manipulated inputs (feedforward). Thus we can state easily that:

   The maximum number of independent controlled variables in a processing system is equal to the number of degrees of freedom minus the number of externally specified variables:

   (number of controlled variables)
   $= f -$ (number of externally specified inputs) $\qquad (23.2)$

   This relationship was used in Examples 5.7 and 5.8 to determine the number of controlled outputs in a binary distillation and a mixing process, respectively.

Having determined the number of independent controlled outputs, the following question arises: *How many independent manipulated inputs do we need in order to keep the controlled outputs at their desired values (set points)?* To answer this question, let us consider a process with the following specifications:

N controlled outputs $(y_1, y_2, \cdots, y_N)$
M independent manipulations $(m_1, m_2, \cdots, m_M)$ *with* $M \geq N$
L disturbances externally specified $(d_1, d_2, \cdots, d_L)$

Let the following N equations represent the relationships between the controlled outputs, the manipulations, and the disturbances,

$$y_1 = f_1(m_1, m_2, \cdots, m_M; d_1, d_2, \cdots, d_L)$$
$$y_2 = f_2(m_1, m_2, \cdots, m_M; d_1, d_2, \cdots, d_L)$$

(23.3)

$$\vdots$$

$$y_N = f_N(m_1, m_2, \cdots, m_M; d_1, d_2, \cdots, d_L)$$

As the values of the disturbances change (specified by the external world), the values of the controlled outputs must remain the same. This is possible if N of the M manipulated variables are free to change so as to satisfy the system of eq. (23.3). Therefore:

For the design of a control system the number of required independent manipulated variables is equal to the number of independent controlled variables.

(number of independent manipulated variables)
$$= \text{(number of controlled variables)} \qquad (23.4)$$
$$= f - \text{(number of externally specified inputs)}$$

Remarks

1. Let k be the number of controlled variables given by eq. (23.2). Then if the actually controlled variables are fewer than k, say $l < k$, there are $(k - l)$ process variables which change "wildly" in an uncontrolled manner and may cause problems to the operation of the process. But if the effects of these "uncontrolled" variables on the operation of the process are acceptable, it is perfectly legitimate to have fewer controlled variables than the number dictated by eq. (23.2).
2. It is impossible to design a control system that can regulate more controlled variables than the number given by eq. (23.2).
3. *The degrees of freedom of a process at dynamic state are equal in number or more than those at steady state.* This is due to the fact that the dynamic balance equations contain the accumulation terms, whereas for steady-state balances the accumulation is zero. An incorrect estimate of the number of degrees of freedom can have a profound effect on the design of the appropriate controller. Consider the simple liquid holding tank of Example 10.1. The dynamic mass balance yields

$$A \frac{dh}{dt} = F_i - F_o$$

Here we have *three independent variables* (h, F_i, F_o) and *one equation*. The cross-sectional area A is a parameter with given value. Therefore, we have *two degrees of freedom*. Since F_i is specified by the external world, we can have *only one controlled variable*. This suggests the conventional feedback loop between h and F_o. Had we examined the steady-state balance, where $dh/dt = 0$, we would have concluded (erroneously) that there is only one degree of freedom and consequently no controlled output.

4. Recall Examples 20.6 and 20.7 on split-range control. Notice that the number of manipulated variables used for control is larger than the number of controlled outputs. Therefore, *eq. (23.4) determines the minimum number of required manipulations.*

Example 23.1: Determining the Number of Controlled and Manipulated Variables for a Flash Drum

Consider the flash drum shown in Figure 23.1a. The feed is composed of N components with molar fractions z_i, $i = 1, 2, \cdots, N$. As the liquid

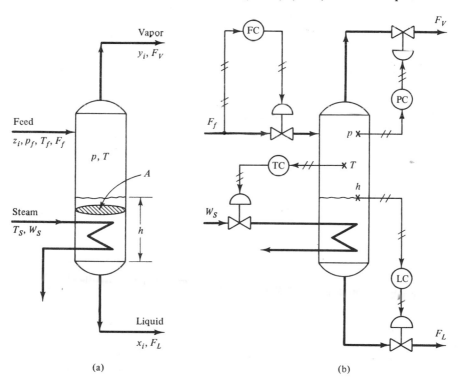

(a)   (b)

Figure 23.1 (a) Flash drum unit; (b) best configuration of control loops.

feed is "flashed" from the high pressure p_f to the lower pressure p of the drum, vapor is produced and reaches equilibrium with the remaining liquid. Steam flowing through a coil supplies the necessary heat for maintaining the desired temperature in the drum, despite any variations in the operating conditions. For this process we would like to (1) identify the controlled variables, (2) identify the manipulated inputs, and (3) generate all feasible loop configurations.

Let us first determine the degrees of freedom for the flash drum. The modeling equations are:

Total mass balance (assuming constant molar density and insignificant vapor holdup):

$$A\rho \frac{dh}{dt} = F_f - (F_V + F_L)$$

Component balances:

$$A\rho \frac{d(hx_i)}{dt} = F_f z_i - (F_V y_i + F_L x_i) \qquad i = 1, 2, \cdots, N-1$$

Heat balance:

$$c_{p,L} A \frac{d(hT)}{dt} = c_{p,f} F_f T_f - (c_{p,V} F_V T + c_{p,L} F_L T) + UA_S(T_S - T)$$

Vapor–liquid equilibrium relationships:

$$y_i = K_i(T, p) x_i \qquad i = 1, 2, \cdots, N$$

Consistency constraints:

$$\sum_{i=1}^{N} x_i = 1 \qquad and \qquad \sum_{i=1}^{N} y_i = 1$$

All the relationships above constitute a system of $2N + 3$ equations with $4N + 14$ variables. These variables are classified as follows:

Constants $(N + 7)$: A, A_S, ρ, U, $c_{p,f}$, $c_{p,V}$, $c_{p,L}$, and $K_i(T, p)$ for $i = 1, 2, 3, \cdots, N$

Externally specified (N): T_f, and z_i for $i = 1, 2, \cdots, N-1$

Unspecified $(2N + 7)$: F_f, F_V, F_L, p, T, h, T_S (or W_S), and x_i, y_i for $i = 1, 2, \cdots, N$

Therefore, the number of controlled variables is equal to

$$(2N + 7) - (2N + 3) = 4$$

But *which four of the $(2N + 7)$ unspecified variables will be selected as controlled outputs?* The operating requirements dictate that T and p should be kept constant to achieve the desired separation. Furthermore, for constant production the flow rate of the liquid feed should be maintained at the desired value. Finally, the liquid level should remain within certain bounds. Thus *T, p, F_f, and h are the controlled variables.* All four controlled variables can be measured directly, using simple and reliable sensors (thermocouples, differential pressure cells, etc.) with fast responses. Therefore, *the measured variables for the control system are T, p, F_f, and h.* From the set of $(2N + 7)$ unspecified variables we can select the required four manipulated variables. Clearly, these are F_f, F_V, F_L, and W_S.

23.3   Generation of Alternative Loop Configurations

After the identification of the controlled and manipulated variables we need to determine the control configuration (i.e., specify the manipulated variable that will control a given controlled variable). In other words, determine the configuration of the control loops.

For a system with N controlled and N manipulated variables there are $N!$ different loop configurations. Figure 23.2 shows the two possible loop configurations for a process with two manipulations and two controlled outputs. As the number N increases, the number of different loop configurations increases very rapidly; for example,

for $N = 3$ there are $3! = 6$ different configurations

for $N = 4$ there are $4! = 24$ different configurations

for $N = 5$ there are $5! = 120$ different configurations

The selection of the "best" among all possible loop configurations is a difficult problem. Various criteria can be used to select the "best"

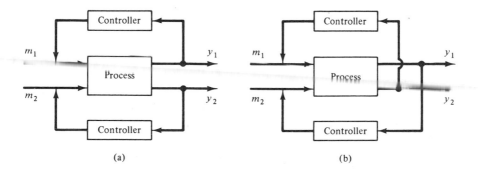

(a)   (b)

Figure 23.2 Alternative loop configurations for a 2 × 2 process.

couplings among the controlled and manipulated variables, such as:

1. Choose the manipulation that has a direct and fast effect on a controlled variable.
2. Choose the couplings so that there is a little dead time between every manipulation and the corresponding controlled variable.
3. Select the couplings so that the interaction of the control loops is minimal, etc.

In subsequent chapters we will develop more precise quantitative criteria for the selection of the loops.

Example 23.2: Alternative Loop Configurations for the Flash Drum

In Example 23.1 we identified the controlled and manipulated variables for the flash drum. But how are these interconnected to form the control loops? Table 23.1 shows the 24(= 4!) possible loop configurations

TABLE 23.1.
ALTERNATIVE LOOP CONFIGURATIONS FOR THE FLASH DRUM OF EXAMPLE 23.1

Configuration number	F_f Control by:	p Control by:	T Control by:	h Control by:
1	F_f	F_L	W_S	F_V
2	F_f	F_L	F_V	W_S
3	F_f	F_V	W_S	F_L
4	F_f	F_V	F_L	W_S
5	F_f	W_S	F_L	F_V
6	F_f	W_S	F_V	F_L
7	F_L	F_f	W_S	F_V
8	F_L	F_f	F_V	W_S
9	F_L	F_V	W_S	F_f
10	F_L	F_V	F_f	W_S
11	F_L	W_S	F_f	F_V
12	F_L	W_S	F_V	F_f
13	F_V	F_f	F_L	W_S
14	F_V	F_f	W_S	F_L
15	F_V	W_S	F_f	F_L
16	F_V	W_S	F_L	F_f
17	F_V	F_L	F_f	W_S
18	F_V	F_L	W_S	F_f
19	W_S	F_f	F_L	F_V
20	W_S	F_f	F_V	F_L
21	W_S	F_L	F_f	F_V
22	W_S	F_L	F_V	F_f
23	W_S	F_V	F_f	F_L
24	W_S	F_V	F_L	F_f

resulting from all possible combinations among the controlled and manipulated variables. The "best" among the 24 can be found using the following qualitative arguments:

1. The effects of F_f, F_V, and F_L on the temperature T are indirect and rather slow, while that of W_S is direct and faster. Therefore, from the 24 loop configurations of Table 23.1, only numbers 1, 3, 7, 9, 14, and 18 look promising for efficient temperature control.
2. The effects of W_S and F_L on the pressure p are also indirect and slow. Therefore F_f and F_V are better manipulated variables for controlling p and from the previously selected loop configurations only numbers 3, 7, 9, and 14 remain valid candidates.
3. Among numbers 3, 7, 9, and 14, the loop configuration 3 seems to be the best because it uses F_L to achieve fast level control and manipulates F_f directly. Furthermore, it uses F_V for direct and fast regulation of the pressure p. This loop configuration is shown in Figure 23.1b.

Remarks

1. To select the most promising control configuration for the flash drum we employed qualitative arguments. In subsequent sections we will study quantitative techniques for selecting the optimal coupling between controlled and manipulated variables.

2. It should be emphasized that the four loops of the control configuration in Figure 23.1b interact with each other. Thus, increasing the steam flow rate to control the temperature will affect and thus decontrol the pressure. The interaction among the control loops is an important design consideration. In Chapter 24 we will examine the *relative gain array* method which determines how the manipulated variables should be coupled with the controlled variables in such a way as to minimize the interaction among the resulting loops.

23.4   Extensions to Systems with Interacting Units

In Sections 23.2 and 23.3 we studied the determination of the necessary controlled and manipulated variables, as well as the generation of all feasible loop configurations for single processing units. In the present section we extend these results to systems composed of several interacting processing units, since such are the systems encountered in a chemical plant.

Consider a process composed of N units which interact with each other through material or energy flows. To determine all feasible control configurations for the overall process, we can adopt the following systematic procedure:

Step 1: Divide the process into separate blocks. Every block may contain a single processing unit or a small number of processing units with an inherent, common operational goal. For example, the block containing a distillation column should also contain the condenser and reboiler attached to the column; two neighboring heat exchangers in series or in parallel should be contained in the same block; a reactor and its feed preheater could be in the same block; and so on.

Step 2: Determine the degrees of freedom and the number of con-

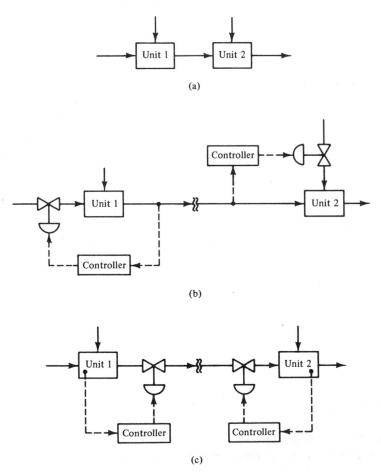

Figure 23.3 (a) Process with two units in series; (b), (c) incompatible control loop configurations.

trolled and manipulated variables for each block. To do this, follow the procedure described in Section 23.2.

Step 3: Determine all feasible loop configurations for each block. Having specified the controlled and manipulated variables for each block, it is easy to generate all possible configurations, following the approach described in Section 23.3. Using qualitative or quantitative arguments, retain a small number of the "best" loop configurations for each block.

Step 4: Recombine the blocks with their loop configurations. It is clear that the number of the generated loop configurations for the overall process is equal to the product of the retained configurations for all blocks.

Step 5: Eliminate conflicts among the control systems of the various blocks. The control configurations resulting in step 4 usually lead to an overspecification of the overall controlled process. This can be explained as follows. Consider two units connected by a common flow (Figure 23.3a). When we design the loops for each unit separately, it is possible to select the interconnecting flow as a controlled variable for both units but in different loops (Figure 23.3b). Also, it is possible to have the common interconnecting flow as the manipulated variable in two different control loops (Figure 23.3c). Both situations correspond to overspecified systems and lead to conflicts among the control loops. Such conflicts must be erased before we can have a feasible control configuration for the overall process.

Let us now demonstrate this procedure on two specific processing systems composed of several interacting units.

Example 23.3: Generate the Control Loop Configuration for a Simple Chemical Process

The heart of the process shown in Figure 23.4 is the continuous stirred tank reactor (CSTR) where the simple exothermic reaction, A → B, takes place. The reactor feed is preheated, first by the hot reactor effluent and then by steam. Coolant, flowing through a jacket around the reactor, removes the heat generated by the reaction in order to maintain the temperature of the reacting mixture at the maximum allowable (for highest conversion). The coolant is provided with two branches, one of which is cooled while the other is heated. The rates of cooling and heating (i.e., Q_c and Q_h) are constant. With this configuration we can fine-tune the temperature of the coolant (increase or decrease it) before it enters the jacket of CSTR. The reactor effluent is first cooled by the feed in the feed-effluent heat exchanger and subsequently it is "flashed" in a flash drum.

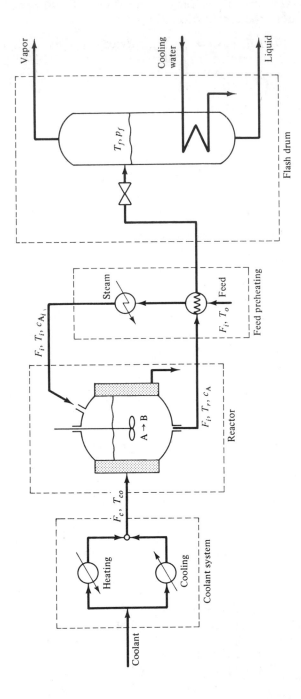

Figure 23.4 Chemical plant of Example 23.3.

There, it is separated into two streams, a vapor and a liquid, which are further processed in separate units. Cooling water is provided to regulate the temperature in the drum. We would like to develop alternative loop configurations for this process which satisfy the following operating objectives:

1. Keep the conversion in the reactor at its highest permissible value.
2. Maintain a constant production rate.
3. Achieve constant composition in the liquid product of the flash drum.

Step 1: Divide the process into four blocks (Figure 23.4): coolant system, feed preheating, reactor, and flash drum with its cooler.

Steps 2 and 3: Determine the degrees of freedom as well as the controlled and manipulated variables for each block. Also, generate all possible loop configurations for each block and retain the "best."

Coolant system (Figure 23.5a). Table 23.2 summarizes all the characteris-

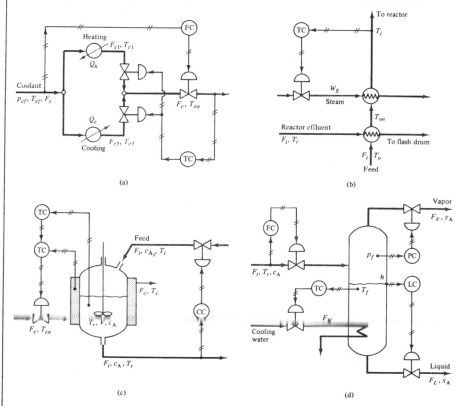

Figure 23.5 Control loops for individual segments of plant in Example 23.3: (a) coolant system; (b) feed preheating; (c) reactor; (d) flash drum.

TABLE 23.2.
CONTROL SYSTEM CHARACTERISTICS FOR THE COOLANT SYSTEM OF EXAMPLE 23.3

Total number of variables
  (excluding constant parameters): 8 $[p_{cf}, T_{cf}, F_{c1}, T_{c1}, F_{c2}, T_{c2}, F_c, T_{co}]$
Externally specified variables: 2 p_{cf} = coolant feed pressure
                                     T_{cf} = coolant feed temperature

Unspecified variables: 6
Number of modeling equations: 4 Heat balance on cooling branch
                                  Heat balance on heating branch
                                  Heat balance on the mixing junction
                                    of the two branches
                                  Mass balance on mixing junction
Number of controlled and
  manipulated variables: 2

| | *Loop Configurations* | |
Configuration	F_c Control by:	T_{co} Control by:
1	F_c	F_{c1} and F_{c2} (split-range control)
2	$F_{c1} + F_{c2}$	F_{c1}/F_{c2}
3	F_{c2}	F_{c1}
4	F_{c1}	F_{c2}
5	$F_{c1} + F_{c2}$	F_{c1}
6	$F_{c1} + F_{c2}$	F_{c2}
etc.		

tics of the coolant system. There are two controlled variables requiring two manipulations. From the operating requirements we can easily determine that:

F_c and T_{co} are the two controlled variables.

F_c and T_{co} are also the two measured variables.

The two manipulated variables can be selected from the set

$$F_c, \quad F_{c1}, \quad F_{c2}, \quad F_{c1} + F_{c2}, \quad F_{c1}/F_{c2}$$

Table 23.2 also indicates a few of the possible loop configurations. Configuration 1 seems to be the simplest and is selected for the control of the coolant system (Figure 23.5a).

Feed preheating system (Figure 23.5b). This block requires one controlled variable (see Table 23.3), which is the temperature T_i. The only available manipulated variable is the steam flow rate W_S, thus yielding only one loop configuration (Figure 23.5b).

TABLE 23.3.
FEED PREHEATING CHARACTERISTICS OF EXAMPLE 23.3

Total number of variables (excluding constant parameters):	6	$[W_S, T_o, T_i, T_r, T_{int}, F_i]$
Externally specified:	3	$[T_o, T_r, F_i]$
Unspecified variables:	3	
Number of modeling equations:	2	Heat balance on steam heater Heat balance on feed-effluent heat exchanger
Number of controlled and manipulated variables:	1	

TABLE 23.4.
CONTROL SYSTEM CHARACTERISTICS FOR THE REACTOR OF EXAMPLE 23.3

Total number of variables (excluding constant parameters):	9	$[V, T_r, c_A, c_{A_i}, T_i, F_i, F_c, T_c, T_{co}]$
Externally specified:	4	$[F_i, c_{A_i}, T_i, T_{co} \text{ (or } F_c)]$
Unspecified variables:	5	
Number of modeling equations:	3	Component A balance around reactor Energy balance on reacting mixture Energy balance on the coolant in the jacket
Number of controlled and manipulated variables:	2	

	Loop Configurations	
Configuration	c_A Control by:	T_r Control by:
1	F_i	F_c (or T_{co})
2	F_c (or T_{co})	F_i
3	F_i	F_c, with T_c as a secondary measurement in a cascade configuration

Reactor (Figure 23.5c). Table 23.4 shows that there should be two controlled variables for the reactor, which are easily identified as the temperature T_r and the concentration c_i of the reactor effluent stream. Available manipulations are

$$F_i, \quad T_{co} \text{ (or } F_c)$$

Table 23.4 shows the three possible loop configurations. Number 3 corresponds to cascade temperature control and, as we have seen in Section 20.1 and Example 20.1, it provides fast compensation. Thus configuration 3 is selected for the reactor.

Flash drum (Figure 23.5d). This is similar to the flash drum system analyzed in Examples 23.1 and 23.2 with one difference; instead of steam heating (see Figure 23.1a) there is a water cooling system (Figure 23.5d). Therefore, following the same procedure as in Example 23.1 we conclude that there should be

Four controlled variables [F_i, p_f, T_f, h]
Four manipulated variables [F_i, F_V, F_L, F_W]

We can generate 24 possible loop configurations, similar to those tabulated in Table 23.1. The configuration shown in Figure 23.5d is selected as the "best" because it provides direct and fast regulation of all controlled variables.

Step 4: Recombine the four blocks with their control configurations. Considering that the four blocks (coolant system, feed preheating, reactor, flash drum) possess 6, 1, 3, and 24 possible loop configurations, we can generate in principle 432 ($= 6 \times 1 \times 3 \times 24$) control configurations for the overall process. Not all of them need to be examined for consistency because some are obviously bad. Figure 23.6a shows the resulting control system if the "best" loop configurations are selected for each block.

Step 5: Eliminate conflicts among the control loops of the various blocks. Consider the control system for the overall process shown in Figure 23.6a. We notice quickly two overspecifications which create conflicts among the control loops.
   (a) The coolant flow rate is used as a manipulated variable by two different loops; the temperature cascade loop of the reactor and the feedforward flow control loop of the coolant system.
   (b) The flow rate F_i (feed to the reactor and reactor effluent) is controlled by two different loops: the feedback concentration control loop in the reactor and the feedforward flow control loop in the flash drum.
To eliminate conflict (a) we can delete the feedforward flow control loop in the coolant system. To erase conflict (b) we delete the flow control loop in the flash drum. Thus the final control configuration for the overall process is shown in Figure 23.6b. It has no conflicts among the loops and the process is exactly specified.

Example 23.4: Generate the Control Loop Configuration for an Integrated Chemical Plant

Consider the process shown in Figure 23.7. An exothermic reaction A + B → C takes place in the gas phase. The product C is taken from the top of a distillation column. The unreacted raw materials A and B are both recycled to the reactor from the flash drum and the bottom of the distillation column, respectively. Compressors (C-1, C-2) are used to increase the pressure of the feed and recycled gas A. The liquid B is vaporized in a series of two heat exchanger (E-1, E-2). The reactor is a

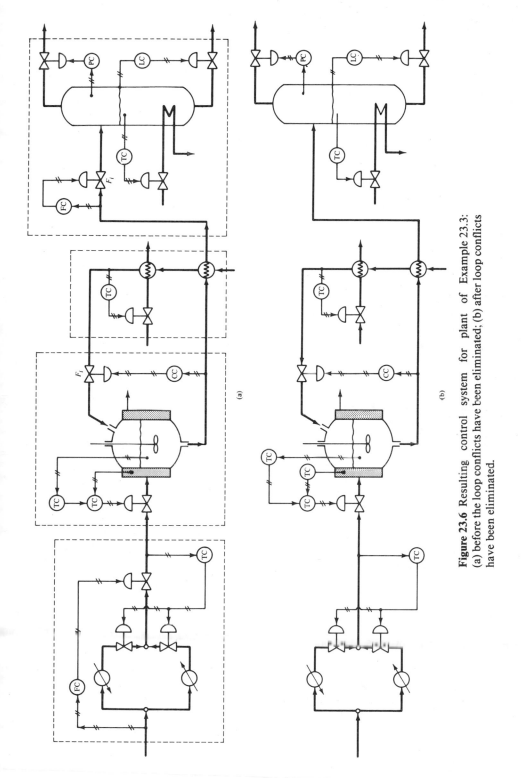

Figure 23.6 Resulting control system for plant of Example 23.3: (a) before the loop conflicts have been eliminated; (b) after loop conflicts have been eliminated.

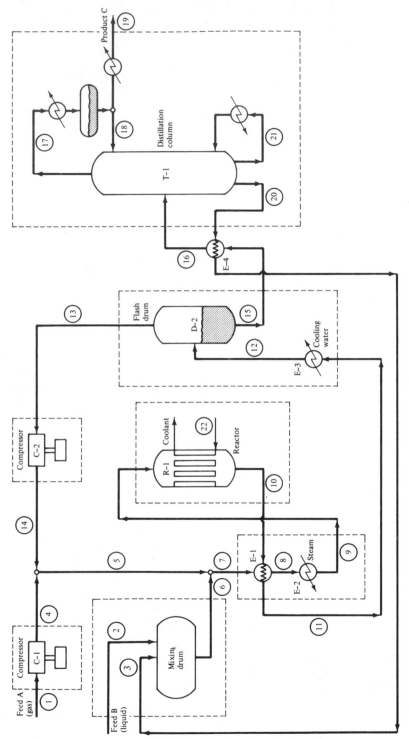

Figure 23.7 Chemical plant of Example 23.4. Reproduced from Umeda, T. et al., "A Logical Structure for Process Control System Synthesis," in *A Link Between Science and Applications of Automatic Control*, A. Niemi (editor), by permission.

tube-and-shell heat exchanger with the reaction taking place in the tubes and the coolant flowing in the shell around the tubes.

The basic control objectives are: maintain the desired steady-state production rate and quality of product C for a long period.

Step 1: Divide the process to seven blocks as shown in Figure 23.7:
 (a) Compressor for the fresh feed gas A
 (b) Compressor for gas A recycled from the flash drum
 (c) Mixing drum for the fresh feed B and the recycled from the bottom of the distillation column
 (d) Feed vaporizing and preheating
 (e) Reactor
 (f) Flash drum with its feed cooler
 (g) Distillation column with its condenser and reboiler

Steps 2 and 3: To simplify the presentation of this example, we have

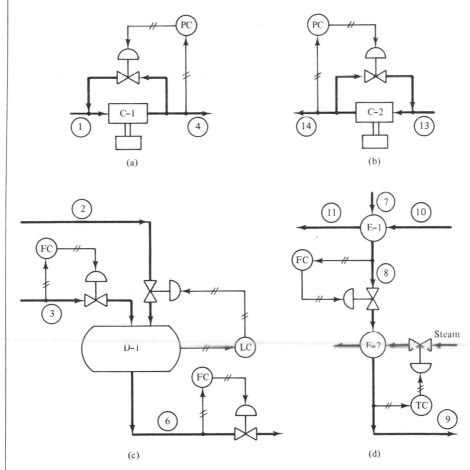

(a)   (b)

(c)   (d)

Figure 23.8 Control loops for individual segments of plant in Example 23.4.

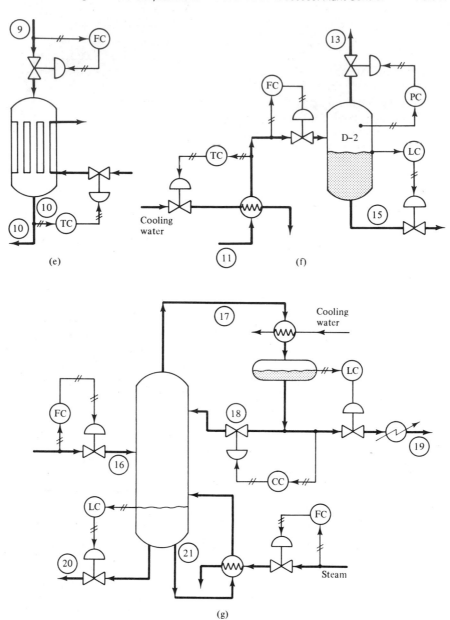

Figure 23.8 (Continued)

omitted the details of modeling, accounting for variables, determination of controlled and manipulated variables, and so on. Figure 23.8a through g shows the selected "best" loop configurations for each block.

Step 4: Figure 23.9 presents the control structure for the overall plant resulting from the particular loop configurations shown in Figure 23.8.

Step 5: Close observation of the control configurations in Figure 23.9 reveals the following conflicts among the various control loops:

    (a) The flow rate of the recycle stream B from the bottom of the distillation column is controlled by two loops: the feedback level control at the bottom of the distillation column (stream 20) and the feedforward flow control in the mixing drum (stream 3). To erase this conflict, eliminate the feedforward loop in the mixing drum because the level controller is absolutely necessary for good operation of the distillation column.

    (b) The flow rate of the reactor feed is controlled by three loops: the flow control loop in the mixing drum (stream 6), the flow control loop in the feed preheating block (stream 8), and the flow control loop in the reactor block (stream 9). Eliminate the conflict by retaining the flow control on stream 6 only.

    (c) Delete the flow control on the feed of the flash drum (stream 11) because its flow is determined by the flow rate of the reactor feed and consequently by the flow of stream 9.

    (d) The flow rate of the feed to the distillation column is controlled by two loops: the level control of the flash drum (stream 15) and the flow control on stream 16. Retain only the second loop and eliminate the first.

Step 6: Improve the control configuration generated in step 5. After the elimination of the four conflicts among the control loops, which we described above, we can make two additional modifications which improve the quality of the resulting control.

    (a) The pressure control of both gaseous streams 1 and 13 may be excessive. Since the pressure of stream 5 is the one of practical importance, we can replace the two pressure control loops by one, which measures the pressure of stream 5 and manipulates the bypass flow around compressor C-1.

    (b) For the pressure control in the flash drum we use the flow rate of the vapor (stream 13) as the manipulated. But the variations in stream 13 are fed back to the main process and may cause additional disruptions in the operation. For better pressure control introduce a purge stream (stream 23) and manipulate its flow into. Figure 23.10 shows the final configuration of the control loops after eliminating any conflicts (step 5) and making the two modifications described in step 6.

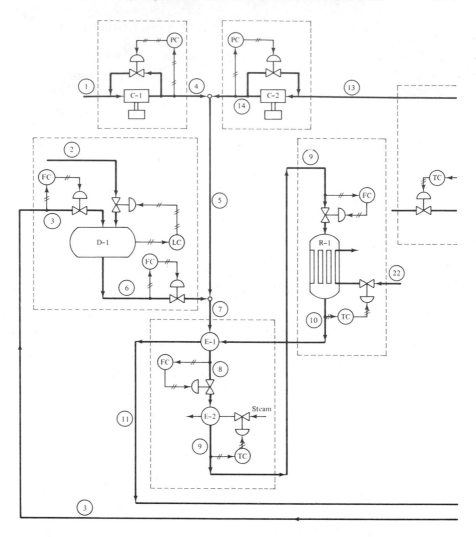

Figure 23.9 Control loops for plant of Example 23.4, before loop conflicts have been eliminated.

THINGS TO THINK ABOUT

1. What is a MIMO process, and in what sense is the design of a control system for a MIMO process different from that for a SISO process?

2. Discuss the design questions related to a MIMO control system.

3. Why do we assume that all manipulated variables are measurable? Is this

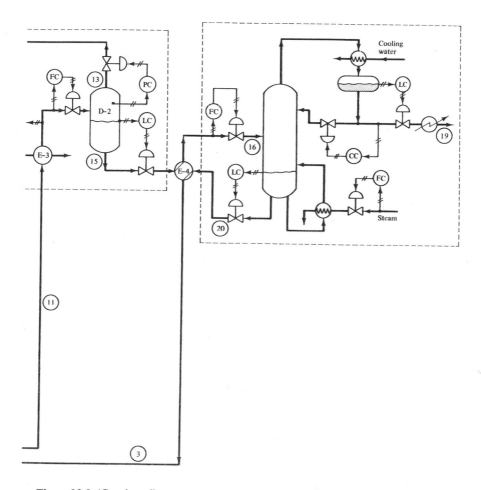

Figure 23.9 (Continued)

assumption correct? How would you use the values of manipulated variables in a control system?

4. Why do we claim that there are a large number of control configurations for a MIMO process? Find the number of alternative loop configurations for a process with N controlled variables and M manipulations, where $M > N$.

5. "Prove" eq. (23.2) and (23.4), which determine the number of necessary controlled and manipulated variables.

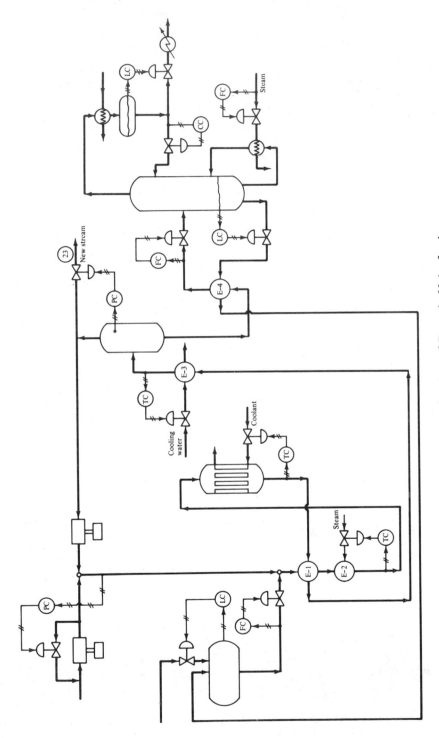

Figure 23.10 Control loops for plant of Example 23.4, after loop conflicts have been eliminated.

6. Construct a physical example where we can have fewer controlled variables than dictated by eq. (23.2) and the operation of the process is acceptable.

7. Equation (23.4) determines the minimum number of required manipulations for a process. Why is it minimum? Could you use more manipulations, and how? Construct a physical example with more manipulations than those dictated by eq. (23.4).

8. Is it sufficient to control the temperature and pressure in a flash drum in order to have vapor and liquid products of desired composition? Elaborate on your answer and explain why yes or no.

9. Consider the stirred tank heater example. Show that for the control of temperature and level there are two distinct loop configurations. One of them is unacceptable because it renders an uncontrollable system. Which one is this configuration, and why is it unacceptable? (*Hint*: Consider the effects of manipulations on the controlled variables.)

10. Extend the observation made for the uncontrollability of the stirred tank heater to other systems. State qualitatively a test for rejecting loop configurations leading to uncontrollable systems.

11. Assuming that all variables can be measured, how many measurements do you need for the design of a control system with N controlled variables?

12. Consider the process examined in Example 23.3. Are there more degrees of freedom when the processing units are considered together in an integrated whole or when the various units are considered separately, detached from each other? Explain.

13. Why is the overall process overspecified when the various blocks with their corresponding loop configuration are recombined to yield the control configuration for the overall process? (See steps 4 and 5 in Section 23.4). How does the overspecification manifest itself in the configuration of control loops? Discuss how it can be eliminated.

14. Determine the number of controlled and manipulated variables for the flash drum (Example 23.1) assuming steady-state operation. Why are the results different from those of Example 23.1? State the danger involved when we consider steady-state models to design a MIMO control system.

15. What controlled variables remain unidentified when we use steady-state models to determine controlled and manipulated variables? How can you overcome this drawback and still use steady-state models?

Interaction **24**
and Decoupling
of Control Loops

From the discussion in Chapter 23, two characteristics should be clear concerning the design of control systems for processes with multiple inputs and multiple outputs:

1. A control system is composed of several interacting control loops.
2. The number of feasible, alternative configurations of control loops is very large.

For example, to control the operation of a flash drum we need a config-

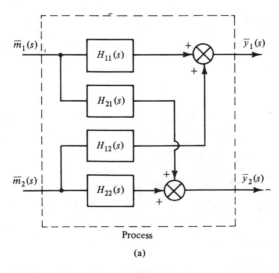

Process

(a)

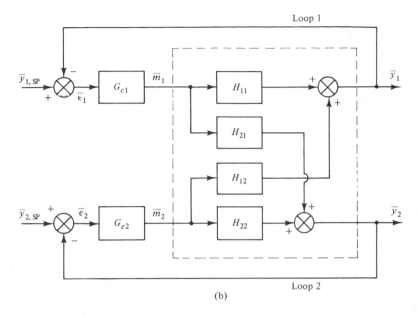

Figure 24.1 Block diagram of process with two controlled outputs and two manipulations: (a) open-loop; (b) closed-loop.

uration of four loops which must be selected from 24 possible such loop configurations (Example 23.2).

These two characteristics dictate the content of this chapter. In particular, we will study;

1. The interaction among the control loops of a MIMO process
2. The relative-gain array method, which determines how the controlled and manipulated variables should be coupled to yield control loops with minimal interaction
3. The design of special control systems with noninteracting loops

24.1   Interaction of Control Loops

Consider a process with two controlled outputs and two manipulated inputs (Figure 24.1a). The input–output relationships are given by

$$\overline{y}_1(s) = H_{11}(s)\overline{m}_1(s) + H_{12}(s)\overline{m}_2(s) \qquad (24.1)$$

$$\overline{y}_2(s) = H_{21}(s)\overline{m}_1(s) + H_{22}(s)\overline{m}_2(s) \qquad (24.2)$$

$H_{11}(s)$, $H_{12}(s)$, $H_{21}(s)$, and $H_{22}(s)$ are the four transfer functions relating the two outputs to the two inputs (see Section 9.2). Equations (24.1) and (24.2) indicate that a change in m_1 or m_2 will affect both controlled outputs.

Let us form two control loops by coupling m_1 with y_1 and m_2 with y_2 as shown in Figure 24.1b. To simplify the presentation, we have assumed that the transfer functions of the measuring devices and final control elements in both loops are equal to 1. If $G_{c1}(s)$ and $G_{c2}(s)$ are the transfer functions of the two controllers, the values of the manipulations are given by

$$\overline{m}_1(s) = G_{c1}[\overline{y}_{1,SP}(s) - \overline{y}_1(s)] \qquad (24.3)$$

$$\overline{m}_2(s) = G_{c2}[\overline{y}_{2,SP}(s) - \overline{y}_2(s)] \qquad (24.4)$$

To understand the nature of interaction between two control loops and how it arises, we will study the effects of input changes on the outputs when (1) one loop is closed and the other is open, and (2) both loops are closed.

1. *One loop closed*: Assume that loop 1 is closed and loop 2 is open (Figure 24.2a). Also assume that $m_2 =$ constant [i.e., $\overline{m}_2(s) = 0$],

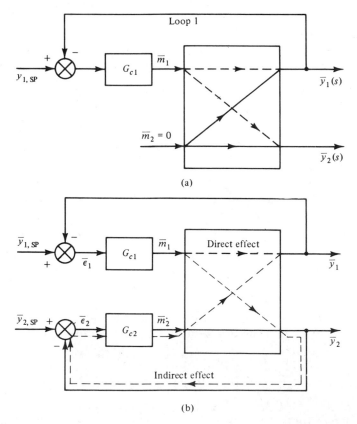

(a)

(b)

Figure 24.2 Interactions among control loops: (a) one loop closed; (b) both loops closed.

and make a change in the set point $y_{1,SP}$. After substituting eq. (24.3) into eqs. (24.1) and (24.2), we take

$$\bar{y}_1 = \frac{H_{11}G_{c1}}{1 + H_{11}G_{c1}} \bar{y}_{1,SP} \tag{24.5}$$

$$\bar{y}_2 = \frac{H_{21}G_{c1}}{1 + H_{11}G_{c1}} \bar{y}_{1,SP} \tag{24.6}$$

It is clear, then, that *any change in the set point $y_{1,SP}$ will affect not only the behavior of the controlled output y_1, but also the uncontrolled output y_2.* The dashed lines in Figure 24.2a indicate schematically the routes through which $y_{1,SP}$ affects the two outputs. Similar conclusions are drawn if we consider loop 1 open and loop 2 closed. The situation becomes more complex when both loops are closed.

2. *Both loops closed* (Figure 24.2b): Initially, the process is at steady state with both outputs at their desired values. Consider a change in the set point $y_{1,SP}$ only, and keep the set point of loop 2 the same (i.e., $\bar{y}_{2,SP} = 0$). Then the following things happen:
   (a) The controller of loop 1 will change the value of m_1 in such a way as to bring the output y_1 to the new set point value. This is the *direct effect* of m_1 on y_1 through loop 1, and is shown schematically by the dashed line in Figure 24.2b.
   (b) The control action of m_1 will not only attempt to bring y_1 to the new set point, but will also disturb y_2 from its steady-state value. Then the controller of loop 2 attempts to compensate for the variations in y_2 by changing appropriately the value of the manipulated variable m_2. But a change in m_2, in turn, affects output y_1. This is an *indirect effect* of m_1 on y_1, through loop 2, and is shown schematically by the dashed line in Figure 24.2b. It constitutes the essence of interaction between two control loops.

An analogous conclusion as to the loop interaction will be drawn if we consider a change in $y_{2,SP}$ while keeping $y_{1,SP}$ the same. In summary, we can make the following statement, which describes the interaction between two control loops:

The regulatory action of a control loop deregulates the output of another loop (in the same process), which in turn takes control action to compensate for the variations in its controlled output, disturbing at the same time the output of the first loop.

Having completed a qualitative presentation, let us now examine the quantitative ramifications of the interaction between two control loops. Substitute eqs. (24.3) and (24.4) into eqs. (24.1) and (24.2) and take

$$(1 + H_{11}G_{c1})\bar{y}_1 + (H_{12}G_{c2})\bar{y}_2 = H_{11}G_{c1}\bar{y}_{1,SP} + H_{12}G_{c2}\bar{y}_{2,SP} \quad (24.7)$$

$$(H_{21}G_{c1})\bar{y}_1 + (1 + H_{22}G_{c2})\bar{y}_2 = H_{21}G_{c1}\bar{y}_{1,SP} + H_{22}G_{c2}\bar{y}_{2,SP} \quad (24.8)$$

Solve eqs. (24.7) and (24.8) with respect to the controlled outputs y_1 and y_2 and take the following *closed-loop input–output relationships*:

$$\bar{y}_1 = P_{11}(s)\bar{y}_{1,SP} + P_{12}(s)\bar{y}_{2,SP} \quad (24.9)$$

$$\bar{y}_2 = P_{21}(s)\bar{y}_{1,SP} + P_{22}(s)\bar{y}_{2,SP} \quad (24.10)$$

where

$$P_{11}(s) = \frac{H_{11}G_{c1} + G_{c1}G_{c2}(H_{11}H_{22} - H_{12}H_{21})}{Q(s)}$$

$$P_{12}(s) = \frac{H_{12}G_{c2}}{Q(s)}$$

$$P_{21}(s) = \frac{H_{21}G_{c1}}{Q(s)}$$

$$P_{22}(s) = \frac{H_{22}G_{c2} + G_{c1}G_{c2}(H_{11}H_{22} - H_{12}H_{21})}{Q(s)}$$

and

$$Q(s) = (1 + H_{11}G_{c1})(1 + H_{22}G_{c2}) - H_{12}H_{21}G_{c1}G_{c2} \quad (24.11)$$

Remarks

1. Equations (24.9) and (24.10) describe the response of outputs y_1 and y_2 when both loops are closed (i.e., they have accounted for the interaction between the two loops).
2. When $H_{12} = H_{21} = 0$, there is no interaction between the two control loops. The closed-loop outputs are given by the following equations:

$$\bar{y}_1 = \frac{H_{11}G_{c1}}{1 + H_{11}G_{c1}}\bar{y}_{1,SP} \qquad \bar{y}_2 = \frac{H_{22}G_{c2}}{1 + H_{22}G_{c2}}\bar{y}_{2,SP}$$

The closed-loop stability of the two noninteracting loops depends on the roots of their characteristic equations. Thus if the roots of the two equations

$$1 + H_{11}G_{c1} = 0, \qquad 1 + H_{22}G_{c2} = 0 \quad (24.12)$$

have negative real parts, the two noninteracting loops are stable.

3. The stability of the closed-loop outputs of two interacting loops is determined by the roots of the characteristic equation

$$Q(s) \equiv (1 + H_{11}G_{c1})(1 + H_{22}G_{c2}) - H_{12}H_{21}G_{c1}G_{c2} = 0 \quad (24.13)$$

Thus if the roots of eq. (24.13) have negative real parts, the two interacting loops are stable.

4. Suppose that the two feedback controllers G_{c1} and G_{c2} are tuned separately (i.e., keeping the loop under tuning closed, and the other open). Then *we cannot guarantee stability for the overall control system, where both loops are closed*. The reason is simple: Tuning each loop separately, we force the roots of the characteristic equations (24.12) for the individual loops to acquire negative real parts. But the roots of these equations are different from the roots of the characteristic equation (24.13), which determines the stability of the overall system with both loops closed.

5. Normally, we tune the two controllers in such a way that the roots of all eqs. (24.12) and (24.13) have negative real parts. Such tuning guarantees stability when both loops are closed [roots of eq. (24.13)], or only one is closed while the other is open [roots of eqs. (24.12)], due to a hardware failure.

6. The previous discussion indicates that the interaction between control loops is a significant factor and affects in a very profound manner the "goodness" of a control system. For this reason, a control designer attempts to couple the manipulated variables with the controlled outputs in such a way as to minimize the interaction of the resulting control loops. If strong interactions persist for any possible pairing, he or she will design a special control system which eliminates the interaction (decoupling the loops).

Example 24.1: Interaction of Control Loops in a Stirred Tank Heater

Consider once more the stirred tank heater (Example 4.4). Figure 24.3 shows the two control loops: loop 1 controls the liquid level by manipulating the effluent flow rate, and loop 2 regulates the temperature by manipulating the steam flow rate. Let us see how the two loops interact:

1. When the inlet flow rate (load) or the desired value of liquid level (set point) change, loop 1 attempts to compensate for the changes by manipulating the value of the effluent flow rate. This, in turn, will disturb the temperature of the liquid in the tank and loop 2 will compensate by adjusting appropriately the value of the steam flow rate.

2. If, on the other hand, the temperature of the inlet stream (load) or the desired value of the temperature (set point) change, loop 2 will

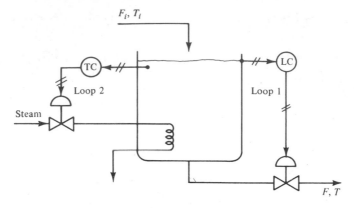

Figure 24.3 Control loops of a stirred tank heater.

adjust the steam flow rate to compensate for the changes. This will leave the liquid level undisturbed.

Thus we notice that *loop 1 affects loop 2, but not vice versa*. In other words, the interaction is in a single direction.

Example 24.2: Interaction of Control Loops in a Stirred Tank Reactor

In the CSTR of Figure 24.4, the temperature is controlled by the flow of coolant in the jacket while the effluent concentration is controlled by the inlet flow rate. Assume that initially both effluent concentration and temperature are at their desired values.

1. Consider a change in the inlet concentration (load) or the desired effluent concentration (set point). Loop 1 will compensate for these changes by manipulating the feed flow rate. However, this change in the feed rate also disturbs the reactor temperature away from the

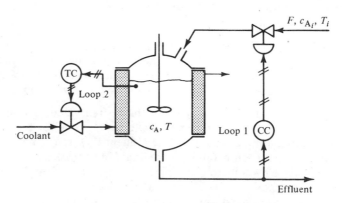

Figure 24.4 Control loops of a CSTR.

desired value. Then loop 2 attempts to compensate for the change in temperature by varying the coolant flow rate, which in turn affects the effluent concentration.

2. On the other hand, an attempt to compensate for changes in feed temperature (load) or the desired set point of reactor temperature, it also causes the effluent concentration to vary. Then loop 1 attempts to compensate for the change in effluent concentration by varying the feed rate, which in turn disturbs the reactor temperature.

It is clear from the above that *loop 1 interacts with loop 2 in both directions* (unlike the loops of the stirred tank heater, which interact in a single direction).

Example 24.3: Tuning the Controllers of Two Interacting Loops

Assume that the input–output relationships of a process with two controlled outputs and two manipulated inputs are given by

$$\bar{y}_1 = \frac{1}{0.1s + 1}\, \bar{m}_1 + \frac{5}{s + 1}\, \bar{m}_2$$

$$\bar{y}_2 = \frac{1}{0.5s + 1}\, \bar{m}_1 + \frac{2}{0.4s + 1}\, \bar{m}_2$$

Form two loops by coupling m_1 with y_1 and m_2 with y_2. The closed-loop input–output relationships are given by eqs. (24.9) and (24.10), where

$$H_{11} = \frac{1}{0.1s + 1} \qquad H_{12} = \frac{5}{s + 1} \qquad H_{21} = \frac{1}{0.5s + 1} \qquad H_{22} = \frac{2}{0.4s + 1}$$

Let the two controllers be simple proportional controllers with

$$G_{c1} = K_{c1} \qquad \text{and} \qquad G_{c2} = K_{c2}$$

1. *Tuning each loop separately*: The characteristic equation of loop 1 (when loop 2 is open) is given by

$$1 + H_{11}G_{c1} = 1 + \frac{K_{c1}}{0.1s + 1} = 0$$

and yields the closed-loop pole

$$s = -10(1 + K_{c1}) < 0$$

Therefore, *when loop 2 is open, loop 1 is stable for any value of gain K_{c1}*. Similarly, the closed-loop pole for loop 2 when loop 1 is open is given by

$$s = -2.5(1 + 2K_{c2}) < 0$$

and consequently, *loop 2 is stable for any value of K_{c2} when loop 1 is open*.

2. *Tuning with both loops closed*: When both loops are closed, the

characteristic equation is given by eq. (24.13) and for this example takes the following form:

$$\left(1 + \frac{K_{c1}}{0.1s + 1}\right)\left(1 + \frac{2K_{c2}}{0.4s + 1}\right) - \frac{5}{s + 1}\frac{1}{0.5s + 1}K_{c1}K_{c2} = 0$$

or

$$0.02s^4 + 0.1(3.1 + 2K_{c1} + K_{c2})s^3 + (1.29 + 1.3K_{c2} + K_{c1}$$
$$+ 0.8K_{c1}K_{c2})s^2$$
$$+ (2 + 3.2K_{c2} + 1.9K_{c1} + 0.5K_{c1}K_{c2})s$$
$$+ (1 + 2K_{c2} + K_{c1} - 3K_{c1}K_{c2}) = 0 \tag{24.14}$$

According to the first test of the Routh–Hurwitz criterion for stability (see Section 15.3), eq. (24.14) has at least one root with positive real part if any of its coefficients are negative. Thus the closed-loop behavior of the process is unstable if K_{c1} and K_{c2} take on such values that make the last term of eq. (24.14) negative [all other terms in eq. (24.14) are always positive]:

$$1 + 2K_{c2} + K_{c1} - 3K_{c1}K_{c2} < 0 \tag{24.15}$$

Inequality (24.15) places restrictions on the values that K_{c1} and K_{c2} can take, to render a stable performance when both loops are closed. This is in direct contrast to our earlier result [see (1) above], whereby all values of K_{c1} and K_{c2} were acceptable if each loop were tuned separately.

Note.   The allowable range of values for K_{c1} and K_{c2}, which render stable responses when both loops are closed, can be found by applying the second test of the Routh–Hurwitz criterion. The reader is encouraged to complete this example and find the range of values of K_{c1} and K_{c2}, which render stable responses.

24.2   Relative-Gain Array and the Selection of Loops

In Chapter 23 we recognized that for a process with N controlled outputs and N manipulated variables there are $N!$ different ways to form the control loops. Which one is best? One way to answer this question is to consider the interactions among the loops for all $N!$ loop configurations and select the one where the interactions are minimal. The *relative-gain array* provides exactly such a methodology, whereby we select pairs of input and output variables in order to minimize the amount of interaction among the resulting loops. It was first proposed by Bristol and today is a very popular tool for the selection of control loops. Let us now study the logic of the method and present some examples describing its usage.

Definitions

Consider a process with two outputs and two inputs (Figure 24.1a). Then do the following two experiments:

1. Assume that m_2 remains constant (Figure 24.5a). Introduce a step change in the input m_1 of magnitude Δm_1 and record the new steady-state value of output y_1. Let Δy_1 be the change from the previous steady state. It is clear that it has been caused *only* by the change in m_1. The *open-loop static gain between y_1 and m_1 when m_2 is kept constant* is given by [see Section 10.4 and eq. (10.20)]

$$\left(\frac{\Delta y_1}{\Delta m_1}\right)_{m_2}$$

2. In addition to the static gain computed above, there is another open-loop gain between y_1 and m_1, when m_2 varies by a feedback loop controlling the other output, y_2 (Figure 24.5b). Thus, introducing a step change Δm_1 we record a change $\Delta y_1'$ in the steady-state value of y_1. In general, $\Delta y_1'$ will be different from Δy_1 for the following reason: The input change Δm_1 affects not only y_1 but also y_2. Then the control loop attempts to keep y_2 constant by varying m_2, which in turn affects the steady-state value of y_1. Therefore, $\Delta y_1'$ is the compound result of the effects from m_1 and

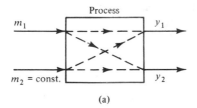

(a)

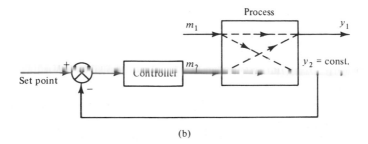

(b)

Figure 24.5 Schematic of two experiments in order to compute relative gains.

m_2. Let the new open-loop gain between y_1 and m_1, when y_2 is kept constant by the control loop, be given by

$$\left(\frac{\Delta y'_1}{\Delta m_1}\right)_{y_2}$$

The ratio of the two open-loop gains computed above defines the *relative gain*, λ_{11}, between output y_1 and input m_1:

$$\lambda_{11} = \frac{(\Delta y_1/\Delta m_1)_{m_2}}{(\Delta y_1/\Delta m_1)_{y_2}} \tag{24.16}$$

The relative gain provides a useful measure of interaction. In particular:

1. If $\lambda_{11} = 0$, then y_1 does not respond to m_1 and m_1 should not be used to control y_1.
2. If $\lambda_{11} = 1$, then m_2 *does not affect* y_1 and the control loop between y_1 and m_1 *does not interact* with the loop of y_2 and m_2. In this case we have *completely decoupled loops*.
3. If $0 < \lambda_{11} < 1$, then an *interaction exists* and as m_2 varies it affects the steady-state value of y_1. The smaller the value of λ_{11}, the larger the interaction becomes.
4. If $\lambda_{11} < 0$, then m_2 causes a strong effect on y_1 and in the opposite direction from that caused by m_1. In this case, the interaction effect is very dangerous.

In a similar manner as above, we can define the following three remaining relative gains between the two inputs and two outputs of the process we have been considering:

$$\lambda_{12} = \frac{(\Delta y_1/\Delta m_2)_{m_1}}{(\Delta y_1/\Delta m_2)_{y_2}} \qquad \text{relative gain between } y_1 \text{ and } m_2$$

$$\lambda_{21} = \frac{(\Delta y_2/\Delta m_1)_{m_2}}{(\Delta y_2/\Delta m_1)_{y_1}} \qquad \text{relative gain between } y_2 \text{ and } m_1$$

$$\lambda_{22} = \frac{(\Delta y_2/\Delta m_2)_{m_1}}{(\Delta y_2/\Delta m_2)_{y_1}} \qquad \text{relative gain between } y_2 \text{ and } m_2$$

The values of these gains can also be used as measures of interaction for the corresponding cases, in a similar way as it was done above for λ_{11}.

Selection of loops

For a process with two inputs and two outputs there are two different loop configurations, shown in Figure 23.2. Let us see how we can use the relative gains to select the configuration with minimum interaction between the loops.

Arrange the four relative gains λ_{11}, λ_{12}, λ_{21}, and λ_{22} into a matrix form, which is known as the *relative-gain array*:

$$\begin{array}{cc} m_1 & m_2 \end{array}$$
$$\Lambda = \begin{bmatrix} \lambda_{11} & \lambda_{12} \\ \lambda_{21} & \lambda_{22} \end{bmatrix} \begin{array}{c} y_1 \\ y_2 \end{array}$$

It can be shown that the sum of the relative gains in any row or column of the array is equal to 1. Thus

$$\begin{array}{ccc} \lambda_{11} + \lambda_{12} = 1 & & \lambda_{11} + \lambda_{21} = 1 \\ & \text{and} & & \text{(24.17)} \\ \lambda_{21} + \lambda_{22} = 1 & & \lambda_{12} + \lambda_{22} = 1 \end{array}$$

Therefore, we need to know only one of the four relative gains, while the other three can be easily computed. For example, if $\lambda_{11} = 0.75$, then $\lambda_{12} = \lambda_{21} = 0.25$ and $\lambda_{22} = 0.75$.

Depending on the value of λ_{11}, we can distinguish the following different situations:

$\lambda_{11} = 1$. Then the relative-gain array is

$$\Lambda = \begin{bmatrix} 1 & 0 \\ 0 & 1 \end{bmatrix}$$

and it is obvious that we can have two noninteracting loops: m_1 coupled with y_1 and m_2 coupled with y_2 (Figure 23.2a).

$\lambda_{11} = 0$. Then the relative-gain array is given by

$$\Lambda = \begin{bmatrix} 0 & 1 \\ 1 & 0 \end{bmatrix}$$

The 1's in the off-diagonal elements indicate that we can form two noninteracting control loops by coupling m_1 with y_2 and m_2 with y_1 (Figure 23.2b).

$\lambda_{11} = 0.5$. Then

$$\Lambda = \begin{bmatrix} 0.5 & 0.5 \\ 0.5 & 0.5 \end{bmatrix}$$

and the amount of interaction between the two loops is the same in both configurations of Figure 23.2. In other words, it does not matter how we couple inputs and outputs; the degree of interaction remains the same.

$0 < \lambda_{11} < 0.5$, say $\lambda_{11} = 0.25$. Then

$$\Lambda = \begin{bmatrix} 0.25 & 0.75 \\ 0.75 & 0.25 \end{bmatrix}$$

The two larger numbers (i.e., 0.75) indicate the recommended coupling with the smaller amount of interaction. Thus we couple m_1 with y_2 and m_2 with y_1 (Figure 23.2b).

$0.5 < \lambda_{11} < 1$, say $\lambda_{11} = 0.8$. Then

$$\Lambda = \begin{bmatrix} 0.8 & 0.2 \\ 0.2 & 0.8 \end{bmatrix}$$

and the recommended coupling is the opposite of the previous case: couple m_1 with y_1 and m_2 with y_2 (Figure 23.2a).

$\lambda_{11} > 1$. Then $\lambda_{22} = \lambda_{11} > 1$ and $\lambda_{12} = \lambda_{21} = 1 - \lambda_{11} < 0$. Situations with relative gains outside the range 0 to 1 create difficult control problems. Let us see why.

1. Suppose that you couple y_1 with m_1 and y_2 with m_2. The corresponding relative gains, λ_{11} and λ_{22}, are larger than 1. Then from the definition of the relative gains, we conclude that

$$\left(\frac{\Delta y_1}{\Delta m_1}\right)_{m_2} > \left(\frac{\Delta y_1}{\Delta m_1}\right)_{y_2} \quad \text{and} \quad \left(\frac{\Delta y_2}{\Delta m_2}\right)_{m_1} > \left(\frac{\Delta y_2}{\Delta m_2}\right)_{y_1}$$

   In other words, the response of the outputs is *held back* by the interaction from the other loop, and the larger the values of the relative gains above unity, the larger the "holding back" effect will be. Thus we need larger values for the controller gains.

2. If you couple y_1 with m_2 and y_2 with m_1, the corresponding gains λ_{12} and λ_{21} are negative. In this case, the interaction will take the controlled outputs in the opposite direction from that desired by the control effort and control will be lost altogether. Therefore, *never form loops by coupling inputs to outputs with negative relative gains*.

We can summarize all the foregoing observations with the following rule for selecting the control loops:

Select the control loops by pairing the controlled outputs y_i with the manipulated variables m_j in such a way that the relative gains λ_{ij} are positive and as close as possible to unity.

Remarks

1. The relative gains provide a measure of interaction based on steady-state considerations. Therefore, the rule given above for the selection of loops does not guarantee that the dynamic interaction between the loops will be also minimal.

2. The relative-gain array is a square matrix, which implies that the number of manipulated variables is equal to the number of controlled outputs. Now, suppose that we have a process with two outputs and three possible manipulations, m_1, m_2, and m_3. There are three possible pairs of manipulated variables: (m_1, m_2), (m_2, m_3), and (m_3, m_1). Therefore, we can form three different relative-gain arrays:

$$
\begin{array}{ccc}
\quad m_1 \quad m_2 & \quad m_2 \quad m_3 & \quad m_3 \quad m_1 \\[4pt]
\Lambda_1 = \begin{bmatrix} \lambda_{11} & \lambda_{12} \\ \lambda_{21} & \lambda_{22} \end{bmatrix} \begin{matrix} y_1 \\ y_2 \end{matrix} &
\Lambda_2 = \begin{bmatrix} \lambda'_{12} & \lambda_{13} \\ \lambda'_{22} & \lambda_{23} \end{bmatrix} \begin{matrix} y_1 \\ y_2 \end{matrix} &
\Lambda_3 = \begin{bmatrix} \lambda'_{13} & \lambda'_{11} \\ \lambda'_{23} & \lambda'_{21} \end{bmatrix} \begin{matrix} y_1 \\ y_2 \end{matrix}
\end{array}
$$

and we need to examine all of them before we can select the set of two loops with minimal interaction. (*Note*: In general, $\lambda'_{11} \neq \lambda_{11}$, $\lambda'_{12} \neq \lambda_{12}$, etc.)

3. There are two ways of obtaining the relative gains of a process: a computational approach using a steady state input–output model for the process and an experimental approach. When a steady-state model is available, we can obtain the numerator and denominator of the relative gain [see eq. (24.16)] in a very simple manner. This way we can express the relative gains in terms of the controlled and manipulated variables themselves, which enables us to evaluate the interaction across a range of operating conditions.

4. For an existing process we can evaluate the relative gains experimentally by performing the following two experiments:

(a) *Experiment 1 (all loops open)*: Keeping all loops open, make a small step change Δm_1 in m_1, keeping m_2 constant. Record the changes in the steady-state values of y_1 and y_2 (i.e., Δy_1 and Δy_2). Then compute

$$
\left(\frac{\Delta y_1}{\Delta m_1} \right)_{m_2} \quad \text{and} \quad \left(\frac{\Delta y_2}{\Delta m_1} \right)_{m_2}
$$

Return the system to the initial steady state and repeat the same experiment by varying m_2 by Δm_2. Record the changes Δy_1 and Δy_2 and compute

$$
\left(\frac{\Delta y_1}{\Delta m_2} \right)_{m_1} \quad \text{and} \quad \left(\frac{\Delta y_2}{\Delta m_2} \right)_{m_1}
$$

(b) *Experiment 2 (one loop closed)*: Make a small change Δm_1 in m_1 while keeping y_2 constant by feedback control through m_2. Record the change Δy_1 in the steady state of y_1 and compute the gain

$$\left(\frac{\Delta y_1}{\Delta m_1}\right)_{y_2}$$

Repeat the same experiment, but now keep y_1 constant through a control loop with m_2. Record the change Δy_2 and compute the gain

$$\left(\frac{\Delta y_2}{\Delta m_1}\right)_{y_1}$$

Similarly, we can compute the following two gains:

$$\left(\frac{\Delta y_1}{\Delta m_2}\right)_{y_2} \quad \text{and} \quad \left(\frac{\Delta y_2}{\Delta m_2}\right)_{y_1}$$

Taking the ratios of the corresponding gains in Experiments 1 and 2, we can compute the relative gains λ_{11}, λ_{12}, λ_{21}, and λ_{22}. [*Note*: Remember that you do not need to compute all relative gains, since they are related by eqs. (24.17).]

5. The definition of the relative gains and their use in selecting the control loops are not limited to systems with two inputs and two outputs. The extension to general processes is straightforward. Thus the relative gain between an output y_i and a manipulation m_j is defined by

$$\lambda_{ij} = \frac{(\Delta y_i/\Delta m_j)_m}{(\Delta y_i/\Delta m_j)_y}$$

The subscript m denotes constant values for all manipulations except m_j (i.e., all loops open), while subscript y indicates that all outputs except y_i are kept constant by the control loops (i.e., all loops are closed). Similarly, the relative-gain array is given by

$$
\Lambda =
\begin{array}{c}
\quad m_1 \quad m_2 \qquad\quad m_N \\
\left[
\begin{array}{cccc}
\lambda_{11} & \lambda_{12} & \cdots & \lambda_{1N} \\
\lambda_{21} & \lambda_{22} & \cdots & \lambda_{2N} \\
\multicolumn{4}{c}{\cdots\cdots\cdots\cdots\cdots} \\
\lambda_{N1} & \lambda_{N2} & \cdots & \lambda_{NN}
\end{array}
\right]
\begin{array}{c}
y_1 \\ y_2 \\ \vdots \\ y_N
\end{array}
\end{array}
$$

The entries of Λ satisfy the following two properties:

$$\sum_{i=1}^{N} \lambda_{ij} = 1 \quad \text{for } j = 1, 2, \ldots, N \qquad \text{summation by columns}$$

$$\sum_{j=1}^{N} \lambda_{ij} = 1 \qquad \text{for } i = 1, 2, \ldots, N \qquad \text{summation by rows}$$

The loop selection rule remains the same.

Example 24.4: Select the Loops Using the Relative-Gain Array

Consider a process with the following input–output relationships:

$$\overline{y}_1 = \frac{1}{s+1}\, \overline{m}_1 + \frac{1}{0.1s+1}\, \overline{m}_2 \tag{24.18}$$

$$\overline{y}_2 = \frac{-0.2}{0.5s+1}\, \overline{m}_1 + \frac{0.8}{s+1}\, \overline{m}_2 \tag{24.19}$$

Let us compute the relative gains:

1. Make a unit step change in m_1 (i.e., $\overline{m}_1 = 1/s$) while keeping m_2 constant (i.e., $\overline{m}_2 = 0$). Then from eq. (24.18) we take

$$\overline{y}_1 = \frac{1}{s+1}\frac{1}{s}$$

Recall the final-value theorem (Section 7.5) and find the resulting new steady state in y_1:

$$y_{1,s} = \lim_{s \to 0}\, [s\overline{y}_1(s)] = \lim_{s \to 0} \left(\frac{1}{s+1} \right) = 1$$

Therefore, $(\Delta y_1/\Delta m_1)_{m_2} = 1/1 = 1$.

2. Keep y_2 constant under control by varying m_2. Introduce a unit step in m_1. Since y_2 must remain constant (i.e., $\overline{y}_2 = 0$), eq. (24.19) will tell us by how much m_2 should change;

$$\overline{m}_2 = \frac{0.2}{0.8}\frac{s+1}{0.5s+1}\, \overline{m}_1$$

Substitute this value in eq. (24.18) and find

$$\overline{y}_1 = \frac{1}{s+1}\, \overline{m}_1 + \frac{1}{0.1s+1}\frac{0.2}{0.8}\frac{s+1}{0.5s+1}\, \overline{m}_1$$

Then, the resulting new steady state for y_1 is given by

$$y_{1,s} = \lim_{s \to 0}\, [s\overline{y}_1] = \lim_{s \to 0} \left[s \left\{ \frac{1}{s+1}\frac{1}{s} + \frac{1}{0.1s+1}\frac{0.2}{0.8}\frac{s+1}{0.5s+1}\frac{1}{s} \right\} \right] = 1.25$$

Therefore, $(\Delta y_1/\Delta m_1)_{y_2} = 1.25/1 = 1.25$ and

$$\lambda_{11} = \frac{(\Delta y_1/\Delta m_1)_{m_2}}{(\Delta y_1/\Delta m_1)_{y_2}} = \frac{1}{1.25} = 0.8$$

Using eqs. (24.17), we find $\lambda_{12} = \lambda_{21} = 0.2$ and $\lambda_{22} = 0.8$. It is easy now to conclude that we should pair m_1 with y_1 and m_2 with y_2 to form two loops with minimum interaction. It should be noted that had we selected the loops differently (i.e., coupled m_1 with y_2 and m_2 with y_1), the interaction of the loops would have been four times larger (i.e., $0.8/0.2 = 4$).

Example 24.5: Selecting the Loops in a Mixing Process

Two streams with flow rates F_1 and F_2 and compositions (mole percent) $x_1 = 80\%$ and $x_2 = 20\%$ in a chemical A are mixed in a vessel (Figure 24.6a). We would like to form two control loops to regulate the product composition x and flow rate F. Let $F \equiv y_1$ and $x \equiv y_2$ be the two controlled outputs, while $F_1 \equiv m_1$ and $F_2 \equiv m_2$ are the two available manipulated variables. There are two possible control configurations with different pairings between the inputs and outputs, and they are shown in Figure 24.6b and c. Which one should we prefer?

The steady-state mass balances yield

$$F = F_1 + F_2 \tag{24.20}$$

$$Fx = F_1 x_1 + F_2 x_2 \tag{24.21}$$

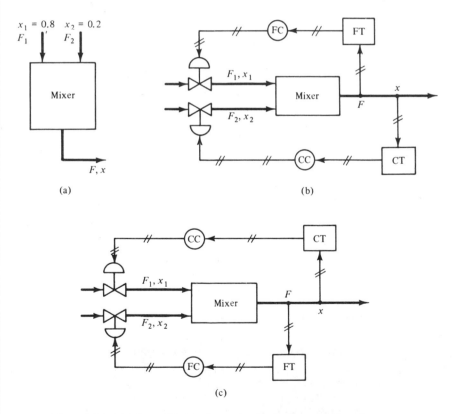

(a)

(b)

(c)

Figure 24.6 (a) Mixer; (b), (c) alternative loop structures for mixer.

(*Note*: We have neglected the energy balance because the temperature of the product stream is not in our operating requirements.) The desired steady state for operational purposes is

$$F = 200 \text{ mol/hr} \quad \text{and} \quad x = 60\% \text{ (by moles)}$$

With these values we find the following steady-state solution of eqs. (24.20) and (24.21):

$$F_1 = 133.4 \quad \text{and} \quad F_2 = 66.6$$

To compute the relative gain between F and F_1, do the following:

1. Change F_1 by one unit (i.e., $F_1 = 134.4$) while holding $F_2 = 66.6$ (the same). Solve eqs. (24.20) and (24.21) for F and x and find the following new steady states:

$$F = 201 \quad \text{and} \quad x = 0.6012$$

   Therefore,

$$\left(\frac{\Delta F}{\Delta F_1}\right)_{F_2} = \frac{1}{1} = 1 \qquad \left(\frac{\Delta x}{\Delta F_1}\right)_{F_2} = \frac{0.0012}{1} = 0.0012$$

2. Change F_1 by one unit (i.e., $F_1 = 134.4$) while holding $x = 60\%$ constant. Solve eqs. (24.20) and (24.21) and find

$$F = 201.67 \quad \text{and} \quad F_2 = 67.27$$

   Therefore,

$$\left(\frac{\Delta F}{\Delta F_1}\right)_x = \frac{1.67}{1} = 1.67$$

Consequently, the relative gain between F and F_1 is

$$\lambda_{11} = \frac{(\Delta F/\Delta F_1)_{F_2}}{(\Delta F/\Delta F_1)_x} = \frac{1}{1.67} = 0.6$$

It follows easily that the complete relative-gain array is

$$\begin{array}{cc} & F_1 \quad F_2 \\ \Lambda = \begin{bmatrix} 0.6 & 0.4 \\ 0.4 & 0.6 \end{bmatrix} & \begin{array}{c} F \\ x \end{array} \end{array}$$

We can draw two main conclusions:

1. The two loops with minimum interaction are formed when we couple F with F_1 and x with F_2 (Figure 24.6b).
2. Although the interaction between the two selected loops is smaller than that of the alternative configuration (Figure 24.6c), it is still significant. Thus any control action to regulate F will seriously disturb x, and vice versa.

24.3   Design of Noninteracting Control Loops

The relative-gain array indicates how the inputs should be coupled with the outputs to form loops with the smaller amount of interaction. But the persisting interaction, although it is the smallest possible, may not be small enough. Example 24.5 demonstrated this aspect clearly. In such a case, the two control loops still affect each other's operation very seriously, and the overall control system is characterized as unacceptable.

When the designer is confronted with two strongly interacting loops, he or she introduces in the control system special new elements called *decouplers*. The purpose of decouplers is to cancel the interaction effects between the two loops and thus render two *noninteracting* control loops. Let us now study how we can design the decouplers for a process with two strongly interacting loops.

Consider the process whose input–output relationships are given by eqs. (24.1) and (24.2). Form two control loops by coupling m_1 with y_1 and m_2 with y_2 (see Figure 24.1b).

Assume that initially both outputs are at the desired set point values. Suppose that a disturbance or set point change causes the controller of loop 2 to vary the value of m_2. This will create an undesired disturbance for loop 1 and will cause y_1 to deviate from its desired value. However, *we could change m_1 by such an amount as to cancel the interaction effect from m_2*. But the question arises: How much should we change m_1?

From eq. (24.1) we find that in order to keep y_1 constant (i.e., $\overline{y}_1 = 0$), m_1 should change by the following amount:

$$\overline{m}_1 = -\frac{H_{12}(s)}{H_{11}(s)}\,\overline{m}_2 \tag{24.22}$$

Equation (24.22) implies that we can introduce a dynamic element with a transfer function

$$D_1(s) = -\frac{H_{12}(s)}{H_{11}(s)} \tag{24.23}$$

which uses the value of m_2 as input and provides as output the amount by which we should change m_1, in order to cancel the effect of m_2 on y_1. This dynamic element is called *decoupler* and when it is installed in the control system (Figure 24.7a) it cancels any effect that loop 2 might have on loop 1, but not vice versa.

To eliminate the interaction from loop 1 to loop 2, we can follow the

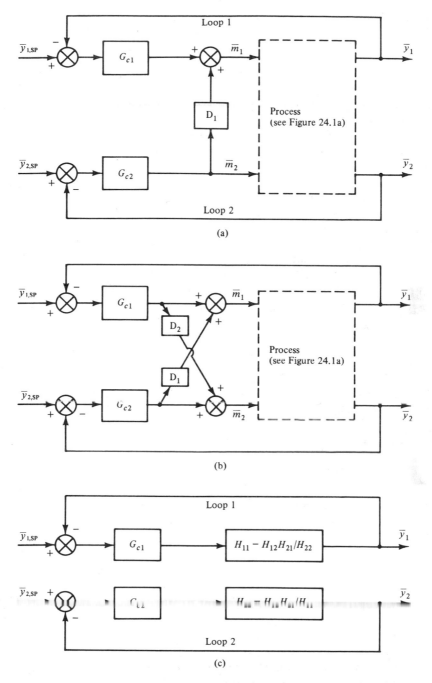

Figure 24.7 (a) A 2 × 2 process with one decoupler; (b) with two decouplers; (c) equivalent block diagram with complete decoupling.

same reasoning as above and we find that the transfer function of the second decoupler is given by

$$D_2(s) = -\frac{H_{21}(s)}{H_{22}(s)} \tag{24.24}$$

The block diagram of the process with two feedback control loops and two decouplers is given in Figure 24.7b.

From the block diagram of Figure 24.7b it is easy to develop the following two closed-loop input–output relationships:

$$\bar{y}_1 = \frac{G_{c1}[H_{11} - H_{12}H_{21}/H_{22}]}{1 + G_{c1}[H_{11} - H_{12}H_{21}/H_{22}]} \, \bar{y}_{1,SP} \tag{24.25}$$

$$\bar{y}_2 = \frac{G_{c2}[H_{22} - H_{12}H_{21}/H_{11}]}{1 + G_{c2}[H_{22} - H_{12}H_{21}/H_{11}]} \, \bar{y}_{2,SP} \tag{24.26}$$

The last two equations demonstrate complete decoupling of the two loops since the controlled variable of each loop depends only on its own set point and *not* on the set point of the other loop. Figure 24.7c shows the net block diagram of the two noninteracting loops described by eqs. (24.25) and (24.26). It is completely equivalent to that of Figure 24.7b.

Remarks

1. Two interacting control loops are perfectly decoupled only when the process is perfectly known, because only in this case are the transfer functions H_{11}, H_{12}, H_{21}, and H_{22} known exactly. Since this requirement is rarely satisfied in practice, the decouplers offer only partial decoupling, with some weak interaction still persisting between the two loops.

2. As we have mentioned repeatedly, chemical processes are mostly nonlinear and nonstationary (i.e., the values of their parameters change). Therefore, even if the decoupling is initially perfect, as the desired operating conditions change, the decoupling deteriorates. One solution to this problem is to use *adaptive decouplers*. Thus as the process changes, we estimate the new transfer functions H_{11}, H_{12}, H_{21}, and H_{22} and compute new decouplers. How to design adaptive decouplers is quite complex and goes beyond the scope of this text.

3. Perfect or very good decoupling allows independent tuning of each controller without risking the stability of the overall system.

4. A close examination of Figure 24.7b reveals that for all practical purposes the *decouplers are essentially feedforward control elements*. Thus decoupler D_1 measures the changes in m_2 and takes

appropriate action to cancel the effect that m_2 would have on y_1 before it has been felt by y_1.

5. If the decouplers are designed using steady-state models for the process, we talk of *steady-state or static decoupling*. Equations (24.23) and (24.24) provide the design of *dynamic decouplers*. It should be emphasized that for severely interacting loops, static decoupling is better than no decoupling at all.

6. For a general process with two inputs and two outputs, we need two decouplers to produce noninteracting loops. Whenever we use only one decoupler, despite the fact that two are needed, we talk about *partial* or *one-way decoupling*. Such systems allow the interaction to travel in one direction. Figure 24.7a shows a partial decoupling of the loops. Thus disturbances entering loop 2 cannot disturb loop 1 due to the decoupler D_1. On the other hand, disturbances originating in loop 1 may enter loop 2 but cannot be returned.

Example 24.6: Partial Decoupling

Let us return to the mixing process we studied in Example 24.5. Suppose that the operating requirements allow small variations in the product flow rate F, while dictating very tight control on the concentration x of the product. Then we can use partial or one-way decoupling to cancel any effects that interaction might have on x, leaving the simple feedback loop to regulate the value of the product flow rate, F.

Assuming that x is kept at the desired value of 0.6, eq. (24.21) yields

$$(F_1 + F_2)0.6 = 0.8F_1 + 0.2F_2$$

or

$$F_2 = \frac{F_1}{2}$$

The last equation describes the necessary steady-state decoupler which cancels any effects that the flow control loop might have on the composition control loop.

Example 24.7: Physically Unrealizable Decouplers

Consider a process whose input–output relationships are given by

$$\bar{y}_1 = \frac{0.5e^{-1.5s}}{s + 1} \bar{m}_1 + \frac{e^{-0.5s}}{2s + 1} \bar{m}_2$$

$$\bar{y}_2 = \frac{2e^{-1.0s}}{0.5s + 1} \bar{m}_1 + \frac{1}{s + 1} \bar{m}_2$$

Form the two control loops by coupling y_1 with m_1 and y_2 with m_2. Then the transfer functions of the two decouplers are given by eqs. (24.23) and (24.24):

$$D_1(s) = -\frac{H_{12}(s)}{H_{11}(s)} = 2\frac{s+1}{2s+1}e^{+1.0s}$$

$$D_2(s) = -\frac{H_{21}(s)}{H_{22}(s)} = 2\frac{s+1}{0.5s+1}e^{-1.0s}$$

Decoupler D_1 is physically unrealizable because the term $e^{+1.0s}$ implies that we need a *future* value of m_2, which is not available at the present time point.

THINGS TO THINK ABOUT

1. Explain in your own words the interaction among the control loops of a flash drum (Figure 23.1b). Do the same for the loops of a distillation column (Figure 5.6).

2. Can you tune two interacting loops separately and retain the stability of the overall process? Explain why or why not.

3. Consider the process of Figure 24.1a. Couple y_1 with m_2 and y_2 with m_1 to form the two loops. Draw the corresponding block diagram. Develop the resulting closed-loop input–output relationships, similar to those given by eqs. (24.9) and (24.10). Has the closed-loop characteristic equation changed or not?

4. Define the two open-loop gains used in the definition of the relative gain λ_{12}. Give two different ways of computing λ_{12}. Why is λ_{12} a good measure of loop interaction? Can you compute λ_{11}, λ_{21}, and λ_{22} when you only know λ_{12}? If yes, show how; if not, explain why.

5. Repeat item 4 for the relative gain λ_{ij} of a general process with N inputs and N outputs. What do the subscripts i and j denote?

6. Define the relative-gain array for a process with two inputs and two outputs. Extend the definition to a process with N inputs and N outputs.

7. Consider a process with the following transfer functions: $H_{12}(s) = H_{21}(s) = 0$ and $H_{11}(s)$, $H_{22}(s) \neq 0$. Show that the relative-gain array is given by

$$\begin{bmatrix} 1 & 0 \\ 0 & 1 \end{bmatrix}$$

8. What are the properties of a relative-gain array? How many relative gains do you need to compute in order to specify completely the relative-gain array of a process with (a) three inputs and three outputs, and (b) N inputs and N outputs?

9. Explain how you can use the relative-gain array to select the loops with minimum interaction. Why would you avoid coupling an output y_i with a manipulated variable m_j if $\lambda_{ij} < 0$? Does $\lambda_{ij} < 0$ imply that another relative gain is larger than 1 or not? Explain.

10. In Example 24.5, let $x_1 = 0.3$ and $x_2 = 0.7$ and select the control loops. Have they remained the same or not? Explain your result on physical grounds. Has the interaction between the two loops increased or decreased? Explain why.

11. Define an interaction index as follows:

$$\frac{1 - \lambda_{ij}}{\lambda_{ij}}$$

Consider the relative-gain array

$$\Lambda = \begin{bmatrix} 0.2 & 0.8 \\ 0.8 & 0.2 \end{bmatrix}$$

and take the interaction index array (using the definition above)

$$I = \begin{bmatrix} 4 & \frac{1}{4} \\ \frac{1}{4} & 4 \end{bmatrix}$$

Which one of the two arrays Λ and I shows more clearly the amount of relative interaction between the corresponding loops?

12. What do we mean by the term "decoupling" two control loops? Do the two loops of the process in item 7 need decoupling? Why?

13. Consider the process of Figure 24.1a. Form the two loops by coupling y_1 with m_2 and y_2 with m_1. Find the transfer functions of the two decouplers required.

14. Find the steady-state decouplers for the two control loops selected in the process of Example 24.4.

15. What is one-way decoupling of two control loops, and why could it be acceptable?

16. Explain in your own words the feedforward control nature of a decoupler. When do you have perfect decoupling, and when not?

17. After introducing the necessary decouplers, can you tune the controllers of two loops separately so that the stability of the overall process is guaranteed? (*Hint*: Examine closely the closed-loop characteristic equations of two decoupled loops.)

18. What do we mean when we say that a decoupler is physically unrealizable? Explain why decoupler D_1 in Example 24.7 is physically unrealizable.

Design of Control **25**
Systems for Complete
Plants

Designing control systems for complete chemical plants is the ultimate goal of a control designer. The problem is quite large and complex. It involves a large number of theoretical and practical considerations such as the quality of controlled response; stability; the safety of the operating plant; the reliability of the control system; the range of control and ease of startup, shutdown, or changeover; the ease of operation; and the cost of the control system.

The difficulties are aggravated by the fact that chemical processes are largely nonlinear, imprecisely known, multivariable systems with many interactions. The measurements and manipulations are limited to a relatively small number of variables, while the control objectives may not be clearly stated or even known at the beginning of the control system design.

Since we cannot cover all aspects of plant control within the limits of one chapter, we focus our attention on describing its essential features. Also, we present a rather general methodology for the control system design, using a particular plant as an example.

25.1 Process Design and Process Control

Traditionally, one undertakes the design of a control system for a chemical plant only after a process flowsheet has been synthesized and designed or even constructed tó a significant detail. This allows the control designer to know (1) what units are in the plant and their sizes; (2) how they are interconnected; (3) the range of operating conditions; (4) possible disturbances, available measurements, and manipulations; and (5) what problems may arise during startup or shutdown of the plant. Such information is necessary for the design of effective control systems.

On the other hand, when the design of a chemical plant has been finalized, the control designer is forced to work with a largely specified system. His or her task is to design a control system that ensures satisfactory operation of the plant. This may not be always possible because (1) the process does not possess a sufficient number of manipulated variables; (2) very strong interactions exist among the processing units; (3) it is not possible to cope with all external disturbances; (4) the time lags (dead time) may be significant or the process gains may be too low or too high; (5) the process is inherently unstable; and so on.

It is clear, therefore, that *a certain degree of cooperation and coordination between process designers and control designers is required*. Ideally, process designers should have a good exposure in process control problems and process control designers should have a profound understanding of process design aspects. It is this close and intricate interplay between design and control that makes *a chemical engineer the most suitable person to undertake the design of a control system for a chemical plant.*

When the chemical processes were simpler, with very few interactions among the units, mostly serial in architecture and without strong requirements for operational optimality (in terms of energy or raw material utilization), the design of control systems was much simpler. Today, the picture is different. Rising costs in energy and raw materials, strong competition in the market, improved safety and pollution standards, and so on, have forced designers to construct complex plants that are highly integrated, with many and strong interactions among the processing units. Under such conditions the interaction between process design and control is more acute and *plants that are optimally designed (from an economic point of view) may be inoperable.*

We close this section with a series of examples that dramatize the interaction between process design and control. These examples also demonstrate that the foundations for effective control systems are placed during the design of a process.

Example 25.1: Effect of Intermediate Storage Tanks on Process Control

Consider the process shown in Figure 25.1a. The reactor effluent stream is condensed and stored in a large tank. Material from the tank is fed to the distillation column, where the desired product is separated

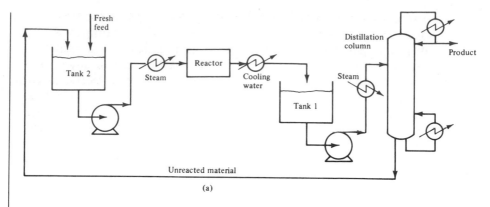

(a)

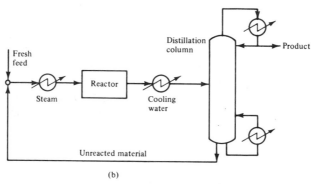

(b)

Figure 25.1 Intermediate storage tanks decouple the processing units of a plant.

from the unreacted reactant, which flows back to a second storage tank before it is again fed to the reactor. *The two intermediate storage tanks decouple the two processing units (i.e., reactor and distillation column).* Thus any variations in the composition, temperature, or flow of the reactor's effluent stream are "damped out" in tank 1 and do not disturb the operation of the column. Similarly, tank 2 absorbs any changes in the flow or composition of column's bottom product, leaving the operation of reactor undisturbed. The *decoupling between reactor and column allows the control designer to synthesize a control system for each unit independently.*

There are several drawbacks in this design. First, the reactor effluent is cooled and then it is heated up again. Second, the two intermediate storage tanks increase significantly the fixed capital expenditure. Figure 25.1b shows an alternative design with lower capital and operating cost. The elimination of the tanks leads to a direct interaction between the two units. Now the control problem is different and certainly more difficult. If the interaction between the two units is very strong, leading to an inoperable plant, we would retain tank 1 but reduce its size significantly.

Example 25.2: Effect of Feed-Effluent Heat Exchange on Reactor Control

In many chemical plants the main reaction is a highly exothermic one. Consequently, the hot reactor effluent is often used to preheat the feed to the reactor (Figure 25.2a). Such feed-effluent heat exchange may lead to serious stability problems in the operation of the reactor. In particular, if T_f is the temperature of the feed to the reactor, the temperature of the reactor effluent T_r follows the sigmoidal curve shown in Figure 25.2b. Furthermore, a heat balance around the heat exchanger shows that T_f depends linearly on T_r (Figure 25.2b). Therefore, the steady-state operation will correspond to one of the three possible points P_1, P_2, or P_3, where the two lines intersect.

As we have seen in Example 1.2, operation at P_2 is inherently unstable and we need to design a stabilizing controller, which should be very robust to maintain stable operation. The design of such controller may not always be possible because either the reaction is extremely sensitive or its mechanism and parameters are poorly known. In such a case we need to redesign the reactor system and make it inherently stable (Figure 25.2c) or eliminate the feed-effluent heat exchange.

One final comment is worth making in this example. Suppose that we want to maintain constant T_f in order to control the conversion at a desired level. *There is no manipulated variable available* to do this. We

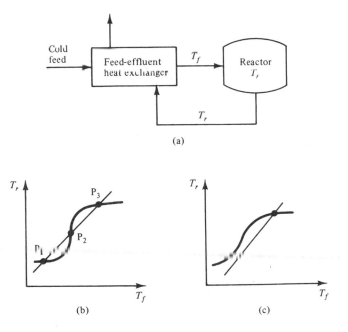

(a)

(b)   (c)

Figure 25.2 Effect of feed-effluent heat exchanger on steady-state patterns of a reactor.

LIVERPOOL JOHN MOORES UNIVERSITY
LEARNING SERVICES

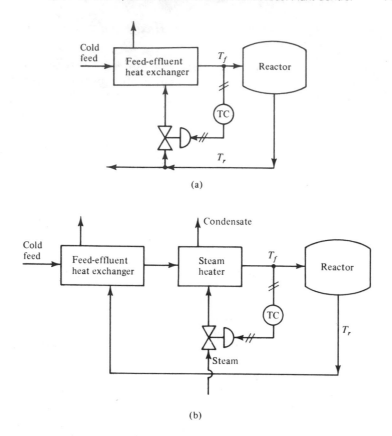

Figure 25.3 Alternative control configurations needed for a feed-efflu-ent heat exchanger system around a reactor.

need to change the design of the reactor system either by splitting the effluent stream (25.3a) or by introducing a second heater with steam (Figure 25.3b).

Example 25.3: Poor Process Designs Leading to Control Problems

When a process is designed solely on economic terms (minimize total cost), it may exhibit serious operating control problems. Let us demon-strate that with three simple examples.

1. A liquid feed is vaporized before it enters the reactor. The heat required for vaporization is supplied by the hot reactor effluent stream, which flows through a coil, immersed in the liquid feed (Figure 25.4a). Suppose that the vaporization tank is poorly designed with large height and small diameter, which makes the

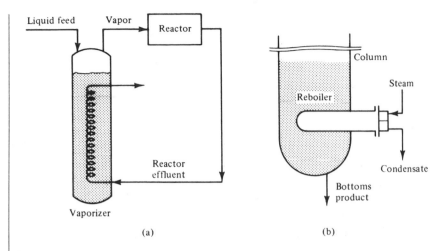

Figure 25.4 Systems with poor dynamics due to bad designs: (a) vaporizer with very large gain; (b) reboiler with very large time constant.

gain of liquid level's response very large (see Example 10.3). In such a case, even small variations in the liquid feed flow rate will have a pronounced effect on the liquid level of the vaporization tank. Thus when the liquid flow rate decreases a little, the level drops significantly, leaving a small portion of the coil immersed in liquid. The amount of heat transferred decreases, leading to a decrease in vapor production. Similar things happen when the liquid flow rate increases. To remedy this "wildly fluctuating" operation we should change the tank's design (i.e., make it shorter with larger diameter). The liquid level will vary by a little and the feed flow rate to the reactor will exhibit smaller fluctuations which are easily controllable.

2. Suppose that we use cooling water to condense the overhead stream of a distillation column. Also, suppose that the heat exchanger area of the condenser has been computed using 90°F as the highest possible temperature of fresh cooling water. On a very hot day when the water temperature is 100°F, no control system can maintain the desired production of overhead product. We need to redesign the condenser by increasing its heat transfer area.

3. Suppose that the reboiler of a distillation column (Figure 25.4b) has been erroneously designed to have a very large volume. From Example 10.2 we know that for energy-storing systems, large volumes imply large time constants. Consequently, the response of the reboiler to any control action will be very slow, in compensating for any disturbance changes. We can improve the controlled response by redesigning the reboiler.

25.2   Hydrodealkylation of Toluene Plant to Produce Benzene: A Case Study

There is no unified and widely acceptable theory for the design of control systems for complete chemical plants. Therefore, instead of presenting an abstract exposition of the various factors that affect control system design, we will use a particular plant as a reference for our discussion. Although a case study has a rather limited scope, it describes clearly the solution methodology and demands concrete answers to specific questions rather than generalities.

The plant we will use for our example *produces benzene* from the *hydrodealkylation of toluene* and is adapted from a similar case study by Douglas [Ref. 21]. The basic reactions are;

$$\text{Toluene} + \text{hydrogen} \longrightarrow \text{benzene} + \text{methane}$$

$$2(\text{benzene}) \longrightarrow \text{diphenyl} + \text{hydrogen}$$

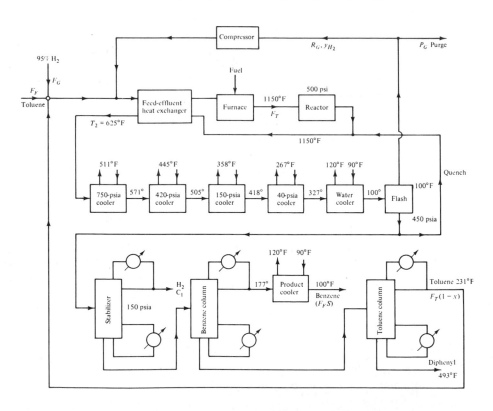

Figure 25.5 Flowsheet of the hydrodealkylation plant. Reproduced from Douglas, J.M., "A Preliminary Design Procedure for Steady-State Control for Complete Chemical Plants," in *Proceedings of Chemical Process Control II*, D. E. Seborg and T. F. Edgar (editors), by permission.

TABLE 25.1.*

ASSUMPTIONS FOR THE PRELIMINARY DESIGN OF THE TOLUENE HYDRODEALKYLATION PLANT

Process operability
 Use all available toluene as feed. All of the benzene produced can be sold.
 The heating value of the fuel for the furnace is constant.
 The hydrogen-to-toluene ratio at the reactor inlet should be 5:1.
 Cooling water is available at 90°F and must be returned at 120°F.
 Water fed to each of the waste heat boilers is at its boiling point.
 The toluene recycle is pure toluene at its boiling point.
 The desired benzene product composition is 99.9%.

Process alternatives
 Distillation is the cheapest way of separating the products.
 Use conventional distillation columns with two product streams instead of
  columns with side streams.
 A stabilizer is required to achieve the desired benzene product purity.
 Aromatics losses in the purge stream and stabilizer overhead are insignificant.
 The heat recovery in the feed-effluent heat exchanger is 50%. The remaining is
  recovered by a battery of steam coolers. The steam lines in the plant are
  available at 750, 420, 150, and 40 psia.
 Liquid from the flash drum is used as a quench stream.

Process optimization
 Conversion of toluene in the reactor is 0.75 and the corresponding selectivity from
  eq. (25.1) is $S = 0.97$.
 The molar fraction of hydrogen in the recycle stream is 0.4.
 The reflux ratios in the distillation columns are 1.2 of the corresponding
  minimum reflux.
 The fractional recovery of benzene in the benzene column is 99%.
 The fractional recoveries of toluene and diphenyl in the toluene column are 99%.

---

*Reproduced from Douglas, J. M., "A Preliminary Design Procedure for Steady-State Control for Complete Chemical Plants," in *Proceedings of Chemical Process Control II*, D. E. Seborg and T. F. Edgar (editors), by permission.

The first is the main reaction; the second leads to diphenyl as a by-product and cannot be avoided. If x is the total conversion of toluene (moles of toluene reacted per mole of toluene fed) and S the selectivity toward benzene (i.e., moles of benzene produced per mole of toluene reacted), experimental data have indicated that these two variables are related by the equation

$$S = 1 - \frac{0.0036}{(1 - x)^{1.544}} \tag{25.1}$$

We notice that as x increases, the selectivity decreases and more diphenyl is produced. For the two raw materials we have the following specifications:

1. Toluene is available at ambient conditions (77°F, 1 atm) and a rate of 2000 bbl/day.

2. Hydrogen stream is available at 100°F and 550 psig, containing 95% by mole H_2 and 5% methane.

The reactor inlet temperature should be at least 1150°F for the reaction rate to be sufficiently high, while the reactor temperature must be kept below 1300°F to prevent coking. Also, the reactor pressure is 500 psig.

Figure 25.5 shows the flowsheet for the hydrodealkylation plant resulting from a preliminary design and optimization of the process. This is a starting flowsheet, not necessarily the best one. It is based on a series of assumptions which can be grouped in the following categories (see Table 25.1):

Process operability assumptions
Assumptions on process alternatives
Assumptions resulting from preliminary process optimization

A quick description of the plant follows:

1. The feed streams (toluene and hydrogen) are mixed with two recycle streams: the gas recycle stream, which contains primarily H_2 and CH_4, and the liquid recycle stream, which is pure toluene.
2. The mixture above is first preheated by the reactor effluent in the feed-effluent heat exchanger, and then passes through a furnace, where its temperature is raised to 1150°F before it enters the reactor.
3. The reaction is exothermic and the reactor effluent leaves the reactor with 1265°F. It is quenched quickly with liquid, down to 1150°F.
4. The reactor effluent is further cooled from 1150°F down to 100°F, first by preheating the feed (feed-effluent heat exchanger) and then through a series of steam coolers, where saturated water at various pressures is vaporized. The steam lines in the plant are available at 750, 420, 150, and 40 psia, and consequently the steam coolers operate at the same pressures. The cooling is completed with cooling water.
5. The reactor effluent is now at 100°F, and in a flash drum operating at 450 psia we can separate the liquid mixture of aromatics (unreacted toluene, produced benzene, and diphenyl) from the gaseous stream of H_2 and CH_4. Part of the flash liquid is used to quench the reactor effluent and the rest goes to a sequence of distillation columns.
6. The gaseous stream from the flash drum is first recompressed and then mixed with fresh feeds. A small portion is purged to avoid a buildup of inert CH_4 in the recycle stream.

7. The sequence of distillations is composed of three columns. The first, called stabilizer, is needed to stabilize the content of the liquid stream by removing the noncondensible gases (H_2 and CH_4). Without a stabilizer we cannot achieve the desired benzene composition in the second column. The toluene from the third column is recycled back to the beginning of the plant, while the diphenyl of the bottom product is used as fuel.

Now that we have completed a description of the hydrodealkylation plant, we can start the design of its control system.

25.3   Material Balance Control for the Hydrodealkylation Plant

The purpose of the control system for the hydrodealkylation plant is to ensure production of benzene with (1) the desired quality (99.9% by mole) and (2) at the desired rate. The product quality must meet the desired specifications; otherwise, it must be reprocessed or wasted. The part of the control system which ensures that the product quality is maintained at the desired levels is known as *product quality control* and will be examined in Section 25.4.

Although a plant is usually designed for a nominal production rate, a design tolerance is always incorporated because the market conditions may require an increase or decrease from the current rate. The control system is then called to ensure a smooth and safe transition from the old to the new production level. This is known as *material balance control*, because its purpose is to direct the control action in such a way as to *make the inflows equal to the outflows and achieve a new steady-state material balance for the plant.* Let us now see how to design the material balance control system for the hydrodealkylation plant.

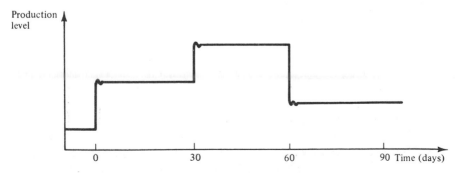

Figure 25.6 Pseudo-steady-state operation of a plant.

TABLE 25.2.*

STEADY-STATE MATERIAL BALANCES FOR THE HYDRODEALKYLATION PLANT

Overall balances

  Toluene fresh feed $= F_F$
  Benzene produced $= F_F S$
  Methane produced $= F_F$
  Diphenyl produced $= (1 - S)F_F/2$
  Hydrogen consumed $= F_F - (1 - S)F_F/2 \approx F_F$

Balances on the gas phase (purge, hydrogen makeup)

$$\text{Hydrogen makeup (fresh)} = F_G$$

$$\text{Purge} = P_G$$

For $S = 0.97$,

$$\text{Hydrogen consumed} = F_F - (1 - S)F_F/2 \approx F_F = \text{methane produced}$$

Assume perfect split at the flash drum.

Overall balance of gases:

$$F_G + (\text{methane produced}) = P_G + (\text{hydrogen consumed})$$

or

$$F_G = P_G \tag{25.2}$$

Overall hydrogen balance:

$$0.95 F_G = (\text{hydrogen consumed}) + y_{H_2} P_G = F_F + y_{H_2} P_G$$

or

$$F_G = P_G = \frac{F_F}{0.95 - y_{H_2}} \tag{25.3}$$

Balances at the recycle mixing points

  Toluene balance at the liquid mixing point:

$$\text{Toluene to reactor} = F_T = F_F + F_T(1 - x)$$

or

$$F_T = F_F/x \tag{25.4}$$

$$\text{Hydrogen to reactor} = 5F_F/x$$

Hydrogen balance at the gas mixing point:

$$0.95 F_G + y_{H_2} R_G = 5F_F/x$$

Therefore,

$$\text{recycle flow rate} = R_G = \frac{F_F}{y_{H_2}} \left(\frac{5}{x} - \frac{0.95}{0.95 - y_{H_2}} \right) \tag{25.5}$$

where y_{H_2} is the molar fraction of hydrogen in the gas recycle stream.

---

*Reproduced from Douglas, J. M., "A Preliminary Design Procedure for Steady-State Control for Complete Chemical Plants," in *Proceedings of Chemical Process Control II*, D. E. Seborg and T. F. Edgar (editors), by permission.

Changes in the production rate come from the plant's management and are rather infrequent (e.g., once every two weeks, or a month, or longer period). The time required by the plant to reach the new operating level is much shorter than the periods noted above. Consequently, the transient dynamic behavior is very short-lived and not very important. Therefore, we can assume that the plant "always" operates at steady state. Figure 25.6 demonstrates this "pseudo-steady-state" behavior. It is clear, then, that *steady-state balances are sufficient for material balance control*.

Table 25.2 shows the steady-state material balances for the hydrodealkylation plant. They will guide the design of material balance control system.

Control the production rate

From Table 25.2 we know that

$$\text{benzene produced} = F_F S$$

Therefore, if we want to control the production rate we need to manipulate the fresh toluene feed rate F_F, or the conversion x, which affects the selectivity S.

1. The conversion x is not a real manipulated variable and in order to change its value we can manipulate the heat input to the furnace, which will affect the reactor temperature and thereby the conversion. But *this is not a good control strategy* because a decrease in S to effect a decrease in the production rate leads to a higher production rate of diphenyl, which is valued only as fuel. Thus to control the production of benzene, we convert valuable toluene to low-value diphenyl.

2. The fresh toluene flow rate is a real manipulated variable. Decreasing F_F we decrease the production rate, leading the excess of toluene to a feed storage tank. This alternative control scheme is preferred for economic and operating reasons. Lower F_F implies less material to process and consequently lower operating cost. Furthermore, the adjustment of F_F is easier and faster than the adjustment of reactor inlet temperature with the furnace heat input. Therefore, we *choose to control production rate by manipulating the fresh toluene feed rate* under constant conversion (Figure 25.7).

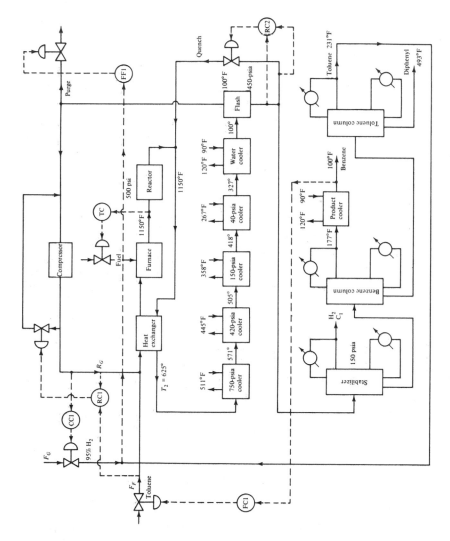

Figure 25.7 Material balance controls for the hydrodealkylation plant.

Material balance control for the other streams

When the benzene production rate changes, the flow rates of the following streams must change appropriately in order to achieve the new steady-state material balance: (1) hydrogen fresh feed F_G, (2) purge P_G, (3) recycled gas R_G, (4) recycled toluene, (5) product benzene, and (6) quench stream. From the material balances of Table 25.2 we can easily see that the flow rates of the first five streams must change proportionally to the change in F_F. From an energy balance at the quenching point we can find that the liquid quench flow rate must change proportionally to F_F. So let us specify the resulting material balance controllers.

1. *Adjust the recycled gas flow rate R_G.* In Table 25.2 we can easily see that the value of R_G is computed so that the ratio of hydrogen to toluene at the reactor inlet is equal to 5:1. Therefore, we can use a *ratio controller* to adjust R_G by the following relationship [see eq. (25.5)]:

$$\frac{R_G}{F_F} = \frac{1}{y_{H_2}}\left(\frac{5}{x} - \frac{0.95}{0.95 - y_{H_2}}\right) = \frac{1}{0.4}\left(\frac{5}{0.75} - \frac{0.95}{0.95 - 0.4}\right) = 12.35$$

2. *Adjust the hydrogen makeup flow rate F_G.* From eq. (25.3) of Table 25.2 we can see that F_G depends on y_{H_2} (i.e., the value of the recycled gas composition in hydrogen), which must be maintained at 0.4 (see the assumptions resulting from process optimization in Table 25.1). Therefore, we can use a *feedback controller* which measures the hydrogen composition in the recycled gas and adjusts F_G in order to maintain $y_{H_2} = 0.4$.

3. *Adjust the purge rate P_G.* From Eq. (25.2) of Table 25.2 we conclude that we can use a *feedforward controller* which adjusts P_G in accordance with the value of F_G.

4. *Adjust the benzene product flow rate.* This is the overhead product of the benzene distillation column and depends on the value of the feed flow rate. In the next section we will see how to adjust the overhead flow rate in order to maintain the desired benzene purity and fractional recovery by a material balance controller which manipulates the distillate to feed ratio (D/F control).

5. *Adjust the recycled toluene flow rate.* This is again the overhead product of the toluene column, and in Section 25.4 we will study how to adjust its value by manipulating the ratio of distillate to feed.

6. *Adjust the quench flow rate.* The easiest way to adjust the quench flow rate is to set it initially as a fraction of the total liquid coming out from the flash drum. Thus when the fresh toluene flow rate

TABLE 25.3.
MATERIAL BALANCE CONTROLLERS FOR THE HYDRODEALKYLATION PLANT

1. Control the production rate by manipulating the fresh toluene feed rate (flow controller, FC1 in Figure 25.7).
2. Control the recycle gas flow rate R_G through a ratio controller. Keep the ratio $R_G/F_F = 12.35$ (ratio controller, RC1).
3. Use a feedback loop which measures the recycled gas hydrogen composition to adjust the value of the hydrogen fresh feed rate (composition controller, CC1).
4. Through a feedforward controller adjust the purge rate so that it is equal to the fresh hydrogen feed rate (feedforward flow controller, FF1).
5. Use a ratio controller to adjust the quench flow rate by maintaining a constant ratio of the quench to the total flash liquid stream (ratio controller, RC2).

changes, so does the liquid from the flash drum, and consequently the quench flow rate is adjusted automatically to the proper value. A *ratio controller* can be used here.

Table 25.3 summarizes the material balance controllers for the hydrodealkylation plant.

Remark. A good material balance control system is an imperative requirement for satisfactory operation of a chemical plant. Therefore, if a process design cannot accommodate efficient material balance control, we must change the design of the plant. This is not the case for the assumed design of the hydrodealkylation plant, because the control system of Table 25.3 is acceptable and operationable.

25.4   Product Quality Control for the Hydrodealkylation Plant

The material balance control system is designed for a single purpose: to adjust the plant's production rate at the level dictated by management and keep it at this level. In other words, the material balance control is not designed to maintain the operating variables in certain units at constant values (set points) against changes in various disturbances. This is accomplished by an additional control system which remains to be designed.

The basic purpose of the additional control system is *to ensure the desired product quality*. This can be accomplished by:

1. Regulating the operation of the benzene column against disturbance changes entering the column
2. Canceling the effects of disturbance changes in upstream units,

which when left unattended will propagate and affect product quality.

Therefore, we will design the *product quality control system* starting with the benzene column and continuing with each of the other units. The general methodology described in Chapter 23 will be used here to develop and screen the alternative loop configurations for each unit.

Benzene column control

Consider the feed as a pseudo-binary mixture composed of benzene ($\equiv$ A) and toluene + diphenyl ($\equiv$ B). From Example 5.7 we know that we can specify four controlled variables for a binary distillation column. In our case these are:

Benzene product purity
Fractional recovery of benzene in the overhead product (distillate rate)
Liquid level in the overhead accumulator
Liquid level at the bottom of the column

Four manipulated variables are also available: distillate flow rate (D), reflux rate (R), steam flow rate (S), and bottoms flow rate (B). The basic disturbances are; feed flow rate and composition, and the cooling water temperature in the overhead condenser. Table 25.4 shows all possible control-loop configurations for the benzene distillation column. Let us now screen these alternatives and select the best.

1. Trying to control the liquid level at the bottom of the column with the reflux flow or distillate flow rate involves *very long time responses* because the action of the manipulated variable must travel the whole length of the distillation column before it is felt by the controlled variable. Therefore, control-loop configurations 4, 5, 10, 12, 14, 16, 17, 18, 20, 21, 23, and 24 are ruled out.

2. A long time response is also involved when we try to control the level in the overhead accumulator by manipulating the bottoms flow rate for the same reason as above. This rules out configurations 1 and 8.

3. A long time response (not quite as long as above) is involved when we try to control the accumulator's level with the steam flow rate. This rules out configurations 2 and 7.

4. It is quite complicated to control the distillate composition or flow rate with the bottoms flow rate. It also involves long time

TABLE 25.4*.
POSSIBLE CONTROL LOOP CONFIGURATIONS FOR THE BENZENE DISTILLATION COLUMN

Configuration number	Benzene product composition	Distillate rate	Overhead accumulator level	Column bottom level	Method for composition control[b]	Comment[a]
1	D	R	B	S	Direct M.B.	(2)
2	D	R	S	B	Direct M.B.	(2)
3	D	S	R	B	Direct M.B.	
4	D	S	B	R	Direct M.B.	(1)
5	D	B	S	R	Mixed	(1)
6	D	B	R	S	Mixed	(3)
7	R	D	S	B	V/F	(2)
8	R	D	B	S	V/F	(2)
9	R	S	D	B	Indirect M.B.	
10	R	S	B	D	Indirect M.B.	(1)
11	R	B	D	S	V/F	(3)
12	R	B	S	D	V/F	(1)
13	B	D	R	S	Mixed	(3)
14	B	D	S	R	Mixed	(1)
15	B	R	D	S	Direct M.B.	(4)
16	B	R	S	D	Direct M.B.	(1)
17	B	S	R	D	Direct M.B.	(1)
18	B	S	D	R	Direct M.B.	(1)
19	S	D	R	B	V/F	(3)
20	S	D	B	R	V/F	(1)
21	S	R	B	D	Indirect M.B.	(1)
22	S	R	D	B	Indirect M.B.	
23	S	B	D	R	V/F	(1)
24	S	B	R	D	V/F	(1)

*Adapted from McCune, L.C. and Gallier, P.W., "Digital Simulation: A Tool for the Analysis and Design of Distillation Controls", *ISA Transactions*, *12*, 193 (1973), by permission.

[a] (1) Long time response; (2) long time response and increasing complexity; (3) V/F control is not desirable in general; (4) column bottom level control by steam flow is desirable. (Comments are due to T. Umeda).

[b] Direct M.B. = direct material balance control
Indirect M.B. = indirect material balance control
V/F = vapor to feed ratio control

responses. Consequently, configurations 6, 11, 13, and 15 are ruled out.

At this point we realize that only configurations 3, 9, 19, and 22 are left for further consideration. Which one of these four will be the "best"?

We must note that the foregoing discussion and elimination of 20 out of 24 control structures applies to any distillation column. Which one we will select among the remaining four depends on the characteristics of the particular column we consider. Thus for the benzene column we select configuration 3 as the "best", for the following reasons.

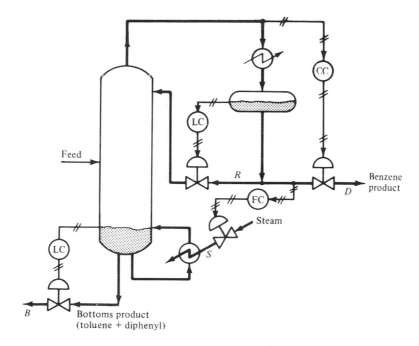

Figure 25.8 Control loops around a distillation column.

First, on a cold day or during a rainstorm the temperature of the cooling water in the overhead condenser drops and the overhead vapor passing through the condenser produces subcooled liquid. When the subcooled liquid returns back (through the reflux flow) to the top tray of the column, it causes less vapor to go overhead. Less vapor overhead causes the liquid level in the accumulator to drop. If the accumulator level is controlled by the reflux flow, the latter will decrease. Thus the disturbance caused by the cooling water temperature drop does not propagate down the column in terms of increased liquid overflows. A new equilibrium takes place quickly on the top tray, thus isolating the effect of the disturbance from the rest of the column. Configurations 3 and 19 control the level of the overhead accumulator by the reflux rate and thus are retained while 9 and 22 are rejected.

Second, in configuration 3 the benzene product composition is controlled directly by the distillate flow, whereas in configuration 19 it is controlled indirectly by the steam rate. Therefore, the first is more responsive and is retained as better.

Figure 25.8 shows the four control loops for the benzene column.

Stabilizer and toluene column control

A similar discussion as above leads to the same control loops as for the toluene column.

Reactor control. Maintain constant conversion

Since the production rate is controlled by the toluene fresh feed rate, the conversion in the reactor must be kept constant. This can be achieved by manipulating the heat duty to the furnace. As measurement we can use either the temperature of the reactor inlet stream or the temperature in the reactor. The first is preferred because it allows us to *take control action earlier* (Figure 25.7).

Flash drum control

We examined this system in Examples 23.1 and 23.2. We will not repeat the same discussion here.

Furnace control

The heat duty to the furnace is manipulated in order to control the conversion in the reactor. The same control loop can be used to compensate for any changes in the heating value of the fuel supply. Finally, a programmed adaptive ratio controller can be used to regulate the fuel/air ratio at its best value for maximum efficiency (see Example 22.2).

Other controllers?

Additional disturbances that may affect the operation of the hydro-dealkylation plant are the following:

1. *Changes in the fresh toluene feed composition:* Most likely, these will be of very small size and should not have an appreciable effect on the material balances of the plant. Therefore, we may not need special control loops to eliminate the effect of this disturbance.
2. *Changes in the hydrogen composition of the fresh hydrogen stream:* This disturbance will affect the hydrogen composition of the reactor inlet stream. To eliminate its effect we can use the feedback composition controller, CC1, which was installed for material balance control. It measures the hydrogen recycle composition and adjusts the fresh hydrogen flow rate.
3. *Fluctuations in the temperature of fresh toluene or fresh hydrogen feeds which may be caused by fluctuations in the ambient temperature:* These temperature changes will cause the exit temperatures from the feed-effluent heat exchanger to vary. Therefore, the heat

load exchanged in the steam coolers will vary, producing more or less steam. The variable heat duty of the furnace will compensate for any changes in the temperature of the reactor feed. Therefore, no additional controllers are needed to compensate for temperature changes in the toluene or hydrogen fresh feeds.

4. *Do we need to control the operation of the steam coolers (waste-heat boilers) in the presence of changes in the production rate?* It is clear that as the production rate decreases, less steam will be produced at the steam coolers. If we maintain the flow rate of the boiler-feed water to the unit constant, lower-quality steam will be produced. However, the steam can be flashed away and the excess water recirculated through the steam production system. Therefore, we do not need controllers for waste-heat boilers (steam coolers). We only need to change the initial design of the plant so that we provide a unit for flashing away the steam and appropriate piping for the return of the remaining water to the boiler.

25.5 Some Comments on the Control Design for Complete Plants

If we review the methodology we followed for the control system design of the complete hydrodealkylation plant, we can easily identify the following central and critical point:

> We have decomposed the control system into two parts: the material balance control and the product quality control. Each has been designed separately, as if they do not interact with each other.

Is this basic assumption true? Let us try to analyze it further.

The material balance control system is called on to compensate for very low frequency disturbances (e.g., plant management decisions for changes in the production rate, once every two weeks, month, or longer). On the other hand, the product quality control system is called on to ensure the operation of the plant against higher-frequency disturbances (once every half a minute, minute, hour, or day). This difference is the basis for the decomposition of the material balance and product quality control systems. It works as follows: Figure 25.9 shows the amplitude ratio versus frequency for the material balance and product quality control systems.

(a) At low frequency the amplitude ratio for the product quality control system is very low and thus every low-frequency disturbance (such as a change in the production rate) leaves it almost

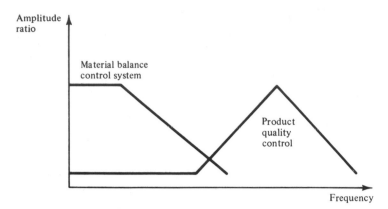

Figure 25.9 Amplitude ratio versus frequency for material balance and product quality control systems.

unaffected. This is not the case for the material balance control, whose amplitude ratio is high and thus is strongly affected.

(b) At high frequencies we have the reverse picture. The product quality control system is affected while the material balance remains undisturbed.

Therefore, each part of the overall control system is designed for a different set of disturbances whose typical frequencies of occurrence differ by several orders of magnitude. Consequently, they can be designed separately.

An additional observation is also important. Figure 25.7 shows that the management decision for a change in the production rate is directed at the material balance control of the toluene feed storage tank. Then the material balance control actions for the other streams follow in the *direction of process flow*. This is the most common type for implementing material balance control actions and is depicted in general form by Figure 25.10a. Sometimes it may be advantageous to implement the material balance control in the *direction opposite to process flow* (Figure 25.10b).

Finally, it must be emphasized that:

The material balance control system can be designed using steady-state balances only, while

The product quality control system should be designed by also considering the dynamic behavior of the various units.

The material balance and product quality control systems do not cover all operating aspects of a complete plant. Several additional

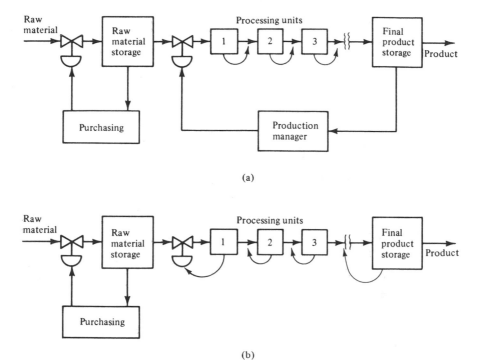

Figure 25.10 Alternative structures of information flow for material balance control.

features need to be considered and they dictate the introduction of additional control systems. Let us examine these features in more detail.

Startup and shutdown of a plant

The control system of a complete plant must permit smooth, safe, and relatively fast startup and shutdown of the plant's operation. Normally, this cannot be accomplished by the material balance or product quality controllers alone, and additional control loops are needed. How to design a good startup or shutdown control system is beyond the scope of the present text. It suffices to mention that it relies heavily on the practical experience of the control designer, who must anticipate all operating contingencies during startup or shutdown. Furthermore, it must be emphasized that quite often we change the material balance or product quality control system, in order to improve the overall capability of the system to startup or shutdown the process.

Safe operation of the plant

This is a primary objective for a good control system, which must ensure that all temperatures, pressures, flows, and concentrations remain within prespecified safe limits. Thus the control system for a complete plant must also include the following elements:

1. Redundant measurements (not used directly in a control loop) which indicate approaching unsafe conditions (e.g., slowly increasing composition, temperature, or pressure in an oxidation reactor leading to explosion; increased pressure in a vessel due to plugging in the pipes, etc.). These measurements allow early corrective actions to avoid unsafe operation.
2. Override control loops (see Section 20.2). Whenever an operating variable tends to exceed an upper or a lower allowable limit, the override control loops transfer the control action from the normal operation to an emergency operation, thus preventing the development of unsafe conditions.
3. An emergency shutdown control procedure. If the developing unsafe conditions seem to be very critical and the override control loops cannot cope with the developing emergency, the control system must be capable of shutting down the plant safely.

Control required by environmental regulations

Federal, state, or local laws and regulations require that the effluents of a plant (liquid, solid, or gaseous) satisfy certain specifications. Thus their temperatures, compositions, and flow rates must be within prespecified limits. Failure to abide by these regulations subjects the plant to additional economic burdens (fines) and makes it a threat to the well-being of its personnel and the surrounding communities. Therefore, additional controllers must be added to ensure that environmental regulations are satisfied.

Optimizing the operation of a plant

It is desirable that a plant operate, almost always, at the point of minimum production cost or maximum profit. This can be achieved by an optimizing control strategy which:

1. Identifies *when* the plant must be moved to a new operating point in order to reduce the operating cost
2. Makes the appropriate set point changes to bring the plant to the new optimum operating point

Let us consider a simple system (Figure 25.11a) with two distur-bances d_1 and d_2, two manipulated variables m_1 and m_2, and two controlled outputs y_1 and y_2. The steady-state input–output relation-ships are

$$y_1 = f_1(m_1, m_2, d_1, d_2)$$
$$y_2 = f_2(m_1, m_2, d_1, d_2)$$

Let the operating cost be

$$\text{cost} = F(y_1, y_2)$$

When the disturbances have the values d_1' and d_2', the minimum opera-ting cost is achieved for $y_1 = y_1'$ and $y_2 = y_2'$, where

$$\left(\frac{\partial F}{\partial y_1}\right)_{y_1',y_2'} = \left(\frac{\partial F}{\partial y_2}\right)_{y_1',y_2'} = 0$$

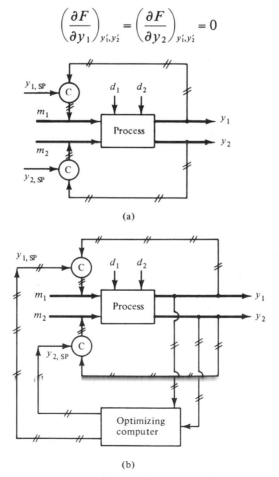

(a)

(b)

Figure 25.11 (a) Simple process with its control loops; (b) supervisory implementation of optimizing control strategy.

The values y_1' and y_2' are the current set point values for the two control loops (Figure 25.11a). But when the disturbances change, say $d_1 = d_1''$ and $d_2 = d_2''$, the values of y_1 and y_2 that minimize the cost function also change. Let the new optimum values be y_1'' and y_2''. Again they satisfy the following necessary conditions for optimality:

$$\left(\frac{\partial F}{\partial y_1}\right)_{y_1'',y_2''} = \left(\frac{\partial F}{\partial y_2}\right)_{y_1'',y_2''} = 0$$

The optimizing control system must be able to:

1. Identify that the current set points y_1' and y_2' no longer correspond to optimum operating conditions.
2. Find the new optimum operating conditions (i.e., the values y_1'' and y_2'').
3. Bring the process to the new optimum operation by changing the set point values of the two control loops.

The introduction of digital computers for process control has allowed the implementation of optimizing control strategies in chemical plants. The usual mode of implementation is that of supervisory control (see Sections 3.2 and 26.3). Figure 25.11b shows the supervisory control implementation of an optimizing control strategy for the simple process of Figure 25.11a. Notice that the computer calculates the new optimum set point values and communicates them to the two control loops.

Optimizing control is used extensively today and its future looks even more promising. It is particularly beneficial for large plants where even small improvements in the operating cost are multiplied by large production throughputs, thus leading to significant economic savings on a yearly basis.

Normally, there will be a tremendous number of alternative control systems for a complete plant. Initially, the control designer uses empirical arguments to eliminate the bad ones. Progressively, he or she uses more and more complex analyses for further screening and elimination of alternatives. The designer may need to resort to steady-state and/or dynamic simulation of the plant and its controllers in order to evaluate the alternative control systems.

THINGS TO THINK ABOUT

1. Identify all the objectives that a control system for a complete chemical plant is called upon to satisfy. Describe the objectives qualitatively and be as exhaustive as you can in enumerating these control objectives.

2. In what sense is the control of a complete chemical plant different from the control of a single processing unit?

3. How does the design of a chemical plant affect its control system? Enumerate all possible effects and provide examples to describe them.

4. When would you introduce intermediate storage tanks in a chemical plant? What is their effect in designing control systems for a plant?

5. Describe processing examples with long response times or significant dead times in their dynamic behavior. What would you prefer to do: attempt to design complicated control systems that try to speed up the response and eliminate dead times, or change the design of the plant? Elaborate on your answer.

6. Consider the distillation systems shown in Figure Q25.1a and b. In the first the columns are thermally uncoupled, whereas in the second they are thermally coupled (i.e., the overhead of the first column is cooled by the bottoms of the second column). For which system is the control design easier? Elaborate on your answer.

7. Identify a control system for the thermally coupled distillation columns of Figure Q25.1b. Do you see any serious problems with your control system? How would you start up the plant? Make the necessary design modifications so that you can start it up.

8. What is material balance control for a complete plant? In what sense is it different from the set point control we have seen in earlier chapters? Discuss the two approaches for the implementation of material balance control. Which one is preferable, implementation in the direction of process flow or opposite to it? Elaborate on your answer. Would you use steady-state or dynamic models for material balance control, and why? (*Note:* You will find Chapter 13 of Ref. 16 very useful in answering these questions.)

9. Consider the plant of Example 23.4 (Figure 23.7). Synthesize a material balance control system for this plant. Use only qualitative arguments because you are not given quantitative information about the plant.

10. Examine the control system of Figure 23.10, which was synthesized for the plant of Example 23.4. Which of the control loops correspond to the material balance control system, and which are used for product quality control?

11. What is product quality control? In what sense is it different from material balance control? Explain why you can synthesize the product quality control system independently of the material balance control system?

12. Discuss again the arguments that were used in the text to select control loop configuration 3 for the benzene distillation column.

13. Why do we not need additional controllers for the waste-heat boilers of the hydrodealkylation plant?

14. Is material balance and product quality control sufficient for the good

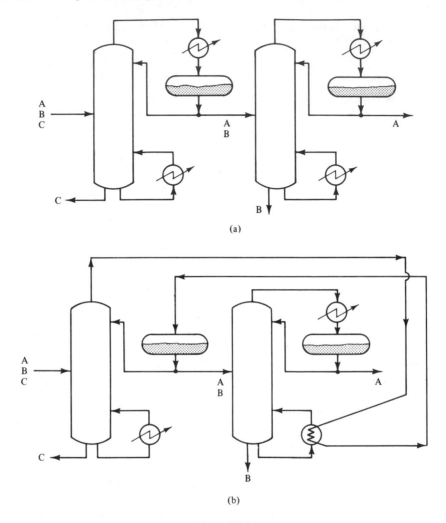

(a)

(b)

Figure Q25.1

operation of a chemical plant? If not, explain why and indicate what more is needed.

15. Can you identify where in the hydrodealkylation plant you may need an override control system? (*Note*: What about the pressure in the gaseous recycle loop?)

16. What is the purpose of an optimizing control system, and how is it implemented in practice? Why is it needed?

17. Consider a distillation column separating A from B. Both products must be produced with desired purities. Do you need to develop an optimizing

control strategy for this system? If yes, explain why and describe qualitatively how such a system would work. (*Note*: You will find Chapter 11 in Ref. 9 very useful in answering this question.)

18. Would you recommend an optimizing control system for the hydrodealkylation plant? Why?

REFERENCES FOR PART VI

Chapter 23.   For additional study on the degrees of freedom and their use for determining the number of controlled and manipulated variables, the reader can consult the following sources:

1. *Automatic Control of Processes*, by P. W. Murrill, Intext Educational Publishers, Scranton, Pa. (1967).
2. *Handbook of Chemical Engineers*, 5th ed, by J. H. Perry (ed.), McGraw-Hill Book Company, New York (1974).
3. "Selection of Control Measurements," by M. E. Findley, in *Process Control* (Module 31b), T. F. Edgar (ed.), AIChE Modular Instruction, New York (1982).

In these references the reader can also find useful examples on the use of degrees of freedom, either for single units or complete plants. Example 23.4 is adapted from a paper by Umeda et al. [Ref. 4], where they discuss a decomposition strategy for the control design of complete chemical plants:

4. "A Logical Structure for Process Control System Synthesis," by T. Umeda, T. Kuriyama, and A. Ichikawa, in *A Link between Science and Applications of Automatic Control*, A. Niemi (ed.), Pergamon Press Ltd., Oxford (1979).

Chapter 24.   For additional reading and examples on loop interaction and design of noninteracting loops, the reader can consult the following references:

5. *Process Dynamics and Control*, Vol. 2, by J. M. Douglas, Prentice-Hall, Inc., Englewood Cliffs, N.J. (1972).
6. *Digital Computer Process Control*, by C. L. Smith, Intext Educational Publishers, Scranton, Pa. (1972).
7. *Process Modeling, Simulation, and Control for Chemical Engineers*, by W. L. Luyben, McGraw-Hill Book Company, New York (1973).

A thorough and very useful discussion on interaction and decoupling with many valuable insights can be found in the following two books by Shinskey [Refs. 8 and 9]. The first treats the subject in a general manner, while the second focuses on interaction and decoupling of loops in distillation columns.

8. *Process Control Systems*, 2nd ed., by F. G. Shinskey, McGraw-Hill Book Company, New York (1979).

9. *Distillation Control*, by F. G. Shinskey, McGraw-Hill Book Company, New York (1977).

For a more advanced theoretical treatment of loop interaction and decoupling, the following book by Rosenbrock is an excellent source:

10. *Computer-Aided Control System Design*, by H. H. Rosenbrock, Academic Press, Inc., New York (1974).

The idea of the relative-gain array and its use for selecting the control loops can be found in the original paper by Bristol:

11. "On a New Measure of Interaction for Multivariable Process Control," by E. H. Bristol, *IEEE Trans. Autom. Control, AC-11*, 133 (1966).

Bristol's relative-gain array initiated an extensive and vigorous research effort in interaction and decoupling of control loops. There is a large number of papers with applications of Bristol's array, as well as several theoretical treatments. Today it has been shown that if dynamic interaction is important Bristol's method may lead to the wrong couplings. The following paper demonstrates this point:

12. "Analysis of Control Input–Output Interactions in Dynamic Systems," by L. S. Tung and T. F. Edgar, *AIChE J., 27*(4), 690 (1982).

In Chapter 24 we have been concerned only with the design of noninteracting multivariable control systems. However, there is a large number of multivariable control design techniques. The interested reader can consult Douglas [Ref. 5] and Rosenbrock [Ref. 10] for further study. Also, the following texts are very good references for advanced study:

13. *Linear Optimal Control Systems*, by H. Kwakernaak and R. Sivan, John Wiley & Sons, Inc. (Interscience Division), New York (1972).

14. *Advanced Process Control*, by W. H. Ray, McGraw-Hill Book Company, New York (1981).

Finally, it should be noted that Example 24.5 was inspired from the following source:

15. "Interaction Matrix Analysis," by T. J. Ward, in *Process Control*, Volume 4, T. F. Edgar (ed.), AIChE Modular Instruction, American Institute of Chemical Engineers, New York (1982).

Chapter 25.   The synthesis of control systems for complete plants has attracted tremendous research interest during the last few years. The earlier systematic treatment of the subject was in Buckley's text [Ref. 16]. There the reader can find a thorough discussion on material balance and product quality control systems, their characteristics, and treatments of the interaction between design and control or between process development and control and between process operation and control.

16. *Techniques of Process Control*, by P. S. Buckley, John Wiley & Sons, Inc., New York (1964).

Other efforts include the work of Boyd and Bakanowski [Ref. 17], who synthesized a control system for a nitric acid plant, and the work of Bettes and Wright [Ref. 18], who designed the control system for a solvent-dewaxing plant. Both works employed analog computers to solve the differential equations which described the dynamic behavior of the plants.

17. "*Electronic Computation in the Chemical Industry,*" by R. N. Boyd and V. J. Bakanowski, *Eng. J., 43*(12), 47 (1960).
18. "Analog Computer Leads the Way to Better Control," by R. S. Bettes and L. T. Wright, *Oil Gas J., 58*(17), 202 (1960).

Recent developments on the synthesis of control systems for complete plants are described in the following papers;

19. "Studies in the Synthesis of Control Structures for Chemical Processes," Parts I, II, III, by M. Morari, Y. Arkun, and G. Stephanopoulos, *AIChE J., 26*, 220 (1980).
20. "Control System Synthesis Strategies," by R. Govind and G. J. Powers, *AIChE J., 28*(1), 60 (1982).
21. "Preliminary Design Procedure for Steady-State Control Systems for Complete Chemical Plants," by J. M. Douglas, in *Chemical Process Control—II*, D. E. Seborg and T. F. Edgar (eds.), Proceedings of Engineering Foundation Conference, Sea Island, Ga. (1981), American Institute of Chemical Engineers, New York (1982).

Two very useful review articles on the various aspects of complete plant control system design with lists of relevant references are the following:

22. "Integrated Plant Control. A Solution at Hand or a Research Topic for the Next Decade?" by M. Morari in *Chemical Process Control—II*, T. F. Edgar and D. E. Seborg (eds.), Proceedings of Engineering Foundation Conference, Sea Island, Ga. (1981), American Institute of Chemical Engineers, New York (1982).
23. "Synthesis of Control Systems for Chemical Plants. A Challenge for Creativity," by G. Stephanopoulos, *Proceedings of International Symposium on Process Systems Engineering*, Kyoto, Japan (1982).

PROBLEMS FOR PART VI

Chapter 23

VI.1 (a) Determine the number of degrees of freedom for the system of three storage tanks shown in Figure PII.2.
(b) Determine the number of controlled and manipulated variables.
(c) Construct all alternative control loop configurations for this system.
(d) Screen the alternative loop configurations using qualitative arguments such as speed of response, low static gain, and possible dead time.

VI.2 Repeat Problem VI.1, but consider the system of two noninteracting heaters shown in Figure PII.3.

VI.3 Repeat Problem VI.1, but consider the system shown in Figure PII.4 (see also Problem II.4).

VI.4 Repeat Problem VI.1, but consider the system of two CSTRs shown in Figure PII.8. The reaction is irreversible A → B. Assume isothermal conditions.

VI.5 Repeat Problem VI.1, but consider the drum boiler system described in Problem II.11 and shown in Figure PII.10.

VI.6 Repeat Problem VI.1, but consider the gas absorption column described in Problem II.23 and shown in Figure PII.13.

VI.7 Problem II.20 describes the simple plant shown in Figure PII.11.
(a) Determine the degrees of freedom and the number of controlled and manipulated variables for the whole plant.
(b) Using the decomposition technique presented in Section 23.4, develop all alternative loop configurations.
(c) Take the most "promising" loop configurations for each unit and develop a compatible control structure for the complete plant by erasing any conflicts among the loops.

VI.8 The plant shown in Figure PVI.1 produces gasoline from the dimerization of olefins. The feed, which is composed of propane (24 mole %), propene (20%), isobutane (26%), n-butane (20%), isobutane (3%), and 1-butene (7%), is preheated and then enters the reactors. Each reactor operates at 360°F and 515 psia and is composed of four catalytic beds with intermediate cooling because the reaction is strongly exothermic. Propane from the depropanizer column is used as coolant. The reactor effluent preheats the feed, and after its pressure is reduced to 200 psia, enters the depropanizer. Propane is the top product and is used partly for interstage cooling in the reactor; the other part is removed to storage tanks. The bottom product is sent to the debutanizer, which operates at 90 psia. The top product is butane and after condensation, is stored; the bottom product is gasoline, the desired product.

Develop a "good" control loop configuration for this plant using the

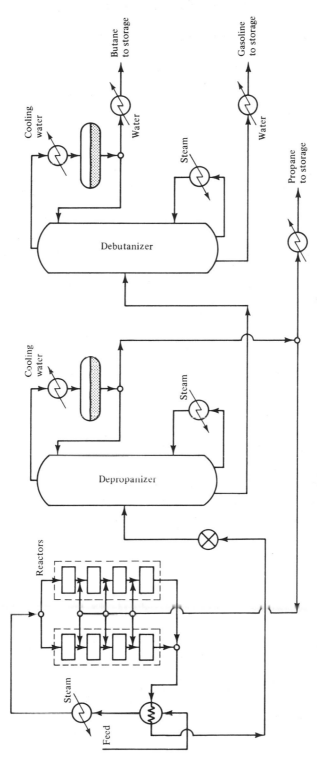

Figure PVI.1

541

decomposition technique discussed in Section 23.4. [Since the amount of quantitative information given in this problem is extremely little, you should take the initiative and make all the qualitative judgements needed. This problem is not intended for rigorous analysis but as an exercise in intuitive and qualitative design of control systems for complete plants.]

VI.9   Consider the two distillation columns shown in Figure Q25.1b. These columns are heat integrated (i.e., the overhead vapor of the first column is used instead of steam as the heating medium in the reboiler of the second column.
(a) Determine the degrees of freedom for the overall system.
(b) Determine the number of controlled and manipulated variables for the system.
(c) Develop a table with all possible control-loop configurations.
(d) Select the seemingly "best" control-loop configuration using qualitative arguments.

VI.10 Figure PVI.2 shows a process for the recovery of acetone from an air–acetone mixture. The air–acetone gaseous mixture enters the absorber at the bottom, and as it goes up the column the acetone is absorbed by the falling liquid water. The solution of acetone in water is subsequently preheated, first by the bottom stream of the distillation column and then by steam. Distillation is used to separate acetone from water.
(a) Determine the degrees of freedom for the overall process.
(b) Determine the number of controlled and manipulated variables for the overall system.
(c) Develop a table with all possible control-loop configurations.
(d) Select the seemingly "best" control-loop configuration using qualitative arguments.

Chapter 24

VI.11 The following sets of linear differential equations describe the dynamic behavior of various multiple-input, multiple-output processes.

(1) $\dfrac{dy_1}{dt} + 2y_1 - y_2 = m_1 + 3m_2$ $y_1(0) = 0$

$\dfrac{dy_2}{dt} - y_1 + 4y_2 = 2m_1 - m_2$ $y_2(0) = 0$

(2) $2\dfrac{dy_1}{dt} + \dfrac{dy_2}{dt} + y_1 + y_2 = m_1 + m_2$ $y_1(0) = 0$

$\dfrac{dy_1}{dt} - 5\dfrac{dy_2}{dt} + 2y_1 - y_2 = m_1 - 2m_2$ $y_2(0) = 0$

(3) $\dfrac{d^2y_1}{dt^2} + 2\dfrac{dy_1}{dt} + y_1 = m_1$ $y_1'(0) = y_1(0) = 0$

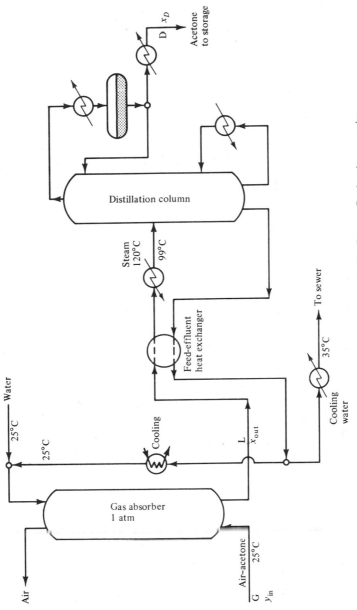

Figure PVI.2 Adapted from Douglas, J. M., *Process Design* (manuscript in preparation), by permission.

$$\frac{dy_2}{dt} - 3\frac{dy_1}{dt} - y_1 - 2y_2 = m_2 \qquad\qquad y_2(0) = 0$$

For each of these processes, assume that manipulated variable m_i is coupled with the controlled output y_i ($i = 1, 2$) through proportional controller.

(a) For which of the processes above do the control loops interact, and for which do they not?

(b) Draw the block diagrams with the control loops included. Indicate the direct and indirect effects of loop interaction.

(c) How would you tune the proportional controllers of the interacting loops so that the overall process is stable?

VI.12 Repeat Problem VI.11 but use the following input–output relationships describing the dynamics of multiple-input, multiple-output processes:

(1) $\bar{y}_1(s) = \dfrac{1}{s+1}\,\overline{m}_1(s) + \dfrac{5}{3s+1}\,\overline{m}_2(s)$

$\bar{y}_2(s) = 10\overline{m}_1(s) - \dfrac{0.8}{10s+1}\,\overline{m}_2(s)$

(2) $\bar{y}_1(s) = \dfrac{2}{s^2+5s+1}\,\overline{m}_1(s) + \dfrac{1}{0.1s+1}\,\overline{m}_2(s)$

$\bar{y}_2(s) = \dfrac{-3}{0.5s+1}\,\overline{m}_1(s) + \dfrac{10}{7s+1}\,\overline{m}_2(s)$

(3) $\bar{y}_1(s) = \dfrac{5e^{-s}}{10s+1}\,\overline{m}_1(s) + \dfrac{2}{s+1}\,\overline{m}_2(s)$

$\bar{y}_2(s) = \dfrac{-4}{(s+1)(2s+1)}\,\overline{m}_1(s) + \dfrac{1}{2s+1}\,\overline{m}_2(s)$

VI.13 Identify the proper couplings between inputs and outputs of the systems in Problem VI.11. Use the relative-gain array methodology so that the resulting loops have the minimum possible steady-state interaction.

VI.14 Consider the processes with input–output relationships given in Problem VI.12. Using Bristol's relative-gain array, select the loops with the minimum steady-state interaction.

VI.15 Consider the following process with three inputs and three outputs.

$$\bar{y}_1(s) = \frac{1}{s+1}\,\overline{m}_1(s) + \frac{2}{(s+2)(s+3)}\,\overline{m}_2(s) - \frac{0.1}{3s+1}\,\overline{m}_3(s)$$

$$\bar{y}_2(s) = \frac{5}{s^2+2s+1}\,\overline{m}_2(s) + \frac{1}{s+1}\,\overline{m}_3(s)$$

$$\bar{y}_3(s) = \frac{10}{5s + 1}\,\overline{m}_1(s) + \frac{7}{(4s + 1)(0.1s + 1)}\,\overline{m}_2(s)$$

$$+ \frac{15}{(s + 1)(s + 5)}\,\overline{m}_3(s)$$

Using Bristol's relative-gain array, select the control loops with minimum steady-state interaction.

VI.16 Select the control loops with minimum steady-state interaction for the following system with two outputs and three inputs:

$$\bar{y}_1(s) = \frac{3}{2s + 1}\,\overline{m}_1(s) - \frac{0.5}{(s + 1)(s + 3)}\,\overline{m}_2(s) + \frac{1}{s^2 + 3s + 2}\,\overline{m}_3(s)$$

$$\bar{y}_2(s) = -10\,\overline{m}_1(s) + \frac{2}{s + 1}\,\overline{m}_2(s) + \frac{4}{(s + 1)(3s + 1)}\,\overline{m}_3(s)$$

VI.17 We like to regulate the composition of distillate and bottom products from a distillation column, using the reflux flow R and vapor flow in the column V (or equivalently the steam flow rate) as manipulated variables. Select the pairings between controlled outputs and manipulated inputs so that the resulting loops offer minimum steady-state interaction. The following input–output relationships for the distillation column have been determined experimentally.

$$\bar{x}_D(s) = \frac{0.60e^{-1.1s}}{(5s + 1)(2s + 1)}\,\overline{R}(s) - \frac{0.5e^{-1.0s}}{(6s + 1)(3s + 1)}\,\overline{V}(s)$$

$$\bar{x}_B(s) = \frac{0.30e^{-1.3s}}{(5s + 1)(s + 1)}\,\overline{R}(s) - \frac{0.50e^{-1.0s}}{(5s + 1)(s + 1)}\,\overline{V}(s)$$

If the best couplings still possess significant interaction between the control loops, design two decouplers that produce two noninteracting control loops.

VI.18 Repeat Problem VI.17, but consider a different distillation column with the following transfer functions:

$$\bar{x}_D(s) = \frac{0.35e^{-1.0s}}{(20s + 1)(25s + 1)}\,\overline{R}(s) + \frac{0.5e^{-2.5s}}{(15s + 1)(10s + 1)}\,\overline{V}(s)$$

$$\bar{x}_B(s) = \frac{0.85e^{-2.0s}}{(40s + 1)(50s + 1)}\,\overline{R}(s) - \frac{0.90e^{-1.0s}}{(30s + 1)(40s + 1)}\,\overline{V}(s)$$

VI.19 Design two noninteracting control loops for the processes of Problem VI.11 if the "best" couplings between inputs and outputs still possess significant loop interaction. Draw the resulting block diagrams.

VI.20 For each of the processes in Problem VI.12, design two noninteracting control loops. Draw the resulting block diagram.

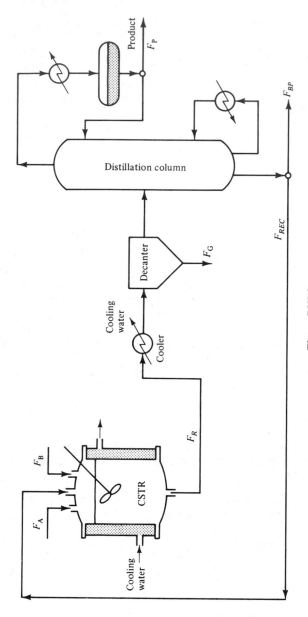

Figure PVI.3

Chapter 25

VI.21 Using qualitative arguments, develop the material balance control system for the process described in Problem II.20 and shown in Figure PII.11.

VI.22 Using qualitative arguments, develop the material balance control system for the process described in Problem VI.8 shown in Figure PVI.1.

VI.23 Using qualitative arguments, develop the material balance control system for the process described in Problem VI.10 and shown in Figure PVI.2. The following information on material balances in also provided:

(1) $Gy_{in} \approx Lx_{out} \approx Dx_D$

(2) $L = 1.4mG$

where G, L, and D are the flows of the gaseous mixture in the absorber, the flow of water in the absorber, and the flow of the distillate product from the top of the distillation column. y_{in} is the concentration of acetone in the air–acetone mixture, x_{out} is the concentration of acetone in the water leaving the absorber, and x_D is the concentration of acetone in the distillate product. m is the slope of the equilibrium curve at the operating conditions of the column.

VI.24 Figure PVI.3 shows the Williams–Otto plant, named after T. I. Williams and R. E. Otto, who originally described it in *Trans. Am. Inst. Electr. Engr., 79*, 458 (1960). Raw materials A and B are mixed in a CSTR and yield the following reactions:

$$A + B \longrightarrow C \qquad C + B \longrightarrow P + E \qquad P + C \longrightarrow G$$

where P is the desired product, and E and G are undesired by-products. The three reactions have an overall exothermic effect and the CSTR is cooled with water. The effluent of the reactor is cooled with cooling water and enters a decanter, where the by-product G is removed. The remaining mixture enters a distillation column where the desired product P is removed as the top distillate stream. The bottom product contains unreacted A, B and by-products C and E. Part of it is removed while the rest is recycled back to the CSTR.

(a) Identify the number of degrees of freedom for each unit and for the overall plant.

(b) Develop a material balance control system for the overall process.

(c) Develop a product quality control system for the process.

(*Note*: If you feel that you need the equations describing each unit, these can be found in the original paper.)

Process Control **VII**
Using Digital
Computers

As one control engineer put it to me in a recent correspondence, "the success of a computer control application depends as much upon the ability of the designer as anything else."

M. Fjeld*

Rapid technological developments in digital computing systems coupled with significant reduction in their cost have had a profound effect on how chemical plants can and should be controlled. Already large plants such as petroleum refineries, ethylene plants, ammonia plants, and many others are under digital computer control. The benefits have been substantial, both in terms of operating cost and in terms of operational smoothness and safety.

The future of digital computers for chemical process control looks even brighter. The high speed of computations, together with the large information storage capacity possessed by digital computers, provides virtually unlimited intelligence which allows the use of quite advanced control techniques, such as adaptive, inferential, multivariable, and supervisory control. Presently, all the new large plants are designed so that they can accept computer control.

*"On Closing the Gap: With Modern Control Theory That Works" in *Chemical Process Control—II*, D. E. Seborg and T. F. Edgar (eds.), Proceedings of Engineering Foundation Conference, Sea Island, Ga. (1981). American Institute of Chemical Engineers, New York (1982)

The subsequent six chapters of Part VII will address the various questions related to the design of digital computer process control systems. In particular:

1. Chapter 26 will present the various hardware elements required by a computer control loop and will identify the new control design questions.
2. Digital computers can process only discrete-time signals, not continuous-time signals. For this reason, in Chapter 27 we will examine how to convert continuous- to discrete-time signals, and vice versa. In the same chapter we will also study the discretization of continuous systems.
3. In Chapter 28 we will introduce z-transforms, which constitute the main tool for the analysis of discrete-time systems and play the same role as Laplace transforms for continuous systems.
4. In Chapter 29 we will analyze the dynamic response of discrete-time systems under open- or closed-loop conditions. In addition, we will develop new criteria in the z-domain for analyzing the stability characteristics of such systems.
5. The design of digital control systems will be the subject of Chapter 30, and Chapter 31 will treat the question of experimentally modeling a process. Finally, the on-line coordination of an experimental modeling procedure with a control algorithm will be examined in Chapter 31, in an attempt to develop on-line adaptive control systems.

Digital Computer **26**
Control Loops

A typical control loop consists of the following components (Figure 26.1): (1) process, (2) measuring sensor and the accompanying transducer (if necessary), (3) controller, (4) final control element with associated electropneumatic converter (if necessary), and (5) transmission lines for the process measurement and the control command signal. As long as the controller is an analog device (pneumatic or electronic) it can (1) process continuously the analog signals generated by the sensors and/or the corresponding transducers, and (2) produce continuous, analog command signals for the final control element. It is obvious that in such a case *all transmission lines carry continuous, analog signals.*

The situation described above has been the basis for all control systems we have examined so far. However, the introduction of a digital

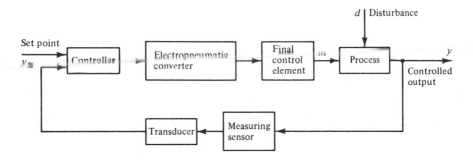

Figure 26.1 Schematic of feedback loop with analog controller.

computer in the place of an analog controller creates the need for new hardware elements and new control design problems. Before we examine what they are, let us briefly review the characteristics of a digital computer and how it is interfaced to the external world.

26.1 The Digital Computer

Despite the differences in capacity, speed, and architecture, all digital computers designed for process control have much the same functions. Figure 26.2 indicates the basic components of such a typical digital computer and the associated peripherals. Let us briefly describe their basic features.

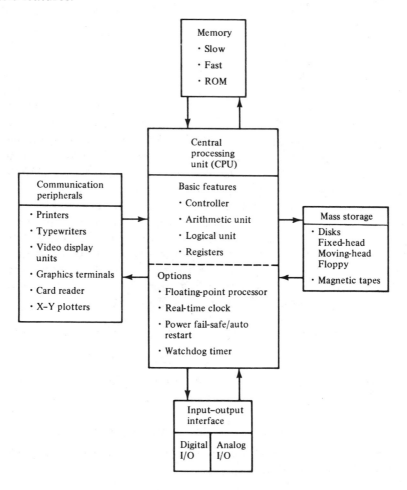

Figure 26.2 Components of a digital computer and associated peripherals.

Central processing unit

Usually the central processing unit is referred to simply as the CPU. The CPU is the heart of a computer system and maintains control over all its functions. Thus it is the CPU that:

1. Determines the next operation to be performed by the computer
2. Executes the various arithmetic or logic operations
3. Retrieves or stores information from or to the memory
4. Directs various other operations, such as data transfer between memory and peripheral devices

The smallest unit of information that the computer uses for communication or computations is the *bit* (from *bi*nary digi*t*), which can assume the value 0 or 1. A specified number of bits together form a *word*. In addition to the bit and word, an intermediate unit called *byte* is also used to characterize information of 8-bit length (i.e., 1 byte = 8 bits). The byte is a useful unit because all alphabet letters, numbers, other typing characters, control characters, and so on, can be fully specified by one byte according to the ASCII Code (industry standards). The CPU stores information in the memory or retrieves information from it in terms of words. The length of a word varies among various computers, with 8-, 16-, or 32-bit words being the most common. Consider the following 16-bit word; 1001010011101111. A usual 16-bit word computer arranges the binary digits in an *octal* system (composed of three binary digits):

$$1 \quad 001 \quad 010 \quad 011 \quad 101 \quad 111$$

The first digit is reserved for the sign (e.g., $0 = +$, $1 = -$) and the other 15 form five groups of three digits each and are used to represent an integer number. Thus the binary form above is equal to the following octal integer:

$$-12357_8 \equiv -\{1 \times 8^4 + 2 \times 8^3 + 3 \times 8^2 + 5 \times 8^1 + 7 \times 8^0\} \equiv -5359_{10}$$

The reader can easily verify that from the binary to the octal system,

$$001_2 = 1_8 \quad 010_2 = 2_8 \quad 011_2 = 3_8 \quad 101_2 = 5_8 \quad 111_2 = 7_8$$

Therefore, the range of integers in a 16-bit word computer is:

$$\text{From: } 1 \quad 111 \quad 111 \quad 111 \quad 111 \quad 111 = -77777_8 - 1 = -32,768_{10}$$

$$\text{To: } 0 \quad 111 \quad 111 \quad 111 \quad 111 \quad 111 = +77777_8 = +32,767_{10}$$

The CPU is equipped with a set of hardware instructions to perform some very basic operations, such as:

Addition, subtraction, and logical comparisons between integer numbers

Storing data in and recalling data from the memory
Transferring data between CPU and the various peripheral devices

More complicated operations can be performed using combinations of these basic hardware instructions.

To expedite and facilitate operations, modern CPUs are equipped with various hardware options. Among all possible options the following are of great value to process control computers.

1. *Hardware floating-point processor:* Performs with very high speed floating-point arithmetic operations and expands tremendously the computational speed of the machine.
2. *Real-time clock:* Every digital computer used for process control must have a real-time clock. This is the device that keeps track of the real world's time and allows the computer to schedule its functions at time intervals in coordination with the various needs of the real world. Thus it is the real-time clock that determines *when* the computer should take data from measuring sensors or change the values of manipulated variables.
3. *Power fail-safe/automatic restart:* In the event that power to the computer is lost, this option senses the power failure and executes a prespecified set of instructions before the machine becomes inoperable. These instructions may transfer control of the process from the digital computer to another back up control system and/or save information necessary for an orderly and automatic restart of the control programs when electrical power has been restored to the computer. This option enhances the safety of computer-controlled processes.
4. *Watchdog timer:* This is another valuable option for process control computers. It allows the computer to determine if the control program is being executed smoothly or if the program is "hung up" in a never-ending loop. In the second case an alarm alerts the operator that the computer has lost control of the process, due to software problems.

Memory

This is the place where the computer stores (1) the instructions of the program it executes, and (2) the values of the initial data, intermediate data, and final results from computations. The smallest unit of storage is the bit, but the memory is organized in terms of words. Thus 4K words memory is composed of 4096 16-bit words (for 16-bit word computers). Each memory word is characterized by a unique address and during the execution of a program the CPU keeps track of the memory address which contains the data or the instruction under execution.

There are two general types of memory: the *random access memory* (RAM) allows data and instructions to be "written" and "read" at any location (address) in the memory. On the contrary, the *read-only memory* (ROM), as its name implies, does not allow alterations of its content (i.e., a program can "read" information from the locations of the ROM but cannot "write" in it). The RAM is used for the storage of any size general-purpose programs, while ROM is employed for the execution of highly specific, small programs. Most of the ROM is "programmed" in the factory and it is used to store basic instructions for starting up the computer or basic input/output commands, and so on. Recently, field-programmable ROMs have been introduced.

The *cycle time* of a computer is the time required by the CPU to read the content of one word from memory and restore its content. According to the value of the cycle time, we can distinguish the following types of RAM:

1. *Core memory*, with a typical cycle time ~ 1 μsec (slow), is low in cost. It is constructed with ferrite rings which retain the stored information when power fails.
2. *Metal-oxide silicon memory* (MOS), with a typical cycle time ~ 500 nsec, is faster and cheaper than core memory and is based on a simple semiconductor device.
3. *Bipolar transistor memory*, with a typical cycle time ~ 300 nsec, is still faster but more expensive. It is contructed from complex integrated circuits.

The cycle time is not the only factor that determines how fast a computer is. Various additional determinants, such as the number and type of basic instructions in CPU, the number of general-purpose registers, and so on, affect the speed with which a computer executes a program.

Mass storage devices

Mass storage devices are used to store large amounts of data and/or instructions. Various types of mass storage devices are available with different (1) capacities for storage, (2) purchase costs, and (3) speeds for accessing and retrieving information. The most common units are:

1. *Disks:* With very large capacity for storage (1 to 300 million 16-bit words), low access time (5 to 100 μsec), and high cost. The disks are distinguished into *fixed-head* and *moving-head* disks. The first have capacities in the range 1 to 20 million 16-bit words and access time ~ 5 μsec, while the second have longer access times

($50 \div 75 \, \mu$sec) but higher capacities (up to 300 million words). *Floppy disks* are low-cost, small-capacity devices and are the most common mass storage facilities for microcomputers.

2. *Magnetic tapes:* These are low-speed mass storage devices with significant capacity (10 to 20 million words). They are seldomly found on process control computers and they are used to store off-line large programs and large amounts of data.

Communication peripherals

These equipments are used for communication between the operator and the computer and include typewriter terminals, line printers, video display units, storage scope graphics terminals, card readers, X-Y plotters, and so on. With such devices the computer can display data describing the current state in the operation of the process it controls, or inform the operator about the current control actions taken by the computer. Furthermore, the communication peripherals allow the operator to intervene and change set points, gains, and other characteristic parameters of a control loop, or switch control from the computer to manual or other backup control systems. The communication peripherals must be supported by easy to use, highly informative, and well-organized software. If this is not the case, the operator may become frustrated or lose confidence in the computer control system, thus rendering it useless.

Input/output (I/O) interface

This is the device which allows the communication between the computer and the process to be controlled. In particular, the I/O interface performs the following functions:

1. It receives the signals from the measuring sensors and transducers associated with the various measured process variables. These signals may be continuous, analog electrical voltages (e.g. thermocouple output, pressure transducer signal), or simple digital information (on–off), from various relays, on–off valves, and so on.
2. It sends out command signals to the various manipulated variables in either analog or digital form.
3. It allows communication with other computers, which are used either as process controllers or number "crunchers." This feature permits the use of several computers in a "distributed digital control systems" architecture.

A digital computer without an input/output interface cannot function as a process controller. But what are the features of such an interface, how does it operate, and how does one select the appropriate interface for a given process control application? These questions are quite central in the design of a computer-based control system and will be covered in Section 26.2 in some detail.

Remark.   It is common practice to characterize the digital computers as large or *maxicomputers, minicomputers,* or *microcomputers.* The standards for such classification are often obscure. Generally, though, a large computer has words of 32 or 64 bits, memory larger than 1,000,000 words, and a large number of associated peripherals. It is used primarily for scientific or business purposes and physically it occupies a large number of cabinets. Its cost is normally larger than $1,000,000. A minicomputer is a 12-, 16-, or 18-bit machine (16 bits is the most popular) with 64,000 to 512,000 words of memory. It has several peripherals and its cost may go up to $200,000, depending on the size of memory and associated peripherals. Microcomputers have been traditionally characterized as machines with 8-bit words, although today one sees typical 16-bit processors characterized as micros. They possess from 1000 to 64,000 words of memory and a few peripheral devices. The cost of the basic CPU is less than $200 and goes up depending on the memory size and peripherals. Maxicomputers arc very costly for process control purposes and are not used. Minicomputers are well suited to control a large number of control loops. But due to their low cost and tremendous abilities, the future in process control applications belongs to micros.

26.2   Computer–Process Interface for Data Acquisition and Control

Return to Figure 26.1, which shows all necessary hardware elements in a loop with an analog controller. Replace the controller by a digital computer. Then the control functions (e.g., feedback P, PI, PID laws) will be performed by an executable program (in BASIC, FORTRAN, assembly language, etc.), which resides in the memory of the computer. It is obvious that such control program requires as data (input) the values of the measured outputs and produces as results (output) the values that the manipulated variables should have, in order to keep the controlled variables at the desired set points. *For a digital computer both input (data) and output (results) are in digital form and correspond to discrete-time values.* Here is where problems of incompatibility arise and dictate the necessary hardware elements for an input/output interface between a digital computer and the controlled process.

Samplers

The process measurement data (flow rates, pressures, liquid levels, temperatures, etc.) are provided *continuously in time* by the various measuring sensors and transducers. However, the computer can handle information on a *discrete-time* basis (i.e., at given time instants) for the following reason: The time taken by the computer to "read" the measured value, calculate the error, and make a control correction is finite. If during this period the measured value changed, this is not recognized by the computer. Then the computer "reads" in effect at discrete-time intervals. This is denoted through the use of a *sampler*, which is simply a switch closing at specified time intervals. In other words, a sampler takes in values of a continuous signal and produces a sequence of sampled values at particular time instants (Figure 26.3a).

Hold elements

On the other hand, most of the final control elements (pneumatic valves in particular) are actuated by continuous-time signals (e.g., compressed air). Therefore, the control commands produced by the computer program should be converted from discrete-time to continuous-time signals. This is accomplished by the *hold elements*. Figure 26.3b shows schematically the conversion of a sequence of discrete-time signals to a stair-step-like continuous signal. Here the hold element keeps the value of a discrete-time signal constant for the entire period until the next signal comes along.

Analog-to-digital converters

The measurement data are not only provided continuously in time, but they are also *analog* electrical signals in nature. They cannot be used directly by the control program, which requires data in a *digital* form (e.g., information coded in 16-bit words, for a 16-bit word machine). Therefore, the input interface should contain an *analog-to-digital converter* (ADC or A/D converter).

The analog signals coming from measuring devices and sensors are modified so that they fall within a prespecified voltage range (e.g., 0 to 10 V, 0 to 5 V, −10 to +10 V, or −5 to +5 V, etc.). The digital signal produced by an A/D converter is expressed by an integer number in a binary form. The resolution of the conversion depends on the number of bits used by the converter to encode an analog value in digital form. The most common converters use 8-bit or 12-bit resolution, with the second providing smaller error and being more costly.

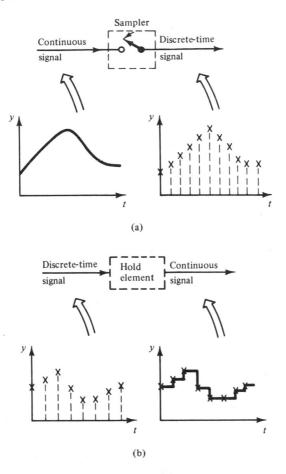

(a)

(b)

Figure 26.3 (a) Conversion of continuous to discrete-time signal; (b) reconstruction of continuous signal from its discrete-time equivalent.

Consider a voltage range of 0 to 1V and an n-bit converter. The n bits define 2^n integer numbers (including zero), which in turn define $2^n - 1$ voltage intervals between 0 and 1. Thus the accuracy of the conversion expressed by the value of *resolution* is given by

$$\text{resolution} = \frac{1}{2^n - 1} \qquad (26.1)$$

For a 12-bit converter (one bit for sign and eleven for discretization) the resolution is about 0.05%; that is, when two voltage values differ by more than 0.05% of the prespecified voltage range, the converter will distinguish the two signals and assign two different integers for them. For an 8-bit converter the resolution is smaller, about 0.4%. Usually, both 8-bit and 12-bit converters are satisfactory for process control

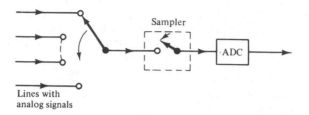

Figure 26.4 A/D converter with multiplexer.

purposes. Converters with more than 12 bits are used only when extremely high precision is required and are quite costly.

The conversion speed is very high and typical A/D converters used for process control allow 20,000 to 100,000 conversions per second. Higher rates can be achieved by high-performance converters and are useful only for very special problems.

To avoid the need for a large number of A/D converters handling the conversion of a large number of different analog signals, it is common practice to use a *multiplexer*. This is an electronic switch with several ports, which can serve sequentially several lines carrying analog signals (Figure 26.4).

Digital-to-analog converters

The control commands produced by the control program are in digital form, but most of the final control elements, pneumatic valves in particular, are actuated by analog signals (e.g., compressed air). To erase this incompatibility, the output interface should include a *digital-to-analog converter* (DAC or D/A converter).

D/A converters function in the reverse manner to A/D converters. Thus for a 12-bit converter we can have $2^{12} = 4096$ integer numbers defining 4095 intervals of the prespecified voltage range, say 0 to 10 V. Then the integer number 516 causes an analog output of

$$\frac{516}{4095} \times 10 = 1.26 \text{ V}$$

Digital I/O

A digital computer control system may be required to handle digital inputs or outputs for a variety of reasons. Typical examples are:

1. Information concerning:
    (a) Status of relays turning pumps, valves, lights, and other devices on or off

(b) Status of multiplexers

(c) Settings of various switches

(d) Status of communication peripherals and various digital logic devices

2. Control commands to:

(a) Relays, switches, solenoids, digital logic devices

(b) Stepping motors

3. Communication between:

(a) Several computers

(b) A computer and its peripherals, etc.

The digital signals are fully compatible with a computer so that no special converters are needed in the I/O interface. The transmission (input or output) of digital signals by the I/O interface can be done either in *parallel* (two-way, in and out, simultaneous transmission) or in *series* (one-way, in or out transmission). The length of a digital information transmitted in or out is one word (i.e., 16 bits for a 16-bit machine). The transmission rates vary from very low to very high, and are expressed in terms of *baud rates*, where

baud rate = 10 × (number of characters transmitted/second)

Remark. When the prespecified range of voltages involves negative and positive values, the first bit of an A/D or D/A converter is used to denote the sign. Consider the range −5 to +5 V. For a 12-bit converter we have $2^{12-1} = 2048$ positive integer numbers (including zero) to represent voltage values in the positive range $0 \leq$ volts ≤ 5. Also, there are 2048 negative integer numbers (excluding zero) covering the range $-5 \leq$ volts < 0.

26.3   Computer Control Loops

In the preceding two sections we gave a brief description of a digital computer and its characteristics, of the associated peripherals, and of the I/O interface required for data acquisition and control. In this section we examine the various types of control loops that result when a digital computer is used as the main controller, as well as the necessary hardware components.

Single-loop control

Figure 26.1 shows the hardware elements of a single-loop control system using an analog controller. When we replace the analog controller by a digital computer, the following changes take place:

1. The measurement signal from the sensor or transducer is *sampled* at prespecified intervals of time using a simple sampler. Thus it is converted from a continuous- to a discrete-time signal. This in turn is converted from analog to digital by an A/D converter and enters the computer.
2. The hardwired analog logic of an analog controller is replaced by the *software of the control program*, which resides in the memory and is executed by the computer whenever it is called.
3. The control commands produced by the control program are digital and discrete-time signals. They are first converted to analog by a D/A converter and then to continuous-time signals by simple hold elements before they actuate the final control elements.

Figure 26.5 summarizes the foregoing changes and indicates all hardware components present in a single computer control loop. We observe that both continuous and discrete-time signals are present in the loop. They are denoted by c: and d:, respectively. It should also be noted that the set point values, as well as the values of the adjustable control parameters (e.g., gains, reset or rate-time constants, etc.), are now introduced by the operator through a typewriter terminal.

Multiple-loop control

A digital computer can be used to control simultaneously several outputs, not only one as discussed above. We still need an interface between the computer and the process, but it is now somewhat different. Thus:

1. Instead of using one A/D converter for every measured variable, we employ a single A/D converter which serves all measured variables sequentially through a multiplexer.
2. A multiplexer can also be used to obtain several outputs from a single D/A converter.
3. The control program is now composed of several subprograms, each one used to control a different loop. Furthermore, the control program should be able to coordinate the execution of the various subprograms so that each loop and all together function properly.

Figure 26.6 shows the use of a single computer (CPU) to control two outputs.

When a digital computer has assumed all control actions of a conventional controller, we talk of *direct digital control* (DDC). Both systems in Figures 26.5 and 26.6 are examples of direct digital control.

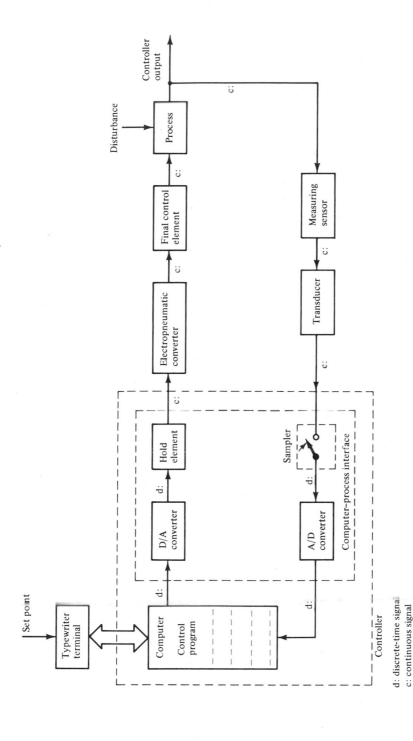

Figure 26.5 Hardware components of digital computer control loop.

d: discrete-time signal
c: continuous signal

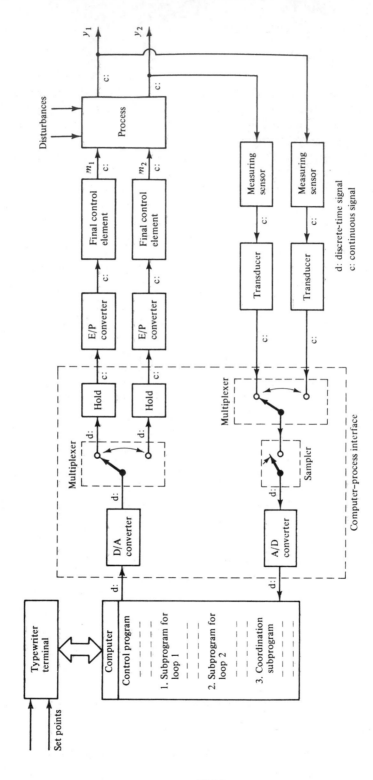

Figure 26.6 Digital computer used in two distinct control loops.

564

Supervisory control

Unlike the situation of direct digital control, we may use the computer to change only the set points or the values of the adjustable control parameters of the local controllers. The resulting system is known as *supervisory control* (**Figure 26.7a**).

The local controllers may be conventional analog devices or digital computers implementing direct digital control. An I/O interface is needed to inform the supervising computer about the state of the local control loops and for the computer to provide the set point or other information to the local controllers. When the local controllers are digital computers, the I/O interface carries only digital signals, allowing communication between the supervising computer and the local DDCs.

Supervisory control has been applied extensively in chemical processes, to optimize their operation (minimize operating cost, maxi-

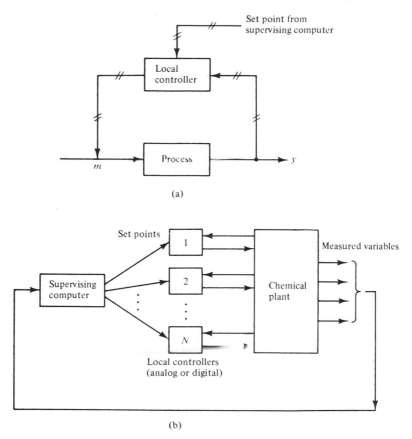

(a)

(b)

Figure 26.7 Supervisory control: (a) single loop; (b) several loops.

mize efficiency in energy or raw materials utilization, maximize production profit, etc.). Thus one computer supervises and coordinates the operation of several control loops, deciding what are the best set point values for the various loops. Figure 26.7b shows schematically the architecture of a supervisory control system for a chemical plant.

Remarks

1. In some cases, the voltage signal produced by the measuring sensor or transducer is very low and easily corrupted by noise. A typical example is the output of a thermocouple which is of the order of milivolts. Low-voltage signals are normally amplified to the prespecified voltage range for A/D conversion. If several signals need amplification, we may use a multiplexer first, followed by a single common amplifier.
2. The computer is physically located some distance from the controlled process. We can put the computer-process I/O interface (a) close to the computer or (b) close to the controlled process.

   In the first case we have analog signals transmitted over some distance between the process and the interface. This approach yields satisfactory results if the analog signals are transmitted over short distances (i.e., no longer than 200 to 300 ft). For longer distances there is significant deterioration of the transmitted signal, due to voltage losses and cable capacitance. Furthermore, external noise may seriously corrupt the transmitted analog signal.

   Alternative (b) is preferred when the transmission distance is long and there are strong sources of external noise. In such case the proximity between process and interface allows conversion of "clean" and "strong" analog signals to digital, which can then be transmitted to the computer. Digital signals are less susceptible to external noise and can be transmitted over long distances by telephone lines.
3. Microcomputers are normally used for local direct digital control of several loops (5 to 10). Minicomputers are usually employed as supervising computers in a supervisory control architecture.

26.4   New Control Design Problems

The introduction of a digital computer for process control raises some design questions which were not covered in earlier chapters.

1. The digital computer uses and produces information in discrete-time form. Therefore, the continuous process models that we have

used for the design of analog controllers are not appropriate. We need to develop a mechanism that will convert the differential equations describing the process to difference equations which are convenient for discrete-time representation.

2. How fast should we sample a measured variable to produce its discrete-time equivalent? Does the sampling rate affect the quality of control?

3. How should we reconstruct a continuous signal from its discrete-time equivalent so that we can actuate the final control elements? How does the type of reconstruction (i.e., type of hold element) affect the quality of control?

4. The Laplace transforms allowed us to develop simple input–output relationships for a process and provided the framework for easy analysis and design of loops with continuous analog controllers. For discrete-time systems we need to introduce new analytical tools. These will be provided by the z-transforms.

5. Does the design of a control loop change when we use digital computer control? What about the stability conditions and the tuning of a loop?

6. How can we use the tremendous computational power of a computer to implement some advanced notions of process control, such as feedforward, adaptive, inferential, optimizing, and so on?

In the following chapters we will address all the above and other questions related to the design of computer control systems.

Before closing this section, let us make a few remarks on the *software* required to implement the control laws, which constitutes a new control design problem introduced by the use of digital computers.

There are two classes of software programs needed for computer process control applications: the *computer system programs* and the *application programs*.

1. *Computer system programs* are supplied by the manufacturer of process control computers or specialized software houses. They include:

   (a) *Operating systems*, which deal with the real-time operation of the computer control system. They supervise the execution of the control programs and to this end organize the various operations of the hardware components in an orderly manner, providing efficient use of the CPU, memory, communication peripherals, and I/O interface.

   (b) *Utility programs*, such as assemblers, editors, debuggers, compilers, and so on, which support the development of the application programs written by the users.

2. *Application programs:* These are written by the user and perform

the specific functions required by the control problem, such as:

(a) Monitoring the measured process variables at specified time intervals
(b) Executing the algorithms of the control laws
(c) Coordinating the control actions to the various final control elements
(d) Computing and changing set points
(e) Computing and changing the values of the adjustable controller parameters
(f) Calling alarms if process variables exceed preset limits, and so on.

The application programs may be written in high-level languages such as FORTRAN or low-level languages such as machine language. High-level languages are easily understood by the programmer and allow an easy statement for the solution procedure. However, they require increased memory and slow down the execution because FORTRAN statements, for example, must be translated into machine language before they can be executed. Normally, one writes the complex part of a control program in a high-level language because it is an overwhelming task for machine language programming, while the latter is used only to encode those functions that are performed at high speeds and repeatedly (data acquisition, implementation of control commands, etc.)

THINGS TO THINK ABOUT

1. What is a digital signal and what is an analog signal (information)? Identify their differences and discuss how one can be converted to the other.

2. Define bit, byte, and word. Why are all needed to encode information in a digital system?

3. Describe the structure of a conventional digital computer, and identify the characteristics of each hardware component in this structure.

4. What are the basic and what are the optional features of a central processing unit? Are all of them needed for a process control computer?

5. What is the real-time clock needed for? How does it function? Why is it very difficult to use a computer for process control without a real-time clock? Do you have any suggestions on how you can count time elapsed without a real-time clock?

6. Identify the functions and hardware components of a computer-process I/O interface.

7. Explain in simple physical terms how you can convert a continuous- to a

discrete-time signal, and vice versa. Why are these two operations necessary in a computer-process I/O interface?

8. Describe in physical terms the conversion of a signal from analog to digital, and vice versa. Why are these conversions needed in the I/O interface?

9. What is easier and less costly to do? (a) multiplex N analog signals first and then use a common A/D converter, or (b) convert the N analog signals to digital first and then multiplex them to enter the computer through a single, common word of storage?

10. Repeat question 9, but use N digital signals, a multiplexer, and D/A converters.

11. For a prespecified voltage range of -10 to $+10$ V, find the resolution of a 12-bit A/D converter. What are the integer numbers representing voltages of -2 V and $+5$ V? What is the possible conversion error in volts?

12. Find the voltages that are represented by the integer numbers -712 and $+1514$ within a prespecified range -10 to $+10$ V and what is the possible conversion error (in volts), for 8-bit and 12-bit converters.

13. Find the number of bits needed for a D/A converter to yield an error of less than 0.0001 V, for a prespecified range of voltages 0 to 5 V.

14. Define direct digital and supervisory control. Which one is used for regulatory control actions and which for servo operations? In a supervisory control mode, what are better as local controllers, analog or digital devices? Discuss relative advantages and disadvantages.

15. What size computers would you use for DDC and supervisory control? Why? How do the local DDCs communicate with the supervising computer?

16. Identify all components of a DDC system using one microprocessor to handle the four loops of a flash drum (see Example 23.2, Figure 23.1b).

17. Describe a supervisory control system for the plant of Example 23.4 (Figure 23.10) using as local controllers: (a) analog devices controlling one loop each, and (b) microcomputers capable of handling four loops each. How would you select the loops to be controlled by each local microprocessor?

18. What is a high-level and what is a low-level language for computer control applications? Which one would you use, and why?

19. Discuss the new design problems raised by the use of a digital computer for process control.

From Continuous-Time **27**
to Discrete-Time Systems

When we use continuous analog controllers, all signals in a loop are continuous in time. Then the dynamic behavior of each component in the loop (process, measuring device, controller, final control element), as well as the response of the overall control system, can be effectively analyzed by continuous models (differential equations in the time domain or transfer functions in the Laplace domain).

The introduction of a digital computer in a process control loop changes the picture because a computer can handle information on a discrete-time basis only (i.e., at particular time instants). As we can see from Figure 26.5, in a computer control loop we have both continuous- and discrete-time signals present. The implication of this feature is twofold:

1. Continuous signals must be converted to discrete-time before they can be "read" by the computer, and the discrete-time control commands produced by the computer must be converted to continuous signals before they can actuate the final control elements.
2. The continuous models (e.g., differential equations in the time domain, or input–output models in the Laplace domain) are not convenient to use to analyze the dynamic behavior of loops with computer control; discrete-time models are needed.

Therefore, before proceeding with the development of design techniques for computer control systems, we should study how to convert continuous signals and models to discrete-time equivalents, and vice versa. This is the subject of Chapter 27.

27.1   Sampling Continuous Signals

Consider a line carrying a measurement signal y, which varies continuously in time as shown in Figure 27.1a. The line is interrupted by a switch, called *sampler*, which closes every T seconds and remains closed for an infinitesimally short period of time (theoretically, a time point). The x's of Figure 27.1b show the value of the signal at the other end of the line, when $T = 1$ second. We notice that the resulting signal has values only when the time is a multiple of T, that is, at time points

$$t = nT, \qquad n = 0, 1, 2, \ldots$$

and is zero for any other times. The signal of Figure 27.1b is called

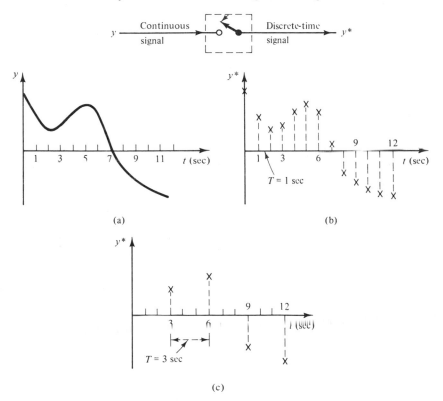

Figure 27.1 Continuous signal and its discrete-time representation with different sampling rates.

discrete-time or *sampled representation* of the continuous signal in Figure 27.1a, with a *sampling period* of 1 second. Figure 27.1c shows another sampled representation of the same continuous signal but with a sampling period of 3 seconds. Two observations are easy to make:

1. As the sampling period tends to zero, the sampled representation comes closer to the continuous signal but requires an explosively larger number of sampled values.
2. On the other hand, as the sampling period increases, fewer sampled values are required, but the sampled representation of a continuous signal deteriorates, and the reconstruction of the original signal becomes poor or impossible.

Thus the obvious question arises:

How does one select the best sampling period so that the sampled representation of a continuous signal is satisfactory without requiring an excessively large number of sampled values?

There exists a mathematical answer to this question whose development is quite complex and goes beyond the scope of this text. We will try to give a practical answer based on the typical dynamic responses encountered in chemical processes.

Example 27.1: Sampling the Response of First-Order Systems

Consider a first-order linear system subject to an input step change. Figure 10.4 shows the response of the system with time. In Section 10.4 we found that the response reaches 63.2% of its final value when the time elapsed is equal to one time constant τ_p. Also, when $t = 2\tau_p$ the response has reached 86.5% of the final value, at $t = 3\tau_p$ 95%, and so on. Therefore, if the sampled representation of the response is going to be of any value, the *sampling period must be smaller than one time constant*. How much smaller? Practical experience suggests that a sampling period between 0.1 and 0.2 of one time constant yields satisfactory results.

Suppose that the first-order system possesses dead time. Then its response to a step input is given by Figure 12.3a. If the dead time t_d is of the same order of magnitude as the time constant τ_p, select the sampling period equal to $0.1t_d$ or $0.1\tau_p$, whichever is smaller. If t_d is much smaller than τ_p, neglect the dead time and take $T = 0.1\tau_p$.

Example 27.2: Sampling the Response of Overdamped Systems

The rule developed in Example 27.1 for the sampling rate of a first-order response can be extended to cover a large class of overdamped systems. Figure 16.9 shows the experimental response of an overdamped process to an input step change. The S-shaped response of Figure 16.9 can be approximated by the response of a first-order plus dead time system, as described in Section 16.5. Thus we can identify the dominant time constant τ_p and the dead time t_d. The sampling period should be $0.1\tau_p$ or $0.1t_d$, whichever is smaller, or $0.1\tau_p$ if t_d is much smaller than τ_p.

Example 27.3: Sampling the Oscillating Response of a System

Oscillatory behavior is exhibited by underdamped open- or closed-loop systems (see Chapter 11) and by the steady-state response of linear systems subject to periodic, sinusoidal input changes (Chapter 17).

To develop a good sampled representation of an oscillating signal, follow the rule:

Sample an oscillating signal more than two times per cycle of oscillation; otherwise, it is impossible to reconstruct the original signal from its sampled values.

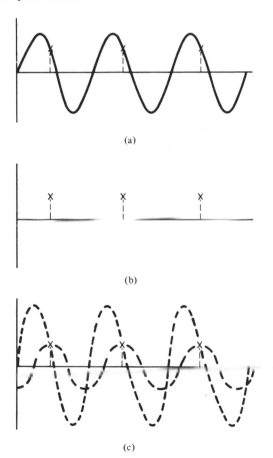

(a)

(b)

(c)

Figure 27.2 Sampling once per cycle does not lead to unique reconstruction of periodic signals.

To demonstrate this rule, consider the sinusoidal signal of Figure 27.2a sampled once per cycle. The sampled values are shown in Figure 27.2b, and Figure 27.2c shows clearly that we cannot reconstruct uniquely the original sinusoid because there exist several waves passing through the sampled values. Therefore, sampling with a period equal to the period of oscillation renders useless sampled values.

The example of Figure 27.3 demonstrates a serious error that can be committed by an improper selection of the sampling period. The sinusoid of Figure 27.3a is sampled with a period equal to three-fourths of the period of oscillation (i.e., 4/3 samples per period or, better expressed, four samples per three cycles of oscillation). The sampled values are shown in Figure 27.3b. When we attempt to reconstruct the sinusoid going through these sampled values, we take the signal of Figure 27.3c, which is clearly different from the original.

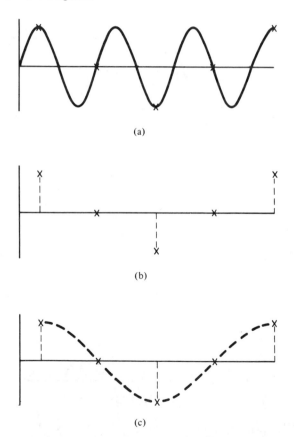

(a)

(b)

(c)

Figure 27.3 Adapted by permission from D.A. Mellichamp (editor), *Real-Time Computing* (*with Applications to Data Acquisition and Control*), Van Nostrand Reinhold Co.

Let us now develop a quantitative description for the sampling operation and the resulting sampled values of a continuous signal.

The sampler is a physical switch that stays closed for a very small but finite period of time, Δt, around the sampling instant. During this time the sampler output takes the value of the continuous signal and has the form shown in Figure 27.4a, or approximately the pulse form of Figure 27.4b. To develop a concise mathematical description we assume that the sampler acts instantly (i.e., $\Delta t \rightarrow 0$). To retain the same area under the pulse as $\Delta t \rightarrow 0$, the height of the pulse goes to infinity and at the limit we take *an impulse of infinite magnitude, zero duration, and an area ("strength") under the impulse equal to the magnitude of the continuous signal at the sampling instant.* Thus the impulse at the sampling point $t = nT$ ($n = 0, 1, 2, \ldots$) is given by

$$y^*(nT) = y(nT)\delta(t - nT) \tag{27.1}$$

where $\delta(t - nT)$ is the unit impulse or Dirac function at $t = nT$ (see Section 7.2 and Figure 7.3b). A sampler with the idealized output given by eq. (27.1) is known as *ideal impulse sampler*.

We can extend eq. (27.1) to apply for any time. Thus the sequence of impulses $y^*(t)$ coming out of an impulse sampler is expressed by the following equation in the time domain:

$$y^*(t) = y^*(0) + y^*(1\,T) + y^*(2T) + \cdots$$

$$= y(0)\delta(t) + y(T)\,\delta(t - T) + y(2T)\,\delta(t - 2T) + \cdots$$

and finally,

$$y^*(t) = \sum_{n=0}^{\infty} y(nT)\delta(t - nT) \tag{27.2}$$

Equation (27.2) is compatible with the idealized physical picture we have considered:

At the sampling instants the "strength" of the impulses is equal to the value of the continuous signal.
Between the sampling instants the "strength" is zero (i.e., no output value).

Remark.   Take the Laplace transforms of both sides of eq. (27.2):

$$\overline{y}^*(s) = \sum_{n=0}^{\infty} y(nT)\mathcal{L}[\delta(t - nT)] = \sum_{n=0}^{\infty} y(nT)e^{-nTs}\mathcal{L}[\delta(t)]$$

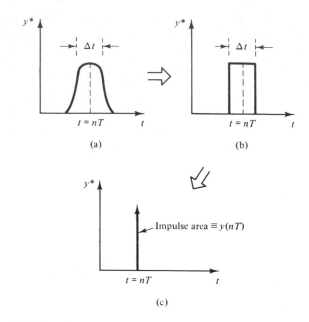

Figure 27.4 From the response of a real sampler to the response of an ideal impulse sampler.

From the last equation we find an expression for the sequence of impulses coming out of the impulse sampler in the Laplace domain (i.e., the s-domain):

$$\overline{y}^*(s) = \sum_{n=0}^{\infty} y(nT)e^{-nTs} \qquad (27.2a)$$

27.2   Reconstruction of Continuous Signals from Their Discrete-Time Values

The discrete-time nature of a digital computer implies that when a computer is used to control a process, the *control commands are given periodically as impulses* at particular time instants and not continuously in time. Such sequence of control impulses cannot maintain a final control element continuously in operation. Thus a valve opens when a control impulse from the computer reaches the valve, but then it closes until the next control impulse arrives at the valve. Such control action is undesirable and the question is: How can we construct a continuous signal from its discrete-time values?

Consider a control signal produced intermittently, every T seconds, by a computer and expressed by a series of impulses (discrete-time values) as shown in Figure 27.5a:

$$m*(0) = m(0)\delta(t), \quad m*(T) = m(T)\delta(t - T),$$
$$m*(2T) = m(2T)\delta(t - 2T), \quad \ldots$$

The simplest way to convert a sequence of discrete-time values into a continuous signal is to keep the discrete-time value of the signal constant, until the next one comes along. Thus if $m(t)$ is the resulting continuous signal, we have

$$m(t) = m(nT) \quad \text{for } nT \le t < (n + 1)T \quad \text{and} \quad n = 0, 1, 2, \ldots \quad (27.3)$$

In particular:

$$\text{for } 0 \le t < T \qquad m(t) = m(0)$$
$$\text{for } T \le t < 2T \qquad m(t) = m(T)$$
$$\text{for } 2T \le t < 3T \qquad m(t) = m(2T), \text{ etc.}$$

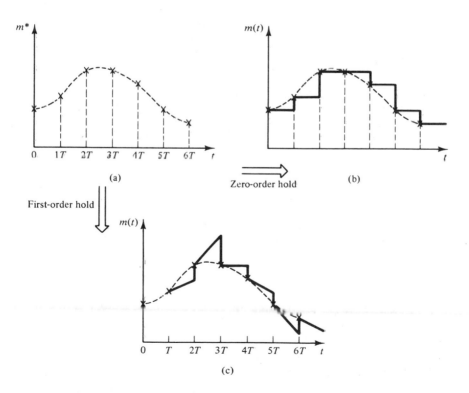

Figure 27.5 (a) Sequence of impulses; (b) reconstruction of continuous signal with zero-order hold; (c) reconstruction with first-order hold.

The resulting stair-step continuous signal is shown in Figure 27.5b. The conversion represented by eq. (27.3) is known as *zero-order hold*. It does not represent the only way to construct a continuous signal from its discrete-time values.

Consider two successive discrete-time values, say $m[(n - 1)T]$ and $m(nT)$. We assume that for the next period $nT \leq t < (n + 1)T$, the continuous signal can be given by a *linear extrapolation of the previous two values*:

$$m(t) = m(nT) + \frac{m(nT) - m[(n - 1)T]}{T}(t - nT)$$

$$\text{for } nT \leq t < (n + 1)T \quad \text{and} \quad n = 1, 2, 3, 4, \ldots$$

(27.4)

Equation (27.4) yields the *first-order hold*, and the continuous signal it produces is shown in Figure 27.5c. Notice that *the first-order hold element needs at least two values to start construction of the continuous signal, whereas the zero-order hold needs only one*.

It is possible to develop second-, third-, or higher-order hold elements. They need three, four, or more initial discrete-time values before they can start the construction of a continuous signal. As the order of a hold element increases, the computational load increases and becomes more complex, with marginal improvements in the quality of the reconstructed signal. Therefore, for most process control applications the zero-order hold element provides satisfactory results with low computational load and is normally used. *To improve the quality of a reconstructed signal, it is better to decrease the period between two successive discrete-time values rather than increase the order of the hold element*.

Example 27.4: Comparing the Results of Zero- and First-Order Hold Elements

We will consider two distinct cases of discrete-time signals: (1) slowly varying with time (Figure 27.6a), and (2) rapidly changing with time (Figure 27.7a):

1. For the slowly varying signal the superiority of the first-order hold is obvious (compare Figure 27.6b and c). This is due to the almost constant slope of the changing signal over large periods of time, which permits a successful linear extrapolation. Nevertheless, the performance of the zero-order hold is also considered satisfactory.
2. For the rapidly changing signal both reconstructions are rather poor (compare Figure 27.7b and c). This is mainly due to the long sampling period between two successive discrete-time values. Any improvement should come from shortening this period (i.e., have more discrete-time values of the signal per unit of time). Nevertheless, Figure 27.7b and c indicate some very useful features:
   (a) The zero-order hold element by its nature never generates

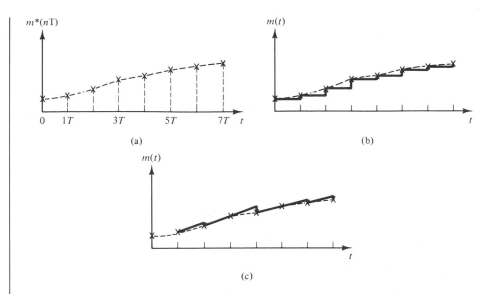

(a)

(b)

(c)

Figure 27.6 Comparison of reconstruction with zero-order and first-order holds, for slowly varying signals.

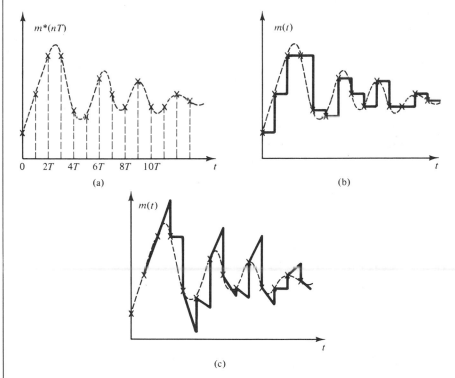

(a)

(b)

(c)

Figure 27.7 Comparison of reconstructions with zero-order and first-order holds, for rapidly changing signals.

"extreme" values outside the range of the discrete-time values. In other words, the zero-order hold produces a "conservative" continuous signal which is satisfactory during periods of slow change and unsatisfactory during periods of fast change in the value of the discrete-time signal.

(b) The first-order hold produces significant excursions beyond the range of the discrete-time values. This could produce undesirably large control actions which may endanger the stability of the controlled process.

Remarks

1. The mathematical basis behind the construction of a hold element, independently of its order, is the following: Consider the continuous signal $m(t)$, which must be constructed from discrete-time values $m(T)$, $m(2T)$, $m(3T)$, and so on. The Taylor series of $m(t)$ around a sampled value $m(nT)$ is given by

$$m(t) = m(nT) + \left(\frac{dm}{dt}\right)_{t=nT}(t - nT) + \frac{1}{2}\left(\frac{d^2m}{dt^2}\right)_{t=nT}(t - nT)^2 + \cdots$$

If we retain only the zero-order term (i.e., the constant) we take the zero-order hold element [eq. (27.3)],

$$m(t) \approx m(nT), \qquad nT \le t < (n + 1)T \qquad (27.3)$$

If we retain the zero- and first-order terms, we take

$$m(t) = m(nT) + \left(\frac{dm}{dt}\right)_{t=nT}(t - nT)$$

The derivative $(dm/dt)_{t=nT}$ can be approximated by

$$\left(\frac{dm}{dt}\right)_{t=nT} = \frac{m(nT) - m[(n - 1)T]}{T}$$

Thus we take the first-order hold element [eq. (27.4)]

$$m(t) = m(nT) + \frac{m(nT) - m[(n - 1)T]}{T}(t - nT) \qquad (27.4)$$

Similarly, by retaining additional terms—second order, third order, and so on—we can develop higher-order hold elements. All necessary derivatives will be numerically approximated as above, but they will require an increasing number of discrete-time values.

2. The output of a zero-order hold element is like a pulse, having a constant height equal to $m(nT)$ and duration T. After recalling that the Laplace transform of a unit pulse is given by eq. (7.12),

then from eq. (27.3) we find that the Laplace transform of a zero-order hold output is given by

$$\overline{m}(s) = m(nT)\frac{1 - e^{-sT}}{s}$$

The last equation implies that the *transfer function of a zero-order hold element* is given by

$$H_0(s) = \frac{1 - e^{-sT}}{s} \tag{27.5}$$

3. In a similar manner we can find the *transfer function of a first-order hold element*:

$$H_1(s) = \frac{1 + sT}{T}\left(\frac{1 - e^{-sT}}{s}\right)^2 \tag{27.6}$$

27.3   Conversion of Continuous- to Discrete-Time Models

We will start by recalling the typical computer loop for direct digital control shown in Figure 26.5. For presentational purposes only we simplify the loop to that shown in Figure 27.8 by retaining its basic four components: process, A/D converter with the associated sampler, digital controller, and D/A converter with the associated hold element. We

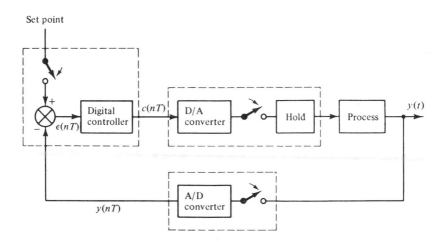

Figure 27.8 Simplified DDC loop.

notice that both continuous- and discrete-time signals are present in the loop. Thus:

1. The process has continuous input and output signals, and consequently it can be described by continuous models (differential equations in the time domain, transfer functions in the Laplace domain).
2. The discrete-time output of the A/D converter can be modeled as a function of the continuous input, by eq. (27.2) in the time domain or by eq. (27.2a) in the Laplace domain.
3. The hold elements can be represented by the corresponding transfer functions [eqs. (27.5) or (27.6)].
4. The digital controller has both input and output signals discrete in time. So far we have not studied any techniques to model such systems, which from now on we will call *discrete*.

Let us go a step further. If the main controller were a continuous feedback PID device, the output of the controller would be given by eq. (13.6) in Section 13.2:

$$c(t) = K_c \epsilon(t) + \frac{K_c}{\tau_I} \int_0^t \epsilon(t)\, dt + K_c \tau_D \frac{d\epsilon}{dt} + c_s \qquad (13.6)$$

A continuous model for the control action such as that of eq. (13.6) is inconvenient for a digital controller which uses error values at particular time instants:

$$\epsilon(0), \quad \epsilon(T), \quad \epsilon(2T), \quad \ldots, \quad \epsilon(nT), \quad \ldots$$

and produces control commands at discrete-time points:

$$c(T), \quad c(2T), \quad \ldots, \quad c(nT), \quad \ldots$$

But how can we convert a continuous-time model to an equivalent discrete-time one? This is the question that we resolve in this section.

Example 27.5: Discrete-Time Model of a Digital PID Controller

Start with the continuous analog of a PID control action, given by eq. (13.6). We will examine each term (proportional, integral, derivative) separately:

1. In every sampling period a sampled value of the process output enters the computer. Let $y(nT)$ be the sampled value at the nth sampling instant. $y(nT)$ is compared to the set point value at the same instant and yields the value of the discrete-time error,

$$\epsilon(nT) = y_{SP}(nT) - y(nT)$$

Then the discrete-time control action produced by the proportional mode is

$$K_c \epsilon(nT)$$

2. The control action produced by the integral mode is based on the integration of errors over a time period. Since the values of the errors are available on a discrete-time basis, the integral $\int \epsilon(t)\, dt$ can only be approximated by numerical integration. Figure 27.9a shows the numerical evaluation of an integral using rectangular integration. It is easy to see that

$$\int_0^t \epsilon(t)\, dt \simeq T \sum_{k=0}^{n} \epsilon(kT)$$

Therefore, the integral mode control action is given by

$$\frac{K_c T}{\tau_I} \sum_{k=0}^{n} \epsilon(kT)$$

3. For the derivative mode action we need a numerical evaluation of the derivative $d\epsilon/dt$. Figure 27.9b shows a first-order difference

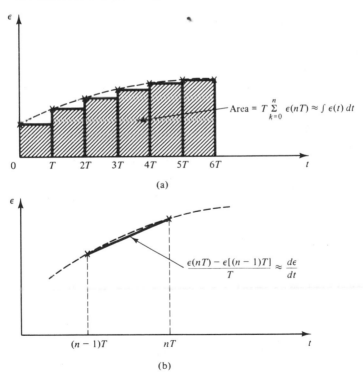

$$\text{Area} = T \sum_{k=0}^{n} \epsilon(nT) \approx \int \epsilon(t)\, dt$$

(a)

$$\frac{\epsilon(nT) - \epsilon[(n-1)T]}{T} \approx \frac{d\epsilon}{dt}$$

(b)

Figure 27.9 Numerical evaluation of: (a) integral; (b) derivative.

approximation of the derivative. Therefore,

$$K_c \tau_D \frac{d\epsilon}{dt} \approx K_c \frac{\tau_D}{T} \{\epsilon(nT) - \epsilon[(n-1)T]\}$$

Consequently, the control action of a digital PID controller is determined by the following discrete-time model:

$$c(nT) = K_c\epsilon(nT) + \frac{K_c T}{\tau_I} \sum_{k=0}^{n}\epsilon(kT)$$

$$+ \frac{K_c \tau_D}{T} \{\epsilon(nT) - \epsilon[(n-1)T]\} + c_s \qquad (27.7)$$

which is nothing other than *a numerical approximation of its continuous counterpart*. Due to the use of finite differences for the approximation of integrals and derivatives, eq. (27.7) is known as the *difference equation*.

Example (27.5) is very instructive on how to develop a discrete-time model from its equivalent continuous one. The procedure can be generalized for any continuous dynamic model as follows:

1. Start with the differential equations describing a continuous model in the time domain.
2. Approximate the derivatives of any order by finite differences.
3. Approximate any integral terms in the model by a scheme of numerical integration.
4. The values of any simple terms are equated to the corresponding discrete-time values at the sampling instants.

The discrete-time modeling equation(s) resulting from the procedure above is known as the *difference equation*(s), in contrast to the term *differential equation*(s) used for the continuous model.

From courses in numerical analysis we know that there exist a variety of techniques to approximate derivatives and integral terms. A detailed exposition of such methods goes beyond the scope of the present text and the interested reader may consult the various references on numerical analysis cited in the references section at the end of Part VII.

Let us close this section with more examples on the time discretization of continuous models.

Example 27.6: Discrete-Time Model of a First-Order Process

A nonlinear first-order process is described by

$$\frac{dy}{dt} = f(y, m) \qquad (27.8)$$

Approximate the derivative by first-order difference:

$$\frac{dy}{dt} \approx \frac{y_{n+1} - y_n}{T}$$

Then, at a given time instant, $t = nT$, eq. (27.8) yields

$$y_{n+1} = y_n + Tf(y_n, m_n) \qquad (27.9)$$

Equation (27.9) is the discrete-time dynamic model of a first-order process and shows what the output of the process will be at the next time instant, using current values of the input m_n and output y_n.

For a linear first-order system we have [see eq. (10.2)]

$$\tau_p \frac{dy}{dt} + y = K_p m$$

and using the foregoing approximation for the derivative we can easily derive the difference equation which is the discrete-time model:

$$y_{n+1} = \left(1 - \frac{T}{\tau_p}\right) y_n + \frac{K_p T}{\tau_p} m_n \qquad (27.10)$$

Note.   In all the expressions above we have used the following simplifying notation: $y(nT) \equiv y_n$ and $m(nT) \equiv m_n$. T is the time period between two successive discrete-time values.

Example 27.7: Discrete-Time Model of a Second-Order Process

A linear second-order system is described by eq. (11.2):

$$\tau^2 \frac{d^2 y}{dt^2} + 2\zeta\tau \frac{dy}{dt} + y = K_p m \qquad (11.2)$$

We have already seen how to approximate the first-order derivative:

$$\frac{dy}{dt} \approx \frac{y_{n+1} - y_n}{T}$$

For the second-order derivative we have

$$\frac{d^2 y}{dt^2} = \frac{d}{dt}\left(\frac{dy}{dt}\right) \approx \frac{d}{dt}\left[\frac{y_{n+1} - y_n}{T}\right] \approx \frac{1}{T}\left\{\frac{y_{n+2} - y_{n+1}}{T} - \frac{y_{n+1} - y_n}{T}\right\}$$

$$= \frac{1}{T^2}(y_{n+2} - 2y_{n+1} + y_n)$$

Replace the derivatives in eq. (11.2) by their approximations and take the following difference equation:

$$\frac{\tau^2}{T^2}(y_{n+2} - 2y_{n+1} + y_n) + 2\zeta\frac{\tau}{T}(y_{n+1} - y_n) + y_n = K_p m_n$$

or

$$y_{n+2} = 2\left(1 - \zeta\frac{T}{\tau}\right) y_{n+1} - \left(\frac{T^2}{\tau^2} - 2\zeta\frac{T}{\tau} + 1\right) y_n + K_p \frac{T^2}{\tau^2} m_n \qquad (27.11)$$

Equation (27.11) represents the discrete-time model of a second-order process. Notice that in order to compute the next value (y_{n+2}) of y, we need its previous *two* values (y_{n+1}, y_n). *For third-, fourth-, and higher-order systems we will need the previous three, four, and more values of y to approximate all derivatives.*

Example 27.8: Discrete-Time Model of a Multivariable Process

Consider the following process with two inputs and two outputs:

$$\frac{dy_1}{dt} + a_{11}y_1 + a_{12}y_2 = b_{11}m_1 + b_{12}m_2$$

$$\frac{dy_2}{dt} + a_{21}y_1 + a_{22}y_2 = b_{21}m_1 + b_{22}m_2$$

With the first-order difference approximation for the derivatives, we take the following difference equations, which represent the discrete-time model of the multivariable process:

$$\frac{y_{1,n+1} - y_{1,n}}{T} + a_{11}y_{1,n} + a_{12}y_{2,n} = b_{11}m_{1,n} + b_{12}m_{2,n}$$

$$\frac{y_{2,n+1} - y_{2,n}}{T} + a_{21}y_{1,n} + a_{22}y_{2,n} = b_{21}m_{1,n} + b_{22}m_{2,n}$$

or

$$y_{1,n+1} = (1 - Ta_{11})y_{1,n} - Ta_{12}y_{2,n} + T(b_{11}m_{1,n} + b_{12}m_{2,n}) \quad (27.12)$$

$$y_{2,n+1} = -Ta_{21}y_{1,n} + (1 - Ta_{22})y_{2,n} + T(b_{21}m_{1,n} + b_{22}m_{2,n}) \quad (27.13)$$

Remarks

1. Numerical differentiation of process measurement data can cause serious problems when there is appreciable process noise (i.e., random effects appearing during the operation but not included in the assumed model). To overcome this difficulty we can use *digital filters*, which filter out any noise and yield "smooth" measurement data.

2. The discretization of continuous models with dead time is rather straightforward. For example, consider a first-order process with dead time t_d between the input $m(t)$ and the output $y(t)$:

$$\tau_p \frac{dy}{dt} + y = K_p \, m(t - t_d)$$

Let $t_d = kT$; that is, the dead time is an integer multiple of the period, T. Then the discrete-time model is easily found to be

$$y_{n+1} = \left(1 - \frac{T}{\tau_p}\right)y_n + \frac{K_p T}{\tau_p} \, m_{(n-k)} \quad (27.14)$$

3. The quality of an approximate discrete-time model improves as the value of the discretization time interval T decreases. Why?

4. The conversion of a discrete to a continuous model in the time domain is possible but not simple. In the next chapter we will see that such conversions are more easily done in the Laplace domain.

THINGS TO THINK ABOUT

1. Explain in your own words why we need to convert continuous- to discrete-time signals, and vice versa. Give a physical example for this need using as a reference the computer control of a stirred tank heater.

2. Using the same computer control system as in item 1, explain why we need to convert continuous- to discrete-time models, and vice versa.

3. What is a sampled signal? Sketch one and indicate how it is related to its continuous counterpart.

4. Define the ideal impulse sampler. How does it differ from a real sampler? Draw two sketches indicating the outputs of an ideal impulse and a real sampler.

5. Develop two mathematical expressions describing the output of an ideal impulse sampler: one in the time domain and the other in the Laplace domain.

6. How would you select the sampling rate for (a) the response of a general underdamped open-loop system, and (b) the oscillating response of a closed-loop system?

7. Is it a satisfactory sampling rate to take three samples every two cycles of an oscillating sinusoidal signal? Demonstrate graphically why or why not.

8. Discuss the mathematical basis for the construction of various orders of hold elements. Develop the time-domain expressions for zero- and first-order hold elements. Describe their functions in physical terms. Can you construct simple electrical circuits that function as zero- and first-order hold elements?

9. Consider the discrete-time signals shown by the two sequences of impulses in Figure Q27.1a and b. What type of hold element would you select to construct the corresponding continuous signals? Elaborate on your answer.

10. Describe two different ways that can be used to improve the quality of a reconstructed continuous signal from its discrete-time values. Outline relative advantages and disadvantages of the two methods.

11. Describe the general procedure for converting a continuous model to a discrete-time model.

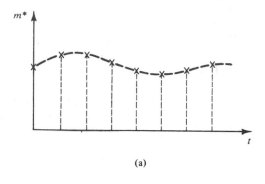

(a)

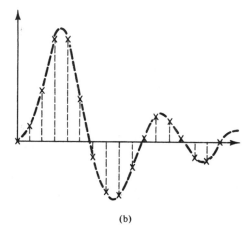

(b)

Figure Q27.1

12. Why is a discrete-time model an approximation of its continuous counter-part?

13. Discretize in time the continuous model of a stirred tank heater.

14. Consider the following two first-order systems:

$$0.01 \frac{dy}{dt} + y = m \qquad \text{and} \qquad 10 \frac{dy}{dt} + y = m$$

With a discretization time interval $T = 1$ second, which one of the two systems will have a better discrete-time representation? Explain why. Also, show how you can improve the quality of the other (worst) discrete-time model.

15. How many sampled output values do you need to construct the discrete-time model for a third-order process? Explain why.

z-Transforms **28**

The use of z-transforms offers a very simple and elegant method for solving linear difference equations which result from the conversion of continuous- to discrete-time models. It allows:

1. The development of input–output models for discrete-time systems, which constitute the basis for the dynamic analysis and design of control loops
2. Simple and straightforward qualitative and quantitative analysis of how discrete-time processes react to external input changes

In other words, z-transforms play the same role for discrete-time systems as that played by Laplace transforms for the dynamic analysis and design of continuous open- or closed-loop systems.

It is for all these reasons that z-transforms have been included in a process control text, although like the Laplace transforms, they constitute a purely mathematical subject.

28.1 Definition of z-Transforms

Consider a continuous function (signal) $y(t)$ sampled at uniform intervals of period T. Let the sequence of sampled values be

$$y(0), \quad y(T), \quad y(2T), \quad \ldots$$

The *z-transform of the sequence of sampled values* is defined by the equation

$$Z\{y(0),\ y(T),\ y(2T),\ \ldots\} = \sum_{n=0}^{\infty} y(nT)z^{-n} \qquad (28.1a)$$

Although the definition above refers to the sequence of sampled values of a continuous function y(t), it is customary to talk about the z-transform of the function y(t), and it is depicted by

$$Z[y(t)] = \hat{y}(z) = \sum_{n=0}^{\infty} y(nT)z^{-n} \qquad (28.1b)$$

From now on, a ^ on top of a variable and the argument z will signify the z-transform of that variable.

Remarks

1. The z-transform maps a discrete-time signal from the time domain to the z-domain (i.e., z is the independent variable).
2. The *z-transform of a given function depends on the selected value of the sampling period T.* This is clear from the defining equation.
3. *If two distinct functions (signals) have the same sampled values at the sampling instants, they also have the same z-transforms and cannot be distinguished.* For example, consider a unit step function and a cosine wave sampled as shown in Figure 28.1. The values of the two functions at the sampling instants are the same and are equal to unity:

$$\cos(\omega \cdot 0) = \cos(\omega T) = \cos(\omega \cdot 2T) = \cdots = 1$$

Then

$$Z[\text{unit step}] = Z[\text{cosine}] = 1 + 1 \cdot z^{-1} + 1 \cdot z^{-2} + \cdots = \frac{1}{1 - z^{-1}}$$

4. A z-transform exists only for those values of the independent

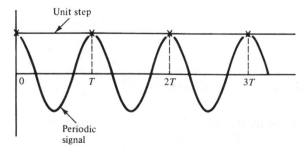

Figure 28.1 Unit step function and periodic signal with same sampled values. (They have the same z-transforms.)

variable z for which the summation of infinite terms in the defining equation (28.1a) or (28.1b) takes finite values.

5. In Section 27.1 we developed the Laplace transform of a sequence of impulses produced by an ideal impulse sampler. Thus if $y(t)$ is the continuous signal entering the sampler and $y^*(t)$ is the sequence of impulses produced by the sampler, the Laplace transform of $y^*(t)$ is given by eq. (27.2a):

$$\overline{y}^*(s) = \sum_{n=0}^{\infty} y(nT)e^{-nTs} \qquad (27.2a)$$

Put

$$z = e^{Ts} \qquad (28.2)$$

and take

$$\overline{y}^*(s) = \sum_{n=0}^{\infty} y(nT)z^{-n} \equiv \hat{y}(z) \qquad (28.3)$$

Equation (28.3) provides an interesting insight: *The z-transform of a sequence of sampled values is a special case of the Laplace transform of the same sequence of impulses* when the independent variables z and s are related through eq. (28.2). We will make frequent use of this result in subsequent sections.

28.2   z-Transforms of Some Basic Functions

Let us now derive the z-transforms of some basic functions which we will use repeatedly in the remaining sections of Part VII.

Unit step function

$$Z[\text{unit step}] = \frac{1}{1 - z^{-1}} = \frac{z}{z - 1} \qquad (28.4)$$

Proof:

$$Z[\text{unit step}] = 1 + 1 \cdot z^{-1} + 1 \cdot z^{-2} + \cdots = \sum_{n=0}^{\infty} 1 \cdot z^{-n} = \frac{1}{1 - z^{-1}}$$

The z-transform exists for $|z^{-1}| < 1$, which makes the series above convergent to a finite value.

Note. In the proof above we have used the following identity, which gives the summation of terms of an infinite geometric series:

$$\sum_{n=0}^{\infty} \lambda^n = \frac{1}{1-\lambda} \qquad \text{with } \lambda < 1 \qquad (28.5)$$

Exponential function

$$Z[e^{-at}] = \frac{1}{1 - e^{-aT}z^{-1}} = \frac{z}{z - e^{-aT}} \qquad (28.6)$$

Proof:

$$Z[e^{-at}] = \sum_{n=0}^{\infty} e^{-anT}z^{-n}$$

This is the summation of terms of an infinite geometric series with ratio $\lambda = e^{-aT}z^{-1}$. Therefore, according to the identity (28.5), eq. (28.6) is proved easily. The transformation exists for those values of z which make the series convergent to finite values (i.e., for $|e^{-aT}z^{-1}| < 1$). From eq. (28.6) it is obvious that

$$Z[e^{at}] = \frac{1}{1 - e^{aT}z^{-1}} = \frac{z}{z - e^{aT}}$$

Ramp function (Figure 7.1a)

$$Z[at] = \frac{aT\,z^{-1}}{(1 - z^{-1})^2} = \frac{aTz}{(z - 1)^2} \qquad (28.7)$$

Proof:

$$Z[at] = 0 + (aT)z^{-1} + (2aT)z^{-2} + (3aT)z^{-3} + \cdots$$
$$= aT(1 + 2z^{-1} + 3z^{-2} + \cdots)z^{-1} = \frac{aTz^{-1}}{(1 - z^{-1})^2}$$

The transformation exists for $|z^{-1}| < 1$.

Trigonometric functions

$$Z[\sin \omega t] = \frac{z^{-1}\sin \omega T}{1 - 2z^{-1}\cos \omega T + z^{-2}} = \frac{z \sin \omega T}{z^2 - 2z \cos \omega T + 1} \qquad (28.8)$$

$$Z[\cos \omega t] = \frac{1 - z^{-1}\cos \omega T}{1 - 2z^{-1}\cos \omega T + z^{-2}} = \frac{z^2 - z \cos \omega T}{z^2 - 2z \cos \omega T + 1} \qquad (28.9)$$

Proof:

$$Z[\sin \omega t] = \sum_{n=0}^{\infty} (\sin \omega nT)z^{-n} = \sum_{n=0}^{\infty} \left\{ \frac{e^{j(\omega nT)} - e^{-j(\omega nT)}}{2j} \right\} z^{-n}$$

In the relationship above we have used Euler's identity (see Section 7.2) to express sin (ωnT):

$$\sin \omega nT = \frac{e^{j(\omega nT)} - e^{-j(\omega nT)}}{2j} \tag{28.10a}$$

Now,

$$\frac{1}{2j} \sum_{n=0}^{\infty} e^{j(\omega nT)} z^{-n} - \frac{1}{2j} \sum_{n=0}^{\infty} e^{-j(\omega nT)} z^{-n} = \frac{1}{2j} Z[e^{j\omega t}] - \frac{1}{2j} Z[e^{-j\omega t}]$$

$$= \frac{1}{2j} \frac{1}{1 - e^{j\omega T} z^{-1}} - \frac{1}{2j} \frac{1}{1 - e^{-j\omega T} z^{-1}} = \frac{1}{2j} \frac{z^{-1}(e^{j\omega T} - e^{-j\omega T})}{1 + z^{-2} - (e^{j\omega T} + e^{-j\omega T})z^{-1}}$$

In the last equation use Euler's identities and replace

$$e^{j\omega T} - e^{-j\omega T} = 2j \sin \omega T \qquad \text{from identity (28.10a)}$$

and

$$e^{j\omega T} + e^{-j\omega T} = 2 \cos \omega T \qquad \text{from identity (28.10b)}$$

Then eq. (28.8) falls out easily.

To prove eq. (28.9), proceed along the same lines as above, using the following Euler's identity:

$$\cos \omega nT = \frac{e^{j(\omega nT)} + e^{-j(\omega nT)}}{2} \tag{28.10b}$$

Translated functions (Figure 28.2)

Consider a function $f(t)$. If this function is delayed by t_d seconds, where t_d is an integer multiple of the sampling period T,

$$t_d = kT \tag{28.11}$$

then the z-transform of the delayed function is given by

$$Z[f(t - t_d)] = \hat{f}(z)z^{-k} \tag{28.12}$$

Proof:

$$Z[f(t - t_d)] = Z[f(t - kT)] = \sum_{n=0}^{\infty} f(nT - kT)z^{-n}$$

Put $\ell = n - k$ and assume that $f(\ell T) = 0$ for $\ell < 0$. Then we have

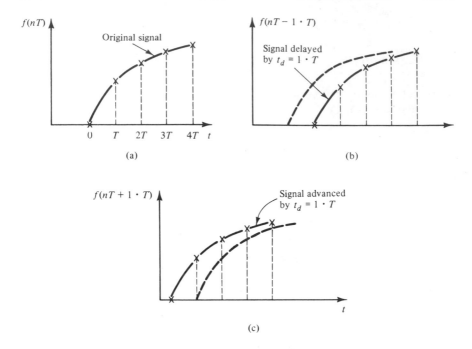

Figure 28.2 Sample values of time-translated functions.

$$\sum_{n=0}^{\infty} f(nT - kT)z^{-n} = \sum_{\ell=-k}^{\infty} f(\ell T)z^{-\ell-k} = \sum_{\ell=0}^{\infty}[f(\ell T)z^{-\ell}]z^{-k} = \hat{f}(z)z^{-k}$$

Similarly, we can show that for a function advanced by $t_d = kT$ seconds, the z-transform is given by

$$\mathcal{Z}[f(t + t_d)] = \hat{f}(z)z^{k} \qquad (28.13)$$

Remarks

1. In Table 28.1 we have tabulated the z-transforms of a number of basic functions that will be encountered often in subsequent sections.
2. It is important to notice from Table 28.1 that *the z-transforms of all functions are ratios of two polynomials in z^{-1} (or z)*.
3. Suppose that a function is translated by a noninteger multiple of the sampling period T (i.e., $t_d = kT + \Delta T$, where $0 < \Delta < 1$). Such systems can be handled by an extension of the z-transforms known as the *modified z-transform*. The interested reader can consult Appendix 28A.
4. By inverting the modified z-transform of a function we can find the value of the function between sampling instants (see Appendix 28A).

28.3   Properties of z-Transforms

Like the Laplace transforms, z-transforms possess certain properties that we will find very useful in dealing with discrete-time systems. Let us examine these properties.

TABLE 28.1.
z-TRANSFORMS OF VARIOUS FUNCTIONS

Function in time domain	Laplace transform	z-Transform
Unit impulse: $\delta(t) = 1$	1	1
Unit step	$\dfrac{1}{s}$	$\dfrac{1}{1 - z^{-1}}$
Ramp: $f(t) = at$	$\dfrac{a}{s^2}$	$\dfrac{aTz^{-1}}{(1 - z^{-1})^2}$
$f(t) = t^n$	$\dfrac{n!}{s^{n+1}}$	$\displaystyle\lim_{a \to 0} (-1)^n \dfrac{\partial^n}{\partial a^n} \dfrac{1}{1 - e^{-aT}z^{-1}}$
Exponential: $f(t) = e^{-at}$	$\dfrac{1}{s + a}$	$\dfrac{1}{1 - e^{-aT}z^{-1}}$
$f(t) = te^{-at}$	$\dfrac{1}{(s + a)^2}$	$\dfrac{Te^{-aT}z^{-1}}{(1 - e^{-aT}z^{-1})^2}$
$f(t) = \sin \omega t$	$\dfrac{\omega}{s^2 + \omega^2}$	$\dfrac{z^{-1} \sin \omega T}{1 - 2z^{-1} \cos \omega T + z^{-2}}$
$f(t) = \cos \omega t$	$\dfrac{s}{s^2 + \omega^2}$	$\dfrac{1 - z^{-1} \cos \omega T}{1 - 2z^{-1} \cos \omega T + z^{-2}}$
$f(t) = 1 - e^{-at}$	$\dfrac{a}{s(s + a)}$	$\dfrac{(1 - e^{-aT})z^{-1}}{(1 - z^{-1})(1 - e^{-aT}z^{-1})}$
$f(t) = e^{-at} \sin \omega t$	$\dfrac{\omega}{(s + a)^2 + \omega^2}$	$\dfrac{z^{-1}e^{-aT} \sin \omega T}{1 - 2z^{-1}e^{-aT} \cos \omega T + e^{-2aT}z^{-2}}$
$f(t) = e^{-at} \cos \omega t$	$\dfrac{s + a}{(s + a)^2 + \omega^2}$	$\dfrac{1 - z^{-1}e^{-aT} \cos \omega T}{1 - 2z^{-1}e^{-aT} \cos \omega T + e^{-2aT}z^{-2}}$

Linearity

The z-transformation is a linear operation. Thus if a_1 and a_2 are constant parameters,

$$\mathcal{Z}[a_1 f_1(t) + a_2 f_2(t)] = a_1 \hat{f}_1(z) + a_2 \hat{f}_2(z) \qquad (28.14)$$

The proof is straightforward.

$$\mathcal{Z}[a_1 f_1(t) + a_2 f_2(t)] = \sum_{n=0}^{\infty} [a_1 f_1(nT) + a_2 f_2(nT)]z^{-n}$$

$$= \sum_{n=0}^{\infty} a_1 f_1(nT)z^{-n} + \sum_{n=0}^{\infty} a_2 f_2(nT)z^{-n}$$

$$= a_1 Z[f_1(t)] + a_2 Z[f_2(t)]$$

Final-value theorem

This is a very useful theorem because it allows us to compute easily the final value of a function from its z-transform. The final-value theorem states that

$$\lim_{t \to \infty} [y(t)] = \lim_{z \to 1} [(1 - z^{-1})\hat{y}(z)] \tag{28.15}$$

The proof is rather easy:

$$(1 - z^{-1})\hat{y}(z) = (1 - z^{-1}) \sum_{n=0}^{\infty} y(nT)z^{-n} = \sum_{n=0}^{\infty} y(nT)z^{-n} - \sum_{n=0}^{\infty} y(nT)z^{-n-1}$$

$$= [y(0) - y(0)z^{-1}] + [y(T) - y(T)z^{-1}]z^{-1}$$

$$+ [y(2T) - y(2T)z^{-1}]z^{-2} + \cdots$$

As $z \to 1$, all the differences above tend to zero, leaving in the limit $y(nT)$ with $n \to \infty$, which is equal to the limit of $y(t)$ as $t \to \infty$.

If $y(nT)$ is the sampled response of a process, *the final-value theorem yields the steady-state value of the process's reponse.*

Initial-value theorem

Given the z-transform of a function, the initial-value theorem allows us to compute the initial value of the function as follows:

$$\lim_{t \to 0} [y(t)] = \lim_{z \to \infty} [\hat{y}(z)] \tag{28.16}$$

The proof of this theorem is also easy.

$$\lim_{z \to \infty} [\hat{y}(z)] = \lim_{z \to \infty} \sum_{n=0}^{\infty} y(nT)z^{-n}$$

As $z \to \infty$, all $z^{-n} \to 0$ for $n = 1, 2, \ldots$. Therefore, the only nonzero term left is that corresponding to $n = 0$ [i.e., $y(0)$], which is the limit of $y(t)$ as $t \to 0$.

z-Transforms of integrals and derivatives

In Sections 7.3 and 7.4 we had developed generalized expressions for the Laplace transforms of integrals and derivatives. *No such general equations exist for the z-transforms of integrals and derivatives* because both have been approximated by various formulas of difference equations. Depending on the approximation formula we use for the integral or derivative, we take different z-transforms for the integral or derivative. Let us demonstrate this feature using two examples.

Example 28.1: z-Transforms of an Integral

Consider the following integral:

$$y(t) = \int_0^{t=nT} f(t)\, dt$$

1. First, approximate the integral using the trapezoidal form of integration (see Figure 28.3a):

$$y(nT) = y[(n-1)T] + \int_{(n-1)T}^{nT} f(t)dt \approx y[(n-1)T]$$
$$+ \frac{f(nT) + f[(n-1)T]}{2} T$$

Taking the z-transform of both sides and noticing that $y\,[(n-1)T]$ and $f[(n-1)T]$ are delayed signals by one T (see Section 28.2), we have

$$\hat{y}(z) = \hat{y}(z)z^{-1} + \frac{T}{2}[\hat{f}(z) + \hat{f}(z)z^{-1}]$$

or

$$\hat{y}(z) = \frac{T}{2}\frac{1 + z^{-1}}{1 - z^{-1}}\hat{f}(z) \qquad (28.17a)$$

2. Second, approximate the integral using Simpson's rule of integration (see Figure 28.3b).

$$y(nT) = y[(n-2)T] + \int_{(n-2)T}^{nT} f(t)\, dt \approx y[(n-2)T]$$
$$+ \frac{T}{3}\{f[(n-2)T] + 4f[(n-1)T] + f(nT)\}$$

Taking the z-transforms of both sides and noticing that $y[(n-2)T]$ and $f[(n-2)T]$ are delayed signals by $2T$, while $f[(n-1)T]$ is delayed only by $1T$, we have

$$\hat{y}(z) = \hat{y}(z)z^{-2} + \frac{T}{3}[\hat{f}(z)z^{-2} + 4\hat{f}(z)z^{-1} + \hat{f}(z)]$$

or

$$\hat{y}(z) = \frac{T}{3}\frac{1 + 4z^{-1} + z^{-2}}{1 - z^{-2}}\hat{f}(z) \qquad (28.17b)$$

Equations (28.17a) and (28.17b) yield two different z-transforms for

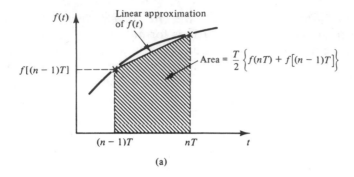

(a)

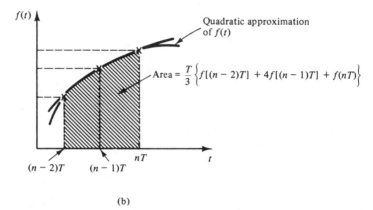

(b)

Figure 28.3 Numerical integration: (a) trapezoidal rule; (b) Simpson's rule.

the same integral. *The difference is caused by the different approximations used to compute the integral from discrete-time values.*

Example 28.2: z-Transforms of a Derivative

Consider the derivative

$$y(t) = \frac{df(t)}{dt}$$

1. Approximate the derivative by a first-order difference

$$y(t) = \frac{df}{dt} \approx \frac{f(nT) - f[(n-1)T]}{T}$$

Then we can easily find

$$\hat{y}(z) = \frac{1}{T}(1 - z^{-1})\hat{f}(z) \qquad (28.18a)$$

2. If we approximate the derivative by a second-order difference

$$y(t) = \frac{df}{dt} = \frac{f(nT) - f[(n-2)T]}{2T}$$

we take

$$\hat{y}(z) = \frac{1}{2T}(1 - z^{-2})\hat{f}(z) \qquad (28.18b)$$

Equations (28.18a) and (28.18b) yield two different z-transforms for the same derivative. *The difference again is due to the different numerical approximations used to characterize the derivative in discrete-time values.*

28.4   Inversion of z-Transforms

With the inverse z-transform we attempt to take back the values of a function at the sampling instants, given its z-transform. The inverse z-transform operation is symbolized as follows:

$$Z^{-1}[\hat{y}(z)] = \{y(0), y(T), y(2T), \cdots\} \qquad (28.19)$$

Before we proceed with the detailed presentation of the two basic methods used for inverting z-transforms, there are a few important points about the inverse transform that we should emphasize.

1. *The inverse z-transform yields the values of a function at the sampling instants only and not the continuous function itself.* This is logical, because we cannot expect to retrieve from the inverse z-transform more information about the function than just its values at the sampling instants, which were used for the computation of the z-transforms in the first place.
2. The inverse z-transform does not even help us to determine the sampling period T for the computed sampled values: $y(0)$, $y(T)$, $y(2T)$, $\cdots$.
3. If we attempt to find a continuous function $y(t)$ which coincides with the sampled values $y(0)$, $y(T)$, $y(2T)$, $\cdots$ produced by the inverse z-transform, we should remember that these sampled values could be derived from two distinct functions (see also Remark 3 in Section 28.1). It follows, then, that *the inverse transform of a function $\hat{y}(z)$ does not necessarily yield a unique continuous function $y(t)$.*

Let us now proceed with the mechanics of two methods used to determine inverse z-transforms. The first is the partial-fractions expansion and the second is based on the long division of two polynomials.

Partial-fractions expansion

This method is completely parallel to the partial-fractions expansion methodology used to invert Laplace transforms and proceeds as follows:

1. Any z-transform should be viewed as the ratio of two polynomials in z^{-1} (or z):

$$\hat{y}(z) = \frac{Q(z^{-1})}{P(z^{-1})}$$

   where $Q(z^{-1})$ and $P(z^{-1})$ are polynomials in z^{-1} of order m and n, respectively.

2. Expand the $\hat{y}(z)$ into a series of fractions:

$$\hat{y}(z) = \frac{Q(z^{-1})}{P(z^{-1})} = \frac{C_1}{r_1(z^{-1})} + \frac{C_2}{r_2(z^{-1})} + \cdots + \frac{C_n}{r_n(z^{-1})} \qquad (28.20)$$

   where $r_1(z^{-1})$, $r_2(z^{-1})$, $\cdots$, $r_n(z^{-1})$ are low-order polynomials in z^{-1} whose inverse z-transform can be computed easily.

3. Compute the values of $C_1, C_2, \cdots, C_n$ from eq. (28.20).

4. Find the inverse z-transform of every partial fraction. Then we can claim that the sequence of sampled values of the unknown function $y(t)$ with the given z-transform is

$$\{y(0), y(T), y(2T), \cdots\}$$

$$= Z^{-1}\left[\frac{C_1}{r_1(z^{-1})}\right] + Z^{-1}\left[\frac{C_2}{r_2(z^{-1})}\right] + \cdots + Z^{-1}\left[\frac{C_n}{r_n(z^{-1})}\right]$$

A few remarks on the application of the method are now in order:

1. The $C_1, C_2, \ldots, C_n$ are computed in a completely similar manner as for the Laplace transforms and the particular computational procedure depends on the nature of the roots of polynomial $P(z^{-1})$ (see Section 8.2).

2. The inverse transforms of the partial fractions can be found easily from Table 28.1 or similar but more inclusive tables. Such tables yield the inverse z-transform in terms of continuous functions. For example, if

$$Z^{-1}\left[\frac{C_1}{r_1(z^{-1})}\right] = f_1(t), \quad Z^{-1}\left[\frac{C_2}{r_2(z^{-1})}\right] = f_2(t), \quad \cdots, \quad Z^{-1}\left[\frac{C_n}{r_n(z^{-1})}\right] = f_n(t)$$

   Then

$$y(t) = f_1(t) + f_2(t) + \cdots + f_n(t)$$

3. We should remember, though, that *the inverse z-transform cannot*

yield complete information about a function, but provides only its values at the sampling instants. Therefore, functions $f_1(t)$, $f_2(t), \ldots, f_n(t)$ are used only to compute the sampled values of $y(t)$ as follows:

$$y(0) = f_1(0) + f_2(0) + \cdots + f_n(0)$$

$$y(T) = f_1(T) + f_2(T) + \cdots + f_n(T)$$

$$y(2T) = f_1(2T) + f_2(2T) + \cdots + f_n(2T), \text{ etc.}$$

Example 28.3: Inverse z-Transform by Partial-Fractions Expansion

Let

$$\hat{y}(z) = \frac{z}{z^2 - 4z + 3}$$

Divide by z^2 and convert into a ratio of two polynomials in z^{-1}:

$$\hat{y}(z) = \frac{z^{-1}}{1 - 4z^{-1} + 3z^{-2}}$$

The polynomial of the denominator has roots $z^{-1} = 1$ and $z^{-1} = 1/3$. Expand into partial fractions:

$$\hat{y}(z) = \frac{z^{-1}}{1 - 4z^{-1} + 3z^{-2}} = \frac{z^{-1}}{(1 - z^{-1})(1 - 3z^{-1})} = \frac{C_1}{1 - z^{-1}} + \frac{C_2}{1 - 3z^{-1}}$$

Compute C_1 and C_2 as follows:

$$C_1 = \frac{z^{-1}}{(1 - 3z^{-1})}\bigg|_{z^{-1}=1} = -\frac{1}{2}$$

$$C_2 = \frac{z^{-1}}{(1 - z^{-1})}\bigg|_{z^{-1}=1/3} = \frac{1}{2}$$

Therefore,

$$\hat{y}(z) = \frac{-1/2}{1 - z^{-1}} + \frac{1/2}{1 - 3z^{-1}}$$

From Table 28.1 we find that:

$\dfrac{-1/2}{1 - z^{-1}}$ is the z-transform of a step function of magnitude $-1/2$

$\dfrac{1/2}{1 - 3z^{-1}}$ has the general form $K/(1 - e^{-aT}z^{-1})$, which is the z-transform of an exponential with $e^{-aT} = 3$ or $-aT = \ln 3 = 1.10$ and a constant multiplying factor $K = 1/2$

Consequently, the sampled values of the unknown function are given by

$$y(nT) = -\frac{1}{2} + \frac{1}{2}\exp(1.1n) \qquad \text{for } n = 0, 1, 2, \cdots$$

or in a tabulated form by

n	0	1	2	3	4 $\cdots$	∞
$y(nT)$	0.	1.0	4.0	13.0	40.0 $\cdots$	∞

Note. Although we can compute the value $-aT = 1.10$, we cannot find the analytic form of the continuous exponential function because T is unknown.

Example 28.4: Inverse z-Transform of a More Complex Expression

Let

$$\hat{y}(z) = \frac{4 + 2.67z^{-1} + 1.56z^{-2} - 1.42z^{-3}}{1 - 0.36z^{-1} + 0.19z^{-2} - 1.03z^{-3} + 0.2z^{-4}}$$

The polynomial of the denominator can be factored into

$$(1 - z^{-1})(1 - 0.2z^{-1})(1 + 0.84z^{-1} + z^{-2})$$

leading to the following partial-fractions expansion

$$\hat{y}(z) = \frac{C_1}{1 - z^{-1}} + \frac{C_2}{1 - 0.2z^{-1}} + \frac{C_3}{1 + 0.84z^{-1} + z^{-2}}$$

Compute C_1, C_2, C_3 and take

$$\hat{y}(z) = \frac{3}{1 - z^{-1}} + \frac{1}{1 - 0.2z^{-1}} + \frac{0.91z^{-1}}{1 + 0.84z^{-1} + z^{-2}}$$

Using the entries of Table 28.1, we observe that:

1. The first term corresponds to a unit step of magnitude 3.
2. The second has the form $1/(1 - e^{-aT}z^{-1})$ and corresponds to the z-transform of an exponential with $e^{-aT} = 0.2$ or $aT = -\ln 0.2 = 1.61$.
3. The third term resembles the z-transform of a sinusoid with $0.91 = \sin \omega T$ and $0.84 = -2 \cos \omega T$. These two equations are compatible for $\omega T = 2.0$ rad = 114.8°.

Therefore, the sampled values of the unknown function are given by

$$y(nT) = 3 + \exp(-1.61n) + \sin 2n \qquad n = 0, 1, 2, 3, \cdots$$

or in tabular form,

n	0	1	2	3	$\cdots$
$y(nT)$	4	4.11	2.28	2.74	$\cdots$

Note. Although we know $aT = 1.61$ and $\omega T = 2.0$, we cannot find analytic expressions for the continuous exponential and cosine functions because T is unknown.

Long division

We have observed that *the z-transform of a function is the ratio of two polynomials in z^{-1} (or z):*

$$\hat{y}(z) = \frac{Q(z^{-1})}{P(z^{-1})} = \frac{\alpha_0 + \alpha_1 z^{-1} + \alpha_2 z^{-2} + \cdots + \alpha_m z^{-m}}{\beta_0 + \beta_1 z^{-1} + \beta_2 z^{-2} + \cdots + \beta_n z^{-n}}$$

The order of the numerator should be lower than or equal to the order of the denominator (i.e., $m \leq n$). Divide the denominator $P(z^{-1})$ into the numerator $Q(z^{-1})$ and take

$$\hat{y}(z) = \gamma_0 + \gamma_1 z^{-1} + \gamma_2 z^{-2} + \cdots \tag{28.21}$$

Recall the defining equation of a z-transform,

$$\hat{y}(z) = y(0) + y(T)z^{-1} + y(2T)z^{-2} + \cdots \tag{28.22}$$

Then, by comparing eqs. (28.21) and (28.22), we find that

$$y(0) = \gamma_0 \qquad y(T) = \gamma_1 \qquad y(2T) = \gamma_2 \qquad \cdots \tag{28.23}$$

The procedure of the long division for computing the inverse z-transform is very simple and can be used for any expression. The expansion to partial fractions, on the other hand, could be a very difficult procedure, if it is very hard to determine the partial fractions or the partial fractions cannot be found among the terms of Table 28.1 or other equivalent.

Example 28.5: Inverse z-Transform by Long Division

Consider the z-transform given in Example 28.3:

$$\hat{y}(z) = \frac{z^{-1}}{1 - 4z^{-1} + 3z^{-2}}$$

Take the long division and find:

$$
\begin{array}{r}
1 \cdot z^{-1} + 4 \cdot z^{-2} + 13 \cdot z^{-3} + 40 \cdot z^{-4} + \cdots \\
1 - 4z^{-1} + 3z^{-2} \overline{\smash{\big)}\ z^{-1} \phantom{+4z^{-2}+13z^{-3}+40z^{-4}}} \\
\underline{z^{-1} - 4z^{-2} + 3z^{-3}} \\
4z^{-2} - 3z^{-3} \\
\underline{4z^{-2} - 16z^{-3} + 12z^{-4}} \\
13z^{-3} - 12z^{-4} \\
\underline{13z^{-3} - 52z^{-4} + 39z^{-5}} \\
40z^{-4} - 39z^{-5} \quad \text{etc.}
\end{array}
$$

or

$$\hat{y}(z) = 1 \cdot z^{-1} + 4 \cdot z^{-2} + 13 \cdot z^{-3} + 40 \cdot z^{-4} + \cdots$$

It is clear from eqs. (28.23) that

n	0	1	2	3	4	$\cdots$	∞
$y(nT)$	0	1	4	13	40	$\cdots$	∞

which is the same result as in Example 28.3.

Example 28.6: Long Division for a More Complex Inversion

Consider the z-transform given in Example 28.4. Take the long division

$$\begin{array}{r} 4 + 4.11z^{-1} + 2.28z^{-2} + 2.74z^{-3} + \cdots \\ 1 - 0.36z^{-1} + 0.19z^{-2} - 1.03z^{-3} + 0.2z^{-4} \; \overline{)\; 4 + 2.67z^{-1} + 1.56z^{-2} - 1.42z^{-3}} \end{array}$$

$$\underline{4 - 1.44z^{-1} + 0.76z^{-2} - 4.12z^{-3} + 0.8z^{-4}}$$

$$4.11z^{-1} + 0.80z^{-2} + 2.70z^{-3} - 0.8z^{-4}$$

$$\underline{4.11z^{-1} - 1.48z^{-2} + 0.78z^{-3} - 4.23z^{-4} + 0.822z^{-5}}$$

$$2.28z^{-2} + 1.92z^{-3} - 3.43z^{-4} - 0.822z^{-5}$$

$$\underline{2.28z^{-2} - 0.82z^{-3} + 0.43z^{-4} - 2.35z^{-5} + 4.56z^{-6}}$$

$$2.74z^{-3} - 3.86z^{-4} + 1.53z^{-5} + 4.56z^{-6}$$

etc.

Therefore,

$$\hat{y}(z) = 4 + 4.11z^{-1} + 2.28z^{-2} + 2.74z^{-3} + \cdots$$

and the sampled values are:

n	0	1	2	3	$\cdots$
$y(nT)$	4	4.11	2.28	2.74	$\cdots$

the same as in Example 28.4.

THINGS TO THINK ABOUT

1. Define the z-transforms. Why is it not correct to talk of the z-transform of a continuous function? What do we mean then when we talk of the z-transform of a continuous function?

2. How is it possible for two functions $f_1(t)$ and $f_2(t)$ to have identical z-transforms?

3. Consider two sinusoidal waves: $\sin \omega_1 t$ and $\sin \omega_2 t$. Under what conditions are the z-transforms of the two waves identical? (*Hint*: Find a condition relating the frequencies ω_1 and ω_2 and the sampling period T.)

4. Discuss the relationship between z-transforms and Laplace transforms.

5. Derive the z-transform of $\cos \omega t$.

6. Discuss the basic properties of z-transforms, and indicate how would you use them. How would you find the steady-state value of the sampled value response of a process?

7. Why is it not possible to develop generalized equations for the z-transforms of integrals or derivatives as it was the case for Laplace transforms? Give three different z-transforms for the integral

$$\int_0^{t=nT} [t^2 + \sin t]\, dt$$

8. Why do we prefer to express z-transforms as ratios of polynomials in negative powers of z instead of positive powers? Is there any fundamental difference, or simply practical reasons?

9. Discuss the two methods available for inverting z-transforms. Describe their relative advantages and disadvantages. Which one would you prefer to use?

10. Show that the inverse z-transform cannot yield a unique continuous function. Why is that so? Demonstrate using a numerical example.

11. Does the inverse z-transform yield unique values at the sampling instants? Explain why or why not.

12. Does the inverse z-transform help you find the sampling period? Explain.

13. Find the final value of $y(t)$ if

$$\hat{y}(z) = \frac{1 + 2z^{-1} + 0.1z^{-2}}{(1 - z^{-1})(1 - 0.2z^{-1})(1 + 2z^{-1} + z^{-2})}$$

14. Using the partial fractions expansion, can you invert the following expression?

$$\hat{y}(z) = \frac{1 + 2z^{-1}}{(1 - z^{-1})^2(1 - 0.2z^{-1})}$$

15. Find the inverse z-transform using the long division

$$\hat{y}(z) = \frac{1 - 0.1z^{-1} + 3z^{-2}}{2 + 3z^{-1} + z^{-2} - 0.1z^{-3} + z^{-4}}$$

APPENDIX 28A

The Modified z-Transform

In Section 28.2 we found that if a function is delayed by an integer number of sampling periods, the z-transform of the delayed function is given by

$$Z[y(t - kT)] = z^{-k}\hat{y}(z) \qquad \text{with } k = \text{integer}$$

What happens, though, when the dead time is a *noninteger multiple* of the sampling period?

Consider the delayed function $y(t - t_d)$, where $t_d = (k + \Delta)T$ with $0 < \Delta < 1$. Then

$$Z[y(t - t_d)] = \sum_{n=0}^{\infty} y(nT - kT - \Delta T)z^{-n} = \sum_{n=0}^{\infty} y[(n - k - 1)T + (1 - \Delta)T]z^{-n}$$

Make the substitutions

$$\ell = n - k - 1 \qquad \text{and} \qquad m = 1 - \Delta$$

and take

$$Z[y(t - t_d)] = \sum_{\ell=0}^{\infty} y(\ell T + mT)z^{-\ell-k-1} = z^{-k}\{z^{-1}\sum_{\ell=0}^{\infty} y(\ell T + mT)z^{-\ell}\}$$

Define the modified z-transform by

$$\hat{y}(z, m) \equiv Z_m[y(t)] = z^{-1}\sum_{\ell=0}^{\infty} y(\ell T + mT)z^{-\ell} \qquad (28A.1)$$

and take

$$Z[y(t - t_d)] = z^{-k} Z_m[y(t)] = z^{-k} \hat{y}(z, m)$$

From the defining eq. (28A.1) we notice that the modified z-transform of a function is characterized by two variables, z and m. The first is a complex variable having the same meaning as in the normal z-transform, while m is a parameter denoting the fraction of a sampling period present in the delay. Table 28A.1 shows the modified z-transforms of several common functions.

We can invert modified z-transforms through long division, taking care to divide separately terms involving m and those that do not include m. This allows us to find the value of a sampled-value function between sampling instants. For example, suppose that we have a signal with

$$\hat{y}(z, m) = \frac{e^{-mT}z^{-1}}{1 - e^{-T}z^{-1}}$$

Invert the last expression for various values of m with $0 < m < 1$ and find the value of the function $y(t)$ for any point between sampling instants.

TABLE 28A.1
MODIFIED z-TRANSFORMS OF VARIOUS FUNCTIONS

Function in time domain	Modified z-transform
Unit step	$\dfrac{z^{-1}}{1 - z^{-1}}$
Ramp: $f(t) = at$	$\dfrac{aTz^{-1}}{1 - z^{-1}} m + \dfrac{aTz^{-2}}{(1 - z^{-1})^2}$
$f(t) = t^n$	$\displaystyle\lim_{a \to 0} (-1)^n \dfrac{\partial^n}{\partial a^n} \left(\dfrac{e^{-amT}z^{-1}}{1 - e^{-aT}z^{-1}} \right)$
Exponential: $f(t) = e^{-at}$	$\dfrac{e^{-amT}z^{-1}}{1 - e^{-aT}z^{-1}}$
$f(t) = te^{-at}$	$\dfrac{Te^{-amT}z^{-1}[m + (1 - m)e^{-aT}z^{-1}]}{(1 - e^{-aT}z^{-1})^2}$
$f(t) = \sin \omega t$	$z^{-1}\left[\dfrac{\sin m\omega T + z^{-1} \sin [(1 - m)\omega T]}{1 - 2z^{-1} \cos \omega T + z^{-2}} \right]$
$f(t) = \cos \omega t$	$z^{-1}\left[\dfrac{\cos m\omega T - z^{-1} \cos [(1 - m)\omega T]}{1 - 2z^{-1} \cos \omega T + z^{-2}} \right]$
$f(t) = 1 - e^{-at}$	$z^{-1}\left(\dfrac{1}{1 - z^{-1}} - \dfrac{e^{-amT}}{1 - e^{-aT}z^{-1}} \right)$
$f(t) = e^{-at}\sin \omega t$	$z^{-1}e^{-amT} \dfrac{\sin m\omega T + z^{-1}e^{-aT} \sin [(1 - m)\omega T]}{1 - 2z^{-1}e^{-aT} \cos \omega T + e^{-2aT}z^{-2}}$
$f(t) = e^{-at}\cos \omega t$	$z^{-1}e^{-amT} \dfrac{\cos m\omega T - z^{-1}e^{-aT} \cos [(1 - m)\omega T]}{1 - 2z^{-1}e^{-aT} \cos \omega T + e^{-2aT}z^{-2}}$

Discrete-Time Response **29**
of Dynamic Systems

In control loops with continuous analog controllers, all dynamic components (process, measuring sensor, controller, final control element) are continuous systems (i.e., respond continuously with time to any changes in their inputs). Therefore, all signals in such loops are continuous in time. Under such conditions we can use the familiar notion of transfer functions in the Laplace domain to analyze the dynamic behavior of control systems in the open- or closed-loop mode. This methodology cannot be used for the dynamic analysis of digital control loops, which possess *discrete elements* (digital control algorithm) and *discrete-time signals*.

Recall the simplified direct digital control loop of Figure 27.8. There are two primary distinct components whose responses we would be interested in analyzing:

1. *Digital control algorithm* (Figure 29.1a): This is the discrete element of the DDC loop with discrete-time input and output signals. The question that we should try to answer is: How can we analyze the discrete-time response $c(nT)$ of digital control algorithms to discrete-time changes in the error signal $\epsilon(nT)$?

2. *Process with the hold element* (Figure 29.1b): These are the continuous elements of the DDC loop. The relationship between the continuous manipulated variable $m(t)$ and continuous output $y(t)$ can be expressed in terms of conventional transfer functions in the Laplace domain. But how can we describe the sampled-

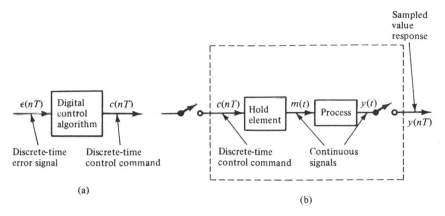

Figure 29.1 Components of a DDC loop with discrete-time inputs and outputs.

value response y(nT) to discrete-time changes in the controller output $c(nT)$?

Once we know how to handle separately the responses of these two systems, we should close the loop and examine the dynamic response of closed-loop digital control systems. The three items above constitute the content of the present chapter.

29.1   Response of Discrete Dynamic Systems

We shall call *discrete* those dynamic systems that process their input signals only at the sampling instants, thus producing output signals that are only defined at particular time instants. In other words, *discrete is a system whose input and output are discrete-time signals*. The input–output relationship for such systems is given by a discrete-time model (i.e., by a difference equation).

Typical examples of discrete systems are all types of digital control algorithms and digital filters. A digital control algorithm uses sampled values of the error signal $\epsilon(nT)$, and produces control commands $c(nT)$ at particular instants of time. As an example, see eq. (27.7) for a digital PID control algorithm. Digital filters are small programs residing in the memory of a process control computer which are used to filter out measurement or process noise from the sampled response of a process. They have replaced earlier analog filters, which were based on electrical *RC* circuits that damped out the high-frequency noise from the measured signal.

In general, any algorithm executed by a process computer is a dis-

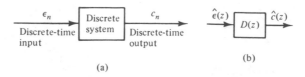

(a)

(b)

Figure 29.2 (a) Discrete-time system; (b) corresponding block diagram.

crete system. Thus algorithms used to estimate the unknown parameters of a process, or to infer the value of an unmeasured process output, or to adapt and tune the controller parameters, and so on, can be considered as discrete systems.

Consider the discrete system shown in Figure 29.2a. Let ϵ_n be the discrete-time input and c_n the discrete-time output. From now on whenever we use a subscript i with a discrete-time signal we will denote the discrete-time value of the signal at the ith sampling instant [e.g., $\epsilon_n \equiv \epsilon(nT)$ and $c_n \equiv c(nT)$].

The input and output signals for a discrete system are related through a linear difference equation of the general form

$$c_n = a_0\epsilon_n + a_1\epsilon_{n-1} + \cdots + a_k\epsilon_{n-k} + b_1 c_{n-1} + b_2 c_{n-2} + \cdots + b_m c_{n-m} \quad (29.1)$$

where $a_0, a_1, a_2 \ldots, a_k$ and $b_1, b_2, \ldots, b_m$ are constant parameters. Let $\hat{c}(z)$ be the z-transform of the output's discrete-time values and $\hat{\epsilon}(z)$ the z-transform of the input's discrete-time values. Then z-transforming both sides of eq. (29.1), we take

$$\hat{c}(z) = a_0\hat{\epsilon}(z) + a_1\hat{\epsilon}(z)z^{-1} + \cdots + a_k\hat{\epsilon}(z)z^{-k} + b_1\hat{c}(z)z^{-1}$$
$$+ b_2\hat{c}(z)z^{-2} + \cdots + b_m\hat{c}(z)z^{-m}$$

or

$$\frac{\hat{c}(z)}{\hat{\epsilon}(z)} = \frac{a_0 + a_1 z^{-1} + a_2 z^{-2} + \cdots + a_k z^{-k}}{1 - b_1 z^{-1} - b_2 z^{-2} - \cdots - b_m z^{-m}} = D(z) \quad (29.2)$$

$D(z)$ is defined as the *transfer function of the discrete system* and we notice that it is completely parallel to our familiar transfer function for continuous systems in the Laplace domain. Equation (29.2) constitutes the *input–output model for the discrete system* and can be used to compute the dynamic response of a discrete system subject to an input change (see Figure 29.2b).

Example 29.1: Discrete-Time Response of a Digital PID Controller

In Example 27.5 we developed the discrete-time model for the digital PID control algorithm [eq. (27.7)]. Consider the control commands at the sampling instants, n and $(n - 1)$, as they are determined by eq. (27.7):

$$c_n = K_c\left[\epsilon_n + \frac{T}{\tau_I}\sum_{k=0}^{n}\epsilon_k + \frac{\tau_D}{T}(\epsilon_n - \epsilon_{n-1})\right] + c_s$$

$$c_{n-1} = K_c \left[\epsilon_{n-1} + \frac{T}{\tau_I} \sum_{k=0}^{n-1} \epsilon_k + \frac{\tau_D}{T} (\epsilon_{n-1} - \epsilon_{n-2}) \right] + c_s$$

Subtract the second from the first equation and take what is known as the *velocity form of the PID algorithm*:

$$c_n - c_{n-1} = K_c [(\epsilon_n - \epsilon_{n-1}) + \frac{T}{\tau_I} \epsilon_n + \frac{\tau_D}{T} (\epsilon_n - 2\epsilon_{n-1} + \epsilon_{n-2})]$$

We will have more to say regarding the velocity form of the PID algorithm in Chapter 30.

Take the z-transforms of both sides of the last equation:

$$\hat{c}(z) - \hat{c}(z)z^{-1} = K_c \left[\hat{e}(z)(1 - z^{-1}) + \frac{T}{\tau_I} \hat{e}(z) + \frac{\tau_D}{T} \hat{e}(z)(1 - 2z^{-1} + z^{-2}) \right]$$

or

$$\frac{\hat{c}(z)}{\hat{e}(z)} = K_c \left[1 + \frac{T}{\tau_I} \frac{1}{1 - z^{-1}} + \frac{\tau_D}{T} \frac{1 - 2z^{-1} + z^{-2}}{1 - z^{-1}} \right] = D(z) \qquad (29.3)$$

Equation (29.3) yields the *discrete transfer function of the velocity PID control algorithm*. For given changes in the sampled values of the error signal, the resulting discrete-time control action can be found from the inverse z-transform:

$$\{c_0, c_1, c_2, \cdots\} = Z^{-1}[D(z)\hat{e}(z)]$$

where $D(z)$ is given by eq. (29.3) and $\hat{e}(z) = \sum_{k=0}^{\infty} \epsilon_k z^{-k}$ with $\epsilon_0, \epsilon_1, \epsilon_2, \ldots$ known.

Example 29.2: Discrete-Time Response of a First-Order Digital Filter

The first-order digital filter is the simplest filter and its input–output relationship is described by a first-order linear difference equation,

$$y_n = (1 - a)y_{n-1} + ax_n \qquad (29.4)$$

where x_n = discrete-time input to the filtering algorithm (i.e., the measurement signal corrupted with process or measurement noise)

y_n = discrete-time output from the filter (i.e., the measurement signal without the noise)

a = a constant, called the *filtering parameter*, with values $0 \le a \le 1$

Take the z-transform of eq. (29.4) and find

$$\hat{y}(z) = (1 - a)\hat{y}(z)z^{-1} + a\hat{x}(z)$$

or

$$\frac{\hat{y}(z)}{\hat{x}(z)} \equiv D(z) = \frac{a}{1 - (1 - a)z^{-1}} \qquad (29.5)$$

Equation (29.5) yields the *discrete transfer function of a first-order digital filter*. The noise-free signal (the output of the filter) is given by

$$\{y_0, y_1, y_2, \cdots\} = Z^{-1}\left[\frac{a}{1 - (1 - a)z^{-1}}\, \hat{x}(z)\right]$$

Remarks

1. It can be easily shown that for N discrete systems in series (Figure 29.3a) the overall discrete transfer function is the product of the discrete transfer functions of the individual systems:

$$D_{\text{overall}}(z) \equiv \frac{\hat{c}_N(z)}{\hat{\epsilon}(z)} = \frac{\hat{c}_1(z)}{\hat{\epsilon}(z)} \frac{\hat{c}_2(z)}{\hat{c}_1(z)} \cdots \frac{\hat{c}_N(z)}{\hat{c}_{N-1}(z)}$$

$$= D_1(z)D_2(z) \cdots D_N(z) \tag{29.6}$$

2. For a discrete system with multiple inputs and outputs, we define the *discrete transfer function matrix* $\mathbf{D}(z)$ as follows:

$$\mathbf{D}(z) = \begin{bmatrix} D_{11}(z) & D_{12}(z) & \cdots & D_{1m}(z) \\ D_{21}(z) & D_{22}(z) & \cdots & D_{2m}(z) \\ \cdot & \cdot & \cdots & \cdot \\ D_{n1}(z) & D_{n2}(z) & \cdots & D_{nm}(z) \end{bmatrix} \tag{29.7}$$

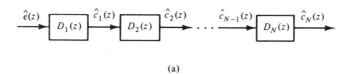

(a)

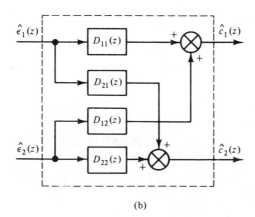

(b)

Figure 29.3 Block diagrams for: (a) N discrete-time systems in series; (b) discrete system with two inputs and two outputs.

where $D_{ij}(z)$ is the discrete transfer function between the ith output and the jth input. Figure 29.3b shows a discrete system with two inputs and two outputs.

29.2 Discrete-Time Analysis of Continuous Systems. The Pulse Transfer Function

In the preceding section the analysis was centered around the response of the discrete components in a direct digital control (DDC) loop with characteristic representative the control algorithm. The use of z-transforms allowed easy and straightforward development of simple input–output models through the discrete transfer functions.

In this section we consider the continuous elements of a DDC loop (i.e., the process and the hold as shown in Figure 29.1b). Although both elements are continuous, the input to the hold is a discrete-time signal $c(nT)$, and the output from the process is sampled by a sampler. We would like to relate the sampled output values $y(nT)$ with the discrete control commands $c(nT)$, through a simple input–output model in the z-domain of the form

$$\frac{\hat{y}(z)}{\hat{c}(z)} = HG_p(z) \qquad (29.8)$$

$HG_p(z)$ is called the *pulse transfer function* and can be related to the continuous transfer functions of the hold and the process, $H(s)$ and $G_p(s)$, respectively. Let us now develop this relationship.

Consider the hold–process combination shown in Figure 29.4a. At time $t = 0$, an impulse $c*(0)$ of "strength" $c(0)$ enters the hold element. From eq. (27.1) we know that

$$c*(0) = c(0)\delta(t)$$

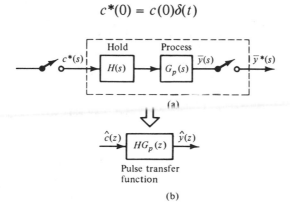

(a)

(b)

Figure 29.4 (a) Process with hold element; (b) corresponding pulse transfer function.

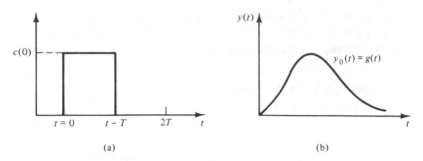

Figure 29.5 (a) Pulse input; (b) response of a continuous system to pulse input.

where $\delta(t)$ is the unit impulse function at $t = 0$. Laplace transform the last equation and take

$$\mathcal{L}[c^*(0)] = c(0) \cdot 1 \qquad (29.9)$$

The output of the hold element to the impulse $c^*(0)$ depends on the order of the hold element. For a zero-order hold element its output is a pulse of height $c(0)$ and duration T (Figure 29.5a). This pulse is the input to the process and produces the output $y_0(t)$ shown in Figure 29.5b. How can we compute this output?

It is clear that

$$\overline{y}_0(s) = [H(s)G_p(s)]\mathcal{L}[c^*(0)]$$

and recalling eq. (29.9),

$$\overline{y}_0(s) = [H(s)G_p(s)]c(0)$$

Invert the last equation and find the response of the process output in the time domain, for an impulse input of "strength" $c(0)$ in the hold element:

$$y_0(t) = \mathcal{L}^{-1}[H(s)G_p(s)]c(0)$$

Define

$$g(t) \equiv \mathcal{L}^{-1}[H(s)G_p(s)]$$

and take

$$y_0(t) = g(t)c(0) \qquad (29.10)$$

Similarly, the process responses to impulses $c^*(T)$, $c^*(2T)$,... entering the hold element at the sampling instants $t = T$, $t = 2T$,... are:

$$y_1(t - T) = g(t - T)c(T)$$
$$y_2(t - 2T) = g(t - 2T)c(2T)$$
$$\vdots$$

Therefore, the *overall response of the process output to a sequence of impulses* $c^*(0)$, $c^*(T)$, $c^*(2T)$,...is the sum of the responses to the sequence of impulses:

$$y(t) = y_0(t) + y_1(t - T) + y_2(t - 2T) + \cdots$$
$$= g(t)c(0) + g(t - T)c(T) + g(t - 2T)c(2T) + \cdots$$
$$.= \sum_{n=0}^{\infty} g(t - nT)c(nT) \tag{29.11}$$

By definition,

$$\hat{y}(z) = \sum_{k=0}^{\infty} y(kT)z^{-k}$$

and from eq. (29.11) we can find the value of the output at the various sampling instants:

$$y(kT) = \sum_{n=0}^{\infty} g(kT - nT)c(nT)$$

Thus

$$\hat{y}(z) = \sum_{k=0}^{\infty} \sum_{n=0}^{\infty} g(kT - nT)c(nT)z^{-k}$$

Put $\ell = k - n$ and remembering that $g(\ell T) = 0$ for $\ell < 0$, take

$$\hat{y}(z) = \sum_{\ell=-n}^{\infty} \sum_{n=0}^{\infty} g(\ell T)c(nT)z^{-\ell-n} = \sum_{\ell=0}^{\infty} g(\ell T)z^{-\ell} \sum_{n=0}^{\infty} c(nT)z^{-n}$$
$$= Z[g(t)]\hat{c}(z)$$

or

$$\frac{\hat{y}(z)}{\hat{c}(z)} = Z[g(t)] = Z\{\mathcal{L}^{-1}[H(s)G_p(s)]\} \tag{29.12}$$

Comparing eqs. (29.12) and (29.8), we conclude that the pulse transfer function is given by

$$HG_p(z) = Z\{\mathcal{L}^{-1}[H(s)G_p(s)]\} \tag{29.13a}$$

which is often denoted by

$$HG_p(z) = Z\{H(s)G_p(s)\} \tag{29.13b}$$

Once the pulse transfer function for a process is known, the sampled values of its output can be found by inverting eq. (29.8):

$$\{y(0), y(T), y(2T), \cdots\} = Z^{-1}[HG_p(z)\hat{c}(z)] \tag{29.14}$$

Example 29.3: Discrete-Time Analysis of Several Linear Continuous Systems

Let us analyze the sampled-value response of (a) a pure integrator, (b) a first-order process, (c) two noninteracting first-order processes in series, and (d) a second-order process, to discrete-time control commands. A zero-order hold with a transfer function

$$H(s) = \frac{1 - e^{-Ts}}{s}$$

will be considered throughout this example.

(a) *Pure integrator* (see Section 10.3): The process transfer function is given by eq. (10.4) (i.e., $G_p = K_p/s$). Therefore, the pulse transfer function (see Figure 29.6a) is

$$HG_p(z) = Z\left\{ \frac{1 - e^{-Ts}}{s} \frac{K_p}{s} \right\} = Z\left\{ \frac{K_p}{s^2} \right\} - Z\left\{ \frac{K_p e^{-Ts}}{s^2} \right\}$$

The e^{-Ts} of the second term denotes delay by one period T and can be replaced by its quivalent z^{-1}. Also, from Table 28.1 we find that the term K_p/s^2 corresponds to a ramp function $K_p t$, which has the following z-transform:

$$\frac{K_p T z^{-1}}{(1 - z^{-1})^2}$$

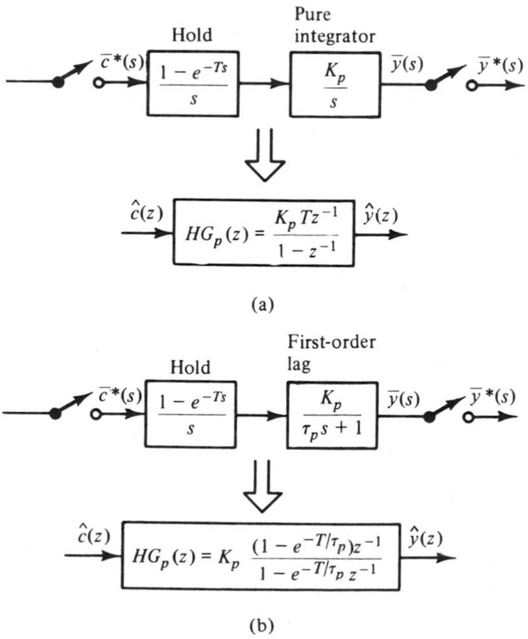

(a)

(b)

Figure 29.6 Pulse transfer functions for: (a) pure integrator; (b) first-order lag system.

Therefore, the pulse transfer function is

$$HG_p(z) = (1 - z^{-1}) \frac{K_p T z^{-1}}{(1 - z^{-1})^2} = \frac{K_p T z^{-1}}{1 - z^{-1}}$$

Consider a sequence of input values $c(nT)$, corresponding to a unit step change:

$$\hat{c}(z) = \frac{1}{1 - z^{-1}}$$

Then the sampled-value process output is given by

$$\hat{y}(z) = HG_p(z)\hat{c}(z) = \frac{K_p T z^{-1}}{1 - z^{-1}} \frac{1}{1 - z^{-1}} = \frac{K_p T z^{-1}}{(1 - z^{-1})^2}$$

and we conclude that $y(nT)$ changes like a ramp function with slope $= K_p$ (Figure 29.7a). The final-value theorem indicates that

$$\lim_{t \to \infty} y(t) = \lim_{z \to 1} [(1 - z^{-1})\hat{y}(z)] = \lim_{z \to 1} \left[\frac{K_p T z^{-1}}{1 - z^{-1}} \right] = \infty$$

and the system reaches no steady state under a unit step input. The same result had been reached for the continuous case in Section 10.3.

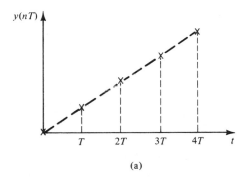

(a)

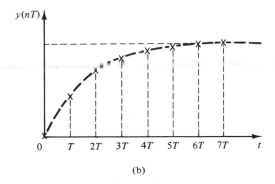

(b)

Figure 29.7 Sampled-value response to unit step inputs of: (a) pure integrator; (b) first-order lag.

(b) *First-order lag process* (see Section 10.4): The process transfer function is given by eq. (10.3) [i.e., $G(s) = K_p/(\tau_p s + 1)$]. Consequently, the corresponding pulse transfer function can be found as follows (see also Figure 29.6b):

$$HG_p(z) = Z\left\{\frac{1 - e^{-Ts}}{s}\frac{K_p}{\tau_p s + 1}\right\} = Z\left\{\frac{K_p}{s(\tau_p s + 1)}\right\} - Z\left\{\frac{K_p e^{-Ts}}{s(\tau_p s + 1)}\right\}$$

$$= (1 - z^{-1})Z\left\{\frac{K_p}{s(\tau_p s + 1)}\right\} = K_p(1 - z^{-1})Z\left\{\frac{1}{s} - \frac{1}{s + 1/\tau_p}\right\}$$

From Table 28.1 we find

$$Z\left\{\frac{1}{s}\right\} = Z\,[\text{unit step}] = \frac{1}{1 - z^{-1}}$$

$$Z\left\{\frac{1}{s + 1/\tau_p}\right\} = Z\,[e^{-t/\tau_p}] = \frac{1}{1 - e^{-T/\tau_p}z^{-1}}$$

and the pulse transfer function is

$$HG_p(z) = K_p(1 - z^{-1})\left[\frac{1}{1 - z^{-1}} - \frac{1}{1 - e^{-T/\tau_p}z^{-1}}\right] = K_p\frac{(1 - e^{-T/\tau_p})z^{-1}}{1 - e^{-T/\tau_p}z^{-1}} \qquad (29.15)$$

Consider a sequence of input values $c(nT)$ corresponding to a unit step. Then

$$\hat{c}(z) = \frac{1}{1 - z^{-1}}$$

and

$$\hat{y}(z) = K_p\frac{1}{1 - z^{-1}}\frac{(1 - e^{-T/\tau_p})z^{-1}}{1 - e^{-T/\tau_p}z^{-1}} = \frac{K_p}{1 - z^{-1}} - \frac{K_p}{1 - e^{-T/\tau_p}z^{-1}}$$

The inverse z-transform yields (see Table 28.1)

$$y(nT) = K_p\left\{1 - \exp\left(\frac{-nT}{\tau_p}\right)\right\} \qquad \text{for } n = 0, 1, 2, \ldots$$

The sampled-value response $y(nT)$ is shown in Figure 29.7b and follows the same pattern as for the continuous case (see Figure 10.4). To find the steady-state value of the process output, invoke the final-value theorem. Thus

$$\lim_{t \to \infty} y(t) = \lim_{z \to 1}[(1 - z^{-1})\hat{y}(z)] = \lim_{z \to 1}\left[(1 - z^{-1})\frac{K_p(1 - e^{-T/\tau_p})z^{-1}}{(1 - z^{-1})(1 - e^{-T/\tau_p}z^{-1})}\right]$$

and

$$\lim_{t \to \infty} y(t) = K_p$$

as expected.

(c) *Two noninteracting first-order lags in series*: The process transfer function is given by

$$G_p(s) = \frac{K_p}{(\tau_1 s + 1)(\tau_2 s + 1)}$$

Therefore, the pulse transfer function is found to be

$$HG_p(z) = Z\left\{\frac{1 - e^{-Ts}}{s} \frac{K_p}{(\tau_1 s + 1)(\tau_2 s + 1)}\right\} = (1 - z^{-1})K_p Z\left\{\frac{1}{s(\tau_1 s + 1)(\tau_2 s + 1)}\right\}$$

$$= (1 - z^{-1})K_p Z\left\{\frac{1}{s} + \frac{\tau_1}{\tau_2 - \tau_1} \frac{1}{s + 1/\tau_1} - \frac{\tau_2}{\tau_2 - \tau_1} \frac{1}{s + 1/\tau_2}\right\}$$

$$= (1 - z^{-1})K_p\left\{\frac{1}{1 - z^{-1}} + \frac{\tau_1}{\tau_2 - \tau_1} \frac{1}{1 - e^{-T/\tau_1}z^{-1}} - \frac{\tau_2}{\tau_2 - \tau_1} \frac{1}{1 - e^{-T/\tau_2}z^{-1}}\right\}$$

It is easy now to proceed as in the previous cases to find the sampled-value response of the process to a sequence of input values corresponding to a unit step.

(d) *Second-order process:* The process transfer function is given by

$$G_p(s) = \frac{K_p}{\tau^2 s^2 + 2\zeta\tau s + 1}$$

In Chapter 11 we found that almost all second-order, open-loop processes are overdamped systems ($\zeta > 1$) composed of two interacting capacities in series. Therefore, the transfer function can be written as

$$G_p(s) = \frac{K_p}{\tau^2 s^2 + 2\zeta\tau s + 1} = \frac{K_p}{(\tau_1' s + 1)(\tau_2' s + 1)}$$

where $-1/\tau_1'$ and $-1/\tau_2'$ are both real roots of the denominator. But the pulse transfer function of a process with $G_p(s)$ given by the last equation was found in case (c).

If the process is inherently second-order (like those in Appendix 11.A), we may have $\zeta = 1$, $\zeta > 1$, or $\zeta < 1$. For each case we will have different expressions for

$$Z\left\{\frac{1 - e^{-Ts}}{s} \frac{K_p}{\tau^2 s^2 + 2\zeta\tau s + 1}\right\}$$

and consequently different pulse transfer functions. The development of these pulse transfer functions is left to the reader as an exercise.

Remarks

1. Consider N continuous processes in series with transfer functions $G_1(s)$, $G_2(s),\ldots, G_N(s)$. The pulse transfer function of the train of continuous systems is characterized by $G_1 G_2 \cdots G_N(z)$ and is computed from the equation

$$G_1 G_2 \cdots G_N(z) = Z[G_1(s)G_2(s)\cdots G_N(s)]$$

2. In general,

$$Z[G_1(s)G_2(s)\cdots G_N(s)] \neq G_1(z)G_2(z)\cdots G_N(z)$$

Thus, unfortunately,

$$HG_p(z) = Z[H(s)G_p(s)] \neq H(z)G_p(z)$$

3. It is interesting to find out what the process output would have been if a sequence of impulses entered a continuous process directly without passing through a hold element. Consider the case of a first-order lag but without the hold element (Figure 29.8a). Then the pulse transfer function is given by

$$G(z) = Z\left\{\frac{K_p}{\tau_p s + 1}\right\} = Z\left\{\frac{K_p/\tau_p}{s + 1/\tau_p}\right\} = \frac{K_p/\tau_p}{1 - e^{-T/\tau_p} z^{-1}}$$

Consider a sequence of discrete input values $c(nT)$ corresponding to a unit step change

$$\hat{c}(z) = \frac{1}{1 - z^{-1}}$$

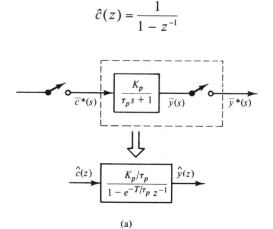

(a)

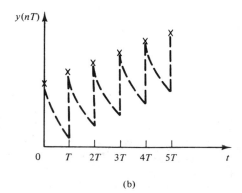

(b)

Figure 29.8 (a) First-order lag without hold element; (b) its discrete-time response to unit step input.

Then

$$\hat{y}(z) = G(z)\hat{c}(z) = \frac{K_p/\tau_p}{(1 - z^{-1})(1 - e^{-T/\tau_p}z^{-1})}$$

and

$$y(nT) = \mathcal{Z}^{-1}\left[\frac{K_p/\tau_p}{(1 - z^{-1})(1 - e^{-T/\tau_p}z^{-1})}\right]$$

$$= \frac{K_p/\tau_p}{1 - e^{-T/\tau_p}} + \frac{K_p/\tau_p}{1 - e^{T/\tau_p}}\exp\left(\frac{-nT}{\tau_p}\right)$$

The sampled value response is shown in Figure 29.8b. Compare the response of Figure 29.8b with that of Figure 29.7b and note the difference that the hold element makes.

29.3 Discrete-Time Analysis of Closed-Loop Systems

Consider the block diagram of a direct digital feedback control loop shown in Figure 29.9. Such loops contain both continuous- and discrete-time signals and dynamic elements. Three samplers are present to indicate the discrete-time nature of the set point $\hat{y}_{SP}(z)$, control command $\hat{c}(z)$, and sampled process output $\hat{y}(z)$. The continuous signals are denoted by their Laplace transforms [i.e., $\bar{y}(s)$, $\bar{m}(s)$, and $\bar{d}(s)$]. Furthermore, the continuous dynamic elements (e.g., hold, process, disturbance element) are denoted by their continuous transfer functions, $H(s)$, $G_p(s)$, and $G_d(s)$, respectively. For the control algorithm, which is the only discrete element, we have used its discrete transfer function, $D(z)$.

Let us now develop the *closed-loop transfer functions* relating the sampled process output $\hat{y}(z)$ to the discrete set point changes $\hat{y}_{SP}(z)$ and to the continuous disturbance $\bar{d}(s)$. The analysis proceeds in a

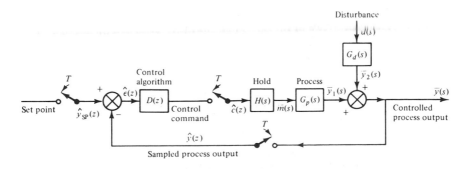

Figure 29.9 Block diagram of DDC loop.

manner analogous to the procedure for completely continuous systems (see Section 14.1).

The continuous controlled output is given by

$$\overline{y}(s) = \overline{y}_1(s) + \overline{y}_2(s) \tag{29.16}$$

where $\overline{y}_1(s)$ is the main process output and $\overline{y}_2(s)$ is the effect of the disturbance. From eq. (29.16) it is easy to find that the z-transform of the sampled process output is given by

$$\hat{y}(z) = Z[\overline{y}_1(s)] + Z[\overline{y}_2(s)] \tag{29.16a}$$

Equation (29.8) implies that

$$Z[\overline{y}_1(s)] = HG_p(z)\hat{c}(z) \tag{29.17}$$

where $HG_p(z)$ is the pulse transfer function for the continuous process of the loop. On the other hand, it is easily seen from the block diagram that

$$Z[\overline{y}_2(s)] = Z[G_d(s)\overline{d}(s)] \tag{29.18}$$

Note that *we cannot define a pulse transfer function $G_d(z)$* for the disturbance element and write

$$Z[G_d(s)\overline{d}(s)] = G_d(z)\hat{d}(z)$$

because *the disturbance is not a sampled-value signal.*

Substitute eqs. (29.17) and (29.18) into eq. (29.16a) and take

$$\hat{y}(z) = HG_p(z)\hat{c}(z) + Z[G_d(s)\overline{d}(s)] \tag{29.19}$$

From the comparator,

$$\hat{\epsilon}(z) = \hat{y}_{SP}(z) - \hat{y}(z)$$

and from the digital controller

$$\hat{c}(z) = D(z)\hat{\epsilon}(z) = D(z)[\hat{y}_{SP}(z) - \hat{y}(z)] \tag{29.20}$$

where $D(z)$ is the discrete transfer function of the control algorithm. Substitute eq. (29.20) into eq. (29.19) and find

$$\hat{y}(z) = HG_p(z)D(z)[\hat{y}_{SP}(z) - \hat{y}(z)] + Z[G_d(s)\overline{d}(s)]$$

or

$$\hat{y}(z) = \frac{HG_p(z)D(z)}{1 + HG_p(z)D(z)} \hat{y}_{SP}(z) + \frac{Z[G_d(s)\overline{d}(s)]}{1 + HG_p(z)D(z)} \tag{29.21}$$

Equation (29.21) yields the *discrete-time (sampled-value) closed-loop response of a direct digital control loop.* Its form is very similar to that for continuous analog control loops.

Remarks

1. The common denominator in eq. (29.21) determines the *characteristic equation* of the DDC loop:

$$1 + HG_p(z)D(z) - 0 \qquad (29.22)$$

The roots of the characteristic equation determine the stability of the closed-loop response of a DDC loop.

2. For the *servo problem*, $\bar{d}(s) = 0$, and the closed-loop response to set point changes is given by

$$\hat{y}(z) = \frac{HG_p(z)D(z)}{1 + HG_p(z)D(z)} \hat{y}_{SP}(z) \qquad (29.21a)$$

3. For the *regulator problem*, $\hat{y}_{SP}(z) = 0$, and the closed-loop response to disturbance changes is characterized by the equation

$$\hat{y}(z) = \frac{Z[G_d(s)\bar{d}(s)]}{1 + HG_p(z)D(z)} \qquad (29.21b)$$

4. What is meant by $Z[G_d(s)\bar{d}(s)]$? Let $\bar{y}_2(s) = G_d(s)\bar{d}(s)$. The inverse Laplace transform yields the function in the time domain:

$$y_2(t) = \mathcal{L}^{-1}[G_d(s)\bar{d}(s)]$$

The z-transform of the sampled values of the function $y_2(t)$ yields

$$Z[y_2(t)] - Z\{\mathcal{L}^{-1}[G_d(s)\bar{d}(s)]\}$$

and for presentational simplification we denote $Z\{\mathcal{L}^{-1}[G_d(s)\bar{d}(s)]\}$ by $Z[G_d(s)\bar{d}(s)]$. Therefore, $Z[G_d(s)\bar{d}(s)]$ *is the z-transform of the disturbance effect on the controlled output*.

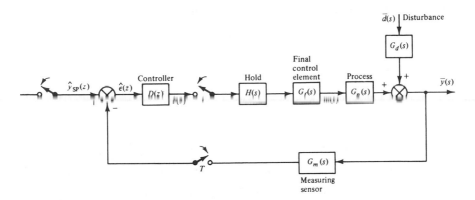

Figure 29.10 Block diagram of DDC loop including dynamics of measuring sensor and final control element.

5. In the feedback control loop of Figure 29.9 we have omitted the dynamics of the measuring sensor and final control element. Thus they are absent from the closed-loop response of eq. (29.21). Consider the loop of Figure 29.10, with the sensor and final control element included. Following the same procedure as above, it is easy to show that the sampled-value, closed-loop response of the loop is given by

$$\hat{y}(z) = \frac{HG_fG_p(z)D(z)}{1 + HG_fG_pG_m(z)D(z)}\,\hat{y}_{SP}(z) + \frac{Z[G_d(s)\overline{d}(s)]}{1 + HG_fG_pG_m(z)D(z)}$$

(29.23)

where $HG_fG_pG_m(z)$ is the pulse transfer function between $\hat{c}(z)$ and $\hat{y}_m(z)$ and it is given by

$$\frac{\hat{y}_m(z)}{\hat{c}(z)} = HG_fG_pG_m(z) = Z[H(s)G_f(s)G_p(s)G_m(s)]$$

Example 29.4: Closed-Loop Response of a First-Order Process under Proportional Digital Control

Let

$$G_p(s) = \frac{K_p}{\tau_p s + 1} \quad \text{and} \quad G_d(s) = \frac{a}{\tau_p s + 1}$$

Assuming zero-order hold, the pulse transfer function for the hold–process combination was developed in Example 29.3, case b, and is given by eq. (29.15):

$$HG_p(z) = K_p \frac{(1 - e^{-T/\tau_p})z^{-1}}{1 - e^{-T/\tau_p}z^{-1}}$$

Also, for the proportional digital controller the discrete transfer function is

$$D(z) = K_c = \text{proportional gain}$$

Put the expressions above in eq. (29.21) and take the following closed-loop response:

$$\hat{y}(z) = \frac{K_p \dfrac{(1 - e^{-T/\tau_p})z^{-1}}{1 - e^{-T/\tau_p}z^{-1}}K_c}{1 + K_p \dfrac{(1 - e^{-T/\tau_p})z^{-1}}{1 - e^{-T/\tau_p}z^{-1}}K_c}\,\hat{y}_{SP}(z) + \frac{Z\left[\dfrac{a}{\tau_p s + 1}\overline{d}(s)\right]}{1 + K_p \dfrac{(1 - e^{-T/\tau_p})z^{-1}}{1 - e^{-T/\tau_p}z^{-1}}K_c}$$

(29.24)

1. *Servo problem*: Put $\overline{d}(s) = 0$. Then eq. (29.24) yields

$$\hat{y}(z) = \frac{K_pK_c(1 - b)z^{-1}}{1 + [K_pK_c - b(1 + K_pK_c)]z^{-1}}\,\hat{y}_{SP}(z)$$

(29.25)

where $b = e^{-T/\tau_p}$. Make a unit step change in the set point:

$$\hat{y}_{SP}(z) = \frac{1}{1 - z^{-1}}$$

Put this value in eq. (29.25) and find

$$\hat{y}(z) = \frac{K_p K_c (1 - b) z^{-1}}{1 + [K_p K_c - b(1 + K_p K_c)] z^{-1}} \frac{1}{1 - z^{-1}}$$

and after inversion we have the sampled-value, closed-loop response,

$$y(nT) = \frac{K_p K_c}{1 + K_p K_c} \left[1 - \exp\left(\frac{-nT}{\tau_p'}\right) \right]$$

where $\tau_p' = \tau_p/(1 + K_p K_c)$. This equation is completely analogous to eq. (14.21), which represents the closed-loop response of a first-order process to a unit step change in the set point under analog proportional control (see Section 14.2).

Let us now find the steady state of the closed-loop response. From the final-value theorem we have

$$\lim_{t \to \infty} y(t) = \lim_{z \to 1}[(1 - z^{-1})\hat{y}(z)]$$

$$= \lim_{z \to 1}\left\{ \frac{K_p K_c (1 - b) z^{-1}}{1 + [K_p K_c - b(1 + K_p K_c)] z^{-1}} \right\} = \frac{K_p K_c}{1 + K_p K_c}$$

Therefore, the offset is

$$\text{offset} = 1 - \frac{K_p K_c}{1 + K_p K_c} = \frac{1}{1 + K_p K_c}$$

the same offset as we have found for analog proportional controllers (see Section 14.2).

2. *Regulator problem*: Put $\hat{y}_{SP}(z) = 0$, and consider a unit step change in the disturbance:

$$\bar{d}(s) = \frac{1}{s}$$

Then

$$Z\left[\frac{a}{\tau_p s + 1}\frac{1}{s}\right] = Z\left[\frac{a}{s} - \frac{a}{s + 1/\tau_p}\right] = \frac{a}{1 - z^{-1}} - \frac{a}{1 - e^{-T/\tau_p}z^{-1}}$$

and the closed-loop response is

$$\hat{y}(z) = \frac{\dfrac{a}{1 - z^{-1}} - \dfrac{a}{1 - bz^{-1}}}{1 + [K_p K_c - b(1 + K_p K_c)] z^{-1}}(1 - bz^{-1}) \qquad (29.26)$$

where $b = e^{-T/\tau_p}$. The inversion of eq. (29.26) is left to the reader as an exercise.

Example 29.5: Closed-Loop Response of a First-Order Process with Dead Time

The transfer function of a first-order process with dead time is

$$G_p(s) = \frac{K_p e^{-t_d s}}{\tau_p s + 1}$$

Assume that the dead time is an integer multiple of the sampling period T:

$$t_d = kT \qquad \text{with } k = \text{integer}$$

The pulse transfer function is

$$HG_p(z) = \mathcal{Z}\left[\frac{1 - e^{-Ts}}{s} \frac{K_p e^{-kTs}}{\tau_p s + 1} \right]$$

The term e^{-kTs} denotes delay by k sampling periods and is equivalent to z^{-k}. Therefore,

$$HG_p(z) = z^{-k} \cdot \mathcal{Z}\left[\frac{1 - e^{-Ts}}{s} \frac{K_p}{\tau_p s + 1} \right] = z^{-k} K_p \left[\frac{(1 - e^{-T/\tau_p}) z^{-1}}{1 - e^{-T/\tau_p} z^{-1}} \right]$$

The rest proceeds in a similar manner as in Example 29.4.

29.4 Stability Analysis of Discrete-Time Systems

In previous sections we examined the dynamic characteristics of discrete-time open- and closed-loop systems and we developed the appropriate transfer functions to describe them. Nowhere, though, did we question the stability of these systems. In this section we extend the previous analysis and derive general rules that will determine the stability characteristics of the response of discrete-time systems.

We again employ the definition of stability outlined in Section 15.1, which is often called bounded input, bounded output stability. Thus:

A discrete-time dynamic system is considered to be stable if for every bounded input it produces a bounded output, regardless of its initial state.

Discrete systems

Consider a discrete system with a discrete transfer function given by

$$D(z) = \frac{\hat{c}(z)}{\hat{e}(z)} = \frac{a_0 + a_1 z^{-1} + a_2 z^{-2} + \cdots + a_m z^{-m}}{1 + b_1 z^{-1} + b_2 z^{-2} + \cdots + b_n z^{-n}}$$

Using the partial-fractions expansion, the output $\hat{c}(z)$ is given by

$$\hat{c}(z) = \left\{ \frac{C_1}{1 - p_1 z^{-1}} + \frac{C_2}{1 - p_2 z^{-1}} + \cdots + \frac{C_n}{1 - p_n z^{-1}} \right\} \hat{e}(z) \quad (29.27)$$

where $p_1, p_2, \ldots, p_n$ are the roots (also known as the *poles*) of the characteristic equation

$$1 + b_1 z^{-1} + b_2 z^{-2} + \cdots + b_n z^{-n} = 0 \qquad (29.28)$$

Take the kth term within the braces from eq. (29.27) and find its inverse z-transform. From Table 28.1 we can easily find that

$$f_k(nT) = Z^{-1}\left[\frac{C_k}{1 - p_k z^{-1}} \right] = C_k \exp\left[n \ln p_k\right] \quad \text{for } n = 0, 1, 2, \cdots \ (29.29)$$

Assume that the root p_k is generally a complex root:

$$p_k = \alpha + j\beta = |p_k| e^{j\omega}$$

where

$$|p_k| = \sqrt{\alpha^2 + \beta^2} \quad \text{and} \quad \omega = \tan^{-1}\left(\frac{\beta}{\alpha}\right)$$

Then, since $\ln p_k = \ln|p_k| + j\omega$, we have

$$\exp\left[n \ln p_k\right] = \exp\left[n \ln|p_k| + nj\omega\right] = \exp\left[n \ln|p_k|\right] \exp\left[nj\omega\right]$$

Now, from Euler's identities we can find that

$$\exp\left[nj\omega\right] = \cos n\omega + j \sin n\omega$$

and we conclude that $\exp\left[nj\omega\right]$ is always bounded for any value of n.

Let us turn our attention to the other term, $\exp\left[n \ln|p_k|\right]$.

If $|p_k| < 1$, then $\ln|p_k| < 0$ and $\exp\left[n \ln|p_k|\right] \to 0$ as $n \to \infty$.
If $|p_k| = 1$, then $\ln|p_k| = 0$ and $\exp\left[n \ln|p_k|\right] = 1$ for every n.
If $|p_k| > 1$, then $\ln|p_k| > 0$ and $\exp\left[n \ln|p_k|\right] \to \infty$ as $n \to \infty$.

From the analysis above we conclude that *the function $f_k(nT)$ is bounded when the root p_k lies inside or on the unit circle in the complex plane* (see Figure 29.11), where its magnitude is smaller or equal to unity. Extending these results to all the terms within the braces of eq. (29.27), we conclude that:

A discrete system is stable if all its poles (i.e., the roots of its characteristic equation) lie inside or on the unit circle in the complex plane (i.e., they have a magnitude less than or equal to unity).

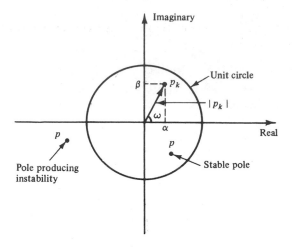

Figure 29.11 Unit circle in complex plane.

Continuous systems with discrete inputs and sampled outputs

With the use of the pulse transfer function, such systems can be converted to discrete systems as above, and consequently the same rule applies for their stability. In particular, if $G_p(s)$ is the transfer function of a continuous process, its pulse transfer function is given by

$$G_p(z) = \mathcal{Z}[G_p(s)]$$

If any of the poles of $G_p(z)$ lie outside the unit circle, the process is unstable. Similarly, for the combination hold–process we examine the location of poles of the combined pulse transfer function

$$HG_p(z) = \mathcal{Z}[H(s)G_p(s)]$$

Digital feedback control systems

The same rule applies here. The characteristic equation is given by eq. (29.22):

$$1 + HG_p(z)D(z) = 0 \qquad (29.22)$$

and if any of its roots lies outside the unit circle, the feedback system is unstable.

Example 29.6: Stability Analysis of a Digital Control Loop

Consider a digital control loop with a block diagram like that given in Figure 29.9. The following specifications are given on the process and controller.

1. The process transfer function is equivalent to two first-order systems in series:

$$G_p(s) = \frac{10}{(0.1s + 1)(2s + 1)}$$

2. For hold we have used the zero-order hold element with

$$H(s) = \frac{1 - e^{-Ts}}{s}$$

3. The controller is a digital PI algorithm in the velocity form. Its discrete transfer function will be developed in Section 30.1, but for the time being consider it is known and given by

$$D(z) = \frac{K_c}{1 - z^{-1}}\left\{\left(1 + \frac{T}{\tau_I}\right) - z^{-1}\right\}$$

where T is the sampling period, K_c the proportional gain, and τ_I the integral time constant.

The pulse transfer function of the combination hold–process is given by

$$HG_p(z) = Z\left[\frac{1 - e^{-Ts}}{s}\,\frac{10}{(0.1s + 1)(2s + 1)}\right]$$

$$= (1 - z^{-1})Z\left[\frac{10}{s} + \frac{50/95}{s + 10} - \frac{1000/95}{s + 0.5}\right]$$

$$= (1 - z^{-1})\left\{\frac{10}{1 - z^{-1}} + \frac{50/95}{1 - e^{-10T}z^{-1}} - \frac{1000/95}{1 - e^{-0.5T}z^{-1}}\right\}$$

Therefore, the stability of the digital loop is characterized by the roots of the following characteristic equation [see eq. (29.22)]:

$$1 + HG_p(z)D(z) = 1 + (1 - z^{-1})$$

$$\left\{\frac{10}{1 - z^{-1}} + \frac{50/95}{1 - e^{-10T}z^{-1}} - \frac{1000/95}{1 - e^{-0.5T}z^{-1}}\right\}\frac{K_c}{1 - z^{-1}}\left\{\left(1 + \frac{T}{\tau_I}\right) - z^{-1}\right\} = 0$$

Note that the roots of the equation above do not depend only on K_c and τ_I but also on the value of the sampling period T. Therefore,

The stability characteristics of a digital loop do not depend only on the controller parameters (e.g., K_c, τ_I, τ_D) but also on the sampling period T.

This is a new feature directly attributable to the discrete-time nature of a digital controller. Therefore, *we need to be very careful not to destabilize an otherwise stable loop by an improper selection of the sampling period.*
    Let us use the following parametric values:

$$K_c = 0.10, \qquad \tau_I = 1, \qquad T = 1$$

Then, after the necessary algebra, the characteristic equation yields

$$1 + HG_p(z)D(z) = 1 - 0.89z^{-1} + 0.316z^{-2} - 0.0321z^{-3} = 0$$

with roots

$$z_1 = 0.162 \qquad z_2 = 0.365 + j\,0.775 \qquad z_3 = 0.365 - j\,0.775$$

Since the magnitude of all roots is smaller than unity, we conclude that the digital control loop is stable.

Parametric analysis

1. If we keep $K_c = 0.10$, $\tau_I = 1$, but change $T = 50$, one of the poles moves outside the unit circle and the system is unstable. *This is a clear manifestation of the destabilizing effect of the sampling period.*
2. Keeping $T = 1$ and changing K_c and τ_I, we find that the loop becomes unstable when $K_c = 1.00$ and $\tau_I = 0.01$.

Remarks

1. It is interesting to point out the relationship between the stability rules for continuous systems in the s-domain (see Section 15.1) and the rule developed in this section for discrete-time systems in the z-domain. Consider a system with a continuous transfer function $G(s)$. This system is stable if the roots of its characteristic equation (poles) lie to the left of the imaginary axis (see Figure 29.12a). Recall the relationship between the variables s and z [eq. (28.2)]:

$$z = e^{Ts}$$

Let s be a complex variable

$$s = \alpha + j\beta$$

Then

$$z = e^{\alpha T} e^{j\beta T}$$

This is a complex number with magnitude

$$|z| = e^{\alpha T}$$

Therefore, *if s lies to the left of the imaginary axis, then $\alpha < 0$ and $|z| < 1$, which is the rule for stability of discrete-time systems.* Figure 29.12 shows the corresponding regions for stability of a system in continuous form (Figure 29.12a) and discrete-time form (Figure 29.12b).
2. It should be emphasized once more that the sampling period affects the stability characteristics of a discrete-time system in a very profound manner. Example 29.6 demonstrated that clearly.

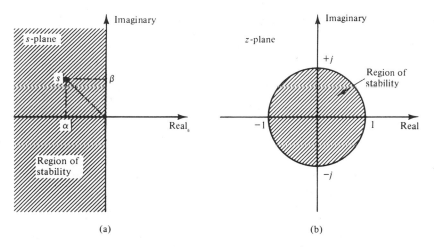

Figure 29.12 Stability regions in s-plane and z-plane.

THINGS TO THINK ABOUT

1. Is it possible to analyze the closed-loop behavior of a DDC loop using continuous transfer functions in the Laplace domain for the various dynamic elements of the loop? Explain why or why not.

2. What types of new transfer functions do you need to define for analyzing a DDC loop, and why?

3. Define a discrete system and indicate how it is different from a continuous system. Give several examples of discrete systems. Why should all algorithms used by computer control systems be considered discrete systems?

4. What is a discrete transfer function, and what is it needed for? Develop the discrete transfer function for (a) a proportional control algorithm, (b) the velocity form of a PI control algorithm, and (c) a second-order digital filter.

5. Define the pulse transfer function for a continuous system. Why is it needed, and why is its continuous transfer function $G(s)$ inadequate?

6. The pulse transfer function of the hold–process combination is given by

$$HG_p(z) = Z\,[H(s)G(s)]$$

What is the right-hand side of the equation above? Indicate a procedure for computing it.

7. Find the pulse transfer function of a pure integrator with and without the hold element.

8. Show that

$$HG_p(z) = Z[H(s)G(s)] \neq H(z)G(z)$$

Under what conditions in $H(s)$ and $G(s)$ can we have

$$Z[H(s)G(s)] = H(z)G(z)?$$

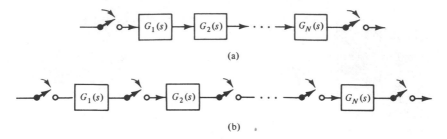

(a)

(b)

Figure Q29.1

[*Hint*: Consider a pure dead-time process with $G(s) = e^{-kTs}$ and compute $HG(z)$. Can you generalize your observation?]

9. Consider N continuous systems in series. For the configuration of Figure Q29.1a, show that the overall pulse transfer function is

$$G_1 G_2 \cdots G_N(z) = Z[G_1(s)G_2(s) \cdots G_N(s)] \neq G_1(z)G_2(z) \cdots G_N(z)$$

while for the system of Figure Q29.1b,

$$G_1 G_2 \cdots G_N(z) = Z[G_1(s)G_2(s) \cdots G_N(s)] = G_1(z)G_2(z) \cdots G_N(z)$$

10. In Figure Q29.2a and b we have drawn the block diagrams of two discrete-

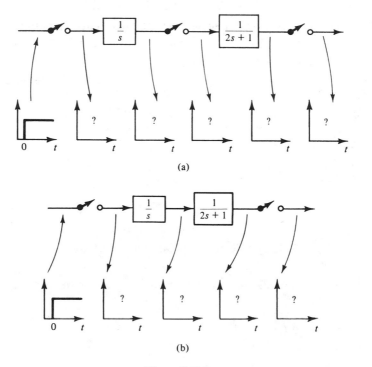

(a)

(b)

Figure Q29.2

time systems. Plot the behavior with respect to time of the requested signals in the appropriate diagrams.

11. Compute the pulse transfer function of the following second-order process with and without a zero-hold element:

$$G_p(s) = \frac{1}{s^2 + 2\zeta s + 1}$$

Examine all cases with $\zeta > 1$, $\zeta < 1$, and $\zeta = 1$.

12. Find the closed-loop response of a first-order process using the velocity form of the PI control algorithm. Show that the steady-state offset of the closed-loop response to a unit step change in the set point is zero.

13. Define the bounded input, bounded output stability of a discrete-time system. What is the rule for characterizing the stability of such a system?

14. Consider the discrete transfer function

$$D(z) = \frac{1}{1 - pz^{-1}} = \frac{\hat{c}(z)}{\hat{e}(z)}$$

and compute the discrete-time response at sampling instants 0, T, $2T$, $3T, \ldots$ for a unit step change in the input $\epsilon(nT)$ and for the following cases:

$$p = -0.9 \quad \text{and} \quad p = +0.9$$

Both systems are stable because $|p| < 1$. But notice the difference in response. (*Note*: The oscillatory response for $p < 0$ is called "ringing" and is undesirable.)

15. Does the sampling period affect the stability of a discrete-time system, and why? What would you expect to happen; a system becomes destabilized when the sampling period increases or decreases? Elaborate on your answer.

Design of Digital **30**
Feedback Controllers

In Chapters 16 and 18 we discussed the design of analog feedback controllers in the time and Laplace domains. Does the introduction of a digital controller change these design methodologies?

The answer is basically no. Most industrial DDC applications use a digital approximation of the analog P, PI, and PID controllers. The sampling rates are usually high, thus making the digital controller approximate very closely the performance of a continuous analog controller. Therefore, for the design of a digital feedback controller we can use:

1. The guidelines discussed in Section 16.4 for the selection of the appropriate type of feedback controller
2. The performance criteria outlined in Sections 16.2 and 16.3
3. The Cohen–Coon or Ziegler–Nichols methodologies for tuning feedback controllers described in Sections 16.5 and 18.3

However, the "intelligence" of a digital computer offers the possibility for more advanced and computationally more complex feedback controllers. In this chapter we will study both: the design of digital approximations to the analog P, PI, and PID controllers, and the design of more complex digital feedback controllers.

$$\frac{0.54}{975s + 30 + 1}$$

$$G(s) \quad \frac{T_p}{M_2} = \frac{0.54\,e^{-15s}}{(15+s)(15+s)} =$$

Fn

M₁

Stam

Steam

M₂

F

To

TT

TT

HSBC ◆◆

with compliments

HSBC Bank plc
2 Bold Street, Liverpool L1 4DT
Tel: 08457 404 404 Fax: 08455 877 431

Registered in England number 14259, Registered Office: 8 Canada Square, London E14 5HQ
Authorised and regulated by the Financial Services Authority.

Ahmed

TUESDAY
18th DEC
3-30 pm

30.1   Digital Approximation of Classical Controllers

The most commonly used analog controller is the three-mode proportional-integral-derivative (PID) controller. Its general form is given by eq. (13.6):

$$c(t) = K_c\left\{ \epsilon(t) + \frac{1}{\tau_I}\int_0^t \epsilon(t)\,dt + \tau_D \frac{d\epsilon}{dt} \right\} + c_s \qquad (13.6)$$

with a continuous transfer function given by eq. (13.7):

$$G_c(s) = K_c\left(1 + \frac{1}{\tau_I s} + \tau_D s \right) \qquad (13.7)$$

Neglecting the derivative control mode, we take the equally popular PI controller,

$$c(t) = K_c\left\{ \epsilon(t) + \frac{1}{\tau_I}\int_0^t \epsilon(t)\,dt \right\} + c_s \qquad (13.4)$$

with

$$G_c(s) = K_c\left(1 + \frac{1}{\tau_I s} \right) \qquad (13.5)$$

Position form of control algorithms

It is easy to develop the digital approximation of the PID controller. This was done in Example 27.5, using rectangular integration to approximate the integral control mode and first-order difference to approximate the derivative mode. The resulting discrete-time approximation is given by eq. (27.7):

$$c_n = K_c\left\{ \epsilon_n + \frac{T}{\tau_I}\sum_{k=0}^n \epsilon_k + \frac{\tau_D}{T}(\epsilon_n - \epsilon_{n-1}) \right\} + c_s \qquad (27.7)$$

where c_s is the controller output signal when the error is zero (i.e., the initial valve position). It is obvious that the difference equation corresponding to the digital PI controller is

$$c_n = K_c\left\{ \epsilon_n + \frac{T}{\tau_I}\sum_{k=0}^n \epsilon_k \right\} + c_s \qquad (30.1)$$

The foregoing two digital approximations for PI and PID controllers are known as the *position form* of the algorithms, because at each sampling instant they compute the actual value (*position*) of the controller output signal. In this form, at the nth sampling instant the PI algorithm saves only the current value of the error, ϵ_n, and the sum of all previous errors, $S_{n-1} = \sum_{l=0}^{n-1} \epsilon_l$, yielding

$$c_n = K_c \left\{ \epsilon_n + \frac{T}{\tau_I} (S_{n-1} + \epsilon_n) \right\} + c_s$$

while the PID algorithm saves the current error, ϵ_n, the previous error, ϵ_{n-1}, and the sum of all previous errors:

$$c_n = K_c \left\{ \epsilon_n + \frac{T}{\tau_I} (S_{n-1} + \epsilon_n) + \frac{\tau_D}{T} (\epsilon_n - \epsilon_{n-1}) \right\} + c_s$$

Velocity form of control algorithms

An alternative form for the PI and PID control algorithms is the so-called *velocity form*. In this form, one does not compute the actual value of the controller output signal at the nth sampling instant, but its change from the preceding period. Thus consider the PID control action at the nth and $(n-1)$th sampling instants. From eq. (27.7) we take

$$c_n = K_c \left\{ \epsilon_n + \frac{T}{\tau_I} \sum_{k=0}^{n} \epsilon_k + \frac{\tau_D}{T} (\epsilon_n - \epsilon_{n-1}) \right\}$$

$$c_{n-1} = K_c \left\{ \epsilon_{n-1} + \frac{T}{\tau_I} \sum_{k=0}^{n-1} \epsilon_k + \frac{\tau_D}{T} (\epsilon_{n-1} - \epsilon_{n-2}) \right\}$$

Subtract the second from the first:

$$\Delta c_n = c_n - c_{n-1} = K_c \left(1 + \frac{T}{\tau_I} + \frac{\tau_D}{T} \right) \epsilon_n - K_c \left(1 + \frac{2\tau_D}{T} \right) \epsilon_{n-1}$$
$$+ K_c \frac{\tau_D}{T} \epsilon_{n-2} \tag{30.2}$$

Equation (30.2) gives the *velocity form of the PID algorithm*. Similarly, we can show that the *velocity form of the PI algorithm* is

$$\Delta c_n = c_n - c_{n-1} = K_c \left(1 + \frac{T}{\tau_I} \right) \epsilon_n - K_c \epsilon_{n-1} \tag{30.3}$$

From eqs. (30.2) and (30.3) we can easily derive the discrete transfer functions for the velocity form of PID and PI control algorithms. Thus we take:

PID algorithm:

$$D(z) = \frac{\Delta \hat{c}(z)}{\hat{\epsilon}(z)} = K_c \left\{ \left(1 + \frac{T}{\tau_I} + \frac{\tau_D}{T} \right) - \left(1 + \frac{2\tau_D}{T} \right) z^{-1} + \frac{\tau_D}{T} z^{-2} \right\} \tag{30.4}$$

PI algorithm:

$$D(z) = \frac{\Delta \hat{c}(z)}{\hat{e}(z)} = K_c \left\{ \left(1 + \frac{T}{\tau_I} \right) - z^{-1} \right\} \qquad (30.5)$$

The velocity form of the algorithms has certain advantages over the position form:

1. *It does not need initialization.* The position form of the algorithms requires the initial value of the controller output c_s, which is not normally known in practice. For example, an operator keeps the control loop in the manual mode until a desired steady-state operation has been reached. At this point the error is zero and the position of the control valve would correspond to the c_s value. Therefore, if the operator would like to transfer the control from manual to automatic, he or she should enter in the position control algorithm the value of c_s, which is not normally known. This difficulty can be bypassed with the velocity form of the control algorithms, which do not need initialization [see eqs. (30.2) and (30.3)].

2. *It is protected against integral "windup."* The integral mode of a controller causes its output to continue changing as long as there is a nonzero error. Often the errors cannot be eliminated quickly, and given enough time they produce larger and larger values for the integral term, which in turn keeps increasing the control action until it is "saturated" (e.g., the valve completely open or closed). This condition is called *integral windup*. Then, even if the error returns to zero, the control action will remain saturated.

   The position form with its continuous summation of errors will produce integral windup and special attention will be required. The velocity form, on the other hand, is protected from integral windup for the following reason: The control action changes continuously until it becomes saturated. But then as soon as the error changes sign, the control action can return within the control range in one sampling period.

3. *It protects the process against computer failure*: With the velocity algorithm we can send out a signal which is used to drive an integrating amplifier or a stepper motor. These devices will retain the last calculated position of the control valve (or other final control element) in case the computer fails, thus avoiding total loss of control of the process.

Tuning digital control algorithms

How do we tune digital control algorithms? In general, the available tuning methodologies are similar to those discussed in Chapters 16 and 18 for continuous controllers.

1. *Cohen–Coon settings* (see Section 16.5): From the process reaction curve we can estimate the process static gain K, the dominant time constant τ, and the process dead time t_d. Then, from eqs. (16.9) through (16.11c), we can compute the parameters K_c, τ_I, and τ_D of a P, PI, or PID control algorithm. The effect of the sampling period T has been accounted for by the nature of the experiment itself, because the reaction curve has been determined using sampled-values of the process output.

2. *Ziegler–Nichols settings* (see Section 18.3): This methodology is completely parallel to that developed for continuous analog controllers. Thus through the digital controller we introduce sinusoidal set point changes of low amplitude and varying frequencies, trying to make the closed-loop response oscillate continuously. Then we can compute the ultimate gain K_u, and ultimate period of sustained oscillation P_u, as described in Section 18.3. Using these values we compute the controller parameters from the given relationships. The effect of the sampling period on determining K_u and P_u has been accounted for by the nature of the experiment itself, because continuous cycling of the process output has been achieved through the use of the digital controller (i.e., using sampled values of the process output).

3. *Tuning through time-integral performance criteria* (see Section 16.3): This approach is identical to that described in Section 16.3 for continuous controllers. We select those values for the adjustable controller parameters which minimize the ISE, IAE, or ITAE.

30.2 Effect of Sampling

The discrete-time sampling of the controlled output is a feature that was not present in analog control systems. The value of the sampling period T affects the quality of the closed-loop response and consequently is considered like a design parameter in a digital control loop.

The intermittent sampling of the controlled output has the following two effects on the control loop:

1. During a sampling interval the digital controller controls the process using "old" and not "current" information.
2. The control loop is in effect "open" during a sampling interval and "closes" only at each sampling instant, when new information arrives.

It is clear, then, that we would like to have a sampling interval as short as possible. However, as the sampling period decreases, the time and

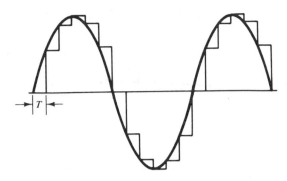

Figure 30.1 Sampling and holding of a continuous signal introduces an effective phase lag.

effort required by a computer increases. What, then, is an optimum sampling rate? Before we attempt to answer this question, let us examine the effect of sampling from another angle.

Consider the continuous, smooth, sinusoidal wave shown in Figure 30.1. Let us sample this signal every T seconds. Figure 30.1 shows the sampling and holding operation, which produces a stair-step-like continuous signal. This is the signal "seen" by the controller. It has the same amplitude and the same frequency as the original wave, but its phase has been shifted. In other words, *the sampling has introduced a phase lag* in the feedback signal, or equivalently, *it has introduced dead time* in the control loop.

Now, from the discussion in Section 19.1 we are well aware of the undesirable consequences that dead time can have on the closed-loop response: sluggishness or, even worse, destabilization of the closed-loop response. Therefore, the sampling period should be selected in such a way that the response of a process does not deteriorate.

There are no firm and rigorous guidelines on how to select the optimal sampling period for a digital control loop. *The optimal choice should always be related to the operating conditions of the controlled process.* From Section 27.1 we know that the sampling period should always be in the range 0.1 to 0.2 of the dominant time constant or dead time, whichever is smaller. For oscillating systems, the sampling period should be smaller than half of the period of oscillation. Therefore, as the operating conditions of the controlled process change and the process output responds faster or oscillates with higher frequency, the sampling rate must increase.

As an initial estimate one could use the following general guidelines which have appeared in various references.

Type of controlled variable (also measured)	Flow	Level and pressure	Temperature
Sampling Period (sec)	1	5	20

These values are only indicative of the sampling periods one should expect for the corresponding loops and simply manifest the fact that flow control loops are faster than level and pressure loops, with temperature loops being the slowest. Thus the faster a loop, the higher its sampling rate. This is in agreement with our discussion in Section 27.1.

30.3 A Different Class of Digital Controllers

In previous sections of this chapter we only considered digital approximations of the familiar analog P, PI, and PID controllers. Such controllers have known structure (e.g., transfer functions) and the only design question is how to select the best values of the adjustable parameters (e.g., K_c, τ_I, τ_D) and the sampling period. In this section we examine a different methodology for designing digital feedback controllers, which makes use of the computational flexibility offered by a digital computer.

Consider the typical DDC loop shown in Figure 29.9. For set point changes only, the closed-loop response is equal to

$$\hat{y}(z) = \frac{HG_p(z)D(z)}{1 + HG_p(z)D(z)} \hat{y}_{SF}(z) \qquad (29.21a)$$

where the pulse transfer function $HG_p(z)$ is known for a given process. *Suppose that we specify the discrete-time character the response should have to a given step change in the set point.* Then both $\hat{y}(z)$ and $\hat{y}_{SP}(z)$ are known and we can solve eq. (29.21a) with respect to the only unknown $D(z)$, the transfer function of the controller we want to design:

$$D(z) = \frac{1}{HG_p(z)} \frac{\hat{y}(z)/\hat{y}_{SP}(z)}{1 - \hat{y}(z)/\hat{y}_{SP}(z)} \qquad (30.6)$$

At this point we must note that $HG_p(z)$, $\hat{y}(z)$, and $\hat{y}_{SP}(z)$ are, in general, ratios of two polynomials in z^{-1}. Therefore, $D(z)$ is generally given by a ratio of two polynomials in z^{-1}: for example,

$$D(z) \equiv \frac{\hat{c}(z)}{\hat{e}(z)} = \frac{\beta_0 + \beta_1 z^{-1} + \beta_2 z^{-2} + \cdots + \beta_k z^{-k}}{1 + \alpha_1 z^{-1} + \alpha_2 z^{-2} + \cdots + \alpha_m z^{-m}} \qquad (30.7)$$

The constant coefficients $\beta_0, \beta_1, \beta_2, \ldots, \beta_k$ and $\alpha_1, \alpha_2, \ldots, \alpha_m$ can be determined from the corresponding coefficients of $HG_p(z)$, $\hat{y}(z)$, and $\hat{y}_{SP}(z)$. Then from eq. (30.7) we find that the control action at the nth sampling instant is given by

$$c_n = \beta_0 \epsilon_n + \beta_1 \epsilon_{n-1} + \beta_2 \epsilon_{n-2} + \cdots + \beta_k \epsilon_{n-k} - \alpha_1 c_{n-1} \\ - \alpha_2 c_{n-1} - \cdots - \alpha_m c_{n-m} \qquad (30.8)$$

Equation (30.8) will constitute the basis of the control algorithm, which will reside in the memory of the process control computer and will be executed every sampling instant.

The control algorithm given by eq. (30.8) is *physically realizable if it requires past or current values for the error*, since future values are not available. This means that *the highest power of z in the numerator of eq. (30.7) cannot be positive*. If this restriction is not satisfied, we have an ill-posed control design problem.

Example 30.1: Physically Unrealizable Digital Controller

Let

$$D(z) \equiv \frac{\hat{c}(z)}{\hat{e}(z)} = \frac{z^{+1} + z^{-1} - 2z^{-2}}{1 + 2z^{-1} + z^{-2}}$$

Then

$$(1 + 2z^{-1} + z^{-2})\hat{c}(z) = (z^{+1} + z^{-1} - 2z^{-2})\hat{e}(z)$$

or

$$c_n + 2c_{n-1} + c_{n-2} = \epsilon_{n+1} + \epsilon_{n-1} - 2\epsilon_{n-2}$$

and

$$c_n = \epsilon_{n+1} + \epsilon_{n-1} - 2\epsilon_{n-2} - 2c_{n-1} - c_{n-2} \qquad (30.9)$$

The control algorithm given by eq. (30.9) is *physically unrealizable* because it requires a future value of the error (i.e., ϵ_{n+1}) to compute the current control action (i.e., c_n).

There are a variety of specifications that we can impose on the closed-loop response $\hat{y}(z)$ for a given step change in $\hat{y}_{SP}(z)$. It is clear that depending on the response specifications, we can derive a series of alternative digital control algorithms. Let us now examine the most commonly used among them.

Deadbeat or minimal prototype response

According to this method, we require that the response of the process to a unit step change in the set point exhibit no error at all sampling instants after the first (see Figure 30.2). Thus for the unit step change in the set point, we have

$$\hat{y}_{SP}(z) = \frac{1}{1 - z^{-1}}$$

If the response is to have zero error at all sampling instants *after the first*, its discrete-time behavior resembles that of a unit step delayed by one sampling instant:

$$\hat{y}(z) = \frac{z^{-1}}{1 - z^{-1}}$$

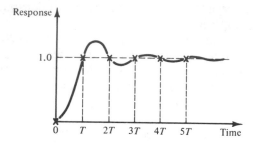

Figure 30.2 Response requirements for the design of a deadbeat controller.

Therefore,

$$\frac{\hat{y}(z)}{\hat{y}_{SP}(z)} = z^{-1} \tag{30.10}$$

and eq. (30.6) yields the following design for a *deadbeat controller*:

$$D(z) = \frac{1}{HG_p(z)} \frac{z^{-1}}{1 - z^{-1}} \tag{30.11}$$

A few remarks on the characteristics of a deadbeat algorithm are now in order.

1. The requirement that the error is zero at the sampling instants does not preclude *large overshoots* (Figure 30.3a) or *highly oscillatory behavior* for the process response (Figure 30.3b). Indeed, as will be explained in the next section, deadbeat controllers suffer from these two weaknesses.

2. The fact that the response reaches its desired value (error = 0) in one sampling instant indicates that the *rise time is minimum*, which in turn implies that the controller *exhibits very strong control action*, not necessarily a desirable feature (causes large overshoots).

3. Nevertheless, it can be shown that *the steady-state error is zero*. Thus

$$\hat{\epsilon}(z) = \hat{y}_{SP}(z) - \hat{y}(z)$$

and using eq. (30.10) yields

$$\hat{\epsilon}(z) = \hat{y}_{SP}(z)(1 - z^{-1})$$

Use the final value theorem and take

$$\lim_{t \to \infty} \epsilon(t) = \lim_{z \to 1} [(1 - z^{-1})\hat{\epsilon}(z)] = \lim_{z \to 1} [\hat{y}_{SP}(z)(1 - z^{-1})^2] = 0$$

4. The deadbeat algorithm *is physically realizable if the time delay in*

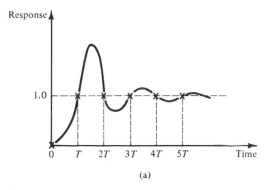

(a)

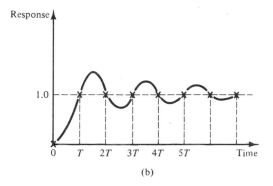

(b)

Figure 30.3 Drawbacks of deadbeat designs: (a) large overshoots; (b) long settling times.

the HG_p (z) is not larger than one sampling period. For example, let

$$HG_p(z) = z^{-k} HG'_p(z)$$

(i.e., the deadtime is equal to k sampling periods). Then from eq. (30.11) we take

$$D(z) = \frac{z^{k-1}}{HG'_p(z)} \frac{1}{1 - z^{-1}}$$

and the controller is not physically realizable if $k > 1$, because $k - 1 > 0$ and the numerator of the design equation has z to a positive power. This implies that we need future error values which are not available.

5. If the dead time is larger than one sampling period, we need to modify the response specifications requiring that the response exhibits zero error at all sampling instants after the first two, three, etc. (Figure 30.4). In such a case we have the following characterizations:

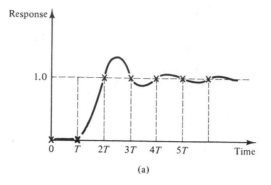

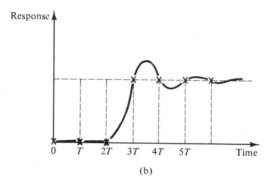

Figure 30.4 Response specifications for deadbeat designs with process dead time: (a) $t_d = T$; (b) $t_d = 2T$.

$$\frac{\hat{y}(z)}{\hat{y}_{SP}(z)} = z^{-2} \qquad \text{or} \qquad \frac{\hat{y}(z)}{\hat{y}_{SP}(z)} = z^{-3} \qquad \text{etc.}$$

yielding the following controller design equations:

$$D(z) = \frac{1}{HG_p(z)} \frac{z^{-2}}{1 - z^{-2}} \qquad \text{or} \qquad D(z) = \frac{1}{HG_p(z)} \frac{z^{-3}}{1 - z^{-3}}$$

Example 30.2: Designing Deadbeat Controllers for a First-Order Process

Consider a process with first-order dynamics:

$$G_p(s) = \frac{10}{0.5s + 1}$$

The pulse transfer function $HG_p(z)$ is given by (assuming that $T = 1$)

$$HG_p(z) = \mathcal{Z}\left[\frac{1 - e^{-Ts}}{s} \frac{10}{0.5s + 1}\right] = 10 \frac{(1 - e^{-T/0.5})z^{-1}}{1 - e^{-T/0.5}z^{-1}} = \frac{8.6z^{-1}}{1 - 0.14z^{-1}}$$

Therefore, from eq. (30.11) we take

$$D(z) = \frac{1 - 0.14z^{-1}}{8.6z^{-1}} \frac{z^{-1}}{1 - z^{-1}} = \frac{1}{8.6} \frac{1 - 0.14z^{-1}}{1 - z^{-1}}$$

and the control action in the time domain is given by

$$c_n = c_{n-1} + \frac{1}{8.6} \epsilon_n - \frac{0.14}{8.6} \epsilon_{n-1}$$

The controller is physically realizable because it uses past and current error values.

Suppose now that the process also had dead time, $t_d = 2T$. This delay introduces the term z^{-2} in $HG_p(z)$:

$$HG_p(z) = z^{-2} \frac{8.6z^{-1}}{1 - 0.14z^{-1}}$$

and the resulting controller is physically unrealizable:

$$D(z) = \frac{z^2}{8.6} \frac{1 - 0.14z^{-1}}{1 - z^{-1}}$$

In this case of increased dead time, we change the response specifications and we require that the error is zero at all sampling instants *after the first three*. Thus $\hat{y}(z)/\hat{y}_{SP}(z) = z^{-3}$ and the controller design equation becomes

$$D(z) = \frac{1}{z^{-2} \dfrac{8.6\,z^{-1}}{1 - 0.14z^{-1}}} \frac{z^{-3}}{1 - z^{-3}} = \frac{1}{8.6} \frac{1 - 0.14z^{-1}}{1 - z^{-3}}$$

The resulting controller is physically realizable:

$$c_n = c_{n-3} + \frac{1}{8.6} \epsilon_n - \frac{0.14}{8.6} \epsilon_{n-1}$$

Dahlin's method

An alternative method suggested by Dahlin requires that the closed-loop response of a DDC loop behave like the response of a first-order system with dead time to a unit step change in the set point:

$$\overline{y}(s) = \frac{e^{-\theta s}}{\mu s + 1} \frac{1}{s} \tag{30.12}$$

where μ is the time constant of the desired response and θ is the dead time of the response assumed to be $\theta = kT$, with k = integer. In discrete form eq. (30.12) yields

$$\hat{y}(z) = z^{-k} \frac{(1 - e^{-T/\mu})z^{-1}}{(1 - z^{-1})(1 - e^{-T/\mu}z^{-1})}$$

Given that for a unit step change in set point,

$$\hat{y}_{SP}(z) = \frac{1}{1 - z^{-1}}$$

we find that

$$\frac{\hat{y}(z)}{\hat{y}_{SP}(z)} = z^{-k} \frac{(1 - e^{-T/\mu})z^{-1}}{1 - e^{-T/\mu}z^{-1}} \tag{30.13}$$

Therefore, the design eq. (30.6) yields Dahlin's control algorithm:

$$D(z) = \frac{1}{HG_p(z)} \frac{(1 - e^{-T/\mu})z^{-k-1}}{1 - e^{-T/\mu}z^{-1} - (1 - e^{-T/\mu})z^{-k-1}} \tag{30.14}$$

The following remarks should be made on Dahlin's control algorithm:

1. The time constant μ is a design parameter whose value is to be selected by the designer according to the desired closed-loop response. Thus we choose low values of μ if we require a faster response, while larger values of μ will yield more sluggish closed-loop response.
2. From eq. (30.14) it is clear that Dahlin's algorithm is physically realizable if the dead time in $HG_p(z)$ is not larger than $(k + 1)T$.
3. With Dahlin's algorithm we can avoid the excessive control action produced by the deadbeat algorithms, thus reducing significantly the undesired large overshoots or highly oscillatory closed-loop response.

Example 30.3: Comparing Dahlin's and Deadbeat Algorithms

A given process has unknown detailed dynamics; that is, it exhibits overdamped open-loop behavior but its exact order and parameter values are poorly known. From the process reaction curve (see Section 16.5) we have approximated its transfer function by the following first-order system with dead time:

$$G_p(s) = \frac{10e^{-2s}}{0.2s + 1} \tag{30.15}$$

Suppose that the sampling period is $T = 1$ second. Then the process dead time is $t_d = 2T$ and the corresponding deadbeat algorithm is given by

$$D_1(z) = \frac{1}{HG_p(z)} \frac{z^{-3}}{1 - z^{-3}}$$

From eq. (30.15) we find that

$$HG_p(z) = 10z^{-2} \frac{(1 - e^{-T/0.2})z^{-1}}{1 - e^{-T/0.2}z^{-1}} = 10 \frac{0.99z^{-3}}{1 - 0.01z^{-1}}$$

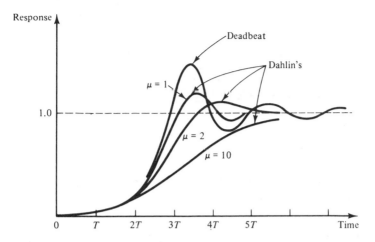

Figure 30.5 Closed-loop responses to unit step change in set point using deadbeat and Dahlin algorithms.

Therefore,

$$D_1(z) = \frac{1}{9.9} \frac{1 - 0.01 z^{-1}}{1 - z^{-3}}$$

To compute Dahlin's algorithm we assume that the closed-loop deadtime is equal to the process dead time plus one sampling period to account for the delay in the sampling and holding operations (i.e., $\theta = 2 + 1 = 3$ seconds). Also, we assume that the desired response has a time constant $\mu = 2$. Then from the design eq. (30.14) we take

$$D_2(z) = \frac{1 - 0.01 z^{-1}}{9.9 z^{-3}} \frac{0.39 z^{-4}}{1 - 0.61 z^{-1} - 0.39 z^{-4}}$$

Figure 30.5 shows the closed-loop response of the process to a unit step change in the set point using the deadbeat and Dahlin's algorithms. We notice that the first produces large overshoot and strong oscillatory response, while the second yields a more moderate controlled response. We also notice the effect that the tuning parameter μ has on the performance of Dahlin's algorithm.

30.4   "Ringing" and the Placement of Poles

While designing a digital controller by the response specification methods (deadbeat, Dahlin's), we should not only be concerned with the closed-loop response, but should pay attention to the resulting movement of the controller output. Excessive control valve movement is unacceptable in industrial practice. Let us examine the movement of a controller's output.

The discrete-time transfer function of a digital controller, which was designed using some specification on the closed-loop response, is given by eq. (30.6). This equation implies that $D(z)$ is the ratio of two polynomials in z^{-1}, as eq. (30.7) demonstrates:

$$D(z) \equiv \frac{\hat{c}(z)}{\hat{e}(z)} = \frac{\beta_0 + \beta_1 z^{-1} + \beta_2 z^{-2} + \cdots + \beta_k z^{-k}}{1 + \alpha_1 z^{-1} + \alpha_2 z^{-2} + \cdots + \alpha_m z^{-m}} \qquad (30.7)$$

The roots of the polynomial in the denominator are the *poles* of the controller and for a stable controller all these poles should lie inside or on the unit circle in the complex plane. Let us examine *what the effect is of a pole's location on the controller output $c(t)$.*

Consider a controller with the following discrete transfer function:

$$D(z) \equiv \frac{\hat{c}(z)}{\hat{e}(z)} = \frac{1}{1 - pz^{-1}}$$

An error equal to a unit impulse enters the controller at $t = 0$. Then from the equation above we take

$$c_n - p\ c_{n-1} = \epsilon_n \qquad (30.16)$$

where $\epsilon_n = 1$ for $n = 0$ and $\epsilon_n = 0$ for $n = 1, 2, 3, \cdots$. Let $p = -0.9$. Then from eq. (30.16) we take

$$c_n + 0.9 c_{n-1} = \epsilon_n$$

and Table 30.1 shows the controller output at the various sampling instants. The behavior of the controller output has been also plotted in Figure 30.6a. *Notice the successive changes in the sign of the controller output.* This phenomenon is known as *ringing* (or *controller ringing*) and is caused by negative poles. *The closer a negative pole is at the origin, the smaller the ringing is.* This is shown in Figure 30.6b by the behavior of the controller output when $p = -0.3$. On the other hand, *for positive poles the controller output exhibits no ringing.* Figure 30.6c and d demonstrate this point for $p = +0.9$ and $p = +0.3$, respectively.

It is clear from the discussion above that:

1. Any negative pole of the controller's discrete transfer function [eq. (30.7)] will cause ringing.
2. The closer a ringing pole is to the unit circle (i.e., the larger its absolute value), the higher the ringing of the controller output will be.

Therefore, *we should design a digital controller so that all its poles are positive*, preferably with medium-size absolute values (e.g., 0.4 to 0.6) (Figure 30.7).

Unfortunately, the deadbeat and Dahlin's algorithms usually contain poles that cause severe ringing of the controller output. This may be the

TABLE 30.1.

DEPENDENCE OF A CONTROLLER OUTPUT ON THE POLE LOCATION

| | | Controller output c_n | |
n	ϵ_n	For $p = -0.9$	For $p = +0.9$
0	1	1	1
1	0	−0.9	0.9
2	0	+0.81	0.81
3	0	−0.729	0.729
4	0	+0.6561	0.6561
5	0	−0.59049	0.59049

case for any response-specification design technique and is not unique
to the foregoing two algorithms.

There exist several design methodologies which eliminate the ringing problem by placing the controller poles on the right-hand side of the
real axis. These methodologies are quite complex and a complete discussion is beyond the scope of the present text. In the following example we discuss one simple method for eliminating a ringing pole and its
effects.

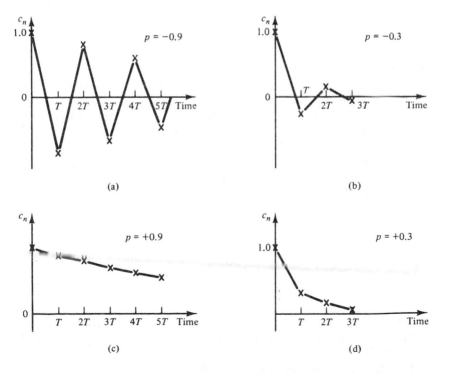

Figure 30.6 Effect of pole location on the behavior of controller output.

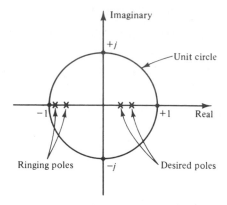

Figure 30.7 Desired location of controller poles.

Example 30.4: Eliminating the Ringing of a Digital Controller

Suppose that for a given process Dahlin's method leads to a controller with the transfer function

$$D(z) = K \frac{1 - 0.5z^{-1}}{(1 + 0.6z^{-1})(1 - z^{-1})(1 - 0.7z^{-1})}$$

The term $(1 + 0.6z^{-1})$ in the denominator implies the presence of a pole at -0.6, which will cause controller ringing. Dahlin has suggested that the ringing pole can be eliminated by multiplying $D(z)$ with $(1 + 0.6z^{-1})$, thus canceling the ringing term from the denominator. But in order to keep the static gain K the same, we should divide $D(z)$ by the term $\lim_{z \to 1}(1 + 0.6z^{-1})$

$= 1 + 0.6 = 1.6$. Thus the ringing-free controller has the transfer function

$$D(z) = \frac{K}{\lim_{z \to 1}(1 + 0.6z^{-1})} \frac{(1 - 0.5z^{-1})(1 + 0.6z^{-1})}{(1 + 0.6z^{-1})(1 - z^{-1})(1 - 0.7z^{-1})}$$

$$= \frac{K}{1.60} \frac{1 - 0.5z^{-1}}{(1 - z^{-1})(1 - 0.7z^{-1})}$$

30.5 Design of Optimal Regulatory Control Systems

The digital control algorithms discussed in Sections 30.2 and 30.3 were designed for set point changes (servo problem). Therefore, the question arises as to how well they perform for load (disturbance) changes. It is a fortuitous coincidence that algorithms such as the deadbeat or Dahlin's perform well for both set point and load changes.

If we want to design a controller primarily for load changes, we can use the response specification procedure outlined in Section 30.2.

Example 30.5: Designing a Digital Controller for Load Changes

Consider the general feedback control loop shown in Figure 29.9. The closed-loop response for load changes only (regulation problem) is given by eq. (20.21b):

$$\hat{y}(z) = \frac{Z[G_d(s)\overline{d}(s)]}{1 + HG_p(z)D(z)} \qquad (29.21b)$$

Solve the last equation with respect to $D(z)$ and take

$$D(z) = \frac{Z[G_d(s)\overline{d}(s)] - \hat{y}(z)}{HG_p(z)\hat{y}(z)} \qquad (30.17)$$

Therefore, we can find the discrete transfer function of a regulatory controller as follows:

1. Assume that the transfer functions $G_d(s)$ and $G_p(s)$ are known.
2. Specify the disturbance change (e.g., step change).
3. Specify the desired closed-loop response, $\hat{y}(z)$.
4. Solve eq. (30.17) and find $D(z)$.

In this section we proceed a step further and develop a methodology which leads to optimal controllers for load changes.

Consider a process described by a kth-order linear difference equation,

$$y_n = a_1 y_{n-1} + a_2 y_{n-2} + \cdots + a_k y_{n-k} + b_1 m_{n-1} + \cdots + b_k m_{n-k} \quad (30.18)$$

where $a_1, a_2, \cdots, a_k; b_1, \cdots, b_k$ are constant parameters with known values. Suppose that the purpose of the regulatory control system is to keep the closed-loop output *as close as possible* to a prescribed set point value y_{SP} in the presence of disturbance changes. The deviation can be specified by one of the following measures:

$$P_1 = \{y_n - y_{SP}\}^2 \qquad (30.19a)$$

$$P_2 = \frac{1}{N} \sum_{n=1}^{N} \{y_n - y_{SP}\}^2 \qquad (30.19b)$$

The first measure, P_1, is referred to as *one-stage control*; the second as *N-stage control*. Thus the controller design problem can be formulated as follows:

Find a controller that minimizes P_1 or P_2 in the presence of load changes.

The control action that minimizes P_1 attempts to keep the output close to the set point by making individual control decisions at each stage. Minimization of P_2, on the other hand, relaxes the restriction above and plans the control action over a longer time horizon. Let us now solve the two design problems above.

Suppose that we are at the nth sampling instant and that we want to compute the control action m_n in such a way that y_{n+1} will be as close as possible to the desired y_{SP}. Using criterion P_1, we take

$$P_1 = \{y_{n+1} - y_{SP}\}^2 = \{a_1 y_n + a_2 y_{n-1} + \cdots + a_k y_{n-k+1}$$
$$+ b_1 m_n + b_2 m_{n-1} + \cdots + b_k m_{n-k+1} - y_{SP}\}^2$$

The minimum of P_1 is found when $\partial P_1 / \partial m_n = 0$. Then we have

$$2\{a_1 y_n + a_2 y_{n-1} + \cdots + a_k y_{n-k+1} + b_1 m_n + b_2 m_{n-1} + \cdots$$
$$+ b_k m_{n-k+1} - y_{SP}\}(b_1) = 0$$

and the optimum regulatory control action at the nth instant is given by

$$m_n = \frac{1}{b_1} \{y_{SP} - a_1 y_n - a_2 y_{n-1} - \cdots - a_k y_{n-k+1}$$
$$- b_2 m_{n-1} - \cdots - b_k m_{n-k+1}\} \qquad (30.20)$$

The controller defined by eq. (30.20) is physically realizable because it uses only current or past information on the manipulated variable and the controlled output.

Now let us turn our attention to the second design criterion, P_2. Consider the situation at the $(N-1)$ sampling instant. The outputs $y_{N-1}, y_{N-2}, \cdots, y_1$ have been measured and the control problem is to determine the value of the manipulated variable m_{N-1}. Since m_{N-1} influences only the last term of P_2, we have

$$\text{minimize } P_2 = \{y_N - y_{SP}\}^2$$

Then the optimal value of m_{N-1} is given by eq. (30.20) with $n = N - 1$.

Consider the situation at $n = N - 2$. The output has been measured for $n = N - 2, N - 3, \cdots, 1$ and the problem is to determine the optimal value of m_{N-2}. Since m_{N-2} influences the last two terms of P_2, we have

$$\text{minimize } P_2 = \{y_N - y_{SP}\}^2 + \{y_{N-1} - y_{SP}\}^2 \qquad (30.21)$$

If the optimum value of m_{N-1} has been used for the last stage, the minimization problem (30.21) yields

$$\text{minimize } P_2 = \{y_{N-1} - y_{SP}\}^2$$

because $y_N - y_{SP} = 0$ for the optimum value of m_{N-1}. But the optimum value of m_{N-2} solving the last problem is given again by eq. (30.20) for $n = N - 2$. Thus we reach the following conclusion:

> The optimal regulatory control action for a system described by a kth-order difference model with constant and known parameters is given by eq. (30.20) independently of which criterion, P_1 or P_2 is used.

The main obstacles in implementing the optimal regulatory control strategy above are:

1. The poor knowledge of the process parameters for many typical chemical processes,
2. The presence of pronounced dead times in the process, leading to a physically unrealizable controller (the reader can show this easily)

Example 30.6: Optimal Regulation of a Process with Dead Time

Consider a process that is modeled by a third-order discrete-time system with dead time. Assume that the dead time is an integer multiple of the sampling period (i.e., $t_d = \ell T$ where ℓ = integer). Then we have

$$y_n = a_1 y_{n-1} + a_2 y_{n-2} + a_3 y_{n-3} + b_1 m_{n-1-\ell} + b_2 m_{n-2-\ell} + b_3 m_{n-3-\ell}$$

and the optimal control action is

$$m_n = \frac{1}{b_1} \{ y_{SP} - a_1 y_{n+\ell} - a_2 y_{n+\ell-1} - a_3 y_{n+\ell-2} - b_2 m_{n-1} - b_3 m_{n-2} \} \quad (30.22)$$

1. If $\ell = 0$ (i.e., no dead time), we take a physically realizable controller:

$$m_n = \frac{1}{b_1} (y_{SP} - a_1 y_n - a_2 y_{n-1} - a_3 y_{n-2} - b_2 m_{n-1} - b_3 m_{n-2})$$

2. If $\ell > 0$, we have physically unrealizable controllers because they require future output values, as depicted by the term $y_{n+\ell}$, which are not available. In this case we can develop *suboptimal* controllers by omitting the future output values. Thus if $\ell = 1$, we can neglect the term $y_{n+\ell}$ from eq. (30.22) and take the following suboptimal control action:

$$m_n = \frac{1}{b_1} (y_{SP} - a_2 y_n - a_3 y_{n-1} - b_2 m_{n-1} - b_3 m_{n-2})$$

If $\ell = 2$, we neglect both terms $y_{n+\ell}$ and $y_{n+\ell-1}$, which correspond to future values, and take

$$m_n = \frac{1}{b_1} (y_{SP} - a_3 y_n - b_2 m_{n-1} - b_3 m_{n-2})$$

THINGS TO THINK ABOUT

1. What are the two classes of methods for designing digital feedback controllers? Discuss their essential characteristics and identify their principal representatives.

2. Describe in your own words the difference between the position and velocity forms of PI or PID control algorithms. Do we have these two forms for a proportional controller?

3. Which one would you prefer to use, the position or velocity form of PI or PID algorithms? Why?

4. Describe in your own words the problem of "initializing" a controller or a controller algorithm.

5. What is integral windup of a controller, and what are its undesired consequences? Does a P or PD controller exhibit integral windup? Why?

6. How can you eliminate the negative consequences of integral windup for the position form of PI or PID control algorithms? [Consult the text by C. L. Smith (see Ref. 7).].

7. Why can you use the classical Cohen–Coon or Ziegler–Nichols techniques for tuning a digital PI or PID controller? What is the additional tuning parameter introduced by the discrete nature of a process control computer?

8. What is the effect of sampling on the response of a closed-loop system? What happens to the process response as the sampling period increases?

9. How would you select the sampling rate in a digital control loop? Does it depend on the operating conditions of a process, and why?

10. Describe the general methodology of designing digital controllers, based on the response specifications. What are two examples of such designs?

11. Discuss the construction of the deadbeat and Dahlin's algorithms. Which one imposes more stringent specifications on the closed-loop response? What are the consequences of such stringent requirements?

12. Design the deadbeat control algorithms for a first-order process with dead time equal to $3T$. Can we have a physically realizable controller if we require that the response exhibit zero error at all sampling instants after the first?

13. What do we mean when we say that a control algorithm is physically unrealizable? What are the necessary conditions for designing physically realizable deadbeat and Dahlin algorithms?

14. Discuss in your own words what ringing is and why it is undesirable in process control. What is the necessary condition in order to have a ringing digital controller?

15. Do the digital PI and PID algorithms (position or velocity forms) produce ringing? Why?

16. How can you eliminate the ringing from a deadbeat or Dahlin algorithm?

17. Design a digital controller with the following specifications. The set point changes like a ramp function, $f(t) = 2t$. The closed-loop response should exhibit a zero error at all sampling instants after the first. What are the conditions for making this algorithm physically realizable?

18. How would you design a digital controller for load changes? Describe two different procedures and implement them to a system with

$$G_d(s) = \frac{1}{0.1s + 1} \quad \text{and} \quad G_p = \frac{10}{0.1s + 1}$$

19. In what sense is the control action given by eq. (30.20) optimal? Under what conditions is it physically realizable?

20. Design a suboptimal controller for load changes when the process can be modeled by a fourth-order discrete-time system with dead time $t_d = \ell T$ (ℓ = integer). Under what conditions is the resulting controller physically realizable?

21. To design digital controllers by response specification techniques, we can use set point or load changes. Set point changes can enter the loop at the sampling instant but load changes cannot because they are not under the designer's control. Do you expect this weakness to affect the quality of the resulting controller? (*Note*: Consult Ref. 7.)

Process Identification 31
and Adaptive Control

It should be clear to the reader by now that the design of effective analog or digital controllers depends very heavily on how well we know the dynamics of the controlled process. Such strong reliance on a process model constitutes a serious weakness of a control design technique because quite often the processes are poorly understood or their physical or chemical parameters are poorly known. In such cases the models that are found on basic principles (see Chapter 4) are quite inadequate to describe the dynamic characteristics of real processes.

Furthermore, even if we have a good initial model for a controlled process, it may not be sufficient for effective control during a long operation, for two reasons. First, chemical processes are nonlinear systems and consequently the characteristics of the corresponding linearized systems change depending on the operating point (point of linearization). Example 10.5 demonstrates a simple such case. Second, chemical processes are nonstationary systems; that is, their dynamic characteristics change with time because several of their important physical or chemical parameters (e.g., overall heat transfer coefficients, catalyst activity, etc.) change values with time (see Example 10.6). Therefore, any attempt to model a process a priori will have at best limited success.

In this chapter we present an experimental approach, known as *process identification*, which can be used to construct a reliable model, either before or after the process has been placed in operation. In addition, we study how process identification can be coupled with various control systems to yield *on-line adaptive control strategies*.

It should be emphasized that process identification and on-line adaptive control require extensive computations, which can be performed rapidly and efficiently only with digital computers. For this reason, extensive use of on-line process identification became a reality only after computers were introduced for process control.

31.1   Process Identification

Consider a process that is poorly known. This may mean that the physical or chemical phenomena in the process are poorly understood or that the various process parameters are imprecisely known. In the first case the model order is not known; the second case is just a parameter estimation problem with known model order.

Let the process be described by the following linear difference equation of order k:

$$y_n = a_1 y_{n-1} + a_2 y_{n-2} + \cdots + a_k y_{n-k} + b_1 m_{n-1} + b_2 m_{n-2}$$
$$+ \cdots + b_k m_{n-k} \tag{31.1}$$

where y_i and m_i are the process output and input values at the ith sampling instant and $a_1, a_2, \cdots, a_k; b_1, b_2, \cdots, b_k$ are constant but imprecisely known parameters. The order k of the model may be known or not.

Introduce to the process a prespecified input change. Let $\tilde{m}_n$ be the measured values of the input variable and $\tilde{y}_n$ the measured values of the resulting process output, at the nth sampling instant with $n = 0, 1, 2, \cdots$. Then we compare the process output values *computed* from the postulated model [eq. (31.1)] with the *measured* output values. Let the error be

$$\epsilon_n = \tilde{y}_n - y_n = \tilde{y}_n - (a_1 \tilde{y}_{n-1} + a_2 \tilde{y}_{n-2} + \cdots + a_k \tilde{y}_{n-k} + b_1 \tilde{m}_{n-1}$$
$$+ b_2 \tilde{m}_{n-2} + \cdots + b_k \tilde{m}_{n-k}) \tag{31.2}$$

The "best" values of the unknown parameters are those which yield the minimum mean error between the theoretical and experimental values of the process output. Thus the best estimate of the process parameters is given by the solution of the following *least-squares problem*:

$$\text{minimize } P = \frac{1}{N} \sum_{n=1}^{N} \epsilon_n^2 = \frac{1}{N} \sum_{n=1}^{N} \{\tilde{y}_n - a_1 \tilde{y}_{n-1} - a_2 \tilde{y}_{n-2} - \cdots - a_k \tilde{y}_{n-k}$$
$$- b_1 \tilde{m}_{n-1} - b_2 \tilde{m}_{n-2} - \cdots - b_k \tilde{m}_{n-k}\}^2 \tag{31.3}$$

There exist several numerical methods for the solution of this minimization problem. One of them is based on the solution of the following set of algebraic equations (necessary conditions, which must be satisfied at the point where P is a minimum):

$$\frac{\partial P}{\partial a_1} = \frac{\partial P}{\partial a_2} = \cdots = \frac{\partial P}{\partial a_k} = \frac{\partial P}{\partial b_1} = \frac{\partial P}{\partial b_2} = \cdots = \frac{\partial P}{\partial b_k} = 0$$

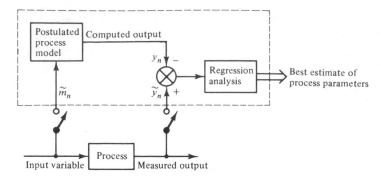

Figure 31.1 Experimental identification of process dynamics.

If the value of the mean squarred error P is significantly larger than the "theoretically possible" value of zero, we conclude that the assumed model order is unacceptably low and that a higher-order model should be used.

Let us now comment on the three steps that constitute the basis for the experimental identification of a process (Figure 31.1).

Step 1: Postulate a model for the process. The unknown process is not completely a black box. Some information about its dynamic behavior is known from basic principles and/or plant experience. Therefore, some estimate of its model's order and some initial values for the unknown parameters will be available. The more we know about the process, the more effective the postulated model will be. Consequently, we should use for its development all available information. Remember, though, that complex models of high order will not necessarily produce better controller designs and will burden the computational effort without tangible results.

The order of the postulated model is a very important factor. For a well-known process such as a stirred-tank heater, it is not a problem. On the other hand, it is not obvious what order of dynamics we should assume for a fluid catalytic cracker (see Example 4.15). Also, it is not obvious what type of low-order model we should use to approximate the high-order models of even simple distillation columns (see Example 4.16). *As a general starting point one could employ first- or second-order models with or without dead time.* There exist a surprisingly large number of processes which could be effectively described by such low-order models.

Step 2: Introduce a known input change to the process and record its output. We could use normal operating data for the values of input and output variables. It should be emphasized, though, that an excitation of the process with an abruptly changing "test" input signal (e.g., step, pulse, sine) will yield more pronounced variations in the output variable. This will uncover more information about the process dynamics. On the other hand, abrupt input changes may disrupt seriously the operation of the process and thus may not be recommended for on-line process identification.

The parameter values computed by the least-squares problem depend on the type of "test" input that was used. Thus we may have different results for step and impulse input changes. This could be a serious drawback of the method for the identification of sensitive systems.

Step 3: Estimate the "best" values for the unknown process parameters. The least-squares methodology is also known as *regression analysis*. If the postulated model is linear, we have *linear regression analysis*; otherwise, it is called *nonlinear regression analysis*.

Example 31.1: On-line Identification in a Jacketed Cooler

Consider a continuous stirred tank where heat is removed by cooling water which flows through a jacket around the tank (see Figure 12.1b). Suppose that we want to develop a control algorithm which regulates the temperature of the tank's content using the coolant temperature or flow rate as the manipulated variable. To do this, we need a "model" that relates the manipulated to the controlled variable. Let us use the experimental modeling procedure described earlier.

Postulate a model. In Example 12.1 we observed that a jacketed cooler is a multicapacity process. For our problem we can identify the following three interacting capacities in series; (1) heat capacity of tank's content, (2) heat capacity of the coolant in the jacket, and (3) heat capacity of the tank's wall. Therefore, our first suggestion is to use a third-order overdamped model without significant dead time. A closer examination of the physical system reveals that the tank's wall does not possess significant capacity for heat storage and could be omitted. Consequently, we suggest a second-order model without dead time of the form

$$G_p(s) = \frac{K_p}{(\tau_1 s + 1)(\tau_2 s + 1)} \tag{31.4}$$

where K_p, τ_1, and τ_2 are poorly known parameters.

Since we will use a digital computer for the process identification, we need to develop the discrete model of the transfer function above. Thus if we use zero-order hold, we take [see Example 29.3, part (c)]

$$HG_p(z) = Z\left\{\frac{1 - e^{-Ts}}{s} \frac{K_p}{(\tau_1 s + 1)(\tau_2 s + 1)}\right\} = (1 - z^{-1})K_p$$

$$\left\{\frac{1}{1 - z^{-1}} + \frac{1}{\tau_2 - \tau_1}\left(\frac{\tau_1}{1 - e^{-T/\tau_1} z^{-1}} - \frac{\tau_2}{1 - e^{-T/\tau_2} z^{-1}}\right)\right\}$$

or

$$HG_p(z) \equiv \frac{\hat{T}(z)}{\hat{F}(z)} = \frac{b_1 z^{-1} + b_2 z^{-2}}{1 - a_1 z^{-1} + a_2 z^{-2}} \tag{31.5}$$

where $\hat{T}(z) = z$-transform of liquid's temperature (controlled output)

$\hat{F}(z) = z$-transform of coolant's flow rate (manipulated input)

and

$$a_1 = e^{-T/\tau_1} + e^{-T/\tau_2} \tag{31.6a}$$

$$a_2 = e^{-T/\tau_1} e^{-T/\tau_2} \tag{31.6b}$$

$$b_1 = K_p(\tau_2 - \tau_1 + \tau_1 e^{-T/\tau_1} - \tau_2 e^{-T/\tau_2})/(\tau_2 - \tau_1) \tag{31.6c}$$

$$b_2 = \frac{K_p[(\tau_2 - \tau_1)e^{-T/\tau_1}e^{-T/\tau_2} + \tau_1 e^{-T/\tau_2} - \tau_2 e^{-T/\tau_1}]}{\tau_2 - \tau_1} \tag{31.6d}$$

From eq. (31.5) we take the following discrete-time model:

$$T_n = a_1 T_{n-1} - a_2 T_{n-2} + b_1 F_{n-1} + b_2 F_{n-2} \tag{31.7}$$

The parameters a_1, a_2, b_1, and b_2 depend on the values of τ_1, τ_2, and K_p, which are imprecisely known.

Regression analysis. Assume that the process is originally at steady state with $T = T_{ss}$ and $F = F_{ss}$. Introduce a unit step change in the coolant flow rate. Then $F_n = 1$ for $n = 1, 2, \cdots, N$. Let $\tilde{T}_n$ be the measured response of the liquid's temperature at the sampling instant $n = 1, 2, \cdots, N$. The objective of the regression analysis is to find the values of τ_1, τ_2, and K_p which minimize the following objective of least squares:

$$\underset{\tau_1, \tau_2, K_p}{\text{Minimize }} P = \frac{1}{N} \sum_{n=1}^{N} \{\tilde{T}_n - a_1 \tilde{T}_{n-1} + a_2 \tilde{T}_{n-2} - b_1 \tilde{F}_{n-1} - b_2 \tilde{F}_{n-2}\}^2$$

where $\tilde{T}_k = T_{ss}$ and $F_k = F_{ss}$ for $k < 0$. There are different numerical methods which could be used to solve the minimization problem. One of these is to find those values of τ_1, τ_2, and K_p which satisfy the following set of algebraic equations (necessary conditions for optimality):

$$\frac{\partial P}{\partial \tau_1} = \frac{\partial P}{\partial \tau_2} = \frac{\partial P}{\partial K_p} = 0$$

Adaptation of the experimental model. During the operation of the cooler the values of the model parameters may change. A typical example is fouling, which reduces the overall heat transfer coefficient, leading to different values for τ_1, τ_2, and K_p. The procedure described above can be repeated during the operation of the cooler, thus yielding updated versions of experimentally determined models.

Example 31.2: Identifying the Order and the Parameters of a Process Model

Consider a process whose dynamics are very poorly known so that we do not have a good estimate for the order of a postulated model. Assuming that the process is initially at steady state, we introduce an input signal whose sampled values are shown in Table 31.1. The recorded response of process output at the various sampling instants is also shown in Table 31.1. Notice that the input and output variables have been expressed in deviation form, having zero values for negative times.

Initially, let us postulate a first-order model:

$$y_n = a_1 y_{n-1} + b_1 m_{n-1}$$

The linear regression analysis will find the values of the parameters a_1 and b_1 which minimize the mean-square error:

$$P = \frac{1}{15} \sum_{n=1}^{15} \{y_n - a_1 y_{n-1} - b_1 m_{n-1}\}^2 \tag{31.8}$$

as follows: The optimum values of a_1 and b_1 must satisfy the necessary conditions for a minimum point:

$$\frac{\partial P}{\partial a_1} = \sum_{n=1}^{15} 2(y_n - a_1 y_{n-1} - b_1 m_{n-1})(-y_{n-1}) - 0 \tag{31.9a}$$

$$\frac{\partial P}{\partial b_1} = \sum_{n=1}^{15} 2(y_n - a_1 y_{n-1} - b_1 m_{n-1})(-m_{n-1}) = 0 \tag{31.9b}$$

Solve eqs. (31.9a) and (31.9b) for a_1 and b_1, using the measured values of Table 31.1 for y_n, y_{n-1}, m_{n-1} ($n = 1, 2, \cdots, 15$) and find

$$a_1 = 0.86 \quad \text{and} \quad b_1 = 0.57$$

These values of a_1 and b_1 minimize the mean of the squared errors given by eq. (31.8) and yield

$$\text{minimum } P = 0.00161$$

This is fairly close to zero so that we can conclude that a first-order model adequately describes the unknown process.

We proceed by postulating a second-order model of the form

$$y_n = a_1 y_{n-1} + a_2 y_{n-2} + b_1 m_{n-1} + b_2 m_{n-2}$$

Then the least-squares objective function becomes

$$P = \frac{1}{15} \sum_{n=1}^{15} \{y_n - a_1 y_{n-1} - a_2 y_{n-2} - b_1 m_{n-1} - b_2 m_{n-2}\}^2$$

TABLE 31.1.
DATA FOR PROCESS IDENTIFICATION (EXAMPLE 31.1)

Sampling instant	Input variable	Output variable
n	m_n	y_n
$n < 0$	0.0	0.0
0	1.0	0.0
1	0.60	0.50
2	0.30	0.90
3	0.10	0.91
4	0.0	0.866
5	0.0	0.732
6	0.0	0.612
7	0.0	0.513
8	0.0	0.430
9	0.0	0.361
10	0.0	0.302
11	0.0	0.253
12	0.0	0.212
13	0.0	0.178
14	0.0	0.149
15	0.0	0.125

Following the same procedure as before, we solve the necessary conditions

$$\frac{\partial P}{\partial a_1} = \frac{\partial P}{\partial a_2} = \frac{\partial P}{\partial b_1} = \frac{\partial P}{\partial b_2} = 0$$

and find

$$a_1 = 0.6, \qquad a_2 = 0.2, \qquad b_1 = 0.5, \qquad b_2 = 0.3$$

These values yield

$$\text{minimum } P = 0$$

Therefore, the postulated second-order model describes exactly the process dynamics and the model that can be used for controller design is

$$y_n = 0.6y_{n-1} + 0.2y_{n-2} + 0.5m_{n-1} + 0.3m_{n-2}$$

31.2 Process Identification and Adaptive Control

The nonlinear and nonstationary nature of a typical chemical process leads to a change in its dynamic characteristics during operation. To cope with this situation, a controller should be able to adjust its parameters in an "optimum" manner. Thus we are naturally led to the adaptive control systems which were discussed in Chapter 22.

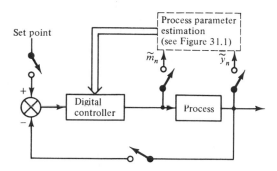

Figure 31.2 Schematic of adaptive control logic.

An optimum adaptation of the controller parameters can be achieved only if a good model is available to describe the process dynamics during the various stages of its operation. Therefore, process identification is an integral part of all adaptive control schemes (Figure 31.2).

Consider a process that is described by the following discrete-time model:

$$y_n = a_1 y_{n-1} + a_2 y_{n-2} + \cdots + a_k y_{n-k} + b_1 m_{n-1}$$
$$+ b_2 m_{n-2} + \cdots + b_k m_{n-k} \tag{31.1}$$

The coefficients $a_1, a_2, \cdots, a_k; b_1, b_2, \cdots, b_k$ are assumed to be constant parameters. Assume that the purpose of the controller is to keep the process output as close as possible to the desired set point y_{SP}. In Section 30.4 we found that the control action, which minimizes either the square error at the next period or the mean square error over N sampling periods, is given by eq. (30.20):

$$m_n = \frac{1}{b_1} \{y_{SP} - a_1 y_n - a_2 y_{n-1} - \cdots - a_k y_{n-k+1} - b_2 m_{n-1}$$
$$- \cdots - b_k m_{n-k+1}\} \tag{30.20}$$

If the parameters $a_1, a_2, \cdots, a_k; b_1, b_2, \cdots, b_k$ were known, then eq. (30.20) describes the optimum control action at the nth sampling instant, which keeps the output as close to the desired point as possible. But the parameters $a_1, a_2, \cdots, a_k; b_1, b_2, \cdots, b_k$ change values either because we move the process operation to a new set point (effect of nonlinearity) or because the process is nonstationary. In both cases we need to estimate new values for the changing parameters. This can be done through linear regression analysis, using experimental input–output data, as described in the preceding section. Consequently, the following on-line adaptive control policy arises for set point changes:

1. Assume that the process is operating at the $y_{SP}^{(i)}$ set point value. The control action given by eq. (31.10) regulates the output at the current $y_{SP}^{(i)}$ set point value against any load changes:

$$m_n = \frac{1}{b_1^{(i)}} \{y_{SP}^{(i)} - a_1^{(i)}\tilde{y}_n - a_2^{(i)}\tilde{y}_{n-1} - \cdots - a_k^{(i)}\tilde{y}_{n-k+1} - b_2^{(i)}\tilde{m}_{n-1}$$
$$- \cdots - b_k^{(i)}\tilde{m}_{n-k+1}\} \qquad (31.10)$$

where the superscript (i) indicates the estimates of the process parameters during the current ith level of operation, and $\tilde{\ }$ indicates measured values.

2. Suppose that we want to move the output to a new set point, $y_{SP}^{(i+1)}$, operation where the process parameters possess different values. Use eq. (31.10) with $y_{SP}^{(i+1)}$ to bring the process to the new set point.

3. During the transfer from the old to the new set point, record the values of the manipulated input and controlled output variables. Use these input–output data to estimate the new values of the parameters through linear regression. Then, the new controller becomes

$$m_n = \frac{1}{b_1^{(i+1)}} \{y_{SP}^{(i+1)} - a_1^{(i+1)}\tilde{y}_n - a_2^{(i+1)}\tilde{y}_{n-1} - \cdots - a_k^{(i+1)}\tilde{y}_{n-k+1}$$
$$- b_2^{(i+1)}\tilde{m}_{n-1} - \cdots - b_k^{(i+1)}\tilde{m}_{n-k+1}\} \qquad (31.11)$$

Figure 31.3 dramatizes the effect of the on-line adaptation on the quality of the closed-loop response versus the case of controlling without controller adaptation.

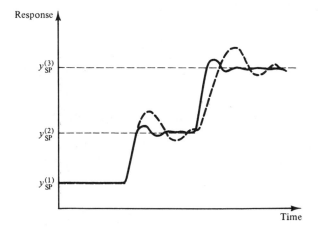

Figure 31.3 Closed-loop process response with (solid lines) and without (dashed line) on-line adaptation.

Example 31.3: Designing a Self-Tuning Regulator for a Second-Order Process

The present example describes how a digital controller can adapt its parameters automatically on-line, during process operation. The resulting system can be called a *self-tuning regulator*. Assume that a poorly known process can be approximated by a second order model:

$$y_n = a_1 y_{n-1} + a_2 y_{n-2} + b_1 m_{n-1} + b_2 m_{n-2}$$

Before the process is placed in operation we have identified the values of its parameters from linear regression analysis of input–output experimental data. Thus we have postulated the following model:

$$y_n = 0.5 y_{n-1} + 0.3 y_{n-2} + 0.6 m_{n-1} + 0.4 m_{n-2}$$

From eq. (30.20) we can find that the optimum control action to keep the process output as close as possible to the desired set point value, $y_{SP} = 1$, is given by

$$m_n = \frac{1}{0.6} \{1 - 0.5 y_n - 0.3 y_{n-1} - 0.4 m_{n-1}\} \tag{31.12}$$

Using the control action given by eq. (31.12), we can optimally regulate the process output at the desired value 1 in the presence of disturbance (load) changes. Suppose now that we want to regulate the process at a new level with $y_{SP} = 5$. We can use the following control algorithm with the old values of the process parameters:

$$m_n - \frac{1}{0.6} \{5 - 0.5 y_n - 0.3 y_{n-1} - 0.4 m_{n-1}\} \tag{31.13}$$

During the transient from the old to the new set point, we record the values of the manipulated variable and the controlled output. These values are shown in Table 31.2. Linear regression analysis using the input–output data of Table 31.2 produces the following values for the process parameters.

$$a_1 = 0.58 \qquad a_2 = 0.35 \qquad b_1 = 0.52 \qquad b_2 = 0.48$$

Then the optimal regulation at $y_{SP} = 5$, against external load changes, is given by

TABLE 31.2.
INPUT–OUTPUT DATA FOR EXAMPLE 31.3

Sampling Instant n	Input m_n	Output y_n
$n < 0$	0.33	1.0
0	0.33	1.0
1	6.63	1.8
2	−0.96	3.7
3	1.09	5.3
4	−0.22	4.9
5	0.25	5.05

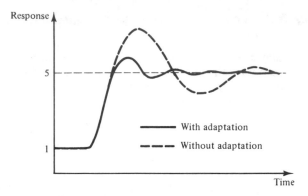

Figure 31.4 Effect of parameter adaptation on the quality of closed-loop response.

$$m_n = \frac{1}{0.52}(5 - 0.58y_n - 0.35y_{n-1} - 0.48m_{n-1}) \qquad (31.14)$$

Figure 31.4 shows the improved quality of the closed-loop response when the adapted control policy [eq. (31.14)] is used in comparison with the closed-loop response using the old process parameters.

Instead of the optimum regulatory control action given by eq. (31.12), we could have used different criteria to design controllers or to tune the adjustable parameters of given control algorithms. Example 31.4 demonstrates the on-line tuning of a process using the "one-quarter decay ratio" criterion.

Example 31.4: On-line Tuning of a PI Controller with One-Quarter Decay Ratio

Consider a first-order process with a PI feedback controller. The closed-loop response to set point changes is given by

$$\overline{y}(s) = \frac{\tau_1 s + 1}{\tau^2 s^2 + 2\zeta\tau s + 1}\,\overline{y}_{SP}(s)$$

where

$$\tau = \sqrt{\frac{\tau_1 \tau_p}{K_p K_c}} \qquad \zeta = \frac{1}{2}\sqrt{\frac{\tau_1}{\tau_p K_p K_c}}\,(1 + K_p K_c)$$

Here τ_p and K_p are the process time constant and static gain, respectively, and K_c and τ_1 are the gain and the integral time constant for the PI controller.

In Example 16.1 we developed a relationship [eq. (16.1)] between τ_p, K_p, τ_1, and K_c which must be satisfied for the response to have one-quarter decay ratio to set point changes:

$$-2\pi\sqrt{\frac{\tau_I}{4\tau_p K_p K_c - \tau_I(1 + K_p K_c)^2}}(1 + K_p K_c) = \ln\left(\tfrac{1}{4}\right) \qquad (16.1)$$

Consequently, as the set point value changes we compute new values for τ_p and K_p using linear regression analysis on the input–output data. Then from eq. (16.1) we can compute the new value of the controller gain. This procedure can be repeated on-line every time we change the set point value.

Remarks

1. In this chapter we have presented a rather simplistic view of the on-line adaptive control systems. There are a number of very important questions which have not been addressed, such as whether the parameter estimates are biased, the interplay between estimation and control, and the stability characteristics of the adaptive controller. A thorough examination of these questions is beyond the scope of this text. The interested reader can consult the relevant references at the end of Part VII.

2. *Under conditions of very good or perfect control we cannot improve the estimates of the poorly known parameters.* This stems from the fact that when the process output is very close to the desired set point, the corresponding values of the manipulated variable change by very little. Consequently, the input–output data are almost "flat" and very poor in new information. Also, quite often the process or measurement noise "covers" the small amounts of information that can be recovered from input–output experimental data during the control phase.

3. On-line adaptation is not limited to feedback systems. On-line process identification can be coupled easily with feedforward, inferential, and other control systems, thus expanding the range of their applicability. Adaptation is particularly valuable for feedforward and inferential systems because they rely heavily on good process models for their successful implementation.

THINGS TO THINK ABOUT

1. Define in your own words the meaning of process identification. Why is it needed in process control? Give some physical examples to demonstrate this need.

2. What is meant by off-line and on-line process identification? Under what circumstances would you use the first or the second? Which one is more valuable in process control?

3. We can use frequency response techniques (see Chapter 17) to identify experimentally a poorly known process. Do you have any ideas on how you could do it? To help you in your thoughts, consider the Bode diagrams of various systems that were examined in Chapter 17. Notice the information provided by characteristics such as the corner frequency (determines the unknown time constant), the level of low-frequency asymptotes (determines the value of static process gains), the slope of high-frequency asymptotes (determines the order of a system), and the behavior of phase lag (keeps increasing for systems with dead time). (*Note*: For further details, consult Ref. 11.)

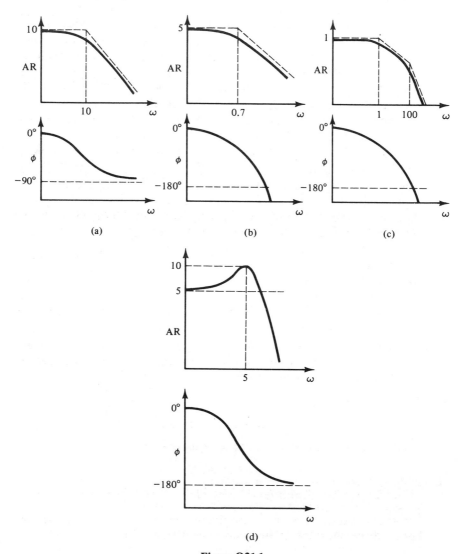

Figure Q31.1

4. Examine the Bode diagrams in Figure Q31.1a through d, which were developed experimentally from the response of real processes to sinusoidal input changes. Identify the models for the unknown processes with the experimental Bode diagrams above; that is, determine (a) the order, (b) whether they possess dead time, and (c) the values of the unknown process parameters (e.g., time constants, gains, dead times, etc.).

5. Describe in your own words the process identification procedure in the time domain using regression analysis. Quantify the procedure and suggest two methods for solving the least mean square error problem.

6. How important is the postulation of a good process model in process identification? Make various suggestions which can help to postulate a good and effective model for an unknown process (recall some general results on the dynamics of various systems from Chapters 10 through 12).

7. How can you decide if a postulated linear model is of acceptable order? What would you do if it were not?

8. Does the type of input change affect the values of the estimated parameters? Explain your answer.

9. Why do you need on-line adaptation of control systems? Can you improve the quality of a process model using input–output data when the process output remains close to the desired set point? Explain your answer.

10. Under what conditions can you improve the model describing the operation of a controlled process? Elaborate on your answer.

11. In what sense is the control action described by eq. (31.12) optimal? Describe in your own words the physical meaning of the control design criteria depicted by eqs. (31.10) and (31.11).

12. Describe an on-line tuning procedure for PI controllers which uses as criterion the ISE (integral square error).

13. Describe an on-line adaptive procedure for a typical feedforward control system (see Chapter 21). Do the same for the inferential control of a distillation column (see Example 22.5).

14. Consider the first-order model

$$y_n = ay_{n-1} + bm_{n-1}$$

Assume that a proportional controller is used;

$$m_n = K_c y_n$$

Show that it is not, in general, possible to estimate both unknown parameters a and b using the least mean square error criterion.
(*Note:* You can find a very helpful discussion of this question in Ref. 18.)
Furthermore, show that if the control law is changed to

$$m_n = K_c y_{n-1}$$

or

$$m_n = K_{c1}y_n + K_{c2}y_{n-1}$$

you can estimate both unknown parameters.

REFERENCES FOR PART VII

Chapter 26. An excellent reference on the various hardware and software aspects of digital computer control systems is the series of monographs edited by D. A. Mellichamp and published by CACHE:

1. *CACHE Monograph Series in Real-Time Computing*, by D. A. Mellichamp (ed.), CACHE, Cambridge, Mass. (1979).

Monograph I:	*An Introduction to Real-Time Computing*
Monograph II:	*Processes, Measurements, and Signal Processing*
Monograph III:	*Introduction to Digital Arithmetic and Hardware*
Monograph IV:	*Real-Time Digital Systems Architecture*
Monograph V:	*Real-Time Systems Software*
Monograph VI:	*Real-Time Applications Software*
Monograph VII:	*Management of Real-Time Computing Facilities*
Monograph VIII:	*Process Analysis, Data Acquisition, and Control Algorithms*

As the titles of the monographs indicate, the reader can find in this series a storehouse of information on every aspect of process computers discussed in Chapters 26 through 30 of the present text. Particular attention is drawn to monographs III and IV, dealing with the hardware components and architecture of digital computers, as well as monographs V and VI, covering all important software aspects for control applications.

Another reference with a large amount of valuable information is:

2. *Computers in Manufacturing*, by U. Rembold, M. K. Seth, and J. S. Weinstein, Marcel Dekker, Inc., New York (1979).

In this text the reader can find material related to computer–instrument–process communication, data acquisition and control, reliability of process control computer systems, management aspects of in-plant computer installations, and a large number of case studies. Additional general references on digital computers and their application in the real-time environment are the following:

3. *Handbook of Industrial Computer Control*, by T. J. Harrison, John Wiley & Sons, Inc., New York (1972).

4. *Minicomputers in Industrial Control*, by T. J. Harrison, John Wiley & Sons, Inc., New York (1978).

5. *Digital Computers in Scientific Instrumentation*, by S. P. Perone and D. O. Jones, McGraw-Hill Book Company, New York (1973).

Measurements, transmission, and signal processing are crucial elements for the design of computer–process interfaces. The reader will find Refs. 4 and 5 very useful in this respect, as well as the following paper:

6. "Universal Process Interfaces," by R. Merritt, *Instrum. Technol., 23*(8), 29 (1976).

Finally, two books from chemical engineers provide useful but limited material on digital computers and their hardware and software features:

7. *Digital Computer Process Control*, by C. L. Smith, Intext Educational Publishers, Scranton, Pa. (1972).
8. *Advanced Process Control*, by W. H. Ray, McGraw-Hill Book Company, New York (1981).

Chapters 27, 28, 29. The material in these chapters has by now become classic and the reader can consult several good references for more details. The following two texts provide an excellent treatment on the mathematical analysis of discrete-time (sampled-data) systems:

9. *Digital and Sampled Data Control Systems*, by J. T. Tou, McGraw-Hill Book Company, New York (1959).
10. *Analysis and Synthesis of Sampled Data Control Systems*, by B. C. Kuo, Prentice-Hall, Inc., Englewood Cliffs, N.J. (1970).

The reader will also find useful Chapter 4 in Ref. 7 and Chapter 14 in:

11. *Process Modeling, Simulation, and Control for Chemical Engineers*, by W. L. Luyben, McGraw-Hill Book Company, New York (1973).

For more details on the mathematical theory of z-transforms as well as their various applications, the following text is a very good reference:

12. *Theory and Application of the z-Transform Method*, by E. I. Jury, John Wiley & Sons, Inc., New York (1964).

In Koppel's book one can find a rigorous theoretical treatment of the sampling of continuous signals and the necessary conditions for their reconstruction from their discrete-time equivalent:

13. *Introduction to Control Theory*, by L. B. Koppel, Prentice-Hall, Inc., Englewood Cliffs, N.J. (1968).

For various questions on numerical integration and differentiation which will be needed during the discretization of continuous systems, the reader can consult several books on numerical analysis, such as:

14. *Digital Computation for Chemical Engineers*, by L. Lapidus, McGraw-Hill Book Company, New York (1962).
15. *Applied Numerical Methods*, by B. Carnahan, H. R. Luther, and J. D. Wilkes, John Wiley & Sons, Inc., New York (1969).

Chapter 30. Chapter 6 of Smith's book [Ref. 7] offers a very useful treatment of the design of digital controllers. It covers the response specification algorithms (deadbeat, Dahlin's), the discrete equivalents of analog controllers, design for load changes, as well as the standard PI and PID algorithms in position or velocity forms. It also discusses and analyzes the phenomenon of controller ringing and provides an extensive coverage of tuning techniques. Finally, it includes a short but useful discussion on how to select the sampling rate in a digital control loop.

For a discussion of the deadbeat algorithms the reader can consult Ref. 10, while Dahlin's method can be found in the original paper:

16. "Designing and Tuning Digital Controllers," by E. B. Dahlin, *Instrum. Control Syst., 41*(6), 77 (1968).

Selecting the appropriate sampling is a very crucial question for digital control systems. Shinskey [Ref. 17] presents a useful discussion on the nature of sampling and its effects on the quality of controlled output. Furthermore, he provides some general guidelines for the selection of an acceptable sampling rate for control purposes.

17. *Process Control Systems*, 2nd ed., by F. G. Shinskey, McGraw-Hill Book Company, New York (1979).

Frequency response or root-locus techniques for the analysis and synthesis of sampled-data control systems have not been included in this text. The procedure is analogous to that for continuous systems. For more information the reader can consult Chapter 15 in Luyben's text [Ref. 11].

The development of the optimal regulatory controller in Section 30.4 is a simplified treatment of the stochastic control problem discussed in the following paper:

18. "Problems of Identification and Control," by K. J. Åström and B. Wittenmark, *J. Math. Anal. Appl., 34*, 90 (1971).

Chapter 31. Several methods exist for the experimental identification of poorly known processes. Luyben in his Chapter 9 [Ref. 11] presents several methods in either the frequency or the time domain and gives some practical tips for their implementation.

For the interested reader the following book by Sage and Melsa constitutes a good start for an in-depth study of process identification theory and techniques:

19. *System Identification*, by A. P. Sage and J. L. Melsa, Academic Press, Inc., New York (1971).

The frequency-domain techniques are not convenient for on-line process identification; regression analysis in the time domain is the main tool. Chapter 7 of Smith's book [Ref. 7] provides a very good start for further reading on time-domain experimental identification. It includes a useful discussion on linear and nonlinear regression and provides some guidance for solving numerical problems. An article by Marquardt offers an excellent algorithm for parameter estimation using the criterion of least squares:

20. "An Algorithm for Least-Squares Estimation of Non-Linear Parameters," by D. W. Marquardt, *J. Soc. Ind. Appl. Math., 11*, 431 (1963).

The simultaneous identification and control is a very difficult problem and is treated in a stochastic framework. The discussion in Section 31.2 is very simplistic. The interested reader can start a careful study of the subject by consulting Åström's book:

21. *Introduction to Stochastic Control Theory*, by K. J. Åström, Academic Press, Inc., New York (1970).

Ref. 18 is also very useful. Here, the interaction between identification and control is discussed and the various idiosyncracies of the two problems are pointed out.

In the following papers the reader can find some typical examples of on-line adaptive control for various processes of interest to chemical engineers:

22. "Self-Tuning Computer Adapts DDC Algorithms," by P. W. Gallier and R. E. Otto, *Instrum. Technol., 15*(2), 65 (1968).

23. "Automatic Tuning of Nonlinear Control Loops," by T. J. Pemberton, *Instrum. Control Syst., 41*(5), 123 (1968).

24. "Adaptive pH Controller Monitors Nonlinear Process," by F. G. Shinskey, *Control Eng., 21*, 57 (1974).

25. "Self-Tuning Adaptive Control of Cement Raw Material Blending," by L. Keviczky, J. Hetthésy, M. Hilger, and J. Kolostori, *Automatica, 14*, 525 (1978).

26. "Self-Tuning Control of a Fixed Bed Reactor," by F. Buchholt, K. Clement, and S. Bar Jorgensen, Preprints of 5th IFAC Symposium on Identification and Parameter Estimation, Darmstadt (1979), p. 1213.

27. "Application of Self-Tuning Regulator to the Control of Chemical Processes," by A. J. Morris, T. P. Fenton, and Y. Nazer, in *Digital Computer Applications to Process Control*, H. R. VanNauta Lemke and H. B. Verbruggen (eds.), 5th IFAC/IFIP International Conference, The Hague (1977).

PROBLEMS FOR PART VII

Chapter 27

VII.1 Find the sampling rate for the following continuous signals so that the sampled values represent well the corresponding continuous signal.

(a) $\bar{y}(s) = \dfrac{10}{0.1s + 1} \dfrac{1}{s}$ $\tau_p = 0.1$ min

(b) $\bar{y}(s) = \dfrac{5e^{-0.5s}}{2s + 1} \dfrac{1}{s}$ $t_d = 0.5$ min and $\tau_p = 2$ min

(c) $\bar{y}(s) = \dfrac{2}{(3s + 1)(10s + 1)} \dfrac{1}{s}$ $\tau_{p1} = 3$ min and $\tau_{p2} = 10$ min

(d) $\bar{y}(s) = \dfrac{1}{\tau^2 s^2 + 2\zeta\tau s + 1} \dfrac{1}{s}$ $\tau = 2$ min and $\zeta = 0.2$

(e) $\bar{y}(s) = \dfrac{3e^{-t_d s}}{\tau^2 s^2 + 2\zeta\tau s + 1} \dfrac{1}{s}$ $t_d = 0.5$ min, $\tau = 5$ min, $\zeta = 0.1$

(f) $\bar{y}(s) = \dfrac{4}{(0.1s + 1)(10s + 1)(100s + 1)}$ $\tau_{p1} = 0.1$ min, $\tau_{p2} = 10$ min, τ_{p3} $= 100$ all min

VII.2 Find the sampling rate for the continuous signals that satisfy the following differential equations. The sampling rate should be such that the sampled values represent well the corresponding continuous signal. (*Note*: All time constants, dead times, and natural periods are in minutes.)

(a) $5\dfrac{dy}{dt} + y = \cos 2t$ $y(0) = 0$

(b) $\dfrac{dy}{dt} + 0.5y = 1$ $y(0) = 0$

(c) $\dfrac{d^2 y}{dt^2} + 0.4\dfrac{dy}{dt} + 2y = \cos(t - 2)$ $dy(0)/dt = y(0) = 0$

(d) $2\dfrac{dy_1}{dt} + y_1 = 1$, $10\dfrac{dy_2}{dt} + y_2 = y_1$, $30\dfrac{dy_3}{dt} + y_3 = y_2$

with $y_1(0) = y_2(0) = y_3(0) = 0$.

VII.3 The liquid of a tank is heated with saturated steam which flows through a coil immersed in the liquid (see Figure 10.2). The energy balance for the system yields

$$V\rho c_p \dfrac{dT}{dt} = UA_t(T_s - T)$$

where V = volume of liquid in the tank = h (liquid level) × A (cross-sectional area); ρ and c_p are the density and heat capacity of liquid in the tank; U and A_t are the overall heat transfer coefficent and the heat transfer area between steam and liquid in the tank; and T_s is the temperature of saturated steam. Discuss how the parameters A, ρ, c_p, U, and A_t affect the rate of sampling the temperature of the liquid in the tank (i.e., how the "suggested" sampling rate changes when the parameters above change in value).

VII.4 Equations (6.36) and (6.37) in Example 6.4 describe the linearized dynamic behavior of a CSTR. Discuss how parameters τ, k_0, U, A, c_p, and V affect the recommended sampling rates for concentration c_A and temperature T.

VII.5 Reconstruct the continuous signal from the following sampled data using zero- and first-order hold elements.

Time t	0	$1T$	$2T$	$3T$	$4T$	$5T$	$6T$	$7T$	$8T$
Signal $y(t)$	0	1.97	3.16	3.88	4.32	4.59	4.75	4.85	4.91

Which reconstruction yields better results?

VII.6 Repeat Problem VII.5, but use the following sampled data.

Time t	0	$1T$	$2T$	$3T$	$4T$	$5T$	$6T$	$7T$	$8T$	$9T$	$10T$
Signal $y(t)$	10	9.8	9.21	8.25	6.97	5.40	3.62	1.70	−0.29	−2.27	−4.16

VII.7 Prove that the transfer function of a first-order hold element is given by eq. (27.6).

VII.8 Convert the following continuous models to equivalent discrete-time models.

(a) $3\dfrac{d^3y}{dt^3} + \dfrac{d^2y}{dt^2} - 2\dfrac{dy}{dt} + 5y = \cos 2t$ $\qquad y''(0) = y'(0) = y(0) = 0$

(b) $\dfrac{d^2y}{dt^2} + 2\dfrac{dy}{dt} + y - \displaystyle\int_0^t y(t)\,dt = 1$ $\qquad y'(0) = y(0) = 0$

(c) $\dfrac{dy_1}{dt} + \dfrac{dy_2}{dt} - 5y_1 + 3y_2 = 1 + e^{-2t}$ $\qquad y_1(0) = y_2(0) = 0$

$2\dfrac{dy_1}{dt} + 3\dfrac{dy_2}{dt} + y_1 - 2y_2 = \sin(t - T)$

VII.9 Derive the discrete-time model of the linearized CSTR model developed in Example 6.4.

VII.10 Derive the discrete-time equivalent of the nonlinear continuous model for a stirred tank heater developed in Example 4.4 and given by eqs. (4.4a) and (4.5b).

Chapter 28

VII.11 Compute the z-transform of the following sequences of sampled values.

(a)

Time t	0	T	$2T$	$3T$	$4T$	$\cdots$	nT	$\cdots$
Signal $y(t)$	1	2	4	8	16	$\cdots$	2^n	$\cdots$

(b)

Time t	0	T	$2T$	$3T$	$\cdots$	nT	$\cdots$
Signal $y(t)$	1	2	3	4	$\cdots$	$n+1$	$\cdots$

(c)

Time t	0	T	$2T$	$3T$	$\cdots$	nT	$\cdots$
Signal $y(t)$	0.7	1.4	2.1	2.8	$\cdots$	$(n+1)0.7$	$\cdots$

VII.12 Derive analytically the z-transforms of the following functions.
(a) $y(t) = \alpha \cos \omega t$
(b) $y(t) = at^n$
(c) $y(t) = te^{-at}$
(d) $y(t) = 1 - e^{-at}$
(e) $y(t) = e^{-at} \sin \omega t$

VII.13 Using the final value theorem compute the value that each sequence of sampled values given in Problem VII.11 reaches, as n → ∞.

VII.14 Compute the final value for each sequence of sampled values with the following z-transforms.

(a) $\hat{y}(z) = \dfrac{0.39z^{-1}}{(1 - z^{-1})(1 - 0.61z^{-1})}$

(b) $\hat{y}(z) = \dfrac{0.14z^{-1}}{(1 - 0.14z^{-1})^2}$

(c) $\hat{y}(z) = \dfrac{1}{1 - 0.37z^{-1}} - \dfrac{1}{1 - 0.14z^{-1}}$

(d) $\hat{y}(z) = \dfrac{0.22z^{-1}}{1 - 0.44z^{-1} + 0.55z^{-2}}$

VII.15 Find the inverse z-transform of the following expressions, using long division.

(a) $\hat{y}(z) = \dfrac{1 + 2z^{-1} - 3z^{-2}}{3 + 5z^{-1} - z^{-2} + 4z^{-3}}$

(b) $\hat{y}(z) = \dfrac{z^{-1}}{1 + z^{-1} + z^{-2} + z^{-3}}$

(c) $\hat{y}(z) = \dfrac{2}{1 - 0.5z^{-1} + z^{-2}} - \dfrac{1}{1 - 0.2z^{-1}}$

VII.16 Compute the inverse z-transforms of the expressions given in Problem VII.14.

VII.17 Compute the inverse z-transforms of the expressions given in Problem VII.15, using the approach through expansion into partial fractions.

VII.18 Compute the modified z-transform of the following functions.
(a) $y(t) = 1 - \sin(2t - 0.2T) + e^{-t}$
(b) $y(t) = 5e^{-(t-1.5T)} - \delta(t - 3.2T)$

VII.19 Compute the inverse of the following modified z-transforms and find the behavior of $y(t)$ between sampling instants.

(a) $\hat{y}(z, m) = \dfrac{0.5z^{-1}}{1 - z^{-1}} m + \dfrac{0.5z^{-2}}{(1 - z^{-1})^2}$

(b) $\hat{y}(z, m) = \dfrac{z^{-1}}{1 - z^{-1}} - \dfrac{e^{-m} z^{-1}}{1 - 0.37z^{-1}}$

Chapter 29

VII.20 Find the discrete transfer functions of the following discrete dynamic systems.
(a) $y_{n+1} = \alpha y_n - \beta \cos 2nT + \gamma y_{n-2} - \delta m_n$ $\quad \alpha, \beta, \gamma, \delta$ constants
(b) $y_{n+1} + 2y_n - 3y_{n-k} = m_{n-1} + m_{n-2}$ $\quad k = $ integer
(c) $y_{1,n+1} + y_{1,n} - 2y_{2,n+1} + y_{2,n} + 5y_{1,n-2} - 4y_{2,n-3} = m_{1,n} + m_{2,n-1} - m_{2,n-2}$
$\quad 2y_{1,n+1} - 3y_{1,n} + y_{2,n+1} - 4y_{2,n} + y_{1,n-2} + y_{2,n-4} = m_{1,n-1} - m_{1,n-2} + m_{2,n}$

VII.21 Find the pulse transfer functions in the z-domain of the continuous systems with the following transfer functions in the Laplace domain. Assume a zero-order hold element.

(a) $G_p(s) = \dfrac{2}{(0.1s + 1)(5s + 1)(s + 1)}$

(b) $G_p(s) = \dfrac{5e^{-t_d s}}{(3s + 1)(s + 1)}$ $\quad$ where $t_d = 2T$

(c) $G_p(s) = \dfrac{1}{s^2 + 0.9s + 1}$

VII.22 Prove the following two statements.
(a) For N continuous systems in series with transfer functions $G_1(s)$, $G_2(s), \ldots, G_N(s)$,

$$Z[G_1(s)G_2(s) \cdots G_N(s)] \neq G_1(z)G_2(z) \cdots G_N(z)$$

(b) For two systems in series with transfer functions $G_1(s)$ and $G_2(s)$,

$$Z[G_1(s)G_2(s)] = G_1(z)G_2(z)$$

when either $G_1(s)$ or $G_2(s)$ contains only dead-time terms.

(c) How is the result of part (b) extended to N systems in series?

VII.23 (a) Compute the discrete-time responses of the processes given in Problem VII.21 to a unit step change in the input variable. The sampling period is $T = 1$.

(b) Compute the ultimate value of the responses.

(c) Compute the steady-state gains of the processes.

VII.24 Compute the discrete-time closed-loop response to unit step change in set point for the following systems.

(a) $G_p(s) = \dfrac{5}{(s + 1)^2(3s + 1)};$ $G_d(s) = \dfrac{1}{s + 1};$ proportional controller with $K_c = 10$

(b) $G_p(s) = \dfrac{e^{-t_d s}}{s + 1}$ with $t_d = 1\,T;$ $G_d(s) = \dfrac{3}{s + 1};$ proportional controller with $K_c = 0.2$

(c) $G_p(s) = \dfrac{1}{(s + 1)(2s + 1)};$ $G_d(s) = \dfrac{1}{2s + 1};$ PI controller with $K_c = 0.1$ and $\tau_I = 0.5$

(d) $G_p(s) = \dfrac{1}{s^2 + 2s + 1};$ $G_d(s) = \dfrac{2}{3s + 1};$ PI controller with $K_c = 0.1$ and $\tau_I = 0.5$

For all of the cases above, assume zero-order hold and sampling period $T = 1$ sec.

VII.25 Compute the discrete-time, closed-loop response to unit step changes in the load for each of the systems described in Problem VII.24. Assume zero-order hold element and sampling period $T = 1$ sec.

VII.26 (a) Assuming sampling period $T = 1$, find the ultimate controller gain for each of the feedback systems described in Problem VII.24 (i.e., find the largest value of gain that will give stable closed-loop behavior).

(b) Using the value of ultimate gain, compute the closed-loop response to unit step changes in set point for the first 10 sampling instants.

VII.27 Consider the feedback loop with the following dynamic components:

Process: $G_p(s) = \dfrac{10e^{-t_d s}}{2s + 1}$ $t_d = kT$ with k = integer

Hold: $H(s) = \dfrac{1 - e^{-Ts}}{s}$

Controller: $G_c(s) = K_c$

(a) Derive the discrete-time, closed-loop response and identify the characteristic equation.
(b) Compute the ultimate gain when $k = 1, 2, 3$ (i.e., compute the largest gain that will give stable closed-loop behavior). Discuss how the ultimate gain changes with the value of dead time. $T = 1$.

VII.28 Repeat Problem VII.27 but consider the following third-order process:

$$G_p(s) = \frac{2e^{-t_d s}}{(0.1s + 1)(s + 1)(3s + 1)}$$

VII.29 Consider the feedback loop shown in Figure PVII.1.
(a) Derive an expression for the closed-loop response to set point changes.
(b) Set $K_m = 1$ and $\tau_m = 0$ and find the largest value of gain K_c (ultimate gain) that will give stable response.
(c) Change the values of K_m and τ_m, one at a time, and find how they affect the ultimate gain.
(d) On the basis of the results from part (c), discuss how the characteristics of the measuring device (i.e., the values K_m and τ_m) affect the stability of digital feedback loop.
Assume a sampling period $T = 1$.

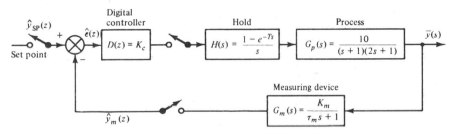

Figure PVII.1

VII.30 Consider the following roots (in the z-domain) of the characteristic equations for various digital feedback loops.
(a) $p_1 = -0.5$ and $p_1 = +0.8$
(b) $p_1 = -1.0$ and $p_2 = +0.5$
(c) $p_1 = -0.5$ and $p_2 = +1.0$
(d) $p_1 = 0.5 - 2j$ and $p_2 = 0.5 + 2j$
(e) $p_1 = 0.5 - 0.7j$ and $p_2 = 0.5 + 0.7j$
(f) $p_1 = -1.5 + j$, $p_2 = -1.5 - j$, and $p_3 = 0.1$
(g) $p_1 = 0.3$, $p_2 = 0.8$, $p_3 = 0.1 + 1.1j$, and $p_4 = 0.1 - 1.1j$
Which of the cases correspond to stable closed-loop systems, and why?

VII.31 Suppose that the closed-loop response of a digital control loop is given by

$$\hat{y}(z) = \frac{1}{1 - pz^{-1}} \hat{y}_{SP}(z)$$

where p is the root of the closed-loop characteristic equation (i.e., the closed-loop pole).

(a) Introduce a unit impulse set point change and compute the response at the first six sampling instants (including $t = 0$) when p takes on the following values: (1) $p = 0.2$, (2) $p = -0.2$, (3) $p = 0.95$, (4) $p = -0.95$, (5) $p = 1.0$, (6) $p = -1.0$, (7) $p = 1.5$, (8) $p = -1.5$.

(b) Summarize your results from part (b) in a few statements, indicating the effect that the closed-loop pole location has on the quality of the closed-loop response.

Chapter 30

VII.32 Which of the following digital controllers are physically realizable and which are not? Explain why a particular controller is physically realizable or not realizable.

(a) $D(z) = \dfrac{\hat{m}(z)}{\hat{e}(z)} = \dfrac{z + 2z^2 + z^3}{1 + 2z - 3z^2 + 4z^3 + z^4}$

(b) $D(z) = \dfrac{\hat{m}(z)}{\hat{e}(z)} = \dfrac{1 + 3z^2 + 5z^4}{2z + z^3}$

(c) $D(z) = \dfrac{\hat{m}(z)}{\hat{e}(z)} = \dfrac{z - z^3 + z^4}{1 + z^{-1} - z^3}$

(d) $D(z) = \dfrac{\hat{m}(z)}{\hat{e}(z)} = \dfrac{1 + z + z^3}{1 + z + z^2 + z^4}$

VII.33 Consider the digital cascade control system shown in Figure PVII.2. Controller D_1 is simple proportional, while controller D_2 is in the velocity form of a PI algorithm with $\tau_I = 0.5$.

(a) Derive the closed-loop response of the cascade system.

(b) Describe how you would tune the two controllers and compute the gains for D_1 and D_2.

(c) Compute the closed-loop response to a unit step change in disturbance d_1 for the first 10 sampling instants using the gains found in part (b).

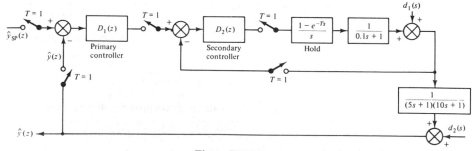

Figure PVII.2

VII.34 Reconsider the control of two noninteracting tanks discussed in Problem IV.17. We would like to use a digital controller instead of a continuous one. Therefore, rework Problem IV.17 using direct digital control. You should select the appropriate sampling. Assume a zero-order hold element.

VII.35 Rework Problem IV.18 using a digital controller. You should select the appropriate sampling. Assume a zero-order hold element.

VII.36 (a) Design a digital PI controller in velocity form, using the Cohen–Coon settings. The experimental process reaction curve was developed with a sampling period of 0.2 min and is given in Table PIV.1.
   (b) Compute the discrete-time, closed-loop response to a unit step in set point using the settings above.
   Assume a zero-order hold element and approximate the process dead time by the nearest larger integer multiple of sampling periods.
   (c) Design a deadbeat controller for this process using a unit step change in set point.
   (d) Compare the closed-loop responses produced by the PI algorithm of part (a) and the deadbeat algorithm of part (c). Which one exhibits (1) a shorter rise time, (2) lower overshoot, and (3) a shorter settling time?

VII.37 Repeat Problem VI.36 but consider the process with the reaction curve data of Table PVI.2. The digital controller should be a PID velocity algorithm. Note from Table PVI.2 that the sampling period is 5 min. Assume zero-order hold.

VII.38 (a) Design a deadbeat sampled-data controller for a first-order process with dead time:

$$G_p(s) = \frac{5e^{-t_d s}}{0.1s + 1} \qquad t_d = kT \text{ with } k = \text{integer}$$

   Consider a step change in set point of magnitude 5. Assume that $k = 0, 1,$ or 2. The controller must be physically realizable. Consider $T = 1$.
   (b) Sketch the closed-loop response specifications that the deadbeat algorithms are called upon to satisfy.
   (c) Identify the location of the controller poles for each case. Do you have ringing poles, and if so, how serious is the ringing problem?

VII.39 (a) Design a deadbeat controller for the following second-order process:

$$G_p(s) = \frac{1}{(s + 1)(5s + 1)}$$

   Use a step change in the set point of magnitude 10 and assume a zero-order hold. The controller must be physically realizable.
   (b) Do you have ringing poles, and if so, how serious is the ringing problem? If the controller possesses ringing poles, show how it can be modified so that the ringing problem is eliminated.
   (c) Using the deadbeat controller designed in part (a), compute the process closed-loop response to a step change in the set point of

magnitude 5. How does the deadbeat algorithm perform for a set point change different from that for which it was designed?

VII.40 Show that the design equation of a deadbeat controller depends on the type of input change but does not depend on the magnitude of the input change.

VII.41 Consider the cascade control system of Figure PVII.2. Design two deadbeat algorithms for the primary and secondary controllers.

VII.42 (a) Design the Dahlin algorithm for the process of Problem VII.39. The algorithm must be physically realizable and should provide no slower response than that of the open-loop process.
 (b) Does the algorithm possess ringing poles, and if so, how serious is the ringing problem?
 (c) If the controller possesses ringing poles, show how it can be modified so that the ringing problem is eliminated.

VII.43 Design a deadbeat and Dahlin's controller for the process with the reaction curve data of Table PIV.1. Assume zero-order hold and approximate the process dead time by the nearest larger integer multiple of the sampling period, which is 0.2 min.

VII.44 Consider a process with the following transfer functions:

$$G_p(s) \equiv \frac{\bar{y}(s)}{\bar{m}(s)} = \frac{5e^{-Ts}}{s+1} \qquad \text{and} \qquad G_d(s) \equiv \frac{\bar{y}(s)}{\bar{d}(s)} = \frac{1}{s+1}$$

Design a minimal prototype algorithm using the following specifications: (1) unit step load change; and (2) the output should return to zero at the fifth sampling instant and remain at zero for every sampling instant after the fifth.

VII.45 Repeat Problem VII.44, but change the output specification so that the response returns to zero as soon as possible and remain at zero for every sampling instant afterward.

VII.46 Consider a process with the following input–output relationship in discrete-time form:

$$y_n = 2y_{n-1} - y_{n-2} + m_{n-1}$$

(a) Design an optimal feedback control algorithm which satisfies the following performance criterion:

$$\text{minimize } P = \sum_{k=1}^{5} (y_k - y_{SP})^2$$

(b) Find the closed-loop response to a step change in the set point of magnitude 2, using the optimal control algorithm of part (a). Compute the value of the performance index P.

VII.47 Consider a process with the following input–output relationship:

$$y_n = 4y_{n-1} + 0.3y_{n-2} + 0.1y_{n-3} + m_{n-1-k} + 0.5m_{n-2-k}$$

where k is an integer and denotes the dead time between input and output.

(a) Derive the optimal feedback control action which minimizes the following performance criterion:

$$P = \sum_{\ell=1}^{10} (y_\ell - y_{SP})^2$$

(b) What values of k make the controller of part (a) physically realizable? Elaborate on your answer.

(c) Modify the optimal controller of part (a) so that it is physically realizable for values of k larger than those identified in part (b).

VII.48 Consider a process with the following transfer functions:

$$G_p(s) \equiv \frac{\overline{y}(s)}{\overline{m}(s)} = \frac{e^{-Ts}}{s + 1} \quad \text{and} \quad G_d(s) \equiv \frac{\overline{y}(s)}{\overline{d}(s)} = \frac{1}{s + 1}$$

(a) Design different controllers using (1) the deadbeat algorithm, (2) Dahlin's algorithm, and (3) the optimal performance criterion of eq. (30.19a). All controllers must be physically realizable.

(b) Using each of the three controllers above separately, compute the closed-loop response to a unit step change in the load d.

(c) Compare the three closed-loop responses found in part (b) and discuss their relative advantages and disadvantages.

VII.49 Repeat Problem VII.48, but consider the following process transfer functions:

$$G_p(s) \equiv \frac{\overline{y}(s)}{\overline{m}(s)} = \frac{5}{(s + 1)(2s + 1)} \quad \text{and} \quad G_d(s) \equiv \frac{\overline{y}(s)}{\overline{d}(s)} = \frac{1}{s + 1}$$

Chapter 31

VII.50 Table PIV.1 gives the open-loop response of a process to a step change in the manipulated input of magnitude 50.

(a) Assuming a first-order model with dead time $t_d = 0.4$ min, estimate the process parameters K (static gain) and τ (time constant) using linear regression.

(b) Compute the value of the least mean square error using the values of the parameters estimated in part (a). Is this value sufficiently close to zero?

(c) How can you improve the parameter estimates without changing the assumed order of the model?

VII.51 Repeat Problem VII.50 but now use the input–output data of Table PIV.2. Initially, assume a first-order model with dead time $t_d = 15$ min.

VII.52 Formulate the linear regression problem for the following assumed orders of unknown processes:

(a) Two first-order systems in series.

(b) A first-order and an overdamped second-order system in series.

VII.53 Consider a stirred tank heater. During the operation of the heater the overall heat transfer coefficient decreases due to fouling. Design a self-tuning regulator which adjusts appropriately the parameters of an optimal feedback controller. (*Note*: Derive the general expressions that can be programmed in a digital computer for the on-line implementation of the self-tuning regulator.)

VII.54 Derive the general expressions for the design of a self-tuning regulator which is to be used for the control of the following second-order process:

$$y_n = 2y_{n-1} + 0.7y_{n-2} + 0.2m_{n-1} + 0.1m_{n-2}$$

Index